Lecture Notes in Control and Information Sciences 388

Editors: M. Thoma, F. Allgöwer, M. Morari

Jean Jacques Loiseau, Wim Michiels,
Silviu-Iulian Niculescu, and Rifat Sipahi (Eds.)

Topics in Time Delay Systems

Analysis, Algorithms and Control

Springer

Editors

Jean Jacques Loiseau
IRCCyN UMR CNRS 6597
Ecole Centrale de Nantes, 1
rue de la Noë, BP 92 101
44321 Nantes cedex 3
France
E-mail: Loiseau@irccyn.ecnantes.fr

Silviu-Iulian Niculescu
L2S (UMR CNRS 8506), CNRSSupélec
3 rue Joliot Curie
91192 Gif-sur-Yvette
France
E-mail: Silviu.Niculescu@lss.supelec.fr

Wim Michiels
Department of Computer Science
Katholieke Universiteit Leuven
Celestijnenlaan 200A
B-3001Heverlee
Belgium
E-mail: Wim.Michiels@cs.kuleuven.be

Rifat Sipahi
Department of Mechanical and
Industrial Engineering
Northeastern University
Boston, MA 02115
USA
E-mail: rifat@coe.neu.edu

ISBN 978-3-642-02896-0 e-ISBN 978-3-642-02897-7

DOI 10.1007/978-3-642-02897-7

Lecture Notes in Control and Information Sciences ISSN 0170-8643

Library of Congress Control Number: Applied for

Typeset & Cover Design: Scientific Publishing Services Pvt. Ltd., Chennai, India.

Printed in acid-free paper

5 4 3 2 1 0

springer.com

‘He is so very much behind the day that occasionally, as things move round in their usual circle, he finds himself, to his bewilderment, in the front of the fashion.’

Behind the Times, A. Conan Doyle (1894)

Preface

Time delays are present in many physical processes due to the period of time it takes for the events to occur. Delays are particularly more pronounced in networks of interconnected systems, such as supply chains and systems controlled over communication networks. In these control problems, taking the delays into account is particularly important for performance evaluation and control system's design. It has been shown, indeed, that delays in a controlled system (for instance, a communication delay for data acquisition) may have an *"ambiguous"* nature: they may stabilize the system, or, in the contrary, they may lead to deterioration of the closed-loop performance or even instability, depending on the delay value and the system parameters. It is a fact that delays have stabilizing effects, but this is clearly conflicting for human *intuition*. Therefore, specific analysis techniques and design methods are to be developed to satisfactorily take into account the presence of delays at the design stage of the control system.

The research on time delay systems stretches back to 1960s and it has been very active during the last twenty years. During this period, the results have been presented at the main control conferences (CDC, ACC, IFAC), in specialized workshops (IFAC TDS series), and published in the leading journals of control engineering, systems and control theory, applied and numerical mathematics. In the control area these results concern either the analysis of systems with delays (stability, controllability, performance evaluation), or the solution of control problems (design a control law that reaches a given stability margin, control objective, or performance specification). Results have been obtained for many specific classes of models (linear, nonlinear, stochastic) and many different controller structures (feedback, feed-forward, predictive, cascaded). In the numerical mathematics area the research has focused on time-stepping and numerical stability, and bifurcation analysis.

We can say that at this point a maturity has been reached *only* regarding the analysis and synthesis of *stand-alone* systems with relatively *small dimensions* and a *small number of delays*. However, the community faces several challenges due to the emergence of important new application fields, mainly in the area of the analysis and control of large-scale interconnected systems and networks (control of communication networks, distributed decision making and control, teleoperation, tele-surgery, cooperative control). This is obviously a *natural* need which arises in parallel to the progress in the technology that demands complicated tasks to be concurrently accomplished. In other words, a typical modern control system is no longer a standalone system, but rather an interconnection of hierarchically organized heterogenous systems (continuous or discrete) with cross-coupled physics and multiple information channels, communication systems and decision-making laws. These new applications are described by highly *complex* models characterized by multiple and time-varying delays with intermittent and stochastic nature, several types of nonlinearities including limited sensor capabilities and saturation of interconnections and actuators, and the presence of different time-scales. These complications clearly trigger the need for advancement; especially the development of new theoretical and mathematical tools, numerically tractable algorithms for real-time control, and implementable controller structures are inevitable (e.g. suitable for a distributed implementation) not only to meet the nominal stability, performance and robustness specifications, but also to scale well with respect to the system's and network size. Furthermore, at various levels in the research effort, it also becomes necessary to consider the particular structure of the system, both in the control design (exploiting the structure of the controller and the network topology) and in the numerical analysis (the sparsity of the large-scale problems needs to be exploited to keep the problems feasible from a computational point of view).

It is crucial to note that not all the above mentioned challenges existed two decades ago and as the field advances along with the technology, new challenges are expected to merge and they will again drive the research in this field in the next decades. The existing difficulties will be overcome by the development of integrated and truly cross-disciplinary approaches that can bridge the applications to control engineering, mathematics, numerical analysis and numerical optimization fields. Since this direction requires a lot to be covered so that it becomes available for practice, the research field in the last decade has exhibited *rapid growth* to close the existing gaps. Therefore, this field of research achieves to renew, regenerate and reproduce in a rapidly growing pace along with the most recent advancements in science and technology. For instance, it is worthy to mention that recently new light has been shed particularly on the approaches based on multi-variable polynomials or polynomial matrices (strongly established in the analysis of time delay systems), semi-definite programming and its application to problems from algebraic geometry, which has led to the availability of new computational tools.

This book is primarily targeted to present the most recent trends in the field of *control and dynamics of time delay systems*. In this rapidly evolving field, the book successfully captures a careful selection of the most recent papers contributed by prominent active researchers in the field and collected under five parts: (i) analysis and methodology, (ii) numerical methods and algorithms, (iii) control and observation, (iv) propagation and flows, (v) delays effects in interconnected systems and networks. The themes of the chapters appropriately categorize the activities in the field as per the motivations discussed above. The contributions in the chapters are a well-balanced combination of two main resources; *invited papers* and the *work presented* at 2007 IFAC Workshop on Time Delay Systems, which is the main specialized meeting venue in the field. In the selection of the topics and the contributors, the editors have not only aimed at maintaining the highest technical quality, but also at achieving an appropriate balance across the chapters (i)-(v). It is worthy to note that the book proposal does not have significant overlap with the contents of 2007 IFAC Workshop on Time Delay Systems. Attendees of the workshop contributing to this book proposal also improved their contributions beyond what they presented at the workshop and the editors comfortably claim that at least 50% of the book content-wise comprises *new* contributions that are closely *relevant* to the most recent trends discussed above.

Structure of the book

As mentioned earlier, this volume is divided into five parts, each of them composed of five to eight chapters. They are respectively devoted to Analysis and Methodology (Part I), Numerical Methods and Algorithms (Part II), Control and Observation (Part III), Propagation and Flows (Part IV), and Delay Effects in Interconnected Systems and Networks (Part V). The first part is concerned with the analysis of the systems with delays, with a special attention paid to their stability properties and characterizations. The second part is devoted to numerical tools and algorithms, and the third part to the control and observation of time delay systems. The fourth and fifth part are concerned with applications ranging from propagation and flow models to interconnected systems and networks. In what follows we present the different chapters and streams of the different parts.

Part I Analysis and Methodology

The first part includes eight chapters. It is devoted to recent trends in stability analysis of time delay systems. Stability is a very important property of dynamical control systems, and it has motivated most of the work on time delay systems since the 90s. The recent research is oriented towards the study of the stability of more general classes of systems with delays that include systems with stochastic delays, multiple delays and distributed delays, distributed systems and time delay systems of neutral type. It is worth to notice that these generalizations are not only of theoretical interest, but they are motivated by important applications in the area of control of systems with spatially distributed parameters, in particular systems distributed over networks.

The first chapter is contributed by ERIK I. VERRIEST and WIM MICHIELS. It provides conditions for the stability of systems with a stochastic delay having the form of a sawtooth with resets to zero at the arrival times of a Poisson process. The models are motivated by an important field of applications, namely systems controlled through a network, where the delays typically have an irregular behavior, but information is available about their probabilistic properties.

In the second chapter, contributed by TUDOR C. IONESCU and RADU ŞTEFAN, the stability of neutral systems is addressed using the comparison method. Two

Padé approximations are used, which lead to two comparison systems and delay-dependent robust stability conditions. The computation of the delay margin is reduced to solving a set of finite dimensional Linear Matrix Inequalities (LMI's).

The third chapter is contributed by VLADIMIR L. KHARITONOV, SABINE MONDIÉ and GILBERTO OCHOA. A new method for studying the stability of a large class of systems with distributed delays is proposed. This is important because control laws involving distributed delays are often very useful for systems with point-wise delays. Furthermore distributed delays can be used to approximate classes of time-varying delays, including stochastically varying delays. A number of examples are presented to illustrate the approach and to show its strength.

The fourth chapter is contributed by KAMRAN TURKOGLU and NEJAT OLGAC. Linear time-invariant minimum phase MIMO plants with multiple control delays are considered. The delay decoupling control is introduced. It aims at decoupling the characteristic matrix in order to facilitate the assessment of stability in each of the delays, independently from each other. For a class of uncertain time delayed dynamics, it is shown that if the feedback control is properly designed, decouplability may still hold, hence, the robustness analysis can be performed efficiently. The result is demonstrated by means of a cart-pendulum system.

The next chapter deals with the case of linear distributed parameter systems of parabolic type. It is contributed by YURY ORLOV and EMILIA FRIDMAN. The analysis of exponential stability via the Lyapunov-Krasovskii method is extended towards linear time delay systems over a Hilbert space. When applied to the heat equation, the obtained stability conditions are represented in terms of standard LMIs, providing an effective tool for the robust control of distributed parameter systems with time delay.

Two procedures for the computation of Lyapunov matrices of time delay systems of neutral type are presented in the sixth chapter, contributed by GILBERTO OCHOA, JUAN E. VELÁZQUEZ, VLADIMIR L. KHARITONOV and SABINE MONDIÉ. This computation is the key step in many analysis problems for time delay systems, such as the determination of robustness bounds or exponential estimates.

The seventh chapter is contributed by YI LIU, HONGFEI LI and KEQIN GU. In this article the stability problem of coupled linear differential-difference equations with block-diagonal uncertainty is discussed. New stability criteria in the form of linear matrix inequality are derived using the discretized Lyapunov-Krasovskii functional method. A number of examples are presented to illustrate the effectiveness of the method. The stability conditions are also extended to the evaluation of H_∞ performance measures.

The last chapter of this part part is provided by RABAH RABAH, GRIGORY M. SKLYAR and ALEXANDER V. REZOUNENKO. It provides a systematic approach to the stabilizability problem of linear infinite-dimensional systems where the infinitesimal generator of the solution operator has an infinite number of unstable eigenvalues. The focus is on strong non-exponential stabilizability by a linear feedback control. The study is based on recent results on the Riesz basis property and a selection of control laws which preserve this property. The results are applied to the linear feedback control of systems described by the wave equation or by delay differential equations of neutral type.

Part II Numerical Methods and Algorithms

The simulation, analysis and control of time delay systems require the development of specific numerical algorithms and computational tools or the adaptation of existing tools for finite dimensional systems. One of the focus points of recent research concerns the location of stability switches and bifurcations. Because the delay operator and the differential operator are algebraically independent, such control problems often boil down to the analysis of multivariable equations, which can be effectively solved using the concept of Ore algebras. For the same reason, the polynomial sum-of-squares algorithms also apply to time delay systems. The results in these directions are rapidly growing. Besides these algebraic approaches the development of fast numerical methods for the simulation (time-integration) and stability analysis are prerequisites for the analysis of complex time-systems.

The first chapter of this part is contributed by MATTHEW M. PEET, CATHERINE BONNET and HITAY ÖZBAY. It is shown how sum-of-squares programming and real algebraic geometry can be used to verify root locations in the complex plane in polynomial time.

In the second chapter, contributed by MIROSLAV HALÁS, a polynomial approach for nonlinear time delay systems is presented, based on the Ore algebra over the field of meromorphic functions. The chapter also introduces transfer functions of nonlinear time delay systems and addresses some basic properties, in particular, the accessibility.

The third chapter is contributed by ALEXANDER M. KAMACHKIN and ALEXANDER V. STEPANOV. They describe sufficient conditions for the existence of stable periodic solutions of time delay control systems containing hysteresis nonlinearities. The cases of stable and perturbed systems are considered.

In the fourth chapter, contributed by ANTONIS PAPACHRISTODOULOU and MATTHEW M. PEET, it is shown how the sum-of-squares approach can be used to understand the stability of biological systems and communication networks.

The fifth chapter, contributed by DIMITRI BREDA, STEFANO MASET and ROSSA-NA VERMIGLIO, aims at presenting the freely available MATLAB package TRACE-DDE, devoted to the computation of characteristic roots and stability charts of linear autonomous systems of delay differential equations with discrete and distributed delays. The main features of the underlying approaches are presented.

Whilst in the control theory field a huge literature on stability analysis of time delay systems exists, simulating such systems by numerical integration often provides the sole approach for studying issues like global stability. In the sixth chapter, contributed by ALFREDO BELLEN and STEFANO MASET, an overview of numerical methods for retarded functional differential equations is presented, including both distributed and discrete delays.

The Maple library OREMODULES is devoted to the symbolic analysis of multidimensional systems, and its subpackage OREMORPHISMS provides constructive tools for the factorization, the reduction and the decomposition problem for general linear functional systems. The Maple library is described in the last two chapters of this part, contributed by THOMAS CLUZEAU and ALBAN QUADRAT. The seventh chapter contains a survey of the main theoretical results on which the constructive homological algebra approach is based, leading to algorithms for computing morphisms between finitely presented left modules over an Ore algebra. In the eighth chapter the library is demonstrated.

Part III Control and Observation

In this part recent results are presented on observation, identification, optimal control and synthesis of controllers such that the closed-loop system is stable and satisfies given performance specifications.

The first chapter is contributed by HITAY ÖZBAY and A. NAZLI GÜNDEŞ. A stabilizing controller for a given plant is considered, in the form of a cascade connected PI (proportional plus integral) controller. The allowable range of the integral action gain is examined, and is discussed how the stabilizing controller can be chosen to maximize this range.

In the second chapter of the part, contributed by MICHAËL DI LORETO, CATHERINE BONNET and JEAN JACQUES LOISEAU, the stabilization problem of time delay systems of neutral type is addressed. For strictly proper systems that are not formally stable, it is shown by means of a parameterization of stabilizing compensators that any stabilizing compensator is necessarily not proper.

LMI conditions for the robust stabilization and robust H_∞ control of uncertain linear systems with distributed systems are presented in the third chapter, contributed by ULRICH MÜNZ, JOCHEN M. RIEBER and FRANK ALLGÖWER.

Chapter four is contributed by ALEXANDRE SEURET, THIERRY FLOQUET, JEAN-PIERRE RICHARD and SARAH SPURGEON. It deals with the design of observers for linear systems with unknown, time-varying, but bounded delays on the state and on the input. For a class of systems the problem is solved by combining the unknown input observer approach with an adequate choice of a Lyapunov-Krasovskii functional.

In chapter five, contributed by MILENA ANGUELOVA and BERNT WENNBERG, the identifiability of delay parameters in nonlinear delay differential equations is considered. The identifiability is directly related to input-output relationships.

The optimal timing and impulsive control for a class of systems described by coupled differential and continuous time difference equations are considered in chapter six, contributed by ERIK I. VERRIEST and PIERDOMENICO PEPE. Necessary conditions for optimality are derived using a streamlined approach based on an appropriately defined Hamiltonian. An application to a model for the regeneration of a deforested area is reported.

Part IV Propagation and Flows

Time delay models are often used as simplified models for spatially distributed processes, where the delays are typically connected to the propagation of a signal or a flow. An important stream of research consists of applying systems theory and control design methods for time delay systems to systems involving diffusion or propagation.

The first chapter, contributed by DANIEL MELCHOR-AGUILAR, addresses the local asymptotic stability of proportional-integral AQM controllers supporting TCP flows. The complete set of stabilizing proportional-integral controllers is obtained. The robustness of the controllers against uncertainties in the network parameters (number of TCP flows, round-trip time and link capacity) is also examined.

Chapter two is contributed by PIERDOMENICO PEPE. A nonlinear state feedback control law for a stirred tank chemical reactor with recycling is studied. The controller drives the output of the system, the reactor temperature, to the desired value with exponential error decay rate. Moreover, the closed loop system is locally input-to-state stable with respect to a disturbance forcing that models sensor and actuator errors, as well as errors due to uncertainty in the delay parameter.

In Chapter three, contributed by VLADIMIR RĂSVAN, a dialectic between propagation systems and the associated functional differential equations is presented. It is focused on two models: the circulating fuel nuclear reactor, where the propagation is no longer lossless, and the overhead crane, where the parameters may be varying in space, inducing distortion in the propagation.

Chapter four, contributed by TOMÁŠ VYHLÍDAL, PAVEL ZÍTEK and KAREL PAULŮ, is devoted to the presentation of a laboratory heating system. A linear time delay model is proposed, the identification of the system parameter is performed, and a state feedback control is designed and implemented. Finally, simulations and measurements are compared.

The fifth chapter is contributed by CÉLINE CASENAVE and GÉRARD MONTSENY. An original method devoted to the optimal identification of a wide class of complex linear operators involving delay components is presented, based on infinite dimensional state formulations of diffusive type. The method is applied to the identification of the acoustic impedance of absorbent materials designed for the noise reduction of aircraft engines.

Part V Delays Effects in Interconnected Systems and Networks

The study of time delay systems is to a large extent motivated by the research fields of teleoperation and control of interconnected systems and networks. The main problems involved concern the assessment of the influence of delays and saturation on performance, and the understanding of the dynamic properties of coupled systems from the properties of the individual agents and the interconnections. The synchronization of coupled systems is of particular importance in teleoperation, cryptography and biological applications.

The first chapter is contributed by AARON CLAUSET, HERBERT G. TANNER, CHAOUKI T. ABDALLAH and RAYMOND H. BYRNE. It presents an overview of problems at the intersection of control theory and complex networks research, thereby supporting the paradigm that issues of interest to the network theorist can have impact on control engineering design.

The second chapter, contributed by ISMAIL ILKER DELICE and RIFAT SIPAHI, aims at establishing connections between system's thinking in operations research and the field of time delay systems. In particular, delays in product deliveries, decision-making and lead times in manufacturing are discussed along with mathematical models exhibiting these delays. The stability maps are seen as promising decision-making tools in supply chain management and they carry rich educational information for training managers.

The third chapter, contributed by JOONO CHEONG, SILVIU-IULIAN NICULESCU, YONGHWAN OH and IRINEL CONSTANTIN MORĂRESCU, addresses the motion synchronization in shared virtual environments in the presence of communication delays. The case considered concerns multiple users interacting with the same dynamics.

The fourth chapter of the part is contributed by EMMANUEL NUÑO, LUIS BASAÑEZ and ROMEO ORTEGA. It deals with the stability analysis of teleoperators with time delay. A general Lyapunov-like function is proposed, from which different control schemes can be derived, for constant and variable time delays, with or without using the scattering transformation and with or without position tracking.

The fifth and sixth chapter are devoted to the synchronization of systems with delay in the coupling. In Chapter five, by FATIHCAN M. ATAY, it is shown that synchronization may happen, in a robust manner with respect to the parameters and in particular to the delays. Delays are often a rule rather than an exception in many practical situations, and it is of interest to investigate how synchronization is affected by their presence. Chapter six, contributed by TOSHIKI OGUCHI, TAKASHI YAMAMOTO and HENK NIJMEIJER, is devoted to the synchronization problem of coupled nonlinear systems with time-varying delays.

The last chapter is contributed by HÉCTOR JAVIER ESTRADA-GARCÍA, LUIS ALEJANDRO MÁRQUEZ-MARTÍNEZ and CLAUDE H. MOOG. A master-slave system consisting of two underactuated inverted pendulums with a constant time delay in the transmission of the measurements is considered. The control objective for the slave is to asymptotically track the reference trajectory imposed by the master. A causal controller is designed.

Last but not least, we would like to thank the editors for professionally handling the volume and the reviewers for their careful suggestions which improved the overall quality of the volume.

Nantes, Leuven, Gif-sur-Yvette, Boston,

May 2009

Jean Jacques Loiseau
Wim Michiels
Silviu-Iulian Niculescu
Rifat Sipahi

Contents

Part I

Analysis and Methodology

On Moment Stability of Linear Systems with a Stochastic Delay Variation

Erik I. Verriest[1] and Wim Michiels[2]

[1] Department of Electrical Engineering, Georgia Institute of Technology, Atlanta, GA 30332-0250, USA, Erik.Verriest@ece.gatech.edu
[2] Department of Computer Science, Katholieke Universiteit Leuven, Celestijnenlaan 200A, B-3001 Heverlee, Belgium Wim.Michiels@cs.kuleuven.be

Summary. Moment stability for a class of linear systems with stochastically varying delay is analyzed. It is assumed that the delay function has the form of a sawtooth with resets to zero occurring at the arrival times of a homogeneous Poisson process. Necessary and sufficient conditions for moment stability are derived. The results are applied to two examples, where the effect on stability of a deterministic approximation is also examined.

1 Introduction

This chapter establishes stability criteria for linear systems subjected to stochastically varying delays. Unlike the mainstream perturbation based approaches, where in the analysis a system with time-varying delay is implicitly seen as a perturbation of a system with a time-invariant delay, we aim at exploiting explicitly the distribution and probabilistic properties of the delay. This leads to the determination of the precise effect of the delay variation on stability, and, in particular, yields a characterization of the situations where the variation of the delay is beneficial and where it is not. >From an application point of view the problem is motivated by networked controlled systems and congestion control problems in communication networks, where the delays typically have an irregular behavior, but information is available about their probabilistic properties [10, 9].

While the existing results on stability of systems with time-varying delays are mostly perturbation based, at the one hand there exist some results on assessing the effects of a delay variation on stability, yet for the special case of a (deterministic) periodic variation, see, e.g., [4, 5] and [8]. On the other hand, several results are available on stability of stochastic systems with delay, see [14] for an overview, but they do not cover the case where precisely the delay variation is stochastic. Exceptions are formed by the papers [13] and [16]. The first extends the Lyapunov-Krasovskii approach in [15]. In the second the available information about the delay is an upper bound which can take two values with a given probability. Finally, jump linear systems with mode dependent delays are briefly discussed in the book [1].

J.J. Loiseau et al. (Eds.): Topics in Time Delay Systems, LNCIS 388, pp. 3–13.
springerlink.com

Because of the difficulty of the important problem of analyzing systems with stochastically varying delays and the fact that this research direction is barely explored, we restrict ourselves in this work to the simplified case where the stochastic delay function has the form of a sawtooth, with discontinuities occurring at the arrival times of a homogeneous Poisson process. The slope of the sawtooth is chosen equal to one, which not only simplifies the analysis, but allows to interpret the problem and the obtained criteria also in the context of stability of products of random matrices, as shall become clear. Furthermore, the inspiring paper [8] suggests to consider a fast delay variation and thus a high slope, but a slope larger than one is debatable from a system's theoretic point of view, as a system with time-varying delay $\tau(t)$ is no longer causal if the function $t - \tau(t)$ is allowed to decrease.

The structure of the chapter is as follows. In Section 2 the model is presented and briefly discussed. Also connections are made with stability of products of random matrices and the concept of the joint spectral radius. The main results are given in Sections 3 and concern moment stability criteria. The obtained results are illustrated in Section 4. Some concluding remarks in Section 5 end the paper.

The following notation will be used. For a square matrix A, $\rho(A)$, respectively $\alpha(A)$ denote its spectral radius, respectively its spectral abscissa, that is, $\alpha(A) := \max_{\lambda\in\mathbb{C}}\{\Re(\lambda) : \det(\lambda I - A) = 0\}$. The expected value of x is denoted by $\mathbf{E}x$. Throughout the paper $\|\cdot\|$ stands for an arbitrary vector norm or its induced matrix norm.

2 Poisson delay model as a random iteration

We consider the equation

$$\dot{x}(t) = A\,x(t) + B\,x(t-\tau(t)),\ x(0) \in \mathbb{R}^n, \tag{1}$$

where the delay τ has a sawtooth form, as shown in Figure 1. The slope is equal to one a.e. and the discontinuities (resets to zero) occur at time-instants

$$0 < t_1 < t_2 < \cdots$$

that correspond to the arrival times of a homogeneous Poisson process with rate λ. The inter arrival times $\delta_k := t_{t+1} - t_k$ are thus independent and identically distributed, and the probability density function of the inter arrival time δ_k is given by

$$P_{\delta_k}(t) = \lambda e^{-\lambda t}.$$

Note that the average inter arrival time is $1/\lambda$ and the average delay $1/2\lambda$. Let

$$\Omega := \mathbb{R}_+ \times \mathbb{R}_+ \times \cdots$$

be the sample space of the inter arrival times $\{\delta_1, \delta_2, \ldots\}$. With each outcome, $\omega \in \Omega$, corresponds a unique sample path for the delay, denoted by $\tau(t;\ \omega)$. Similarly, let the function

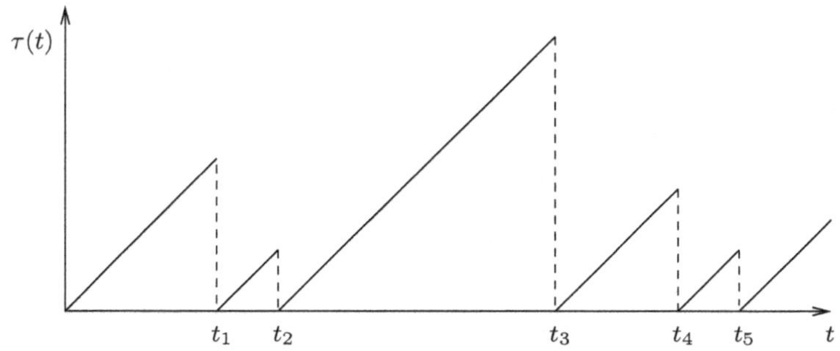

Fig. 1. Sample path for the stochastic delay function.

$$t \in \mathbb{R}_+ \mapsto x(t;\ y, \omega)$$

denote the induced sample path of the forward solution of (1) with initial condition $x(0) = y$. If no confusion is possible, we will write $\tau(t)$ and $x(t)$ instead of $\tau(t;\ \omega)$ and $x(t;\ y, \omega)$ in what follows, in order to simplify the notation.

For $t_k \leq t < t_{k+1}$ the equation (1) reduces to

$$\dot{x}(t) = A\,x(t) + B\,x(t_k), \tag{2}$$

which is readily integrated to

$$\begin{aligned} x(t) &= \mathrm{e}^{A(t-t_k)}x(t_k) + \int_{t_k}^{t} \mathrm{e}^{A(t-\theta)}Bx(t_k)\,\mathrm{d}\theta \\ &= \mathrm{e}^{A(t-t_k)}x(t_k) + \left(\mathrm{e}^{A(t-t_k)} - I\right)A^{-1}Bx(t_k). \end{aligned} \tag{3}$$

At the next Poisson arrival time we get

$$x(t_{k+1}) = \mathrm{e}^{A(t_{k+1}-t_k)}x(t_k) + \left(\mathrm{e}^{A(t_{k+1}-t_k)} - I\right)A^{-1}Bx(t_k). \tag{4}$$

Denoting $x(t_k)$ simply by x_k, we arrive at the discrete model

$$x_{k+1} = M(\delta_k)\,x_k, \tag{5}$$

where the iteration matrix M satisfies

$$M(\delta) = e^{A\delta}(I + C) - C, \;\; \delta \geq 0, \tag{6}$$

with

$$C = A^{-1}B.$$

Remark 1. From (5) we get

$$x_{k+1} = M(\delta_k)\cdots M(\delta_0)\,x_0,$$

hence, the stability problem under consideration can be interpreted as a stability problem of products of random matrices. In this context the work on the so-called joint spectral radius is important (see, e.g. [11] for an overview). There are however two distinct features. First, the results on the joint spectral radius almost exclusively concern products of matrices belonging to a finite set, while in our problem we consider a continuous parameterized set of matrices (6). Second, the joint spectral radius concerns stability in a worst-case scenario, while in what follows we study stability in a probabilistic framework and take into account the distribution of the matrices.

3 Moment stability conditions

Taking expectations of the iteration (5), we get

$$\begin{aligned} \mathbf{E}\, x_{k+1} &= \mathbf{E} M(\delta_k)\, \mathbf{E} x_k \\ &= \int_0^\infty M(t) \lambda e^{-\lambda t} \mathrm{d}t\, \mathbf{E} x_k \\ &= (\lambda I - A)^{-1} (\lambda I + B)\, \mathbf{E} x_k \end{aligned}, \tag{7}$$

under the condition $\lambda > \alpha(A)$, which assures the existence of the integral. Consequently, if the eigenvalues of the bilinear function $(\lambda I - A)^{-1}(\lambda I + B)$ lie within the unit circle, then the first moment converges to zero exponentially:

Theorem 1. *There exist constants $C > 0$ and $\gamma \in (0,\ 1)$ such that*

$$\|\mathbf{E} x_k\| \leq C \gamma^k\ \|x_0\|$$

if and only if $\lambda > \alpha(A)$ and, in addition,

$$\rho\left((\lambda I - A)^{-1}(\lambda I + B)\right) < 1. \tag{8}$$

Next we consider the stability of the second moment. From (5) we have

$$x_{k+1} x_{k+1}^T = M(\delta_k) x_k x_k^T M(\delta_k)^T,$$

hence,

$$\begin{aligned} \mathbf{E}\, x_{k+1} x_{k+1}^T &= \mathbf{E} M(\delta_k) x_k x_k^T M(\delta_k)^T \\ &= \mathbf{E}\left\{ M(\delta_k)(\mathbf{E} x_k x_k^T) M(\delta_k)^T \right\}. \end{aligned}$$

If we define

$$P_k = \mathbf{E} x_k x_k^T, \tag{9}$$

then we arrive at the iteration

$$P_{k+1} = \mathbf{E} M(\delta_k) P_k M(\delta_k)^T. \tag{10}$$

Vectorizing this expression yields:

$$\text{vec } P_{k+1} = \mathbf{E}\left(M(\delta_k) \otimes M(\delta_k)\right) \text{vec } P_k,$$

or

$$\text{vec } P_{k+1} = U \text{ vec } P_k, \tag{11}$$

where

$$U := \mathbf{E}(M(\delta_k) \otimes M(\delta_k)) = \int_0^\infty M(t) \otimes M(t)\ \lambda e^{-\lambda t} dt. \tag{12}$$

Note that the integral exists if $\lambda > 2\alpha(A)$. From (11) we obtain:

Theorem 2. *There exist constants $C > 0$ and $\gamma \in (0,\ 1)$ such that*

$$\|\mathbf{E}x_k x_k^T\| \leq C\gamma^k\ \|x_0 x_0^T\|$$

if and only if $\lambda > 2\alpha(A)$ and, in addition,

$$\rho(U) < 1, \tag{13}$$

where U is defined in (12).

Remark 2. From (10) we obtain

$$\begin{aligned} P_{k+1} &= \mathbf{E}\, M(\delta_k) P_k M(\delta_k)^T \\ &= \mathbf{E}\left(e^{A\delta_k}(I+C) - C\right) P_k \left(e^{A^T\delta_k}(I+C^T) - C^T\right). \end{aligned} \tag{14}$$

By defining

$$X_k := \frac{1}{\lambda}\, \mathbf{E}\, e^{A^T\delta}(I+C)P_k(I+C^T)e^{A\delta},$$

which satisfies the Lyapunov equation

$$AX_k + X_k A^T - \lambda X_k = (I+C)P_k(I+C^T), \tag{15}$$

we can rewrite (14) as

$$P_{k+1} = \lambda X_k - \lambda C P_k (I+C^T)(\lambda I - A^T)^{-1} - \lambda(\lambda I - A)^{-1}(I+C)P_k C^T + C P_k C^T. \tag{16}$$

The expressions (15) and (16) define a *two-term recursion* for the iteration (10) or (11). This iteration can be used in a power method for the computation of $\rho(U)$ in the application of Theorem 2.

Remark 3. In the scalar case,

$$\dot{x}(t) = ax(t) + bx(t - \tau(t)),\ a, b \in \mathbb{R}, \tag{17}$$

explicit expressions for the stability of higher moments can be obtained. Due to the independence of the inter-arrival times δ_i we have for $p \geq 1$,

$$\mathbf{E}\, x_k^p = \prod_{i=0}^{k-1} \mathbf{E}\left[\mathrm{e}^{a\delta_i}(1 + b/a) - b/a\right]^p \mathbf{E}x_0^p, \tag{18}$$

hence, the p-th moment is stable if

$$\left|\mathbf{E}\left[\mathrm{e}^{a\delta}(1 + b/a) - b/a\right]^p\right| = \left|\int_0^\infty \left[\mathrm{e}^{at}(1 + b/a) - b/a\right]^p \lambda \mathrm{e}^{-\lambda t}\, \mathrm{d}t\right| < 1. \tag{19}$$

The integral only converges for $b/a \neq -1$, if $ap < \lambda$. For $b/a = -1$, the expected value is $(-1)^p$. Hence the line $b = -a$ establishes a boundary between stability and instability for all moments. In addition $ap < \lambda$ is necessary for stability of the p-th moment. We note that using the binomial formula the integral in (5) can be computed explicitly as

$$\int_0^\infty \left[\mathrm{e}^{at}(1 + b/a) - b/a\right]^p \lambda \mathrm{e}^{-\lambda t}\, \mathrm{d}t = 1/a^p \sum_{k=0}^{p} \binom{p}{k} \frac{(a+b)^k(-b)^{p-k}}{1 - \frac{ak}{\lambda}}. \tag{20}$$

4 Examples

4.1 Scalar system

We revisit the equation (17), where τ is a sawtooth function, as described in Section 2. Since its switching is determined by a homogeneous Poisson process with rate λ, it is readily seen that the average value of the delay, τ^{av}, satisfies

$$\tau^{\mathrm{av}} = \frac{1}{2\lambda}. \tag{21}$$

In order to analyze the stability in the $(a, b, \tau^{\mathrm{av}})$ parameter space, we consider the dimensionless time

$$t^{(\mathrm{new})} = t^{(\mathrm{old})}/\tau^{\mathrm{av}}, \tag{22}$$

which brings the equation in the form

$$\dot{x}(t) = \alpha x(t) + \beta x(t - r(t)), \tag{23}$$

where

$$\alpha = a\tau^{\mathrm{av}}, \; \beta = b\tau^{\mathrm{av}}$$

and

$$r(t) = \frac{\tau\left(\tau^{\mathrm{av}} t\right)}{\tau^{\mathrm{av}}}. \tag{24}$$

Note that the discontinuities of the normalized function r are now determined by a Poisson process with rate $1/2$.

An application of Theorem 1 leads to the following stability condition for the first moment:

$$\alpha - 1 < \beta < -\alpha.$$

By Theorem 2 the second moment is stable if

$$\left| \int_0^\infty \left(\mathrm{e}^{\alpha t}\left(1 + \frac{\beta}{\alpha}\right) - \frac{\beta}{\alpha} \right)^2 \frac{1}{2}\mathrm{e}^{-\frac{1}{2}t} dt \right| < 1.$$

A simple calculation yields that this condition is equivalent with

$$\alpha - \frac{1}{2} < \beta < -\alpha.$$

The result of this computation is shown on Figure 1(a). The dashed line bounds the stability region for a time-invariant delay, $\tau equiv \tau^{\mathrm{av}}$, which is described by

$$\gamma(\alpha) < \beta < -\alpha,$$

where the function $\alpha \mapsto \gamma(\alpha)$ is implicitly defined by

$$\begin{cases} \alpha = \frac{\omega}{\tan(\omega)}, \\ \beta = -\frac{\omega}{\sin(\omega)}, \; \omega \in (0, \; \pi). \end{cases}$$

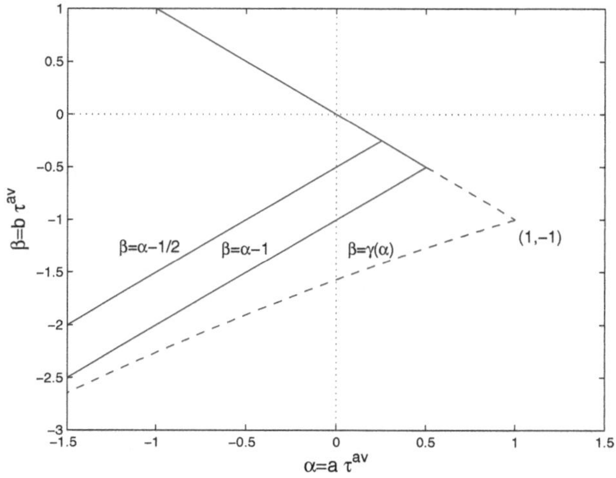

Fig. 2. Stability regions of (23) in the (a, b, τ^{av}) space, obtained from Theorems 1-2 (bounded by the full curve). Stability region for a time-invariant delay τ^{av} (bounded by the dashed curve)

4.2 Characterization of stabilizable second order systems

We consider the stabilizability of the second-order system

$$\dot{x}(t) = Ax(t) + Hu(t - \tau(t)), \quad x(t) = [x_1(t)\ x_2(t)]^T \in \mathbb{R}^2, \quad u(t) \in \mathbb{R}, \tag{25}$$

where τ is as in Section 2 and the pair (A, H) is controllable. Without loss of generality we assume that (25) is in the reachable canonical form, that is,

$$A = \begin{bmatrix} 0 & 1 \\ -a_2 & -a_1 \end{bmatrix}, \quad H = \begin{bmatrix} 0 \\ 1 \end{bmatrix}.$$

The control law takes the form

$$u(t) = -k_2 x_1(t) - k_1 x_2(t),$$

with k_1 and k_2 the controller parameters. Note that the closed-loop system is of the form (1), with

$$B = \begin{bmatrix} 0 & 0 \\ -k_2 & -k_1 \end{bmatrix}.$$

Our goal is to make a complete characterization of the class of stabilizable second-order systems, that is, to determine the values of (a_1, a_2, λ) for which stabilizing values of k_1 and k_2 exist. The stability notions under consideration concern the stability of the first and second moment, for which necessary and sufficient conditions were derived in Section 3.

According to Theorem 1 the first moment is stable if $\lambda > \alpha(A)$ and

$$\rho\left((\lambda I - A)^{-1}(\lambda I + B)\right) < 1.$$

Since

$$(\lambda I - A)^{-1}(\lambda I + B) = \lambda(\lambda I - A)^{-1} - (\lambda I - A)^{-1} H [k_2\ k_1]$$

the stabilization problem can be solved by classical pole placement techniques. Because the corresponding controllability condition is always fulfilled, the first moment is stabilizable for all values of (a_1, a_2, λ).

To characterize the stabilizability of the second moment we first apply the time-transformation (22), followed by a scaling of x_2, to bring the closed-loop system in the form

$$\begin{bmatrix} \dot{x}_1(t) \\ \dot{\xi}_2(t) \end{bmatrix} = \begin{bmatrix} 0 & 1 \\ -\alpha_2 & -\alpha_1 \end{bmatrix} \begin{bmatrix} x_1(t) \\ \xi_2(t) \end{bmatrix} + \begin{bmatrix} 0 & 0 \\ -\kappa_2 & -\kappa_1 \end{bmatrix} \begin{bmatrix} x_1(t - r(t)) \\ \xi_2(t - r(t)) \end{bmatrix},$$

where r is the normalized delay function defined in (21) and (24), and

$$\alpha_1 = a_1 \tau^{\mathrm{av}},\ \alpha_2 = a_2 \left(\tau^{\mathrm{av}}\right)^2,\ \ \kappa_1 = k_1 \tau^{\mathrm{av}},\ \kappa_2 = k_2 \left(\tau^{\mathrm{av}}\right)^2,\ \xi_2 = \tau^{\mathrm{av}} x_2.$$

The advantage of this transformation is that the three plant parameters (a_1, a_2, λ) (or, equivalently, $(a_1, a_2, \tau^{\mathrm{av}})$) are condensed in two parameters (α_1, α_2).

According to Theorem 2, a necessary stability condition for the second moment is given by $\lambda > 2\alpha(A)$, which is equivalent with

$$\alpha_1 > -\frac{1}{2},\ \ \alpha_2 > -\frac{1}{4}\alpha_1 - \frac{1}{16}. \tag{26}$$

Furthermore, if (26) holds, then the second moment is stabilizable if and only if

$$\min_{\kappa_1, \kappa_2} \rho(U) < 1,$$

with U, defined by (12), depending on (κ_1, κ_2). Since the dimensions of U are larger than two, an assignment of all eigenvalues of U is not possible, which suggests to determine stabilizability by solving directly the optimization problem

$$\min_{\kappa_1, \kappa_2} \rho(U).$$

It is well known in the context of eigenvalue optimization that the objective function, $(\kappa_1, \kappa_2) \mapsto \rho(U)$, is typically nondifferentiable, even non Lipschitz, but it is smooth almost everywhere [3]. Therefore, we have used the gradient sampling algorithm of [2] in our computations, which generalizes the steepest descent method to this type of problems and which has already been successfully applied to fixed-order stabilization problems for time-delay system in [12, 6]. It only requires a routine that returns the value of the objective function and its gradient, whenever it exists. It is important to note that it is not necessary to compute explicitly the integral in (12) to evaluate the objective function $\rho(U)$. Instead we have used a Krylov method (more precisely the Arnoldi method) to compute the extreme eigenvalue of U. Such method only relies on products of U with vectors, which can be computed using the formulae (15) and (16). The gradients were computed by means of finite differences. The application of the algorithm leads to a monotone decrease of the objective function towards a (nonsmooth) minimum.

Using the above elements, we have determined the class of stabilizable systems in the (α_1, α_2) parameter space as follows. First, we have minimized $\rho(U)$ using the gradient sampling algorithm for values of (α_1, α_2) chosen on a coarse grid, to estimate the position of the boundary between stabilizable and non-stabilizable systems in the (α_1, α_2) plane. Next we have used a bisection based rootfinder on the equation

$$\min_{(\kappa_1,\kappa_2)} \rho(U) - 1 = 0,$$

where a_1 or a_2 is freed, to determine the exact position of the boundary. Surprisingly, the final result is very simple and expresses that the second moment is stabilizable if and only if

$$\alpha_1 > -\frac{1}{4}, \quad \alpha_2 > -\frac{1}{4}\alpha_1 - \frac{1}{16}.$$

A visualization is given in Figure 3. The solid curve correspond to the stabilizability boundary, that is, the curve separating plant parameters for which the second moment is stabilizable and for which it is not. For comparison, the dashed curve is the stabilizability boundary for the case of a time-invariant delay, $\tau(t) \equiv \tau^{\mathrm{av}}$. For the computation of the latter boundary we refer to [7].

5 Conclusions

The stability properties of a class of linear systems with stochastically varying delays were investigated. The model was restricted to causal form, implying a strict upper bound on the delay variation. Since the delay function itself is not bounded from above and we are particularly interested in the case where the matrix A in (1) is not Hurwitz, methods based on viewing the system as a perturbation of a delay-free system are too restrictive. Likewise, the classical stability notions like uniform asymptotic stability are also too conservative. In contrast to this, stability was studied within a probabilistic framework. Whereas various notions for stochastic stability exist, we singled out moment stability and the corresponding stability criteria were derived. It should be noted that the criteria can also be interpreted in the context of stability of products of random matrices, due to the structure of the equation (5).

The obtained results were illustrated by means of two examples. The first one concerned the stability analysis of scalar systems. The second example concerned a complete characterization of the stabilizable second-order systems. In this, the criteria for moment stability were combined with eigenvalue optimization techniques. The examples further demonstrate the importance of taking the precise properties

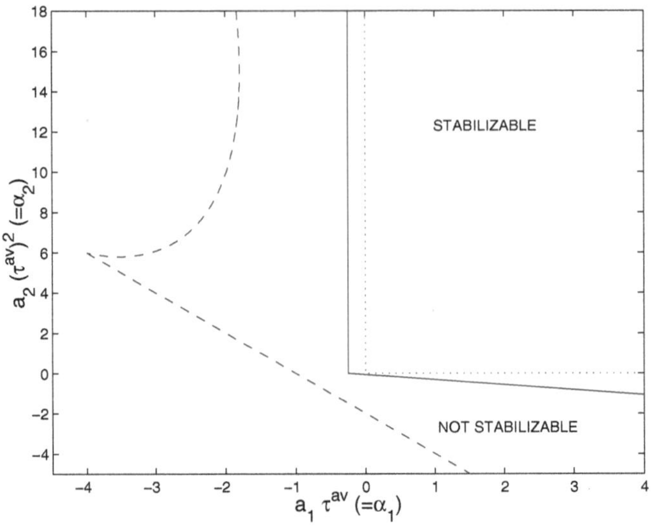

Fig. 3. Characterization of second-order systems, for which the second moment is stabilizable. The systems are parameterized by a_1, a_2 and τ^{av}. The solid curve is the stabilizability boundary for the stochastic delay function $\tau(t)$. The dashed curve corresponds to a time-invariant delay τ^{av}.

of the delay variation explicitly into account. Indeed, they illustrate that an approximation of the stochastic delay with a time-invariant delay may lead to considerably different stability / stabilizability regions in parameter spaces.

Acknowledgements

This paper present results of the Belgian Programme on Interuniversity Poles of Attraction, initiated by the Belgian State, Prime Minister's Office for Science, Technology and Culture, and of OPTEC, the Optimization in Engineering Centre of K.U.Leuven.

References

1. Boukas, E.K., Liu, Z.K.: Deterministic and stochastic time-delay systems, Birkhäuser (2002)
2. Burke, J.V., Lewis, A.S., Overton, M.L.: A robust gradient sampling algorithm for nonsmooth, nonconvex optimization. SIAM Journal on Optimization 15(3), 751–779 (2005)
3. Burke, J.V., Overton, M.L.: Differential properties of the spectral abscissa and the spectral radius for analytic matrix valued mappings. Nonlinear Analysis: Theory, Methods and Applications 23, 467–488 (1994)

4. Butcher, E.A., Ma, H., Bueler, E., Averina, V., Szabo, Z.: Stability of linear time-periodic delay-differential equations via Chebyshev polynomials. International Journal for Numerical Methods in Engineering 59, 895–922 (2004)
5. Insperger, T., Stepan, G.: Stability analysis of turning with periodic spindle speed modulation via semi-discretisation. Journal of Vibration and Control 10, 1835–1855 (2004)
6. Michiels, W., Niculescu, S.I.: Stability and stabilization of time-delay systems. An eigenvalue based approach. In: Advances in Design and Control, p. 12. SIAM, Philadelphia (2007)
7. Michiels, W., Roose, D.: Limitations of delayed state feedback: a numerical study. International Journal of Bifurcation and Chaos 12(6), 1309–1320 (2002)
8. Michiels, W., Van Assche, V., Niculescu, S.I.: Stabilization of time-delay systems with a controlled, time-varying delay and applications. IEEE Transactions on Automatic Control 50(4), 493–504 (2005)
9. Roesch, O., Roth, H., Niculescu, S.I.: Remote control of mechatronic systems over communication networks. In: Proc. 2005 IEEE International Conference on Mechatronics & Automation, Niagara Falls, Canada (2005)
10. Tarbouriech, S., Abdallah, C.T., Chiasson, J.N.: Advances in communication control networks. Lecture Notes in Control and Information Sciences, p. 308. Springer, Heidelberg (2005)
11. Theys, J.: Joint Spectral Radius: Theory and Approximations. PhD thesis, Universié Catholique de Louvain, Louvain-la-Neuve, Belgium (2005), `http://www.inma.ucl.ac.be/~blondel/05thesetheys.pdf`
12. Vanbiervliet, J., Verheyden, K., Michiels, W., Vandewalle, S.: Stabilization of time-delay systems using nonsmooth optimisation techniques. ESAIM Control Optimisation and Calcalus of Variations 14(3), 478–493 (2008)
13. Verriest, E.I.: Stability of systems with state-dependent and random delays. IMA Journal of Mathematical Control and Information 19, 103–114 (2002)
14. Verriest, E.I.: Asymptotic properties of stochastic time-delay systems. In: Niculescu, S.I., Gu, K. (eds.) Advances in Time-Delay Systems. Lecture Notes in Computational Science and Engineering, vol. 38, pp. 389–420. Springer, Heidelberg (2004)
15. Verriest, E.I., Florchinger, P.: Stability of stochastic systems with uncertain time delays. System & Control Letters 24(1), 41–47 (1995)
16. Yue, D., Tian, E., Wang, Z., Lam, J.: Stabilization of systems with probabilistic interval input delays and its applications to networked control systems. IEEE Transactions on Systems, Man and Cybernetics A (2008) (in press)

Stability Analysis of Neutral Systems: A Delay-Dependent Criterion

Tudor C. Ionescu[1] and Radu Ştefan[2]

[1] Dept. of Mathematics and Natural Sciences, University of Groningen, t.c.ionescu@rug.nl
[2] Dept. of Automatic Control and Computers, University Politehnica Bucharest, r.stefan@clicknet.ro

Summary. The main goal of this paper is to investigate the stability of neutral systems, by using the comparison method introduced by Zhang, Knospe and Tsiotras [13] for the standard time-delay case. The central idea consists in replacing the delay by two appropriate Padé approximations, obtaining two comparison systems whose robust stability give a simple delay-dependent stability condition for the (nominal) neutral system. Such a condition guarantees, as in the standard time-delay case, an a priori upper bound for the degree of conservatism induced by the comparison method. This degree depends on the order of the Padé approximation and its estimation makes the difference between this method and other approaches encountered in the literature. The effective computation of the delay margin is reduced to a set of finite dimensional LMI's .

1 Introduction

An important problem in asymptotic stability issues of time delay systems consists of finding the set of values of the delay for which the system achieves asymptotic stability. If this set is $[0,\infty]$, then the system exhibits delay independent asymptotic stability. There are many approaches to tackle this problem, for instance in [1] an analytical solution is given, based on the analysis of a radius function which depends continuously on the delay. A matrix-pencil approach can be found in [6], whereas in [9, 10] a direct method is proposed based on the involved analysis of clusters of infinite roots given by the characteristic equation associated to the system. Another way to treat this problem stems from robustness concepts, [7, 8, 12]. In [13] the problem is turned into one of robust stability analysis of a linear (comparison) system, free of delays, but with uncertain parameters, using "covering" and "under-covering" sets of the delay element $e^{-\tau s}$. The approach relies on exploiting some remarkable properties of the Padé approximation of $e^{-\tau s}$. The robust stability of the system created using the outer covering set turns into a sufficient condition for the asymptotic stability of the original system. The result obtained in this way has a very low degree of conservatism, with an upper bound depending only on the order of the Padé approximation, because the robust stability of the system created using the inner covering set gives a necessary asymptotic stability condition. Moreover, from numerical point of view, the conditions yield an algorithm which is based on the test of a *finite set of LMI*, which can be implemented using e.g. the LMI Toolbox, [2].

J.J. Loiseau et al. (Eds.): Topics in Time Delay Systems, LNCIS 388, pp. 15–24.
springerlink.com

In the multiple delay case, the number of LMI is expected to depend exponentially on the number of delays of the system.

The goal of this paper is to extend this approach to the case of neutral systems (useful for instance in the stability analysis performed in [11]). In Section 2 we will formulate the problem together with some analytical properties peculiar to neutral systems. Using the work of Datko [1], we give an analytical stability criterion for the case of neutral systems. Combining this criterion with the aforementioned comparison method, in Section 3 we analyze the stability of neutral systems with a single delay. The structure of the comparison systems will be more involved than in the standard time-delay case and so will be the corresponding LMI's. Everything is considered under basic assumptions made with respect to the parameters of the neutral system, ensuring that the free of delays system is stable. We conclude with a discussion about the possible extensions to the multiple delay case and we point out some applications of the present approach.

2 Problem Formulation

Consider a neutral system with a single delay:

$$\dot{x}(t) - B\dot{x}(t-\tau) = A_0 x(t) + A_1 x(t-\tau). \tag{1}$$

Its characteristic equation is given by:

$$\det(G_\tau(s)) = 0, \tag{2}$$

where $G_\tau(s) = s\left(I - Be^{-s\tau}\right) - A_0 - A_1 e^{-s\tau}$. Define the spectral limit of $G_\tau(s)$ as

$$\sigma(G_\tau) = \sup\{Re\ s \mid \det\left(G_\tau(s)\right) = 0\}, \quad \text{if} \det\left(G_\tau(s)\right) \text{ not identical } 0.$$

If $\det\left(G_\tau(s)\right) = 0$ for every $s \in \mathbb{C}$, then, by definition, $\sigma(G_\tau) = -\infty$.

The goal is to determine how the stability properties of (1) depend on a change in the delay τ. In particular, if the neutral system (1) is asymptotically stable for $\tau = 0$, we want to find (or at least to approximate) the largest value of τ that preserves this property. In other words, it all boils down to finding the largest delay interval $[0, \overline{\tau})$, $\overline{\tau} > 0$ where the system (1) is asymptotically stable.

Subsequently, we present some properties of the function $\sigma(G_\tau)$. Necessary and sufficient conditions for asymptotic stability - actually exponential stability - have been obtained in [1], based on the continuity of the spectral limit as a real function of the variable τ.
Consider the analytical functions $A_\tau(s) = A_0 + A_1 e^{-s\tau}$ and $B_\tau(s) = I - Be^{-s\tau}$. One can immediately see that

$$G_\tau(s) = sB_\tau(s) - A_\tau(s). \tag{3}$$

The results below - Theorems 1, 2 and 3 - are practically special cases (single delay situation) of Theorems 2.3, 3.2 and 3.4 in [1].

Theorem 1. *The function $G_\tau(s)$ has a spectral limit function $\sigma(G_\tau)$ which is continuous on $(0, \infty)$. If, in addition, B is a Schur matrix (all eigenvalues belong to the unit disc), then $\sigma(G_\tau)$ is continuous on $[0, \infty)$.*

The following results give necessary and sufficient exponential stability conditions that will provide an analytic method for determining the largest delay interval for which (1) remains asymptotically stable.

The method consists in deciding whether the system (1) is asymptotically stable for given τ_0. It is known from the general theory of delayed and neutral differential equations [6], that, if $\sigma(G_{\tau_0}) < 0$, then the system (1) is asymptotically stable for τ_0. Thus, if $\sigma(G_0) < 0$ and $\tau_0 \in [0, \overline{\tau})$, where $\overline{\tau}$ is the smallest number for which $\sigma(G_{\overline{\tau}}) = 0$, it follows by the continuity property in Theorem 1 that $\sigma(G_{\tau_0}) < 0$ and therefore the system (1) is asymptotically stable for τ_0.

Theorem 2. *Let the system (1) be under the assumption that the following conditions are satisfied:*
(i) B is a Schur matrix;
(ii) $\sigma(G_0) < 0$.
Then the following statements all hold:
1. There exists a maximal interval $[0, \overline{\tau})$, $\overline{\tau} > 0$ such that $\sigma(G_\tau) < 0$, $(\forall)\ \tau \in [0, \overline{\tau})$, hence for all τ in this interval the system (1) is exponentially stable.
2. Moreover, if $\overline{\tau} < \infty$, then $\sigma(G_{\overline{\tau}}) = 0$ and the system has a periodic solution for $\tau = \overline{\tau}$.

>From this, the next stability criterion follows:

Theorem 3. *If the system (1) satisfies the following:*
1. B is a Schur matrix;
2. $B_0^{-1}(s)A_0(s)$ is Hurwitz;
3. $\det G_\tau(j\omega) \neq 0$, for all τ, $\omega \geq 0$, then (1) is exponentially stable for all $\tau \geq 0$.
A practical consequence of Theorem 3 is that $\det G_\tau(j\omega) \neq 0$ becomes an essential condition in determining the largest interval $[0, \overline{\tau})$ that ensures asymptotic stability.

Consider the system (1) and its associated characteristic equation (2). For $s = j\omega$, $\omega \geq 0$ the equation (2) becomes:

$$\det[j\omega(I - Be^{-j\omega\tau}) - A_0 - A_1 e^{-j\omega\tau}] = 0 \Leftrightarrow$$
$$\det[j\omega(I - B) - A_0 - A_1 + j\omega B(1 - e^{-j\omega\tau}) + A_1(1 - e^{j\omega\tau})] = 0.$$

Let $\overline{B} = I - B$, $\overline{A} = A_0 + A_1$. We obtain

$$\det[(j\omega\overline{B} - \overline{A}) - (j\omega B + A_1)(e^{-j\omega\tau} - 1)] = 0 \tag{4}$$

We make the following natural assumptions (for the stability of the system free of delays, $\tau = 0$): matrices $\overline{B}$ and $j\omega\overline{B} - \overline{A}$ are invertible. Then, equation (4) can be rewritten as:

$$\det[I - (j\omega\overline{B} - \overline{A})^{-1}(j\omega B + A_1)\Phi(j\omega, \tau)] = 0 \tag{5}$$

where $\Phi(j\omega, \tau) = \text{diag}\left(e^{-j\omega\tau} - 1\right)$.

Let

$$G(s) = (s\overline{B} - \overline{A})^{-1}(sB + A_1). \tag{6}$$

In order to determine a standard state-space realization for $G(s)$ let us notice that since $\overline{B}$ is invertible one has

$$(s\overline{B} - \overline{A})^{-1} = \overline{B}^{-1}\left(sI - \overline{A}\,\overline{B}^{-1}\right)^{-1} \tag{7}$$

and

$$(s\overline{B} - \overline{A})^{-1} s\overline{B} = \left(s\overline{B} - \overline{A}\right)^{-1}\overline{A} + I \tag{8}$$

By choosing now $\widetilde{G}(s) = C_G(sI - A_G)^{-1}B_G + D_G$ where

$$A_G = \overline{A}(\overline{B})^{-1},\ B_G = A_1 + (\overline{A})\overline{B}^{\,-1}B,\ C_G = (\overline{B})^{-1},\ D_G = (\overline{B})^{-1}B \qquad (9)$$

one can immediately check by using (7) and (8) that $G(s) = \widetilde{G}(s)$. Hence equation (5) becomes

$$\det[I - \widetilde{G}(j\omega)\Phi(j\omega,\tau)] = 0. \qquad (10)$$

According to Theorems 1, 2 and 3, if B is a Schur matrix, $(\overline{B})^{-1}\overline{A}$ is Hurwitz and the equality (10) does not hold for any $\omega \geq 0$, then the system (1) is asymptotically stable.
We actually proved the following result:

Lemma 4. *Assume that system (1) satisfies the following conditions:*
(i) B is a Schur matrix;

(ii)$(\overline{B})^{-1}\overline{A}$ is Hurwitz, where $\overline{A} = A_0 + A_1$ and $\overline{B} = I - B$;

(iii) $\det[I - \widetilde{G}(j\omega)\Phi(j\omega,\tau)] \neq 0,\ (\forall)\omega \geq 0,\ \tau \in [0,\overline{\tau})$.
Then the system (1) is asymptotically stable for all $\tau \in [0,\overline{\tau})$ and $\Phi(j\omega,\tau) = \mathrm{diag}\left(e^{-j\omega\tau} - 1\right)$.

Remark 5. We succeeded in fact to turn the initial problem into a particular form which allows to follow closely the comparison method in [13] in order to verify condition (iii). Assumptions (i) and (ii) are standard stability tests, therefore much easier to check.

The main idea of the above mentioned approach is to cover the set $\Phi(j\omega,\tau)$ with another value set which does not depend on τ, say $\underline{\Phi}(\omega)$, such that $\det[I - \widetilde{G}(j\omega)\Delta(j\omega)] \neq 0$, for all $\Delta(j\omega) \in \underline{\Phi}(\omega)$ and for each $\omega \geq 0$. Under these circumstances, condition (iii) in Lemma 4, is automatically implied by the *robust stability* of the feedback interconnection of $\widetilde{G}$ and Δ, denoted by $\Sigma(\widetilde{G},\Delta)$ (here Δ can be regarded as an uncertainty: we only know that Δ it belongs to the value set $\underline{\Phi}(\omega)$). We will subsequently call such an interconnection, a *comparison system*.

The construction of $\underline{\Phi}(\omega)$ and the appropriate Δ relies on rational approximations of the delay element.

Consider the Padé approximation of the delay element $e^{-\tau s}$

$$R_m(s) = \frac{D_m(s)}{D_m(-s)}, \text{ where } D_m(s) = \sum_{k=0}^{m} c_k s^k, \quad c_k = (-1)^k \frac{(2m-k)!m!}{(2m)!k!(m-k)!}.$$

One can introduce the following value sets

$$\Omega_d(\omega, \overline{\tau}) = \left\{ e^{-j\omega\tau} \mid \tau \in [0, \overline{\tau}] \right\},$$

$$\Omega_o(\omega, \overline{\tau}) = \{ R_m(-j\omega\alpha_m\theta) \mid \theta \in [0, \overline{\tau}] \},$$
$$\Omega_i(\omega, \overline{\tau}) = \{ R_m(-j\omega\theta) \mid \theta \in [0, \overline{\tau}] \}$$

where $\alpha_m = \frac{\omega_{cm}}{2\pi}$ and ω_{cm} is the smallest frequency where the phase has the value -2π, i.e.: $\omega_{cm} = \min\{\omega > 0 \,| R_m(j\omega) = 1\}$. According to Lemma 6 in [13] (which is using results from [4]), we have the following remarkable properties:

1. All poles of $R_m(s)$ are in the open left complex half plane.
2. For any $\overline{\tau} > 0$ and any $\omega > 0$, one has $\Omega_i(\omega, \overline{\tau}) \subseteq \Omega_d(\omega, \overline{\tau}) \subseteq \Omega_o(\omega, \overline{\tau})$ and $\lim_{m \to \infty} \alpha_m = 1$.

Thus, Ω_o and Ω_i are the "covering" and the "under-covering" value sets, respectively. Accordingly, let $\Delta_o(s, \theta) := \text{diag}\,(R_m(\alpha_m\theta s) - 1)$ and $\Delta_i(s, \theta) := \text{diag}\,(R_m(\theta s) - 1)$, where $\theta > 0$ is an *independent* variable. The feedback interconnection of $\widetilde{G}$ with Δ_o and Δ_i, will be denoted by $\Sigma_o(\theta)$ and $\Sigma_i(\theta)$, respectively.
The central idea of this methodology is that the asymptotic stability problem of a family of infinite dimensional systems, parameterized by τ, has been transferred to a robust stability problem of two finite dimensional systems. The next result (see Theorems 1 and 2 in [13]) gives necessary and sufficient asymptotic stability conditions for the neutral system, in terms of the robust stability of Σ_o and Σ_i.

Theorem 6. *[13]*
1. If the comparison system Σ_o is robustly stable for all $\theta \in [0, \overline{\tau}]$, then the neutral system (1) is asymptotically stable for all $\tau \in [0, \overline{\tau}]$.
2. If the neutral system (1) is asymptotically stable for all $\tau \in [0, \overline{\tau}]$, then the comparison system Σ_i is robustly stable for all $\theta \in [0, \overline{\tau}]$.

Another important consequence of considering both necessary and sufficient asymptotic stability conditions, is that one can estimate the degree of conservatism induced by checking only the sufficiency: the robust stability of Σ_o.

Definition 7. Let $\mathcal{P}$ be a condition that ensures the stability of the system (1) on $[0, \overline{\tau}]$ and which guarantees the delay margin

$$\overline{\tau}^*_{\mathcal{P}} = \sup \{ \overline{\tau} \mid \mathcal{P} \text{ is true for any } \tau \in [0, \overline{\tau}] \}.$$

The degree of conservatism of the condition $\mathcal{P}$ is defined as:

$$d.o.c.(\mathcal{P}) = \frac{\overline{\tau}^* - \overline{\tau}^*_{\mathcal{P}}}{\overline{\tau}^*},$$

where

$$\overline{\tau}^* = \sup \{ \overline{\tau} \,|(1) \text{is asymptotically stable for any } \tau \in [0, \overline{\tau}] \}$$

is the actual delay margin introduced by Theorem 2.

Remark 8. The degree of conservatism is a quantitative measure. Obviously $0 < d.o.c. \leq 1$. If $d.o.c. = 0$, then $\mathcal{P}$ is a necessary and sufficient condition.
It can be shown (see again [13]) that the degree of conservatism of Theorem 6 depends only on m, the order of the Padé approximation of the delay element, that is,

$$d.o.c.(Theorem\ 6) \leq \frac{\alpha_m - 1}{\alpha_m}.$$

This d.o.c. is independent on the problem data and $d.o.c.(Theorem\ 6) \overset{m\to\infty}{\longrightarrow} 0$. For $m = 3, 4$ and 5, one has $\dfrac{\alpha_m - 1}{\alpha_m} \approx 18.9\%$, 3.05% and 0.361%, respectively.

3 Estimation of $\overline{\tau}$

In this section, based on the previous results - Lemma 4 and Theorem 6 - we give a LMI condition that will allow us to determine whether or not, for a given $\overline{\tau}$, the system (1) exhibits asymptotic stability for all $\tau \in [0, \overline{\tau}]$. Furthermore, exactly as in Theorem 4 in [13], we retrieve an explicit formula for the maximal stability interval.
Let $\{A_P, B_P, C_P, D_P\}$ be a state-space realization of diag $(R_m(\alpha_m \theta s) - 1)$. Obviously, $\{\theta^{-1}A_P, \theta^{-1/2}B_P, \theta^{-1/2}C_P, D_P\}$ is a state-space realization for $\Delta_o(s,\theta)$, $\theta > 0$ being an independent variable. Denote by $\{A_L, B_L, C_L, D_L\}$ a realization of Σ_o - the feedback interconnection of $\widetilde{G}$ and Δ_o. Elementary matrix manipulations yield

$$A_L(\theta) = \begin{bmatrix} A_s & \theta^{-1/2}B_s \\ \theta^{-1/2}C_s & \theta^{-1}\widehat{A}_s \end{bmatrix} \tag{11}$$

where with $S = I - D_G D_P$, $\widehat{S} = I - D_P D_G$ one defines

$$\begin{aligned} A_s &= A_G + B_G D_P S^{-1} C_G,\ B_s = B_P S^{-1} C_G, \\ C_s &= B_G \widehat{S}^{-1} C_P,\ \widehat{A}_s = A_P - B_P D_G \widehat{S}^{-1} C_P. \end{aligned} \tag{12}$$

According to [13], if for every $0 < \theta \leq \overline{\tau}$ there exists $X(\theta) > 0$ for which the Lyapunov inequality

$$A_L(\theta)^T X(\theta) + X(\theta) A_L(\theta) < -\epsilon I < 0,\ \ \epsilon > 0$$

is satisfied, then the system (1) is asymptotically stable for all $\tau \in (0, \overline{\tau}]$.

By choosing $X(\theta)$ to be affine with respect to the variable θ, that is,

$$X(\theta) = \begin{bmatrix} X_0 + \theta X_1 & \theta^{1/2} X_{12} \\ \theta^{1/2} X_{12}^T & X_{22} \end{bmatrix} \tag{13}$$

and since $X(\theta)$ *is bounded* for all $0 < \theta \leq \overline{\tau}$, we obtain - according to Lemma 8 in [13] and Theorem 6 - the following result:

Theorem 9. *The system (1) is asymptotically stable for all* $\tau \in [0, \overline{\tau}]$ *if there exist positive matrices* $X_0, X_1 \in \mathbb{R}^{n \times n}$, *matrix* $X_2 \in \mathbb{R}^{n \times n_{\widehat{A}_s}}$ *and a positive matrix* $X_{22} \in \mathbb{R}^{n_{\widehat{A}_s} \times n_{\widehat{A}_s}}$, *such that:*

$$\Pi(\overline{\tau}) = \begin{bmatrix} \Pi_{11}(\theta) & \Pi_{12}(\theta) \\ \Pi_{12}^T(\theta) & \Pi_{22}(\theta) \end{bmatrix} \Bigg|_{\theta=\overline{\tau}} < 0$$

and (14)

$$X(\overline{\tau}) > 0,$$

where X is given by (13) and

$$\Pi_{11}(\theta) = (X_0 + \theta X_1)A_s + A_s^T(X_0 + \theta X_1) + X_{12}B_s + B_s^T X_{12}^T$$
$$\Pi_{12}(\theta) = (X_0 + \theta X_1)C_s + X_{12}\widehat{A}_s + \theta A_s^T X_{12} + B_s^T X_{22}$$
$$\Pi_{22}(\theta) = \theta X_{12}^T C_s + \theta C_s^T X_{12} + X_{22}\widehat{A}_s + \widehat{A}_s^T X_{22}.$$

Remark 10. As mentioned before, in the single delay case it is possible to obtain an explicit formula for the largest delay margin $\overline{\tau}_{max}$. The LMI "test" (14) is deciding whether or not, the system (1) is asymptotically stable for any $\tau \in [0, \overline{\tau}]$, and this kind of test can also be employed in the multiple delay case.
It is important to mention that the conditions (14) introduce some additional conservatism due to the particular choice of $X(\theta)$ when verifying the robust stability condition in Theorem 6.

Let $\overline{\tau}_0$ be such that the LMI's in Theorem 9 all hold, then (1) is asymptotically stable for all $\tau \in [0, \overline{\tau}_0]$. Under these circumstances, it is easy to see that Theorem 4 in [13] can be adapted in a straightforward manner.

Theorem 11. *The delay margin guaranteed by Theorem 6 is given by*

$$\overline{\tau}_{Theorem6} = \frac{\overline{\tau}_0}{\alpha_m} + \frac{1}{\lambda_{max}^{+}\left(-(M_0 \oplus M_0)^{-1}(M_1 \oplus M_1)\right)} \tag{15}$$

where $M_0 = \begin{bmatrix} \frac{\overline{\tau}_0}{\alpha_m} A_s & B_s \\ \frac{\overline{\tau}_0}{\alpha_m} C_s & \widehat{A}_s \end{bmatrix}$ *and* $M_1 = \begin{bmatrix} A_s & 0 \\ C_s & 0 \end{bmatrix}$.

The numerical procedure deriving from the above mentioned statements is tackling the following problem:

Given A, A_1, B, m (the order of the Padé approximation) and $\overline{\tau}$, check if the system $\dot{x}(t) - B\dot{x}(t-\tau) = Ax(t) + A_1 x(t-\tau)$ is asymptotically stable for all $\tau \in [0, \overline{\tau}]$, under the assumptions to be verified at Step 1.

The algorithm.
Step 1. Determine the matrices $\overline{A} = A + A_1$, $\overline{B} = I - B$ and check if B is a Schur and $(\overline{B})^{-1}\overline{A}$ is a Hurwitz matrix.
Step 2. Compute $P(s) = R_m(s) - 1$ and determine ω_{cm} on its Bode plot. Let $\alpha_m = \omega_{cm}/(2\pi)$.
Step 3. Write a realization $\{A_T, B_T, C_T, D_T\}$ of $R_m(\alpha_m s) - 1$. Then $\Delta_o(s) = (R_m(\alpha_m s) - 1) \cdot I = C_P(sI - A_P)^{-1}B_P + D_P$, where

$$A_P = I \otimes A_T, \quad B_P = I \otimes B_T,$$

$$C_P = I \otimes C_T, \quad D_P = I \otimes D_T,$$

.

Step 4. Construct matrices A_G, B_G, C_G, D_G given by (9) and matrices $A_s, B_s, C_s, \widehat{A}_s$, given by (12).
Step 5. Solve the feasibility problem (14).

Remark 12.

1. Solving the inequality

$$A_L(\theta)^T X(\theta) + X(\theta)A_L(\theta) < -\epsilon I, \ \epsilon > 0 \tag{16}$$

is equivalent to solve the linear programming problem:

$$\min \epsilon$$

with constraints

$$A_L(\theta)^T X(\theta) + X(\theta)A_L(\theta) < 0.$$

The constraints are feasible if the global solution $\epsilon_{\min}$ is negative.
2. If conditions (14) are not verified for given $\overline{\tau}$, then one can take $\overline{\tau} := \overline{\tau}/2$ and start again from Step 1. Otherwise (i.e. the LMI (14) is feasible) one can use formula (15) from Theorem 11.

4 Conclusion

The aim of this paper is to extend the approach in [13] to the case of neutral systems. The presence of the delay in the left hand term of the dynamic equation of such kind of systems makes the stability analysis more involved. In a certain context, established by natural assumptions which lead to a well-posed problem (stability for the delay-free system, continuity of the spectral limit function), we have shown that it is possible to follow the framework in [13]: converting the problem of asymptotic stability of the original system into one of robust stability of a comparison linear system. The structure of the comparison system is more involved than in the standard delay systems case. The result is analogous to the one obtained for the standard case and works in the same way. The degree of conservatism is the same, because we keep its dependency on the order of the Padé approximation, m.

In an immediate perspective, this method could be extended to deal with the delay-dependent stability analysis for systems with characteristic equations of the form

$$\chi(s) := P(s) + e^{-\tau s}Q(s) = 0, \tag{17}$$

where, in the general case, $P(s) = s^n + a_{n-1}s^{n-1} + \cdots + a_0$, and $Q(s) = b_n s^n + b_{n-1}s^{n-1} + \cdots + b_0$, b_n not necessarily equal to 0.
On one hand, if $b_n = 0$, then $\chi(s) = \det(sI - A_0 - A_1 e^{-\tau s})$, where

$$A_0 = \begin{bmatrix} 0 & 1 & \cdots & 0 \\ \vdots & \vdots & \ddots & \vdots \\ 0 & 0 & \cdots & 1 \\ -a_0 & -a_1 & \cdots & -a_{n-1} \end{bmatrix}, \quad A_1 = \begin{bmatrix} 0 & 0 & \cdots & 0 \\ \vdots & \vdots & \ddots & \vdots \\ 0 & 0 & \cdots & 0 \\ -b_0 & -b_1 & \cdots & -b_{n-1} \end{bmatrix}.$$

The stability analysis for the quasi-polynomial $\chi(s)$ can be now performed on the standard time-delay system $\dot{x}(t) = A_0 x(t) + A_1 x(t-\tau)$.
On the other hand, for $b_n \neq 0$, one has $\chi(s) = \det(sI - Be^{-\tau s} - A_0 - A_1 e^{-\tau s})$, where

$$A_0 = \begin{bmatrix} 0 & 1 & \cdots & 0 \\ \vdots & \vdots & \ddots & \vdots \\ 0 & 0 & \cdots & 1 \\ -a_0 & -a_1 & \cdots & -a_{n-1} \end{bmatrix}, \; A_1 = \begin{bmatrix} 0 & 0 & \cdots & 0 \\ \vdots & \vdots & \ddots & \vdots \\ 0 & 0 & \cdots & 0 \\ -b_0 & -b_1 & \cdots & -b_{n-1} \end{bmatrix}, \; B = \begin{bmatrix} 0 & 0 & \cdots & 0 \\ \vdots & \vdots & \ddots & \vdots \\ 0 & 0 & \cdots & 0 \\ 0 & 0 & \cdots & -b_n \end{bmatrix}.$$

This time, the quasi-polynomial $\chi(s)$ is the characteristic polynomial of a neutral system, $\dot{x}(t) - B\dot{x}(t-\tau) = A_0 x(t) + A_1 x(t-\tau)$. Here the present approach can found an useful application. Such models (17) show up in biological phenomena or communication networks.

Recently, several methods have been proposed in the literature for the robustness analysis of arbitrary Hurwitz quasi-polynomials with respect to the time-delays (see for instance [5]). It would be interesting to combine the outcome of the different methods, in order to reach less conservative delay stability margins or to guarantee a certain degree of conservatism for situations of practical importance.

The extension to the multiple delay case could use a similar framework, by adapting the starting point results of R. Datko and finding an appropriate state-space realization for $G(s)$.

References

1. Datko, R.: A procedure for determination of the exponential stability of certain differential-difference equations. Quart. Appl. Math. 36, 279–292 (1978)
2. Gahinet, P., Nemirovski, A., Laub, A.J., Chilali, M.: LMI Control Toolbox for Use with MATLAB. The Math. Works Inc. (1995)
3. Hale, J.: Theory of functional differential equations. Springer, New York (1977)
4. Lam, J.: Convergence of a class of Padé approximations for delay-systems. Int. J. Control 2(4), 989–1008 (1990)
5. Morarescu, C.I.: Qualitative analysis of distributed delay systems: methodology and algorithms. PhD thesis, University of Bucharest (Romania)/Université de Technologie de Compiègne (France) (2006)
6. Niculescu, S.I.: Stability and hyperbolicity of linear systems with delayed state: a matrix-pencil approach. IMA Journal of Mathematical Control & Information 15(4), 331–347 (1998)

7. Niculescu, S.I., Neto, A.T., Dion, J.M., Dugard, L.: Delay-dependent stability of linear systems with delayed state: An LMI approach. In: Proc 34th IEEE Conf. Dec. Contr., New Orleans, LA, pp. 1495–1497 (1995)
8. Niculescu, S.I., Verriest, E.I., Dugard, L., Dion, J.M.: Stability and Robust Control of Time Delay Systems, pp. 1–71. Springer, Heidelberg (1997)
9. Olgac, N., Sipahi, R.: A practical method for analyzing the stability of neutral type lti-time-delayed systems. Automatica, 847–853 (2004)
10. Olgac, N., Sipahi, R.: The cluster treatment of characteristic roots and the neutral type time-delayed systems. Journal of Dynamic Systems, Measurement and Control, Transactions of the ASME, 88–97 (2005)
11. Wang, Z., Lam, J., Burnham, K.J.: Stability Analysis and Observer Design for Neutral Delay Systems. IEEE Transactions on Automatic Control 47(3), 448–483 (2002)
12. Zhang, J., Knospe, C.R., Tsiotras, P.: A unified approach to time-delay system stability via scaled small gain. In: Proc. American Control Conference, San Diego, CA, pp. 307–308 (1999)
13. Zhang, J., Knospe, C.R., Tsiotras, P.: New Results for the Analysis of Linear Systems with Time-Invariant Delays. Int. J. Robust and Nonlinear Control 13, 1149–1175 (2003)

Frequency Stability Analysis of Linear Systems with General Distributed Delays

Vladimir L. Kharitonov[1], Sabine Mondié[2] and Gilberto Ochoa[2]

[1] Applied Mathematics and Control Process Department, St.-Petersburg State University, St.-Petersburg, Russia khar@apmath.spbu.ru
[2] Automatic Control Department, CINVESTAV-IPN, D. F., México smondie,gochoa@ctrl.cinvestav.mx

Summary. A new method that reveals conditions under which the characteristic function of a linear time delay system has a root s_0 such that $-s_0$ is also a root of the function is presented. The method is based on some recent results on the computation of the Lyapunov matrices for time delay systems. A general class of linear systems with distributed delays is studied. A number of examples are given to illustrate the approach and to show its strength.

1 Introduction

It is well known that a linear time invariant delay system of the form

$$\frac{dx(t)}{dt} = A_0x(t) + A_1x(t-h) \tag{1}$$

where $A_0, A_1 \in \mathbb{R}^{n\times n}$, and h is a positive delay, is exponentially stable if and only if all roots of its characteristic quasipolynomial

$$f(s) = \det(sI - A_0 - A_1e^{-hs})$$

have negative real parts [1]. When the system matrices depend continuously on some parameters and/or the delay h is not fixed one may try to detect the values of the parameters for which the system may loose its stability. For such critical values the following should occur: at least one of the roots of $f(s)$ comes across the imaginary axis of the complex plane.

Among the algorithms available for the computation of the critical values one should mention those based on constant matrix test [3], on matrix pencil techniques [7], on properties of the determinant of the resultant matrix see [3], [10] and finally those based on the Rekasius substitution [8], [9]. Another approach for the computation of the critical values of the system parameters has been presented in [5], [6]. It is based on the fact that any pure imaginary root of the characteristic function $f(s)$ is also an eigenvalue of the delay free system of the matrix equations

$$\begin{cases} X_0'(\tau) = X_0(\tau)A_0 + X_{-1}(\tau)A_1 \\ X_{-1}'(\tau) = -A_1^TX_0(\tau) - A_0^TX_{-1}(\tau) \end{cases}$$

J.J. Loiseau et al. (Eds.): Topics in Time Delay Systems, LNCIS 388, pp. 25–36.
springerlink.com

that is used for the computation of the Lyapunov matrices of system (1). In this contribution, we use some recent results on the Lyapunov matrices for time delay systems reported in [2] and [4] to extend our analysis to a class of systems with distributed delays.

The contribution is organized as follows: we start Section 2 with an extension of the connection between the spectrum of the characteristic equation of systems, and that of a special delay free system of matrix equations arising in the computation of the Lyapunov matrices for time delay systems. Then a general methodology on the use of these results for the computation of the critical values of system parameters is explained. In Section 3, we impose additional conditions on the kernel of the distributed delay that lead to a substantial reduction of the size of the delay free system. Several illustrative examples validate the proposed approach in Section 4. The contribution ends with some concluding remarks.

2 Systems with distributed delay

In this section, we extend our approach to the stability analysis of a general class of distributed delay systems of the form

$$\frac{dx(t)}{dt} = A_0x(t) + A_1x(t-h) + \int_{-h}^{0} H(\theta)x(t+\theta)d\theta, \tag{2}$$

where the kernel $H(\theta)$ is a real matrix polynomial of the form

$$H(\theta) = \sum_{j=0}^{m} \eta_j(\theta)B_j. \tag{3}$$

Here $B_j \in \mathbb{R}^{n\times n}$, $j = 1, ..., m$ are given matrices and the scalar functions $\eta_0(\theta), ..., \eta_m(\theta)$ are such that

$$\frac{d}{d\theta}\eta_j(\theta) = \sum_{k=0}^{m} \delta_{jk}\eta_k(\theta) \tag{4}$$

The characteristic matrix of the system is

$$G(s) = \det\left(sI - A_0 - e^{-hs}A_1 - \sum_{j=0}^{m} f^{(j)}(s)B_j\right),$$

where

$$f^{(j)}(s) = \int_{-h}^{0} \eta_j(\theta)e^{s\theta}d\theta, \quad j = 1, ..., m \tag{5}$$

The following result presented in the framework of the study of the Lyapunov matrices for system (2) is the starting point of our proposal, see Remark 9 in [4].

Proposition 1. *Let s_0 be an eigenvalue of the time delay system (2-3) such that $-s_0$ is also an eigenvalue of the system. Then s_0 belongs to the spectrum of the delay free system*

$$Z'(\tau) = Z(\tau)A_0 + V(\tau)A_1 + \sum_{j=0}^{m} X_j(\tau)B_j$$

$$V'(\tau) = -A_1^{\mathsf{T}}Z(\tau) - A_0^{\mathsf{T}}V(\tau) - \sum_{j=0}^{m} B_j^{\mathsf{T}}Y_j(\tau)$$

$$X_j'(\tau) = \eta_j(0)Z(\tau) - \eta_j(-h)V(\tau) - \sum_{k=0}^{m} \delta_{jk}X_k(\tau)$$

$$Y_j'(\tau) = \eta_j(-h)Z(t) - \eta_j(0)V(\tau) + \sum_{k=0}^{m} \delta_{jk}Y_k(\tau) \tag{6}$$

for $j = 0, 1, ..., m$. Furthermore, the spectrum of system (6) is symmetrical with respect to the imaginary axis.

Proof. A complex number s belongs to the spectrum of the delay free system (6) if and only if there exists a nontrivial set of $n \times n$ constant matrices Z, V, X_j, and Y_j, $j = 0, 1, ..., m$, such that

$$\begin{cases} sZ = ZA_0 + VA_1 + \sum\limits_{j=0}^{m} X_jB_j \\ sV = -A_1^{\mathsf{T}}Z - A_0^{\mathsf{T}}V - \sum\limits_{j=0}^{m} B_j^{\mathsf{T}}Y_j \\ sX_j = \eta_j(0)Z - \eta_j(-h)V - \sum\limits_{k=0}^{m} \delta_{jk}X_k \\ sY_j = \eta_j(-h)Z - \eta_j(0)V + \sum\limits_{k=0}^{m} \delta_{jk}Y_k \end{cases} \tag{7}$$

As s_0 and $-s_0$ are eigenvalues of the system then there exist two nonzero vectors γ and μ such that

$$\gamma^{\mathsf{T}}G(s_0) = 0, \quad G^{\mathsf{T}}(-s_0)\mu = 0. \tag{8}$$

Multiplying the first equality in (8) by μ from the left, and the second equality by $-e^{-hs_0}\gamma^{\mathsf{T}}$ from the right we obtain

$$s_0\mu\gamma^{\mathsf{T}} - \mu\gamma^{\mathsf{T}}A_0 - e^{-hs_0}\mu\gamma^{\mathsf{T}}A_1 - \sum_{j=0} f^{(j)}(s_0)\mu\gamma^{\mathsf{T}}B_k = 0, \tag{9}$$

and

$$s_0e^{-hs_0}\mu\gamma^{\mathsf{T}} + A_0^{\mathsf{T}}e^{-hs_0}\mu\gamma^{\mathsf{T}} + A_1^{\mathsf{T}}\mu\gamma^{\mathsf{T}} + \sum_{j=0}^{m} e^{-hs_0}f^{(j)}(-s_0)B_j^{\mathsf{T}}\mu\gamma^{\mathsf{T}} = 0. \tag{10}$$

If we introduce the matrices

$$Z = \mu\gamma^{\mathsf{T}},\ V = e^{-hs_0}\mu\gamma^{\mathsf{T}},\ X_j = f^{(j)}(s_0)\mu\gamma^{\mathsf{T}}, Y_j = e^{-hs_0}f^{(j)}(-s_0)\mu\gamma^{\mathsf{T}} \tag{11}$$

with $j = 0, 1, ..., m$, then these equalities take the form

$$s_0Z - ZA_0 - VA_1 - \sum_{j=0}^{m} X_jB_j = 0,$$

$$s_0V + A_0^{\mathsf{T}}V + A_1^{\mathsf{T}}Z + \sum_{j=0}^{m} B_j^{\mathsf{T}}Y_j = 0.$$

thus for $s = s_0$ these matrices satisfy the first two matrix equations of system (7). Next, we show that for $s = s_0$ they also satisfy the remaining $2(m+1)$ matrix equations of system (7).

Observe that on the one hand,

$$\int_{-h}^{0} \frac{d}{d\theta}(\eta_j(\theta)e^{s\theta})d\theta = \eta_j(0) - \eta_j(-h)e^{-sh} \tag{12}$$

and on the other hand

$$\int_{-h}^{0} \frac{d}{d\theta}(\eta_j(\theta)e^{s\theta})d\theta = s\int_{-h}^{0} \eta_j(\theta)e^{s\theta}d\theta + \sum_{k=0}^{m}\delta_{jk}\int_{-h}^{0} e^{s\theta}\eta_k(\theta)d\theta.$$

Substituting (5) we obtain

$$\int_{-h}^{0} \frac{d}{d\theta}(\eta_j(\theta)e^{s\theta})d\theta = sf^{(j)}(s) + \sum_{k=0}^{m}\delta_{jk}f^{(k)}(s). \tag{13}$$

It follows from (12) and (13) that

$$sf^{(j)}(s) = \eta_j(0) - \eta_j(-h)e^{-sh} - \sum_{k=0}^{m}\delta_{jk}f^{(k)}(s). \tag{14}$$

Now for $s = s_0$ by multiplying (14) by $\mu\gamma^{\mathsf{T}}$ we get

$$s_0 f^{(j)}(s_0)\mu\gamma^{\mathsf{T}} = \eta_j(0)\mu\gamma^{\mathsf{T}} - \eta_j(-h)e^{-s_0 h}\mu\gamma^{\mathsf{T}} - \sum_{k=0}^{m}\delta_{jk}f^{(k)}(s_0)\mu\gamma^{\mathsf{T}},$$

and for $s = -s_0$ by multiplying (14) by $\mu\gamma^{\mathsf{T}}e^{-hs_0}$ we obtain

$$\begin{aligned} s_0 f^{(j)}(-s_0)\mu\gamma^{\mathsf{T}}e^{-hs_0} &= -\,\eta_j(0)\mu\gamma^{\mathsf{T}}e^{-hs_0} + \eta_j(-h)\mu\gamma^{\mathsf{T}} + \\ &\quad + \sum_{k=0}^{m}\delta_{jk}f^{(k)}(-s_0)e^{-hs_0}\mu\gamma^{\mathsf{T}} \end{aligned}$$

Clearly, the substitution of the matrices defined in (11) into these expressions give the last $2m$ matrix equations of system (7).

As the above defined set of matrices Z, V, X_j, and Y_j, $j = 0, 1, ..., m$, is non trivial, s_0 is an eigenvalue of the delay free system of the matrix equations (6). The same is true for $-s_0$.

The fact that the spectrum of system (6) is symmetrical with respect to the imaginary axis follows directly from the observation that if for s there exists a non trivial set of matrices Z, V, X_j, and Y_j, $j = 0, 1, ..., m$, satisfying (7), then, applying the transposition operation to the equalities in (7), one can check that the matrices $\hat{V} = Z^{\mathsf{T}}, \hat{Z} = V^{\mathsf{T}}, \hat{Y}_j = X_j$, and $\hat{X}_j = Y_j$, $j = 0, 1, ..., m$, satisfy (7) for $-s$.

3 A particular case of interest

In this section, we focus our attention on a particular case of the time delay systems studied in the previous section. The additional conditions imposed on the kernels $\eta_j(\theta)$, $j = 0, 1, ..., m$ allow to reduce substantially the dimension of the corresponding delay free system. Notice that this class includes matrix polynomial kernels.

The systems analyzed here are of the form (2) were the kernel defined in (3) satisfies not only the condition (4) but also

$$\eta_j(-\theta - h) = \sum_{k=0}^{m} \gamma_{jk}\eta_k(\theta), \quad j = 0, 1, ..., m, \quad \theta \in [-h, 0]. \tag{15}$$

When this particular situation occurs, it is shown in [4] that this implies that

$$Y_j(\tau) = \sum_{k=0}^{m} \gamma_{jk}X_k(\tau), \; j = 0, 1, ..., m, \tag{16}$$

and one can reduce the complexity of the analysis by excluding the auxiliary matrices $Y_j(\tau)$ of the system (6). We prove first a key technical result.

Lemma 1. *If the conditions (4) and (15) hold, then the matrices defined as*

$$\eta^T(\theta) = \left(\eta_0(\theta) \; \eta_1(\theta) \cdots \eta_m(\theta) \right),$$
$$\Gamma = \begin{pmatrix} \gamma_{00} & \cdots & \gamma_{0m} \\ \vdots & & \vdots \\ \gamma_{m0} & \cdots & \gamma_{mm} \end{pmatrix}, \; \Delta = \begin{pmatrix} \delta_{00} & \cdots & \delta_{0m} \\ \vdots & & \vdots \\ \delta_{m0} & \cdots & \delta_{mm} \end{pmatrix}. \tag{17}$$

satisfy:

$$\Delta\Gamma\eta(\theta) = -\Gamma\Delta\eta(\theta) \tag{18}$$

$$\eta(\theta) = \Gamma^2\eta(\theta). \tag{19}$$

Proof. The conditions (15) and (4) can be respectively rewritten in the above notation as

$$\eta(-\theta - h) = \Gamma\eta(\theta), \; \theta \in [-h, 0], \tag{20}$$

$$\eta'(\theta) = \Delta\eta(\theta), \; \theta \in [-h, 0]. \tag{21}$$

Notice that the condition (20) and (21) imply that the following equalities hold as well:

$$\eta(\theta) = \Gamma\eta(-\theta - h), \; \theta \in [-h, 0], \tag{22}$$

$$\eta'(-\theta - h) = -\Delta\eta(-\theta - h), \; \theta \in [-h, 0]. \tag{23}$$

The property (19) clearly follows from using successively (22) and (20). The property (18) is obtained easily through the differentiation of (20), followed by appropriate substitutions of (21), (23) and (20).

Proposition 2. *Consider a time delay system of the form (2-3) such that conditions (4) and (15) are satisfied. Let s_0 be an eigenvalue of this system such that $-s_0$ is also an eigenvalue of the system. Then s_0 belongs to the spectrum of the delay free system*

$$
\begin{aligned}
Z'(\tau) &= Z(\tau)A_0 + V(\tau)A_1 + \sum_{j=0}^{m} X_j(\tau)B_j \\
V'(\tau) &= -A_1^{\mathsf{T}} Z(\tau) - A_0^{\mathsf{T}} V(\tau) - \sum_{j=0}^{m} B_j^{\mathsf{T}} \sum_{k=0}^{m} \gamma_{jk} X_k(\tau) \\
X_j'(\tau) &= \eta_j(0)Z(\tau) - \eta_j(-h)V(\tau) - \sum_{k=0}^{m} \delta_{jk} X_k(\tau),
\end{aligned}
\tag{24}
$$

with $j = 0, 1, ..., m$. Furthermore, the spectrum of system (24) is symmetrical with respect to the imaginary axis.

Proof. A complex number s belongs to the spectrum of the delay free system (24) if and only if there exists a nontrivial set of $n \times n$ constant matrices Z, V, and X_j, $j = 0, 1, ..., m$, such that

$$
\begin{cases}
sZ = ZA_0 + VA_1 + \sum\limits_{j=0}^{m} X_j B_j \\
sV = -A_1^{\mathsf{T}} Z - A_0^{\mathsf{T}} V - \sum\limits_{j=0}^{m} B_j^{\mathsf{T}} \sum\limits_{k=0}^{m} \gamma_{jk} X_k \\
sX_j = \eta_j(0)Z - \eta_j(-h)V - \sum\limits_{k=0}^{m} \delta_{jk} X_k
\end{cases}
\tag{25}
$$

Observe that the equalities (9), (10) and (14) obtained in the general case remain valid. According to the definition (5) the Laplace transform of (15)

$$
\int_{-h}^{0} \eta_j(-\theta - h)e^{-s\theta} d\theta = \sum_{k=0}^{m} \gamma_{jk} \int_{-h}^{0} \eta_k(\theta)e^{-s\theta} d\theta,
$$

implies that

$$
e^{-hs} f^{(j)}(-s) = \sum_{k=0}^{m} \gamma_{jk} f^{(k)}(s).
$$

If we substitute this equality into (10), we get

$$
s_0 e^{-hs_0} \mu\gamma^{\mathsf{T}} + A_0^{\mathsf{T}} e^{-hs_0} \mu\gamma^{\mathsf{T}} + A_1^{\mathsf{T}} \mu\gamma^{\mathsf{T}} + \sum_{j=0}^{m} \sum_{k=0}^{m} B_j^{\mathsf{T}} \gamma_{jk} f^{(k)}(s_0) \mu\gamma^{\mathsf{T}} = 0, \tag{26}
$$

Finally, if we introduce the matrices

$$
Z = \mu\gamma^{\mathsf{T}},\; V = e^{-hs_0} \mu\gamma^{\mathsf{T}}, X_j = f^{(j)}(s_0)\mu\gamma^{\mathsf{T}},\; j = 0, 1, ..., m,
$$

into (9), (26) and (14), the system (25) is obtained. It is evident that the previously defined set of matrices Z, V, X_j, $j = 0, 1, ..., m$ is not trivial. Therefore, the complex value s_0 belongs to the spectrum of the delay free system of matrix equations (24). The same is true for $-s_0$.

Now we address the second statement of the theorem. The transposition and negation of system (25) gives

$$-sZ^{\mathsf{T}} = -A_0^{\mathsf{T}} Z^{\mathsf{T}} - A_1^{\mathsf{T}} V^{\mathsf{T}} - \sum_{j=0}^{m} B_j^{\mathsf{T}} X_j^{\mathsf{T}}, \tag{27}$$

$$-sV^{\mathsf{T}} = Z^{\mathsf{T}} A_1 V^{\mathsf{T}} A_0 + \sum_{j=0}^{m} \left(\sum_{k=0}^{m} \gamma_{jk} X_k^{\mathsf{T}} \right) B_j, \tag{28}$$

$$-sX_j^{\mathsf{T}} = -\eta_j(0) Z^{\mathsf{T}} + \eta_j(-h) V + \sum_{k=0}^{m} \delta_{jk} X_k, \tag{29}$$

We establish first two key preliminary equalities. We define $X = (X_0, X_1, ..., X_m)$ and we rewrite (29) using Knonecker products and the notation introduced in (17).

$$-sX^{\mathsf{T}} = -\left(\eta(0) \otimes I\right) Z^{\mathsf{T}} + \left(\eta(-h) \otimes I\right) V^{\mathsf{T}} + (\Delta \otimes I) X^{\mathsf{T}}. \tag{30}$$

Pre-multiplication of (30) by $(\Delta\Gamma \otimes I)$ and by $(\Gamma\Delta \otimes I)$ gives respectively

$$\begin{aligned} -s(\Delta\Gamma \otimes I)X^{\mathsf{T}} = &- (\Delta\Gamma \otimes I)\left(\eta(0) \otimes I\right) Z^{\mathsf{T}} + (\Delta\Gamma \otimes I)\left(\eta(-h) \otimes I\right) V^{\mathsf{T}} + \\ &+ (\Delta\Gamma \otimes I)(\Delta \otimes I) X^{\mathsf{T}} \end{aligned} \tag{31}$$

and

$$\begin{aligned} -s(\Gamma\Delta \otimes I)X^{\mathsf{T}} = &- (\Gamma\Delta \otimes I)\left(\eta(0) \otimes I\right) Z^{\mathsf{T}} + (\Gamma\Delta \otimes I)\left(\eta(-h) \otimes I\right) V^{\mathsf{T}} \\ &+ (\Gamma\Delta \otimes I)(\Delta \otimes I) X^{\mathsf{T}}. \end{aligned}$$

Using (18) in the above equation, we get

$$\begin{aligned} -s(\Gamma\Delta \otimes I)X^{\mathsf{T}} = &(\Delta\Gamma \otimes I)\left(\eta(0) \otimes I\right) Z^{\mathsf{T}} - (\Delta\Gamma \otimes I)\left(\eta(-h) \otimes I\right) V^{\mathsf{T}} + \\ &+ (\Gamma\Delta\Delta \otimes I) X^{\mathsf{T}}. \end{aligned} \tag{32}$$

Adding (31) and (32) we get

$$-s\left[(\Delta\Gamma \otimes I)X^{\mathsf{T}} + (\Gamma\Delta \otimes I)X^{\mathsf{T}}\right] = (\Delta\Gamma\Delta \otimes I)X^{\mathsf{T}} + (\Gamma\Delta\Delta \otimes I)X^{\mathsf{T}}$$

hence

$$(\Delta\Gamma \otimes I)X^{\mathsf{T}} = -(\Gamma\Delta \otimes I)X^{\mathsf{T}}. \tag{33}$$

Now, observe that the premultiplication of (30) by $(\Gamma\Gamma \otimes I)$ yields

$$\begin{aligned} -s(\Gamma\Gamma \otimes I)X^{\mathsf{T}} = &- (\Gamma\Gamma \otimes I)\left(\eta(0) \otimes I\right) Z^{\mathsf{T}} + (\Gamma\Gamma \otimes I)\left(\eta(-h) \otimes I\right) V^{\mathsf{T}} + \\ &+ (\Gamma\Gamma \otimes I)(\Delta \otimes I) X^{\mathsf{T}} \end{aligned}$$

and (19) gives

$$-s(\Gamma\Gamma \otimes I)X^{\mathsf{T}} = -\left(\eta(0) \otimes I\right) Z^{\mathsf{T}} + \left(\eta(-h) \otimes I\right) V^{\mathsf{T}} + (\Gamma\Gamma \otimes I)(\Delta \otimes I) X^{\mathsf{T}}.$$

Substracting this equation from (30) we obtain

$$-s\left[X^{\mathsf{T}} - (\Gamma\Gamma \otimes I)X^{\mathsf{T}}\right] = \left[X^{\mathsf{T}} - (\Gamma\Gamma \otimes I)\right](\Delta \otimes I)X^{\mathsf{T}}$$

hence

$$X^{\mathsf{T}} = (\Gamma\Gamma \otimes I)X^{\mathsf{T}}. \tag{34}$$

We are now ready to prove our result. If s is an eigenvalue of system (24), then there exists a non trivial set of constant matrices Z, V, X_0, X_1 satisfying the equalities (25). Let us define the new non trivial set of matrices

$$\tilde{Z} = V^{\mathsf{T}},\ \tilde{V} = Z^{\mathsf{T}}, \tag{35}$$

$$\tilde{X}_j = \sum_{j=0}^{m} \gamma_{jk} X_k^{\mathsf{T}},\ j = 0, 1, ..., m,\ \text{ or } \tilde{X} = (\Gamma \otimes I) X^{\mathsf{T}} \tag{36}$$

If we substitute these matrices into (28), we obtain

$$-s\tilde{Z} = \tilde{V}A_1 + \tilde{Z}A_0 + \sum_{j=0}^{m} \tilde{X}_j B_j. \tag{37}$$

Observe that (34) implies that

$$X^{\mathsf{T}} = (\Gamma \otimes I)\tilde{X}$$

Using this equality and substituting (35) into (27) we get

$$-s\tilde{Z} = -A_0^{\mathsf{T}}\tilde{Z} - A_1^{\mathsf{T}}\tilde{V} - \sum_{j=0}^{m}\left(\sum_{k=0}^{m} \gamma_{jk}\tilde{X}_k\right) B_j. \tag{38}$$

Now, premultiplying (30) by $(\Gamma \otimes I)$ we get

$$\begin{aligned}-s(\Gamma \otimes I)X^{\mathsf{T}} = &- (\Gamma \otimes I)(\eta(0) \otimes I) Z^{\mathsf{T}} + (\Gamma \otimes I)(\eta(-h) \otimes I) V^{\mathsf{T}} + \\ &+ (\Gamma \otimes I)((\Delta \otimes I)X^{\mathsf{T}}.\end{aligned}$$

The properties (16) and (33) imply that

$$-s(\Gamma \otimes I)X^{\mathsf{T}} = -(\eta(-h) \otimes I) Z^{\mathsf{T}} + (\eta(0) \otimes I) V^{\mathsf{T}} - (\Delta \otimes I)(\Gamma \otimes I)X^{\mathsf{T}}$$

and this equality looks in the new matrices $\tilde{V}$, $\tilde{Z}$ and $\tilde{X}$ defined in (35), (36) as

$$-s\tilde{X} = (\eta(0) \otimes I)\tilde{Z} - (\eta(-h) \otimes I)\tilde{V} - (\Delta \otimes I)\tilde{X}. \tag{39}$$

Collecting the equalities (37), (38) and (39) we conclude that $-s$ is also an eigenvalue of the system (24).

Remark 1. The class of distributed delay systems (2-3) with real matrix polynomial kernel, $\eta_j(\theta) = \theta^j$ is of particular interest because it allows polynomial approximation of arbitrary kernels. In this case, the result reduces to the following [4]. Let s_0 be an eigenvalue of the time delay system (2-3) such that $-s_0$ is also an eigenvalue of the system, then s_0 belongs to the spectrum of the corresponding delay free system

$$\begin{aligned}Z'(\tau) &= Z(\tau)A_0 + V(\tau)A_1 + \sum_{j=0}^{m} X_j(\tau)B_j \\ V'(\tau) &= -A_1^{\mathsf{T}}Z(\tau) - A_0^{\mathsf{T}}V(\tau) + \sum_{k=0}^{m}\left[\frac{1}{k!}B^{(k)}(h)\right] X_k(\tau) \\ X_0'(\tau) &= Z(\tau) - V(\tau) \\ X_j'(\tau) &= -(-h)^j V(\tau) - jX_{j-1}(\tau),\ j = 1, 2, ..., m\end{aligned} \tag{40}$$

where $B(h) = \sum_{j=0}^{m} (-h)^j B_j^{\mathsf{T}}$. Furthermore, the spectrum of system (40) is symmetrical with respect to the imaginary axis.

Remark 2. The characteristic polynomial of the delay free system (40) $p(s)$ has degree $(m+3)n^2$. It is important to underline that for $m = 0$, the coefficients of the polynomial de not depend of h, while for $m \geq 1$ they are polynomial functions of h. If $(m+3)n^2$ is odd, $p(s)$ can be written as $sp(\lambda)$ where $\lambda = s^2$, and if $(m+3)n^2$ is even, $p(s)$ can be written as $p_1(\lambda)$. Then the problem of determining the purely imaginary roots of $p(s)$ reduces to finding the negative real roots of the polynomial $p_1(\lambda)$.

4 Illustrative examples

In this section we use this results in order to determine the critical parameters in some illustrative examples.

Example 1. Let us consider the system

$$\dot{x}(t) = A_0 x(t) + A_1 x(t-h) + \int_{-h}^{0} \left[\sin\left(\frac{\pi\theta}{h}\right) B_0 + \cos\left(\frac{\pi\theta}{h}\right) B_1 \right] x(t+\theta) d\theta,$$

where

$$A_0 = \begin{bmatrix} -1 & 0 \\ 0 & -1 \end{bmatrix}, \quad A_1 = \begin{bmatrix} 0 & 1 \\ -1 & 0 \end{bmatrix}, \quad B_0 = \begin{bmatrix} \frac{3}{10} & 0 \\ 0 & \frac{3}{10} \end{bmatrix}, \quad B_1 = \begin{bmatrix} 0 & \frac{3}{10} \\ \frac{-3}{10} & 0 \end{bmatrix},$$

$h > 0$, $\eta_0(\theta) = \sin\left(\frac{\pi\theta}{h}\right)$ and $\eta_1(\theta) = \cos\left(\frac{\pi\theta}{h}\right)$.
Its characteristic function is

$$f(s,h) = \det\left(sI - A_0 - A_1 e^{-sh} + B_0 \frac{h\pi\left(1+e^{-sh}\right)}{(\pi^2 + h^2 s^2)} - B_1 \frac{h^2 s\left(1+e^{-sh}\right)}{(\pi^2 + h^2 s^2)} \right)$$

The characteristic polynomial of the corresponding delay free system (24) is

$$p(s) = \frac{1}{625h^8} q(s) r(s) v(s),$$

where

$$\begin{aligned} q(s) &= s^2 h^2 + \pi^2 \\ r(s) &= s^8 25h^4 + s^6\left(-100h^4 + 50\pi^2 h^2\right) + s^4\left(109h^4 - 30\pi h^3 - 200\pi^2 h^2 + 25\pi^4\right) + \\ &\quad + s^2\left(78\pi h^3 + 200\pi^2 h^2 - 30\pi^3 h - 100\pi^4\right) + 18\pi^2 h^2 + 60\pi^3 h + 100\pi^4 \\ v(s) &= s^6 25h^2 + s^4 25\pi^2 - s^2 30\pi h + 9h^2 \end{aligned}$$

The corresponding pair of pure imaginary roots of $q(s)$, is $\pm i\omega = \pm i\frac{\pi}{h}$. We can observe that $f(i\omega, h)$ is indeterminate, so if we use the rule of L'Hôpital one can observe that $f(i\omega, h)$ does not vanishes for $h > 0$. If we set $\lambda = s^2$, and we analyze $r_1(\lambda)$ with the help of it's Sturm array, we conclude that for $h > 0$, $r_1(\lambda)$ has no negative real roots.

The Sturm array of

$$v_1(\lambda) = \lambda^3 25h^2 + \lambda^2 25\pi^2 - \lambda 30\pi h + 9h^2$$

shows that for $h > 0$, $v_1(\lambda)$ has one negative real root, the pure imaginary roots $\pm\omega(h)$ are sketched on Fig. 1(a).

Now we must check if there exists a pair $\{\omega(h), h\}$ that annihilate the characteristic function. For $h \in (0, 10]$ we can see on Fig. 1(b) that the imaginary and real part of the characteristic function vanish simultaneously at $\omega = 1.0709i$ and $h^* = 5.7345$.

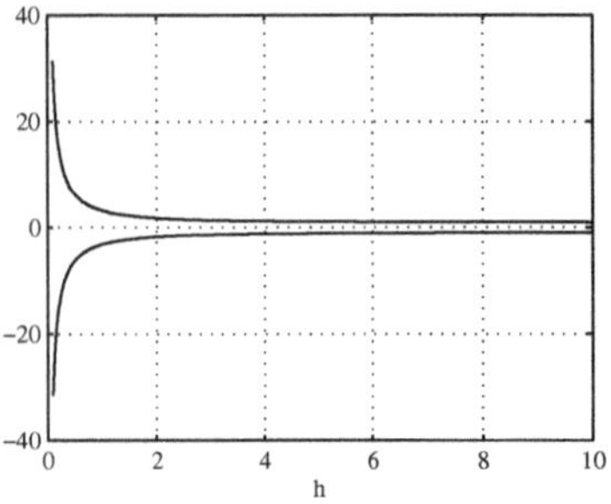

(a) Roots $\pm\omega(h)$

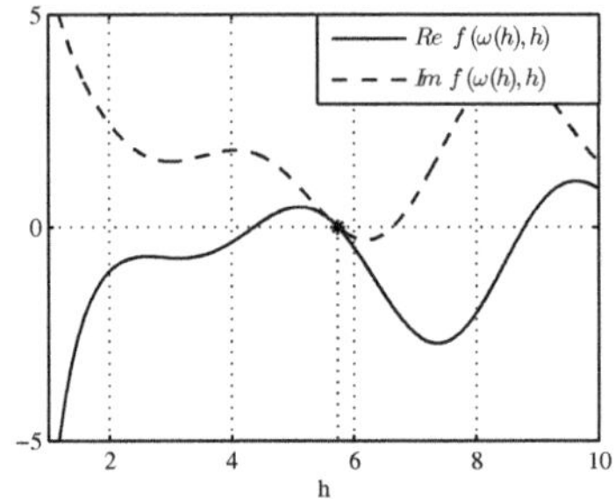

(b) Re **and** Im **of** $f(i\omega(h), h)$

Next, a two dimensional system with a parameter is considered.

Example 2. Consider the following system

$$\dot{x}(t) = A_0 x(t) + \mu A_1 x(t-1) + \int_{-1}^{0} [\sin(\pi\theta) B_0 + \cos(\pi\theta) B_1]\, x(t+\theta) d\theta,$$

where A_0, A_1, B_0 and B_1 are as in example (1), μ is a real parameter of the system, and $h = 1$. Here functions $\eta_0(\theta) = \sin(\pi\theta)$ and $\eta_1(\theta) = \cos(\pi\theta)$. The characteristic function of this system is

$$f(s,h) = \det\left(sI - A_0 - \mu A_1 e^{-s} + \frac{\pi\left(1+e^{-s}\right)}{\pi^2+s^2} B_0 - \frac{s\left(1+e^{-s}\right)}{\pi^2+s^2} B_1 \right). \qquad (41)$$

The characteristic polynomial of the delay free Lyapunov matrix is of the form

$$p(s) = \frac{1}{625} q(s) r(s) v(s),$$

where

$$\begin{aligned}
q(s) &= s^2 + \pi^2 \\
r(s) &= 25s^8 + s^6\left(50\pi^2 - 50\mu^2 - 50\right) + s^4(25\pi^4 - 100\pi^2 - 30\pi + \\
&+ 50\mu^2 + +25\mu^4 - 100\pi^2\mu^2 + 34) + s^2(30\pi + 18\pi\mu + 50\pi^2 - \\
&- 30\pi^3 - 50\pi^4 + +30\pi\mu^2 + 100\pi^2\mu^2 + 50\pi^2\mu^4 - 50\pi^4\mu^2) + \\
&+ (9\pi^2 + 30\pi^3 + 25\pi^4 + 9\pi^2\mu^2 + 30\pi^3\mu^2 + 50\pi^4\mu^2 + 25\pi^4\mu^4) \\
v(s) &= 25s^6 + s^4\left(25\pi^2 + 50\mu^2 - 50\right) + s^2(25\mu^4 - 50\pi^2 - 5\mu^2 - 30\pi + \\
&+ 50\pi^2\mu^2 + 25) + \left(30\pi + 25\pi^2 - 30\pi\mu^2 - 50\pi^2\mu^2 + 25\pi^2\mu^4 + 9\right)
\end{aligned}$$

The roots of $q(s)$ are $\pm\omega i = \pm\pi i$. If we compute $f(\omega i, h)$ it is easy to verify that ωi is not a root of the characteristic function (41). Let us now analyze $r(s)$. If we set $\lambda = s^2$, we can conclude with the help of its Sturm array that $r_1(\lambda)$ has no negative real roots for $\mu > 0$.

The Sturm array of

$$v_1(\lambda) = 25\lambda^3 + \lambda^2\left(25\pi^2 + 50\mu^2 - 50\right) + \lambda(25\mu^4 - 50\pi^2 - 5\mu^2 - 30\pi + \\ + 50\pi^2\mu^2 + 25) + \left(30\pi + 25\pi^2 - 30\pi\mu^2 - 50\pi^2\mu^2 + 25\pi^2\mu^4 + 9\right)$$

shows that for $\mu \in [0, 1.0909)$, $v_1(\lambda)$ has one negative real root, while for $\mu \in [1.0909, \infty)$ it has three negative real roots. Let us consider $\mu \in [0, 2]$. The behavior of the pure imaginary roots $\pm\omega_{1,2,3}(\mu)$ corresponding to the negative real roots of $v_1(\lambda)$ are sketched on Fig. 1(c).

Next we must check if there exist pairs $\{i\omega_k(\mu), \mu\}$, $k = 1, 2, 3$ for which (41) vanishes. Direct calculations show that there are no values $\omega_{1,2}(\mu)$ and μ that annihilate this characteristic function for $\mu \in [0, 2]$. For $\omega_3(\mu)$ and $\mu \in [0, 2]$ one can see on Fig. 1(d) that the real part and the imaginary part of $f(i\omega_3(\mu), \mu)$ vanish simultaneously only for $\mu = 1.4831$ and $\omega = 0.9615$.

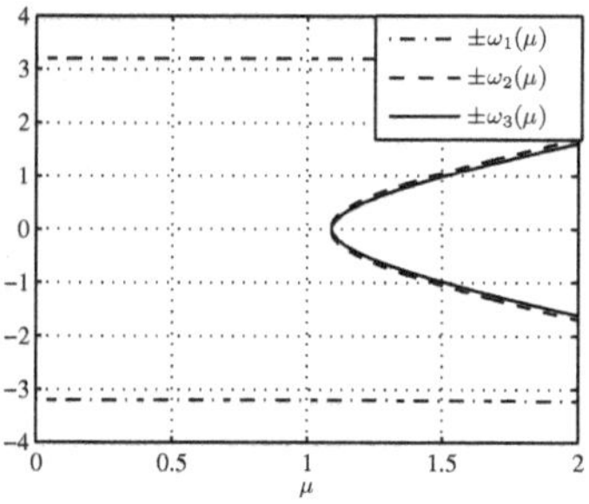

(c) ***Roots*** $\pm\omega_{1,2,3}(\mu)$

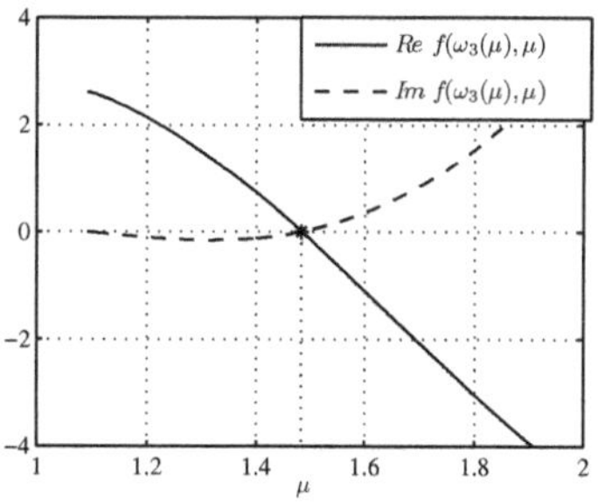

(d) Re **and** Im ***of*** $f(i\omega_3(\mu), \mu)$

5 Concluding remarks

A novel approach for determining the parameters critical values of time delay systems for which the characteristic function of the system has purely imaginary roots is presented. It covers a general class of systems with distributed delay. A special case with further restrictions on the kernel allows a reduction of the size of the associated delay free system. We would like to emphasize the fact that, unlike the classical algebraic tools such as the Kronecker sum and resultant matrix used until now, this new approach establishes an interesting connection between the frequency domains stability analysis of time delay system, and numerical schemes used for the computation of the Lyapunov matrices of the systems.

References

1. Bellman, R., Cooke, K.L.: Differential Difference Equations. Academic Press, New York (1963)
2. García-Lozano, H., Kharitonov, V.L.: Lyapunov matrices for time delay systems with commensurate delays. In: Proceedings of the 2-nd IFAC Symposium on Systems, Structure and Control, Oaxaca, México, pp. 102–106 (2004)
3. Chen, J., Gu, G., Nett, C.: A new method for computing delay margins for stability of linear delay systems. Syst. Contr. Let. 26, 107–117 (1995)
4. Kharitonov, V.L.: Lyapunov matrices for a class of time delay systems. Syst. Contr. Let. 55, 610–617 (2006)
5. Louisell, J.: Numerics of the stability exponent and eigenvalue abcissas of a matrix delay system. In: Duagard, L., Verriest, E.I. (eds.). LNCIS, p. 228. Springer, Heidelberg (1998)
6. Louisell, J.: A matrix method for determining the imaginary axis eigenvalues of a delay system. IEEE Transactions on Automatic Control 48, 2008–2012 (2001)
7. Niculescu, S.I.: Stability and hiperbolicity of linear systems with delay states: a matrix pencil approach. IMA J. Math. Contr. Inf. 15, 331–347 (1998)
8. Olgac, N., Sipahi, R.: An exact method for the stability analysis of time-delayed linear time invariant (lti) systems. IEEE Transactions on Automatic Control 47, 793–797 (2002)
9. Sipahi, R., Olgac, N.: A comparative survey in determining the imaginary characteristic roots of lti time delayed systems. In: Proceedings of the 16th IFAC Worls Congress (2005)
10. Walton, K.E., Marshall, J.E.: Direct method for tds stability analysis. IEEE Proceedings 134, 101–107 (1987)

Robust Control for Multiple Time Delay MIMO Systems with Delay - Decouplability Concept

Kamran Turkoglu[1] and Nejat Olgac[2]

[1] Department of Mechanical Engineering, University of Connecticut, Storrs, Connecticut, 06269 turkogluk@engr.uconn.edu
[2] Department of Mechanical Engineering, University of Connecticut, Storrs, Connecticut, 06269 olgac@engr.uconn.edu

Summary. In this paper, we consider linear time-invariant minimum phase MIMO plants with multiple control delays. The time delays appear at several components of the state. The delay decoupling control (DDC) aims to force the characteristic equation to facilitate the assessment of stability in each of the delays, independently from one another (thus introducing "delay decouplability"). When, however, some system parameters are uncertain, the corresponding characteristic equation exhibits truly coupled delays, which brings the stability assessment problem into N-P hard complexity class. This operation can still be very efficient using a recent paradigm called the Cluster Treatment of Characteristic Roots (CTCR). The main contribution of this study is to show that, for a class of uncertain time-delayed dynamics, if the feedback control is properly designed, decouplability may still hold, consequently the robustness analysis can be performed efficiently. This result is demonstrated for a 2-input, 2-output system, and it is claimed that the findings are scalable to higher dimensional dynamics. An example case study of a cart-pendulum system is examined, considering varying parametric uncertainties.

1 Introduction

In recent years, stability of time delayed systems, both in retarded and neutral classes, is broadly investigated [1-3]. Several complexities arise if there are multiple rationally independent delays in the system dynamics, especially when cross-talk between delays exists (i.e., a linear combination of the delays occurs), [4-6].

The main topic that is addressed in this paper is on the synthesis of a stabilizing controller for the systems, which includes multiple delays in the feedback loop. The suggested control logic must guarantee stability and provide desirable tracking as well as disturbance rejection capabilities. In the literature, several control strategies are encountered for the stability analysis of time-delayed systems. For instance, a PI type feedback stabilization of first-order systems with single delay is studied in [7]. This task becomes very complicated for MIMO systems with multiple delays [8]. The investigation for the design of controllers to remove the cross-interaction between the components of controller and the states, can be found in [9, 10]. In both studies, the small gain theorem is applied for a robust stability analysis. However,

J.J. Loiseau et al. (Eds.): Topics in Time Delay Systems, LNCIS 388, pp. 37–47.
springerlink.com

the exercise is conducted in point-wise fashion, i.e., stability is assured for selected values of delays. As such, the computational cost is very high.

Even though, all of these control strategies have original contributions to the literature, none of them is capable of eliminating the occurrence of delay cross-talk (which are represented by the terms containing multiplicity of delays together) or commensurate delays (which are formed by integer multipliers of individual delays) in the characteristic equation of the system dynamics. As a qualitative remark, even when there is no commensurate and cross-talking delays in the characteristic equation, stability analysis part of the study with multiple delays is very intricate [11, 12].

In this study, Delay Decoupling Control (DDC), from [8] and [13], has been used to avoid the complexities for LTI minimum phase MIMO systems with multiple input delays. The deployment of proposed DDC controller creates a characteristic equation which facilitates the assessment of stability in each of the delays independently from one another. That is, the characteristic equation which is influenced by delays is represented as a product of several parts, each of which has a single time delay. This feature introduces a great convenience from the stability analysis point-of-view. Appropriately, the technique is named as "Delay-Decoupling Control" (DDC). The core action of this control logic is that it results in a factorization of the characteristic equation, which decouples the delays. When, however, some system parameters are uncertain, the characteristic equation exhibits truly coupled delays, which force the stability assessment into an N-P hard complexity class problem. The main contribution of this study is to show that, for some uncertain structures, the feedback control may be properly selected such that with multiple independent delays decouplability may still be achieved and consequently the stability robustness analysis can be done efficiently. The procedure is explained for n-input, n-output systems. As an example case, a fully-actuated cart pendulum dynamics is studied and it is shown that the proposed assessment is efficiently performed, even for uncertain systems, using the Cluster Treatment of Characteristic Roots (CTCR) paradigm.

The outline of the paper is as follows: Section 1 describes the synthesis of DDC problem. In Section 2, the delay-decoupling control law (DDC) is reviewed. In Section 3, we explain the main logic behind the delay decouplability concept. In Section 4, a case study on a fully actuated cart-pendulum plant is presented and Section 5 treats the "delay decouplability conditions" under uncertainty. Finally, the concluding remarks are made. [2].

2 Problem Statement

Consider the closed-loop system in Fig. 1, where the transfer matrices of plant and controller are ***G(s)*** and ***C(s)***, respectively. Both of them are square matrices of appropriate dimensions with rational polynomial entries. The diagonal matrix ***Λ(s)*** represents the input delay dynamics on each control channel which makes different segments of control, ***U***, carry different time delays, as it is common in practice. The elements of the matrices ***G(s)*** and ***C(s)*** are denoted by

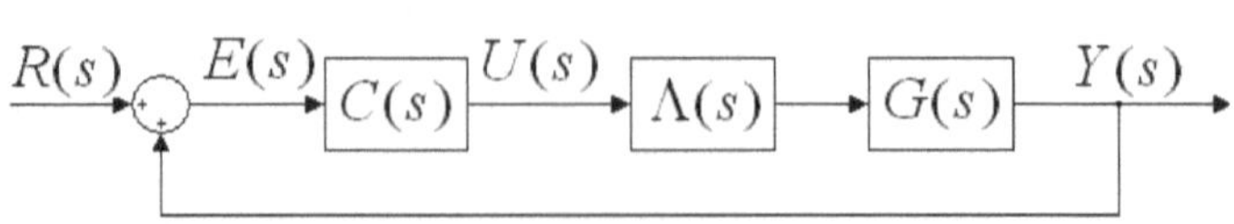

Fig. 1. MIMO system with unity feedback.

$g_{ij}(s)$, $c_{ij}(s)$, i= 1, ..., n, j= 1, ..., n and **Λ(s)**= $diag(e^{-\tau_i s})$ where $\tau_i \in \Re^+$, i= 1, ..., n are the n independent delays. We will use $\tau \in \Re^n$ to represent the delay vector $(\tau_1, \tau_2 \ldots \tau_n)$.

The corresponding transfer matrix from reference input to output of the closed-loop system is expressed as

$$H(s) = (\bar{I} + G\Lambda C)^{-1} G\Lambda C \;, \tag{1}$$

where $\bar{I}$ is the n-dimensional identity matrix. The characteristic equation of this system is

$$\begin{aligned} CE &= \left|\bar{I} + G\Lambda C\right| \times (Non - delayed\, polynomial) \\ &= P_1(s,\tau) \times P_2(s) \;. \end{aligned} \tag{2}$$

In this equation, the only term which contains delays is $\left|\bar{I} + G\Lambda C\right|$, while the non-delayed polynomial $P_2(s)$ is shaped by the denominators of g_{ij} and c_{ij} terms. $P_1(s,\tau)$ typically has the delay cross-coupling elements (which are also called the delay cross-talks). The stability assessment of such a dynamics is an open problem in mathematics [5].

In this paper, we revisit an interesting approach, the Delay Decoupling Control (DDC), through which one can form a control logic that has the capability of decoupling the effects of individual delay terms when it comes to the stability question. The DDC causes the characteristic equation to exhibit a factorized form with only one delay appearing in each factor. The ensuing stability study is straightforward, which simply examines that each factor (which are quasi-polynomials themselves and thus posses infinitely many poles) has all its poles in the stable open left half plane of the complex space. Taking all system parameters as known constants, except the delays, one can utilize the earlier developed technique of the authors group, which is called the Cluster Treatment of Characteristic Roots (CTCR) [1, 4-5]. As briefly described in the text, the CTCR reveals the stable pockets of the only time delay for each factor, where this factor possesses stable poles. Furthermore this result is created exhaustively and precisely in the domain of the particular delay. Searching for the similar results in all n delay spaces one can create n-D prismatic stability regions for the controlled system with DDC. This conclusion was reached by [13]. We introduce a new dimension in this paper from the perspective of robustness of the DDC logic against parametric uncertainties. The following section reviews the derivation and design of such a controller with decouplability concepts.

3 Delay Decoupling Control (DDC)

Next, we briefly review the key steps of DDC following [8, 13]. Delay decoupling process is constructed over a co-prime factorization of the plant transfer matrix ***G(s)*** as a product of two matrices

$$G(s) = A(s)B(s) \;, \tag{3}$$

where $\boldsymbol{A}_{nxn}(s)$ consists of elements which are proper or improper (but not strictly proper) transfer functions, and $\boldsymbol{B}_{nxn}(s)$ is diagonal. This is a right co-prime factorization operation, and it forms the fundamental step for the lemma which is presented next.

Lemma

The control logic for the dynamics in Fig.1 is proposed as

$$C(s) = K(s)A^{-1}(s) \;, \tag{4}$$

where $\boldsymbol{K}_{nxn}(s)$ is a diagonal transfer matrix, and $\boldsymbol{A}^{-1}(s)$ is now a proper matrix. With this selection of ***C(s)***, delays which come from **Λ**'s in Fig.1, appear in the characteristic equation of the closed-loop system in completely decoupled form. In other words, delay terms in $P_1(s,\tau)$ in (2) consist of n factors, each of which contains only one of the delays.

Proof

The proposed control structure of (4) yields the characteristic equation given in (1), of which the delay influenced segment $P_1(s,\tau)$, becomes

$$\begin{aligned} P_1(s,\tau) &= |\bar{I} + AB\Lambda KA^{-1}| = |AA^{-1} + AB\Lambda KA^{-1}| \\ &= \left|A(\bar{I} + B\Lambda K)A^{-1}\right| = \left|\bar{I} + B\Lambda K\right| . \end{aligned} \tag{5}$$

Noting that $\boldsymbol{B\Lambda K}$ is diagonal, the expression $P_1(s,\tau)$ exhibits a factorized form as

$$P_1(s,\tau) = \prod_{i=1}^{n} (1 + B_i K_i e^{-\tau_i s}) \,, \tag{6}$$

where B_i and K_i are the ith diagonal term of the respective matrices. The characteristic equation of the closed-loop system becomes

$$CE(s,\tau) = P_2(s) \prod_{i=1}^{n} (1 + B_i K_i e^{-\tau_i s}) = 0 \,. \tag{7}$$

Therefore, the effects of delays τ_i, $i = 1, 2, ..., n$ on the system stability are all decoupled from each other. In other words, for the quasi-polynomial, (7), to represent stable dynamics, each factor should represent stable dynamics, which can be analyzed separately as single-delay quasi-polynomials. Furthermore, each of these delays appears with no commensuracy, which is a very important advantage when judging the stability.

In summary, the proposed controller in (4), offers the following properties:

1) It eliminates the stability cross-talk among the delays completely. It also reshapes the delayed part of the characteristic equation into a product of "*n*" single-delay expressions, with no commensurate terms. The stability analysis of the closed-loop system becomes substantially simplified. There are many procedures in the literature to handle this class of dynamics. An example is the **Cluster Treatment of Characteristic Roots (CTCR)**, which is a treatment recently introduced by the author's group, to address this class of stability problems [4-6, 14, 16, 17]. For each of the delays, the **CTCR** procedure creates an exact and complete set of delay intervals where the stability is guaranteed. The numerical efficiency of the method is remarkable and it can be executed in real time. The combination of **DDC (Delay-Decoupling Control)** and the **CTCR** procedure creates a very effective methodology to determine the stable operating points of the system within the domain of the *n* delays.

2) $\boldsymbol{K}_{nxn}(s)$ can still be tailored by the designer so that the desired performance specifications could be achieved, as it is described in the example case in the following section. Obviously, the stability and trajectory tracking capabilities of the controlled system (Fig. 1) are directly influenced by the selection of $\boldsymbol{K}_{nxn}(s)$. It influences the stability map in delay domain but does *not* affect the delay decoupling property.

4 Delay Decoupling and Robustness Against Parametric Uncertainties

The above description of DDC logic is comfortably applied when the plant dynamics is certain. When there are uncertainties, however, the procedure becomes problematic. This difficulty needs to be treated in order to introduce robustness to the DDC logic against dynamic uncertainties. For notational simplicity in the discussions of this section, we take the nominal plant dynamics as ***G(s)=A(s)B(s)*** where ***A(s)*** is the uncertain transfer matrix and the feedback control logic based on the nominal plant is **C(s)** $= K(s)\bar{A}(s)^{-1}$, where ***A(s)*** represents the uncertain transfer matrix, and $\bar{A}(s)$ is the nominally known transfer matrix which is used when we form the control logic. Since these two matrices are not the inverses of each other, similarity transformation properties do not hold any more

$$\left|\bar{I} + AB\Lambda K\bar{A}^{-1}\right| \neq \left|\bar{I} + B\Lambda K\right| . \tag{8}$$

As an immediate effect of this, delay decoupling feature of the characteristic equation does not hold.

Consequent to these arguments, the characteristic equation is not expected to display a factorization with one single delay term in each of the factors. However, the particular formation of uncertainty and the resulting equation (8), may be such that the delay decoupling can still be in effect. There are some necessary and sufficient conditions for such decouplability to occur. Instead of giving a full-scale (n-D) treatment of this development here, we will demonstrate it over an example case study within the next section.

The important point is that if a particular uncertainty structure yield the delay decouplability, the control synthesis can still be conveniently treated using the DDC logic. The procedure, in that case, again is reduced to determine the stable delay intervals in each of the independent delay domains, and create the n-dimensional prismatic stability zones. This prism can be determined in a very efficient manner using the CTCR paradigm. One can then superpose the stable delay regions for the corresponding parametric uncertainty range and find the intersection of these stable regions. This would result in the core stable operating zone for all the uncertainties in the delays, as well as the system's parameters uncertainties.

5 Example Case Study

A fully actuated cart pendulum example is taken from [13] in order to demonstrate the functionality of DDC, and decouplability concepts in more detail.

The cart-pendulum system is shown in Fig. 2, which is a 2-input 2-output system. The aim of this investigation is to control the position of the cart, $x(t)$, and the angular displacement of the pendulum, $\theta(t)$, by deploying an appropriate feedback control law which could only be performed with unavoidable time delays. Two delays that appear in the force (F) and torque (T) are independent from each other.

Linearized equations of motion (for small θ angle assumptions) are borrowed from [15] as

$$(M+m)\ddot{x} + b\dot{x} + mL\ddot{\theta} = F(t), \tag{9}$$

$$(I+mL^2)\ddot{\theta} + c\dot{\theta} + mgL\theta + mL\ddot{x} = T(t), \tag{10}$$

where M, m are the masses of the cart and the pendulum; b, c are the viscous friction coefficients for the cart and the pendulum, respectively. L is the half length of the pendulum,

F is the force applied to the cart, I is the moment of inertia of the pendulum about its center of gravity, g is the gravitational acceleration and T is the torque applied to the pendulum. The Laplace Transformed version of these equations are given in (11) and (12)

$$\begin{bmatrix} X(s) \\ \Theta(s) \end{bmatrix} = G(s)U(s) = G(s)\begin{bmatrix} F(s) \\ T(s) \end{bmatrix}, \tag{11}$$

where ***G(s)*** is (2x2) the self-evident transfer matrix,

$$\begin{bmatrix} X(s) \\ \Theta(s) \end{bmatrix} = \frac{1}{Ds}\begin{bmatrix} (I+mL^2)s^2+cs+mgL & -mLs^2 \\ -mLs^2 & (M+m)s^2+bs \end{bmatrix}\begin{bmatrix} F(s) \\ T(s) \end{bmatrix}, \tag{12}$$

$$D = (MI + mI + MmL^2)s^3 + (bI + bmL^2 + cm + cM)s^2$$
$$+(mMgL + m^2gL + bc)s + bmgL.$$

It is assumed that the force (F) and the torque (T) are applied with

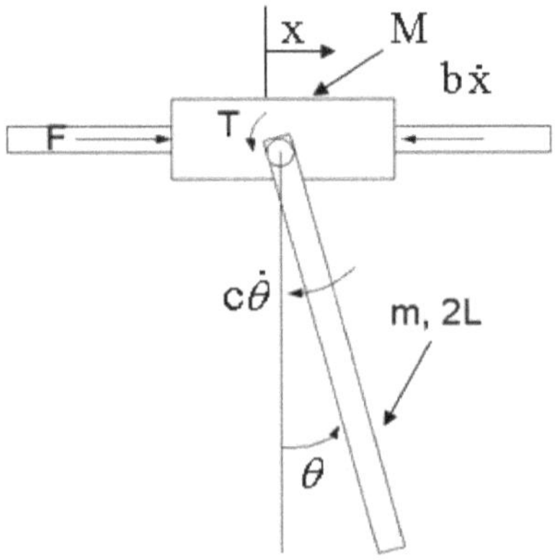

Fig. 2. Fully actuated cart pendulum system.

transmission delays τ_1 and τ_2, respectively. These delays are represented by

$$\Lambda(s) = \begin{bmatrix} e^{-\tau_1 s} & 0 \\ 0 & e^{-\tau_2 s} \end{bmatrix}. \tag{13}$$

Including (13) in the dynamics of (11) one obtains

$$\begin{bmatrix} X(s) \\ \Theta(s) \end{bmatrix} = G(s)\Lambda(s)\begin{bmatrix} F(s) \\ T(s) \end{bmatrix}. \tag{14}$$

To find a stabilizing controller using the DDC method, the right co-prime factorization of the transfer matrix ***G(s)*** is suggested according to (3), in the forms of (15) and (16)

$$A = \begin{bmatrix} (I+mL^2)s^2+cs+mgL & -mLs \\ -mLs^2 & (M+m)s+b \end{bmatrix}, \tag{15}$$

$$B = \frac{1}{D}\begin{bmatrix} \frac{1}{s} & 0 \\ 0 & 1 \end{bmatrix}, \tag{16}$$

$$C(s) = K(s)\left\{\frac{1}{D}\begin{bmatrix}(M+m)s+b & mLs \\ mLs^2 & (I+mL^2)s^2+cs+mgL\end{bmatrix}\right\}. \tag{17}$$

In this factorization process, one should keep in mind that the proposed formations of ***A(s)*** and ***B(s)*** matrices are not unique and could be done in an arbitrary fashion provided that ***A(s)*** is improper and ***B(s)*** is diagonal. Once they are defined, DDC control law is generated as in (17), according to (4). As mentioned earlier, in DDC the designer has the capability of selecting the control matrix ***K(s)*** in order to obtain the desired system performance (such as better tracking and disturbance rejection capabilities). For instance, the ***K(s)*** factor in the control logic of (4) can be chosen as:

$$K(s) = \begin{bmatrix} k_1 & 0 \\ 0 & (1.1+\frac{k_2}{s}+0.13s)\end{bmatrix}, \tag{18}$$

which implies the utilization of P and PID type of controllers on the first and second channels, respectively. k_1 and k_2 are the two arbitrarily chosen (free) parameters to be fine tuned based on the system stability analysis. The corresponding *decoupled* characteristic equation of the system is given in (19),

$$\begin{aligned} CE = {} & [(MI+mI+MmL^2)s^4+(bI+bmL^2+cm+cM)s^3 \\ & +(mMgL+m^2gL+bc)s^2+bmgLs+k_1e^{-\tau_1 s}] \\ & \times[(MI+mI+MmL^2)s^3+(bI+bmL^2+cm+cM)s^2 \\ & +(mMgL+m^2gL+bc)s+bmgL+(1.1+\tfrac{k_2}{s}+0.13s)e^{-\tau_2 s}]=0. \end{aligned} \tag{19}$$

It is obvious that the characteristic equation is a product of two factors *for the nominal plant dynamics*, where each factor contains only one delay element. This formation is the direct result of the DDC logic. It imparts the separation (decoupling) of the stability analysis of the two-delay system into two single-delay systems for the *nominal* system dynamics,

$$\begin{aligned} CE = {} & (2MI^2m+m^2L^4M^2+m^2I^2+2M^2IML^2+2m^2IL^2M \\ & +M^2I^2)s^8+(4MIcm+2mL^2M^2c+m^2L^4Mb_0+m^2L^2bI \\ & +mI^2b_0+m^2IL^2b_0+2m^2L^2Mc+bI^2M+2bI\,mL^2M \\ & +2MI\,mL^2b_0+2m^2Ic+bI^2m+2M^2Ic+MI^2b_0+m^2L^4bM)s^7 \\ & +(m^2L^4bb_0+2c^2mM+cm^2L^2b_0+2bI\,mL^2b_0+2bIcm \\ & +2m^2L^3M^2g+2m^3IgL+2mL^2Mcb_0+m^2L^2bc+2M^2I\,mgL \\ & +4MI\,m^2gL+2mL^2bMc+M^2c^2+2m^3L^3Mg+2mIcb_0 \\ & +2bIMc+c^2m^2+2MIcb_0+bI^2b_0)s^6+(c^2bm+m^3gL^3b_0 \\ & +2m^2L^3bMg+2BI\,m^2gL+2MI\,mgLb_0+c^2mb_0+2m^2IgLb_0 \\ & +2bIMmgL+2bIcb_0+4cm^2MgL+2m^2L^3Mgb_0+2M^2cmgL \\ & +c^2bM+2mL^2bcb_0+m^3L^3bg+Mc^2b_0+2cm^3gL)s^5 \\ & +(M^2m^2g^2L^2+2cbm^2gL+2cm^2gLb_0+2Mm^3g^2L^2 \\ & +2m^2L^3bgb_0+2McmgLb_0+2cbMmgL+m^4g^2L^2 \\ & +2bI\,mgLb_0+c^2bb_0)s^4+(m^3g^2L^2b+m^2g^2L^2bM \\ & +Mm^2g^2L^2b_0+m^3g^2L^2b_0+2cbmgLb_0)s^3+(g^2m^2L^2b_0b)s^2 \\ & +((mL^2k_1M+Ik_1m+Ik_1M)s^4+(ck_1m+ck_1M+Ik_1b_0 \\ & +mL^2k_1b_0)s^3+(ck_1b_0+m^2gLk_1+mgLk_1M)s^2 \\ & +mgLk_1b_0s)e^{-\tau_1 s}+((0.13mL^2M+0.13MI+0.13mI)s^6 \\ & +(1.1mI+0.13Mc+1.1MI+0.13cm+1.1mL^2M+0.13mL^2b \\ & +0.13bI)s^5+(0.13cb+mIk_2+1.1cm+mL^2Mk_2+0.13m^2gL \\ & +1.1mL^2b+1.1Mc+0.13MmgL+IMk_2+1.1bI)s^4+(1.1MmgL \\ & +mL^2bk_2+cMk_2+1.1cb+0.13mgLb+Ibk_2+1.1m^2gL \\ & +mck_2)s^3+(gmLMk_2+cbk_2+1.1mgLb+gm^2Lk_2)s^2 \\ & +gmLbk_2s)e^{-\tau_2 s}+(0.13k_1s^2+1.1k_1s+k_1k_2)e^{-(\tau_1+\tau_2)s}=0. \end{aligned} \tag{20}$$

The novelty of the paper comes at this point. Under the parametric uncertainties, the similarity conditions will not hold as expressed in (8). For the case of single parametric uncertainty, let's say on viscous friction constant, b, the characteristic equation for uncertain plant will become as given in (20), where b_0 is the nominal value of uncertain viscous friction coefficient, b. In what follows, we give a complete and thorough procedure of necessary and sufficient conditions for decouplability of (20) under this particular parametric uncertainty.

Characteristic equation of the uncertain plant, (20), can be expressed as

$$\alpha(s) + \beta(s)e^{-\tau_1 s} + \gamma(s)e^{-\tau_2 s} + \delta(s)e^{-(\tau_1+\tau_2)s} = 0. \tag{21}$$

where the self explanatory terms, $\alpha(s)$, $\beta(s)$, $\gamma(s)$and $\delta(s)$ represent the delay free terms, the terms that are influenced by $e^{-\tau_1 s}$, the terms influenced by $e^{-\tau_2 s}$ and the terms influenced by the cross-talk term $e^{-(\tau_1+\tau_2)s}$, respectively. It is clear that (21) can be rewritten as

$$\left[\frac{\alpha(s)}{\delta(s)} + \frac{\beta(s)}{\delta(s)}e^{-\tau_1 s} + \frac{\gamma(s)}{\delta(s)}e^{-\tau_2 s} + e^{-(\tau_1+\tau_2)s}\right] = 0. \tag{22}$$

At this point let's assume that, our uncertain plant given in (22) is decouplable, and could be expressed as

$$\left[\hat{A}(s) + e^{-\tau_1 s}\right]\left[\hat{B}(s) + e^{-\tau_2 s}\right] = 0. \tag{23}$$

By comparing (22) and (23), we can state

$$\frac{\gamma(s)}{\delta(s)} = \hat{A}(s) \quad \text{and} \quad \frac{\beta(s)}{\delta(s)} = \hat{B}(s). \tag{24}$$

Then the necessary and sufficient conditions for decouplability of (20) becomes as

$$\frac{\alpha(s)}{\delta(s)} = \frac{\gamma(s)}{\delta(s)}.\frac{\beta(s)}{\delta(s)} \quad \text{or} \quad \alpha(s)\delta(s) = \gamma(s)\beta(s). \tag{25}$$

One can demonstrate that this condition holds for the fully-actuated cart dynamics, as represented in (20), with uncertain viscous friction value, b. Therefore the linearized control system which is given in (14-17) remains delay decouplable for an uncertainty on b (the viscous friction coefficient) alone. This result arises due to the special formation of $\boldsymbol{A}$ matrix (15), which has only the A_{22} element affected by b_0. Nevertheless, DDC logic is deployable, and one can obtain the corresponding stability regions of each delay term independently.
The problem is reduced to determine the stable delay intervals in each of the delay domains, and create the rectangular stability zones in (τ_1, τ_2). These rectangles can be evaluated in an efficient manner using the CTCR procedure. Repeated deployment of CTCR for various b values and superposition of the stable delay intervals for the corresponding parametric variations generates the robustness picture in the domain of the delays as well as the uncertain parameters. The intersection of these stable regions would result in the core operating zones which are stable for the delay uncertainty as well as the parametric uncertainty for the given system. The stability analysis of each factor of equation (23) for fixed gains (k_1 or k_2) can easily be analyzed in $\tau \in \Re^{2+}$. Then one can sweep the gains, to create the complete stability picture in 4-D (τ_i, k_i) space (i = 1, 2). Fig.3 is the result of such an analysis for k_1 = k_2 = 6.95.

The rectangular stability shapes indicate the delay decoupling characteristic (i.e., no cross-talk effect exists). It is also possible to see that for increasing values of the parameter b, the stability region is enlarging in τ_1 dimension. The reason of τ_2 delay boundaries to remain static is because of the independence of viscous friction coefficient, b, from τ_2 delay terms, due to the special formation of $\boldsymbol{A(s)}$.

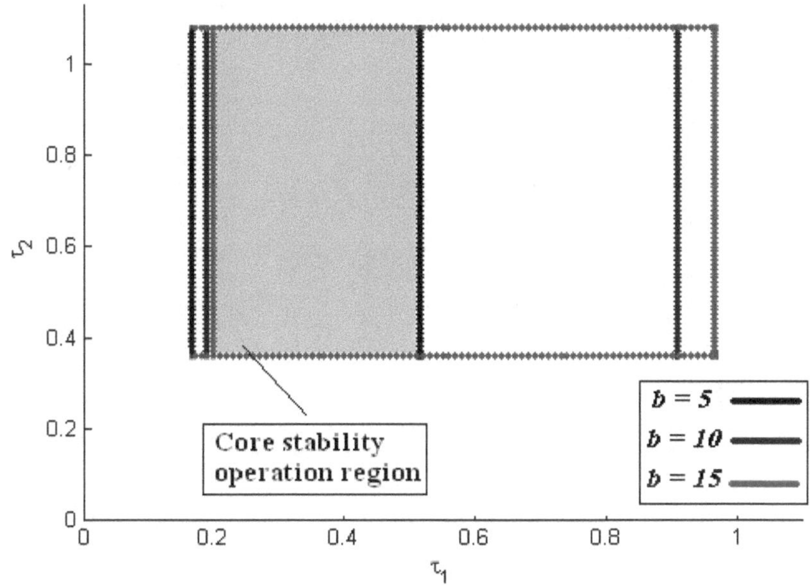

Fig. 3. Stability picture of uncertain plant.

6 Robustness Analysis of the Delay Decouplability Concept

Since condition (25) is satisfied for the system with uncertainty on b only, the "delay decouplability" still holds for all b variations. When, for example, the parameters m and b are uncertain, the decouplability condition (25) is no longer satisfied. Consequently the factorization displayed in (23) cannot be written. Then the characteristic equation (19) has to be studied for its stability, taking τ_1 and τ_2 as independent delays. The CTCR paradigm is again utilized but not with the level of ease that is encountered for the "delay decoupled" case seen above.

For the 2-input, 2-output example case, we consider two parametric uncertainties, for b and m. We leave the details of deploying CTCR to [5] and present the general outlook of stability maps for various pairs of (b, m). We consider the nominal parameters of viscous friction coefficient, b, and mass of the pendulum, m, as $b_0 = 8.4$, $m_0 = 0.231$ and consequently provide some percentage variations of (b, m) from these values. We consider three discrete parametric settings: (b_0+12%, m_0+100%), (b_0+20%, m_0+120%) and (b_0+ 28%, m_0+144%). The respective stability regions are shaded in Fig.4 (with different colors, if viewed in color) for these three discrete cases. We also identify the intersection of these stable zones, which shows the delay composition (τ_1, τ_2) that exhibits stable operating conditions for all 3 pairs of uncertain parameters. Note that, we cannot claim the robustness against the complete ranges of $9 < b < 11$, $0.4620 < m < 0.6006$ due to the non-monotonous and complex dependency of the stability pictures on the parameters b and m. But Fig.4 provides useful information concerning the discrete stable operating points.

Fig.4 is provided also to demonstrate the fact that the DDC procedure offers a remedy even when the decouplability is not preserved. Otherwise the treatment would be again an N-P hard complex and numerically cumbersome.

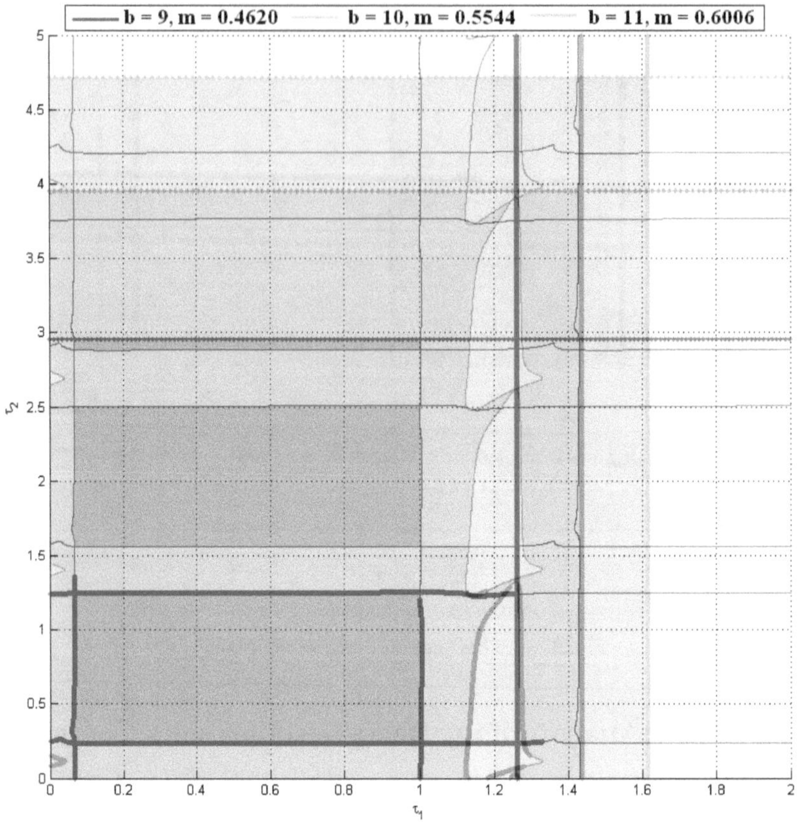

Fig. 4. Stability picture for uncertain b and m.

7 Conclusions

In this study, we show that the deployment of delay decoupling control (DDC) creates a characteristic equation which facilitates the assessment of stability in each of the delays independently from one another. When, however, some system parameters are uncertain, the characteristic equation exhibits truly coupled delays and the main contribution of this study comes at this point. For some special uncertainty structure, if the feedback control is properly formed with independent delays on separate feedback channels, decouplability still holds, and the robustness analysis becomes efficient. Necessary and sufficient conditions for decouplability are derived. The results are demonstrated on a 2-input, 2-output system, and it is claimed that the findings are scalable to higher dimensional dynamics. An example case study was conducted on a fully actuated cart-pendulum system and corresponding stability regions were obtained from the decouplability analysis.

Acknowledgments

Authors express their appreciation for the financial support through the grants from NSF (DMI 0522910) and Army Research Office grant ARO W911NF-07-1-0557.

References

1. Olgac, N., Sipahi, R.: The cluster treatment of characteristic roots and the neutral type time-delayed systems. ASME Journal of Dynamic Systems Measurement and Control 127(1), 88–97 (2005)
2. Gu, K., Kharitonov, V.L., Chen, J.: Stability of Time-Delay systems, 1st edn. Birkhauser, Boston (2003)
3. Niculescu, S.-I.: Delay effects on stability. In: A robust control approach. LNCIS, p. 269. Springer, Heidelberg (2001)
4. Olgac, N., Sipahi, R.: A unique methodology for chatter stability mapping for simultaneous machining. ASME Journal of Manufacturing Science and Engineering 127(4), 791–800 (2005)
5. Sipahi, R., Olgac, N.: A unique methodology for the stability robustness of multiple time delay systems. Systems and Control Letters 55, 819–825 (2006)
6. Fazelinia, H., Sipahi, R., Olgac, N.: Stability analysis of multiple time delayed systems using Building Block Concept. IEEE Transactions on Automatic Control 52(5), 799–810 (2007)
7. Silva, G.J., Datta, A., Bhattacharyya, S.P.: PI stabilization of first-order systems with time delay. Automatica 37, 2025–2031 (2001)
8. Gundes, A.N., Ozbay, H., Ozguler, A.B.: PID controller synthesis for a class of unstable MIMO plants with I/O delays. Automatica 43, 135–142 (2007)
9. Liu, T., Zhang, W., Gao, F.: Analytical decoupling strategy using a unity feedback control structure for MIMO processes with time delays. Journal of Process Control 17, 173–186 (2007)
10. Wang, Q.G., Zhang, Y., Chiu, M.S.: Decoupling internal model control for multivariable systems with multiple time Delays. Journal of Chemical Engineering Science 57, 115–124 (2002)
11. Hale, J.K., Huang, W.: Global geometry of the stable regions for two delay differential equations. Journal of Mathematical Analysis and Applications 178, 344–362 (1993)
12. Gu, K., Niculescu, S.I., Chen, J.: On stability crossing curves for general systems with two delays. Journal of Mathematical Analysis and Applications 311(1), 231–253 (2005)
13. Poursina, M., Olgac, N.: Delay Decoupling Control, A Novel Method for Systems with Multiple Delays. In: IFAC-TDS Conference 2007, Nantes France (2007)
14. Olgac, N., Sipahi, R.: An exact method for the stability analysis of time delayed LTI systems. IEEE Transactions on Automatic Control (47), 793–797 (2002)
15. Franklin, G., Powell, J.D., Emami-Naeini, A.: Feedback Control of Dynamic Systems, 5th edn., pp. 57–59. Prentice Hall, Englewood Cliffs (2005)
16. Olgac, N., Ergenc, A.F., Sipahi, R.: Delay Scheduling a new concept for stabilization in multiple delay systems. Journal of Vibration and Control 11(9), 1159–1172 (2005)
17. Olgac, N., Sipahi, R.: An improved procedure in detecting the stability robustness of systems with uncertain delay. IEEE Transactions on Automatic Control 51(7), 1164–1165 (2006)

On Stability of Linear Retarded Distributed Parameter Systems of Parabolic Type

Yury Orlov[1] and Emilia Fridman[2]

[1] CICESE Research Center, Km.107, Carretera Tijuana-Ensenada, Ensenada, B.C. 22 860, Mexico yorlov@cicese.mx
[2] Department of Electrical Engineering-Systems, Tel-Aviv University, Tel-Aviv 69978, Israel emilia@eng.tau.ac.il

Summary. In this chapter the stability analysis via Lyapunov-Krasovskii method is extended to linear parabolic time-delay systems in a Hilbert space. The operator acting on the delayed state is supposed to be bounded. The system delay is admitted to be unknown and time-varying with an *a priori* given upper bound on the delay derivative, which is less than 1. Sufficient delay-independent asymptotic stability conditions are derived. These conditions are given in the form of Linear Operator Inequalities (LOIs) , where the decision variables are operators in the Hilbert space. Being applied to a heat equation with the Dirichlet boundary conditions, these LOIs are represented in terms of standard Linear Matrix Inequalities (LMIs).

1 Introduction

Time-delay phenomenon is natural for many control systems and it is frequently a source of instability [5]. In the case of distributed parameter systems, even arbitrarily small delays in the feedback may destabilize the system (see e.g. [1]). The stability issue of systems with delay is, therefore, of theoretical and practical value.

By now, a considerable amount of attention has been paid to stability analysis of dynamic systems governed by ordinary differential equations with uncertain, possibly, time-varying delays (see e.g. [4], [10], [13]). The stability analysis of partial differential equations with delay is essentially more complicated. There are only a few works on Lyapunov-based technique for such equations [7, 14].

The analysis of asymptotic stability via the Lyapunov-Krasovskii method is extended towards linear time delay systems over a Hilbert space. The system delay is admitted to be unknown and time-varying with an *a priori* known upper bound on the delay derivative. Sufficient stability conditions are derived in the form of Linear Operator Inequalities (LOIs), where the decision variables are operators in the Hilbert space. Being applied to a parabolic heat equation, these conditions are represented in terms of standard finite-dimensional LMIs. Although the development is confined to linear parabolic systems, the generalization to linear hyperbolic systems is indeed possible and can be found in [3].

J.J. Loiseau et al. (Eds.): Topics in Time Delay Systems, LNCIS 388, pp. 49–59.
springerlink.com

1.1 Background Material

Let $\mathcal{H}$ be a Hilbert space equipped with the inner product $\langle \cdot, \cdot \rangle$ and the corresponding norm $|\cdot|$. $\mathcal{L}(\mathcal{H})$ stands for the space of linear bounded operators from $\mathcal{H}$ to $\mathcal{H}$. The identity operator in $\mathcal{H}$ is denoted by I.

Given a linear operator $P : \mathcal{H} \to \mathcal{H}$ with a dense domain $\mathcal{D}(P) \subset \mathcal{H}$, the notation P^* stands for the adjoint operator. Such an operator P is strictly positive definite, i.e., $P > 0$, iff it is self-adjoint in the sense that $P = P^*$ and there exists a constant $\beta > 0$ such that

$$\langle x, Px \rangle \geq \beta \langle x, x \rangle \text{ for all } x \in \mathcal{D}(P), \tag{1}$$

whereas $P \geq 0$ means that P is self-adjoint and nonnegative definite, i.e.,

$$\langle x, Px \rangle \geq 0 \text{ for all } x \in \mathcal{D}(P). \tag{2}$$

If an infinitesimal operator $-A$ generates a strongly continuous semigroup $T(t)$ on the Hilbert space $\mathcal{H}$ (see, e.g., [2] for details), the domain of the operator A forms another Hilbert space $\mathcal{D}(A)$ with the graph inner product $(\cdot, \cdot)_{\mathcal{D}(A)}$ defined as follows: $(x, y)_{\mathcal{D}(A)} = \langle x, y \rangle + \langle Ax, Ay \rangle$, $x, y \in \mathcal{D}(A)$.

Moreover, the induced norm $\|T(t)\|$ of the semigroup $T(t)$ satisfies the inequality $\|T(t)\| \leq \kappa e^{\delta t}$ everywhere with some constant $\kappa > 0$ and growth bound δ, and the following relation $A^{-1}\mathcal{H} = \mathcal{D}(A)$ holds for the operators A with negative growth bounds δ. Just in case, the norm of $x \in \mathcal{D}(A)$ given by $|Ax|$ is equivalent to the graph norm $\|x\|_{\mathcal{D}(A)}$ of $\mathcal{D}(A)$. The domain $\mathcal{D}(A)$ of such an operator A is thus continuously embedded into $\mathcal{H}$, i.e. $\mathcal{D}(A) \subset \mathcal{H}$, $\mathcal{D}(A)$ is dense in $\mathcal{H}$ and the inequality $|x| \leq \omega|Ax|$ holds for all $x \in \mathcal{D}(A)$ and some constant $\omega > 0$. Apart from this, the square root $\sqrt{A}$ of the operator A is rigorously introduced on $\mathcal{D}(A)$ as a positive definite solution X of the algebraic operator equation $X^2 = A$. Being extended by continuity, this operator is well-posed on the domain $\mathcal{D}(\sqrt{A})$, continuously embedded into $\mathcal{H}$ whereas $\mathcal{D}(A)$ turns out to be continuously embedded into $\mathcal{D}(\sqrt{A})$. In other words, $\mathcal{D}(A)$ and $\mathcal{D}(\sqrt{A})$ are densely embedded into $\mathcal{D}(\sqrt{A})$ and $\mathcal{H}$, respectively, and the following inequalities

$$|x| \leq \omega|\sqrt{A}x| \quad \text{for all } x \in \mathcal{D}(\sqrt{A}) \tag{3}$$

$$|\sqrt{A}x| \leq \omega|Ax| \quad \text{for all } x \in \mathcal{D}(A) \tag{4}$$

hold with a generic constant $\omega > 0$. All relevant background material on fractional operator degrees can be found, e.g., in [11].

The space of the continuous $\mathcal{H}$-valued functions $x : [a, b] \to \mathcal{H}$ with the induced norm

$$\|x\|_{C([a,b],\mathcal{H})} = \max_{s \in [a,b]} |x(s)|$$

is denoted by $C([a, b], \mathcal{H})$. The space of the continuously differentiable $\mathcal{H}$-valued functions $x : [a, b] \to \mathcal{H}$ with the induced norm

$$\|x\|_{C^1([a,b],\mathcal{H})} = \max(\|x\|_{C([a,b],\mathcal{H})}, \|\dot{x}\|_{C([a,b],\mathcal{H})})$$

is denoted by $C^1([a, b], \mathcal{H})$.

$L_2(a, b; \mathcal{H})$ is the Hilbert space of square integrable $\mathcal{H}$-valued functions on (a, b) with the corresponding norm; $L_2(a, b; \mathbf{R}) := L_2(a, b)$.

$W^{l,2}([a, b], \mathcal{H})$ is the Sobolev space of absolutely continuous square integrable $\mathcal{H}$-valued functions on (a, b) with square integrable derivatives up to the order $l \geq 1$ and with the corresponding norm; $W^{l,2}([a, b], \mathbf{R}) := W^{l,2}(a, b)$.

Given $x(\cdot) \in L_2([a,b],\mathcal{H})$, we denote $x^t = x(t+\theta) \in L_2([-h,0],\mathcal{H})$ for $t \in [a+h,b]$; to reduce the notational burden the dependence x^t on θ is subsequently suppressed.

The following result from [14] is instrumental in the subsequent derivation.

Lemma 1 (Wirtinger's Inequality). *Let $u \in W^{1,2}(a,b)$ be a scalar function with $u(a) = u(b) = 0$. Then $\int_a^b u^2(\xi)d\xi \le \frac{(b-a)^2}{\pi^2}\int_a^b (u'(\xi))^2 d\xi$.*

2 Linear Time-delay Systems

Consider a linear infinite-dimensional system

$$\dot{x}(t) + Ax(t) = A_0 x(t) + A_1 x(t-\tau(t)), \tag{5}$$

evolving in a Hilbert space $\mathcal{H}$ where $x(t) \in \mathcal{H}$ is the instantaneous state of the system, the system delay $\tau(t) > 0$ is piece-wise continuous and it is such that

$$\inf_t \tau(t) > 0, \ \sup_t \tau(t) \le h \tag{6}$$

for all t and for some constant $h > 0$; $A_0, A_1 \in \mathcal{L}(\mathcal{H})$ are linear bounded operators, $-A$ is an infinitesimal operator, generating a strongly continuous semigroup $T(t)$, and the domain $\mathcal{D}(A)$ of the operator A is dense in $\mathcal{H}$. Since the operator $A + \lambda I$ is strictly positive definite for the corresponding identity operator I and positive $\lambda \in \mathbf{R}$ large enough [11], without loss of generality we assume throughout that $T(t)$ possesses a negative growth bound δ (otherwise, the bounded operator λI is readily absorbed into A from A_0). We also assume that the operators $\sqrt{A}A_0$ and $\sqrt{A}A_1$ are *subordinate* to $\sqrt{A}$ in the sense that

$$|\sqrt{A}A_i x| \le \omega_i|\sqrt{A}x| \text{ for any } x \in \mathcal{D}(\sqrt{A}) \tag{7}$$

and some constants $\omega_i > 0$, $i = 0,1$. The latter assumption is particularly satisfied if the operators A_0 and A_1 commute to A.

Let the initial conditions

$$x^{t_0} = \varphi(\theta),\ \theta \in [-h,0], \phi \in W \tag{8}$$

be given on a time interval $[t_0 - h, t_0]$ in the space

$$W = C([-h,0],\mathcal{D}(A)) \cap C^1([-h,0],\mathcal{H}). \tag{9}$$

Throughout, solutions of such a system are defined in the classical sense.

Definition 1. *A function $x(t) \in C^1([t_0, t_0+\eta],\mathcal{H})$, $\eta > 0$ with range in $\mathcal{D}(A)$ for all $t \in [t_0, t_0+\eta]$ is said to be a solution of the initial-value problem (5), (8) on $[0,\eta)$ if $x(t)$ is initialized with (8) and it satisfies (5) for all $t \in [t_0, t_0+\eta]$ where $\tau(t)$ is continuous.*

As a matter of fact, along with the above definition, the Caratheodory solutions, satisfying (5) for almost all $t \in (t_0, t_0+T)$ only, and even generalized solutions could be considered. This would particularly allow one to involve initial functions ϕ beyond the space W, thereby capturing a situation where the initial condition (8) is given for $\theta \in [-h,0)$ while the instantaneous value $x(t_0)$ at time instant t_0 does not necessarily coincide with $\phi(0)$ and it is introduced independently of (8). However, in order to facilitate the exposition we prefer to confine the investigation to the classical solutions that captures all the essential features of the general treatment. By the same reason, a potential extension to multiple time-varying delays is also beyond the presentation.

Our aim is to derive stability criteria for linear time-delay systems (5) thus defined. The stability concept under study is based on the initial data norm

$$\|\phi\|_W = \|A\phi\|_{C([-\tau,0],\mathcal{H})} + \|\dot{\phi}\|_{C([-\tau,0],\mathcal{H})} \tag{10}$$

in space (9). Suppose $x(t,t_0,\phi)$ denotes a solution of (5), (8) at a time instant $t \geq t_0$ and $\|x(t,t_0,\phi)\|_{\mathcal{D}(A)} = |Ax(t,t_0,\phi)|$ stands for its instantaneous norm.

Definition 2. *System (5) is said to be stable in $\mathcal{D}(A)$ iff for each $\varepsilon > 0$ and for each $t_0 \in \mathbf{R}$ there exists $\delta > 0$ such that for each solution $x(t,t_0,\phi)$ with the initial condition (8), satisfying $\|\phi\|_W < \delta$, the inequality $\|x(t,t_0,\phi)\|_{\mathcal{D}(A)} < \varepsilon$ holds for all $t \geq t_0$.*

Definition 3. *System (5) is said to be globally asymptotically stable in $\mathcal{D}(A)$ iff it is stable in $\mathcal{D}(A)$, and for all initial data $t_0 \in \mathbf{R}$ and $\phi(\cdot) \in W$ solution $x(t,t_0,\phi)$ of (5), (8) is such that $\|x(t,t_0,\phi)\|_{\mathcal{D}(A)} \to 0$ as $t \to \infty$.*

3 Well-posedness Issues

To begin with, we demonstrate that the solutions of the initial-value problem (5), (8) are well-posed on the semi-infinite time interval $[t_0,\infty)$ and they can be found as mild solutions, i.e., as those of the integral equation

$$x(t) = T(t-t_0)x(t_0) + \int_{t_0}^{t} T(t-s)[A_0x(s) + A_1x(s-\tau(s))]ds, \; t \geq t_0. \tag{11}$$

Moreover, solutions $x(t)$ of (5), (8), uniformly bounded in the state space $\mathcal{H}$, remain uniformly bounded in $\mathcal{D}(A)$ and they possess uniformly bounded time derivatives $\dot{x}(t)$.

Theorem 1. *Let $\mathcal{H}$ be a Hilbert space and let the following assumptions be satisfied:*

1. *the operator $-A$ generates an analytical semigroup $T(t)$ with a negative growth bound δ;*
2. *the linear operators A_0 and A_1 are bounded in $\mathcal{H}$;*
3. *the operators $\sqrt{A}A_0$ and $\sqrt{A}A_1$ are subordinate to $\sqrt{A}$ in the sense of (7);*
4. *the function $\tau(t)$ is piecewise-continuous of class C^1 on the closure of each continuity subinterval and it satisfies (6) for all t and some constant $h > 0$.*

Then there exists a unique solution of the initial-value problem (5), (8). This solution is also a unique solution of the integral initial-value problem (8), (11).

Proof. To begin with, let us choose a positive η_0 small enough to ensure that $\eta_0 < \inf_t \tau(t)$ and the first discontinuity point $t_0^1 > t_0$ of $\tau(t)$ is such that the difference $t_0^1 - t_0$ is multiple to η_0, i.e., $t_0^1 = t_0 + k_0\eta_0$ for some integer $k_0 > 0$. While being viewed over the time segment $[t_0, t_0+\eta_0]$, the initial-value problem (5) is equivalent to

$$\dot{x}(t) + Ax(t) = A_0x(t) + A_1\phi(t-t_0-\tau(t)), \; x(t_0) = \phi(0) \tag{12}$$

where the inhomogeneous term $A_1\phi(t-t_0-\tau(t))$ is of class C^1 on $[0,\eta_0]$. By [2, Theorem 3.1.3], there exists a unique local solution of (12) and this solution satisfies the integral equation (11) on $[t_0, t_0+\eta_0]$.

The same line of reasoning is step-by-step applied to the time segments $[t_{i-1}, t_{i-1} + \eta_0]$, $i = 1, \cdots, k_0$ with $t_i = t_{i-1} + \eta_0$ and $t_{k_0} = t_0^1$. Following this line, the initial-value problem is demonstrated to possess a unique solution $x(t, t_0, \phi)$ for $t \in [t_0, t_0^1]$, which satisfies the integral equation (11) on $[t_0, t_0^1]$. The assertion of Theorem 1 is then concluded by iteration on the time segments $[t_0^j, t_0^{j+1}]$, $j = 1, 2, \dots$ where $t_0^1 < t_0^2 < \dots$ are the successive discontinuity points of the function $\tau(t)$.

Theorem 2. *Let the assumptions of Theorem 1 be satisfied and let a solution $x(t, t_0, \phi)$ of the initial-value problem (5), (8) be uniformly bounded in the Hilbert space $\mathcal{H}$ for all $t \geq t_0$. Then this solution is uniformly bounded in $\mathcal{D}(A)$ for all $t \geq t_0$, and it possesses a uniformly bounded time derivative $\dot{x}(t, t_0, \phi)$ on the semi-infinite time interval $[t_0, \infty)$.*

Proof. Let us first demonstrate that $|\sqrt{A}x(t)|$ is uniformly bounded for all $t \geq t_0$. Indeed, using the solution representation (11), taking into account Assumption 1 of the Theorem 1, and applying the well-known estimate

$$\|T(s)\sqrt{A}\| \leq \frac{ce^{-\delta s}}{\sqrt{s}}, \; s > 0 \tag{13}$$

of the induced operator norm with a positive constant c (cf. [6, Theorem 1.4.3]) yield

$$\begin{aligned}
|\sqrt{A}x(t)| \leq \|T(t-t_0)\||\sqrt{A}\phi(0)| + \int_{t_0}^{t} |T(t-s)\sqrt{A}[A_0x(s)\\
+A_1x(s-\tau(s))]|ds \leq N_1 + \int_{t_0}^{t} \|T(t-s)\sqrt{A}\|[\|A_0\||x(s)|\\
+\|A_1\||x(s-\tau(s))|]ds \leq N_1 + N_2\int_{t_0}^{t} \|T(t-s)\sqrt{A}\|ds\\
\leq N_1 + N_2\int_{t_0}^{t} \frac{ce^{-\delta(t-s)}}{\sqrt{t-s}}ds < N
\end{aligned} \tag{14}$$

for all $t \geq t_0$ and some positive constants N_1, N_2, N.

Now, by employing the operator subordination (7) and by applying estimates (13), (14), the following inequalities

$$\begin{aligned}
|Ax(t)| \leq \|T(t-t_0)\||A\phi(0)| + \int_{t_0}^{t} |T(t-s)A[A_0x(s) + A_1x(s-\tau(s))]|ds\\
\leq M_1e^{-\delta(t-t_0)} + \int_{t_0}^{t} \|T(t-s)\sqrt{A}\| \times |\sqrt{A}[A_0x(s) + A_1x(s-\tau(s))]|ds\\
\leq M_1 + \int_{t_0}^{t} \|T(t-s)\sqrt{A}\|[\omega_0|\sqrt{A}x(s)| + \omega_1|\sqrt{A}x(s-\tau(s))|]ds \leq M_1\\
+M_2\int_{t_0}^{t} \|T(t-s)\sqrt{A}\|ds \leq M_1 + M_2\int_{t_0}^{t} \frac{ce^{-\delta(t-s)}}{\sqrt{t-s}}ds < M
\end{aligned} \tag{15}$$

are derived for some positive constants M_1, M_2, M and all $t \geq t_0$. Hence, $|Ax(t)|$ is uniformly bounded on $[t_0, \infty)$. By taking into account Assumption 3 of Theorem 1 and by virtue of (5), it follows that the time derivative $\dot{x}(t, t_0, \phi)$ is also uniformly bounded on $[t_0, \infty)$. The proof is completed.

4 Lyapunov-Krasovskii Method

For later use we extend the Lyapunov-Krasovskii method to Hilbert space-valued time-delay systems. Given a continuous functional

$$V : \mathbf{R} \times W \times C([-\tau, 0], \mathcal{H}) \to \mathbf{R}, \tag{16}$$

its upper right-hand derivative along solutions of the initial-value problem (5), (8) is defined as follows:

$$\dot{V}(t,\phi,\dot{\phi}) = \limsup_{s\to 0^+} \frac{1}{s}[V(t+s, x^{t+s}(t,\phi), \dot{x}^{t+s}(t,\phi)) - V(t,\phi,\dot{\phi})]. \tag{17}$$

Theorem 3. *Let the assumptions of Theorem 1 be in force. Then (5) is globally asymptotically stable in $\mathcal{D}(A)$ if there exist positive numbers α, β, γ and a continuous functional (16) such that*

$$\beta|\phi(0)|^2 \leq V(t,\phi,\dot{\phi}) \leq \gamma\|\phi\|_W^2, \tag{18}$$

$$\dot{V}(t,\phi,\dot{\phi}) \leq -\alpha|\phi(0)|^2, \tag{19}$$

and the function $\bar{V}(t) = V(t, x^t, \dot{x}^t)$ is absolutely continuous on solutions x^t of (5).

Proof. By Theorem 1, there exists a unique solution of the initial-value problem (5), (8), which is globally defined for all $t \geq t_0$. Integrating (19), where $\phi = x^s$, with respect to s from t_0 to t we have

$$V(t, x^t, \dot{x}^t) - V(t_0, \phi, \dot{\phi}) \leq -\alpha \int_{t_0}^{t} |x(s)|^2 ds. \tag{20}$$

Relations (18) and (20), coupled together, result in

$$\beta|x(t)|^2 \leq V(t, x^t, \dot{x}^t) \leq V(t_0, \phi, \dot{\phi}) \leq \gamma\|\phi\|_W^2 \tag{21}$$

thereby yielding the system stability. Indeed, by applying (4) and (21), Inequality (14) is modified to

$$\begin{aligned}|\sqrt{A}x(t)| \leq \|T(t-t_0)\||\sqrt{A}\phi(0)| + \int_{t_0}^{t} |T(t-s)\sqrt{A}[A_0 x(s) \\ +A_1 x(s-\tau(s))]|ds \leq \|\phi\|_W\{c\omega + \sqrt{\frac{\gamma}{\beta}}[\|A_0\| + \|A_1\|] \int_{t_0}^{t} \frac{ce^{-\delta(t-s)}}{\sqrt{t-s}} ds\} \\ \leq N_0\|\phi\|_W,\ t \geq t_0\end{aligned} \tag{22}$$

where N_0 is some positive constant. With this in mind, Inequality (15) can be strengthen to

$$\begin{aligned}|Ax(t)| \leq \|T(t-t_0)\||A\phi(0)| + \int_{t_0}^{t} \|T(t-s)\sqrt{A}\||\sqrt{A}[A_0 x(s) \\ +A_1 x(s-\tau(s))]|ds \leq \|\phi\|_W\{c + N_0[\omega_0 + \omega_1] \int_{t_0}^{t} \frac{ce^{-\delta(t-s)}}{\sqrt{t-s}} ds \\ \leq M_0\|\phi\|_W,\ t \geq t_0\end{aligned} \tag{23}$$

where M_0 is some positive constant, and subordination (7) has been taken into account. Hence, the solution norm $\|x(t, t_0, \phi)\|_{\mathcal{D}(A)} = |Ax(t)|$ is small for all $t \geq t_0$ if $\|\phi\|_W$ is small. Thus, System (5) is shown to be stable.

It remains to demonstrate that $\|x(t,t_0,\phi)\|_{\mathcal{D}(A)} \to 0$ as $t \to \infty$. For this purpose, let us first demonstrate that $|x(t)| \to 0$ as $t \to \infty$. Similar to (20), the following inequality

$$V(t,x^t,\dot{x}^t) - V(t_0+\eta, x^{t_0+\eta}, \dot{x}^{t_0+\eta}) \le -\alpha \int_{t_0+\eta}^{t} |x(s)|^2 ds \tag{24}$$

is obtained for an arbitrary $\eta > h$. Since

$$V(t_0+\eta, x^{t_0+\eta}, \dot{x}^{t_0+\eta}) \le V(t_0,\phi,\dot{\phi}) < \infty$$

by virtue of (20), the implication $|x(t)| \in L_2[t_0+\eta,\infty)$ is then guaranteed by (24). Apart from this, Theorem 2 ensures that $\dot{x}(t)$ is uniformly bounded on the semi-infinite interval $[\eta,\infty)$. Thus, by applying Barbalat's Lemma [9, p.323], the convergence $|x(t)| \to 0$ as $t \to \infty$ is established.

It follows that for arbitrary $\epsilon_1 > 0$ there exists $t_1 > t_0$ such that $|x(t)| < \epsilon_1$ for all $t \ge t_1$. Then taking into account the assumptions of Theorem 1 and employing (14) – (23) yield

$$|\sqrt{A}x(t)| \le \|T(t-t_1-h)\| |\sqrt{A}x(t_1+h)| + \int_{t_1+h}^{t} \|T(t-s)\sqrt{A}\| [|A_0x(s)| \\ +|A_1x(s-\tau(s))|]ds \le Nce^{-\delta(t-t_1-h)} + \epsilon_1[\|A_0\|+\|A_1\|] \int_0^\infty \frac{ce^{-\delta s}}{\sqrt{s}} ds \tag{25}$$

for $t \ge t_1+h$. By making ϵ_1 small enough, we conclude from (25) that for arbitrarily small $\epsilon_2 > 0$ there exists $t_2 > t_1+h$ such that $|\sqrt{A}x(t)| < \epsilon_2$ for all $t \ge t_2$. Once again taking into account the assumptions of Theorem 1 and employing (14) – (23), we now derive that

$$|Ax(t)| \le \|T(t-t_2-h)\| |Ax(t_2+h)| \\ + \int_{t_2+h}^{t} \|T(t-s)\sqrt{A}\| [|\sqrt{A}A_0x(s)| + |\sqrt{A}A_1x(s-\tau(s))|]ds \\ \le Mce^{-\delta(t-t_2-h)} + \epsilon_2[\omega_0\|A_0\| + \omega_1\|A_1\|] \int_0^\infty \frac{ce^{-\delta s}}{\sqrt{s}} ds \tag{26}$$

for $t \ge t_2+\tau$. By making ϵ_2 small enough, we conclude from (26) that for arbitrarily small $\epsilon > 0$ there exists $T > t_2+h$ such that $|Ax(t)| < \epsilon$ for all $t \ge T$. With this in mind, the desired convergence $\|x(t,t_0,\phi)\|_{\mathcal{D}(A)} = |Ax(t)| \to 0$ as $t \to \infty$ is established. Theorem 3 is thus proved.

5 Linear Operator Inequalities (LOIs) in Hilbert Space

In the sequel, delay-independent asymptotic stability conditions are derived for the case where the system delay $\tau(t)$ is additionally assumed to be differentiable with an *a priori* known derivative bound $d < 1$, i.e., $\dot{\tau}(t) \le d < 1$ for all t.

In the Hilbert space $\mathcal{D}(A)$, let us consider the following Lyapunov-Krasovskii functional

$$V(t,x^t) = \langle x(t), Px(t)\rangle + \textstyle\int_{t-\tau(t)}^{t} \langle x(s), Qx(s)\rangle ds \tag{27}$$

where $P > 0$ and $Q > 0$ are subordinate to the positive definite operator A, i.e.,

$$\langle x, Px\rangle \leq \gamma_0\langle x, Ax\rangle,\ \langle x, Qx\rangle \leq \gamma_0\langle x, Ax\rangle \tag{28}$$

for all $x \in \mathcal{D}(A)$ and some constant $\gamma_0 > 0$. In other words, P and Q are strictly positive definite and bounded on $\mathcal{D}(A)$.

Since by definitions (1) and (2) it follows that

$$\begin{aligned}\beta|x(t)|^2 &= \beta\langle x(t), x(t)\rangle \leq \langle x(t), Px(t)\rangle + \int_{t-h}^{t}\langle x(s), Qx(s)\rangle ds \\ &\leq \gamma_0\langle x(t), Ax(t)\rangle + \gamma_0\int_{t-h}^{t}\langle x(s), Ax(s)\rangle ds \leq \gamma_0|\sqrt{A}x(t)|^2 \\ &\quad +\gamma_0 h \max_{s\in[-h,0]}|\sqrt{A}x(t+s)|^2 \leq \gamma\|x^t\|_W^2\end{aligned} \tag{29}$$

for all $x \in \mathcal{D}(A)$ and some positive β and γ, the Lyapunov-Krasovskii functional (27) appears to satisfy Condition (18) of Theorem 3.

Differentiating V in t along the trajectories of the time-delay system (5), we have

$$\begin{aligned}\dot{V}(t, x^t) &= 2\langle x(t), PA_0x(t)\rangle + 2\langle x(t), PA_1x(t-\tau)\rangle - 2\langle Ax(t), Px(t)\rangle \\ &\quad +\langle x(t), Qx(t)\rangle - (1-\dot{\tau}(t))\langle x(t-\tau(t)), Qx(t-\tau(t))\rangle.\end{aligned} \tag{30}$$

It follows that (19) is satisfied if the Linear Operator Inequality (LOI)

$$\Phi = \begin{bmatrix}(A_0^* - A^*)P + P(A_0 - A) + Q & PA_1 \\ A_1^*P & -(1-d)Q\end{bmatrix} < 0 \tag{31}$$

holds in the Hilbert space $\mathcal{D}(A) \times \mathcal{D}(A)$ in the sense of (1). Thus, the following result is concluded.

Theorem 4. *Let assumptions of Theorem 1 be in force. Then System (5) is globally asymptotically stable in $\mathcal{D}(A)$ for all differentiable delays $\tau(t)$ with $\dot{\tau}(t) \leq d < 1$ if there exist strictly positive definite operators Q and P subject to (28) such that the LOI (31) holds in the Hilbert space $\mathcal{D}(A) \times \mathcal{D}(A)$.*

Proof. First, let us note that by virtue of (29) (which results from Conditions (28), imposed on the Lyapunov-Krasovskii functional (27)), Condition (18) of Theorem 3 is satisfied.

Next let us verify Condition (19) of Theorem 3. Setting

$$\zeta(t) = col\{x(t), x(t-\tau)\}$$

and employing (31) in accordance with (1) yield

$$\dot{V}(t, x^t) \leq \langle\zeta(t), \Phi\zeta(t)\rangle \leq -\alpha\langle\zeta(t), \zeta(t)\rangle \leq -\alpha|x(t)|^2 \tag{32}$$

with some positive α where for the sake of reducing notational overload, the same symbol $\langle,\rangle$ is also used for the inner product in the Hilbert space $\mathcal{H} \times \mathcal{H}$. Condition (19) is thus satisfied.

To this end, we note that while being viewed on solutions of (5), the Lyapunov-Krasovskii functional (27) is absolutely continuous as a function of t because the solutions are absolutely continuous in t.

Thus, Theorem 3 proves to be applicable to System (5). By applying this theorem, the assertion of Theorem 4 is established.

General methods for solving LOIs have not been developed yet. Some finite dimensional approximations were considered in [8]. In the rest of the section, we show that for delay heat equations the LOI (31) becomes feasible if a certain finite-dimensional LMI is feasible.

6 Illustrative Example: Delay Heat Equation

Consider the heat equation

$$u_t(\xi,t) = au_{\xi\xi}(\xi,t) - a_1 u(\xi,t-\tau(t)),\ t \geq 0,\ 0 \leq \xi \leq \pi \tag{33}$$

with constant parameters $a > 0$ and a_1, with a time-varying delay $\tau(t)$ such that $\dot{\tau}(t) \leq d < 1$, and with the Dirichlet boundary condition

$$u(0,t) = u(\pi,t) = 0,\ t \geq 0. \tag{34}$$

If along with the trivial operator $A_0 = 0$ and the bounded operator $A_1 = -a_1$ of the multiplication by the constant $-a_1$, we introduce the operator $A = -a\frac{\partial^2}{\partial\xi^2}$ of double differentiation with the dense domain $\mathcal{D}(\frac{\partial^2}{\partial\xi^2}) = \{u \in W^{2,2}(0,\pi) : u(0) = u(\pi) = 0\}$, then the boundary-value problem (33), (34) can be rewritten as the differential equation (5) in the Hilbert space $\mathcal{H} = L_2(0,\pi)$ with the infinitesimal operator $-A$, generating an analytical semigroup with a negative growth bound (see, e.g., [2] for details).

Let the Lyapunov-Krasovskii functional (27) be simplified to

$$V(t,u^t) = p\int_0^\pi u^2(\xi,t)d\xi + q\int_{t-\tau(t)}^t \int_0^\pi u^2(\xi,s)d\xi ds \tag{35}$$

with some positive constants p and q. Then the operators P and Q in (31) take the form $P = p$, $Q = q$ of the operators of the multiplication by positive constants p and q, respectively, and Condition (28), imposed on these operators, is guaranteed by (3), being specified in the form of the Wirtinger's inequality of Lemma 1.

Integrating by parts and taking into account (34), we find that

$$\langle x, (A^*P + PA)x\rangle = -2a\int_0^\pi puu_{\xi\xi}d\xi = 2a\int_0^\pi pu_\xi^2 d\xi \geq 2a\int_0^\pi pu^2 d\xi \tag{36}$$

for $x \in \mathcal{D}(A)$ where the last inequality has been derived by using the Wirtinger's inequality of Lemma 1. We thus obtain

$$\langle x(t), \Phi x(t)\rangle \leq \int_0^\pi [u(\xi,t)\ u(\xi,t-\tau)] \times \begin{bmatrix} q-2ap & -a_1p \\ -a_1p & -(1-d)q \end{bmatrix} \begin{bmatrix} u(\xi,t) \\ u(\xi,t-\tau) \end{bmatrix} d\xi < 0 \tag{37}$$

provided that the LMI

$$\begin{bmatrix} q-2ap & -a_1p \\ -a_1p & -(1-d)q \end{bmatrix} < 0 \tag{38}$$

is satisfied. Summarizing, the following result is obtained.

Theorem 5. *Let the LMI (38) holds for some positive scalars p and q. Then the Dirichlet boundary-value problem (33), (34) is globally asymptotically stable in the Sobolev space $W^{1,2}(0,\pi)$ for all differentiable delays $\tau(t)$ with $\dot{\tau}(t) \leq d < 1$.*

By Schur complement formula, LMI (38) is feasible iff the following inequality $q^2 - 2(a+a_0)pq + a_1^2p^2/(1-d) < 0$ holds for some $p > 0$ and $q > 0$. The left part of the latter inequality achieves its minimum at $q = (a+a_0)p$ and, thus, the LMI holds iff $a_1^2 < (a+a_0)^2(1-d)$.

It should be noted that the same LMI (38) appears to guarantee the scalar time-delay equation

$$\dot{y}(t) + (a + a_0)y(t) + a_1 y(t - \tau(t)) = 0, \tag{39}$$

to be exponentially stable for all delays with $\dot{\tau} \leq d < 1$. While being interpreted in terms of the modal representation of the Dirichlet boundary-value problem (33), (34), confined to the projection

$$\dot{y}_j(t) + (aj^2 + a_0)y_j(t) + a_1 y_j(t - \tau(t)) = 0, \; j = 1, 2, \ldots, n \tag{40}$$

on the first n eigenfunctions, Theorem 5 essentially utilizes the structure of such a representation to reduce the corresponding LMIs, derived for the diagonal n-th order system (40), to the LMI (38), derived for the first order system (39), that corresponds to the first modal dynamics (40) with $j = 1$. Actually, this interpetation becomes surprising since no similar results on LMI of reduced order are known from the finite-dimensional theory.

To this end, consider the heat equation (33) with a constant delay τ (that corresponds to $d = 0$ in the conditions above) and with the boundary condition (34). The characteristic equations of such a boundary-value problem are given by (see, e.g., [17])

$$\lambda_k + ak^2 + a_0 + a_1 e^{-\lambda_k \tau} = 0, \quad k = 1, 2, \ldots . \tag{41}$$

The exponential stability of (33), (34) is shown in [7] to be determined by (41) with $k = 1$, i.e., the system is exponentially stable if the roots of (41) with $k = 1$ have negative real parts. Since the afore-given root condition is equivalent to the exponential stability of the first order system (39), the constant-delay system proves to be delay-independently exponentially stable iff $a_1 \in (-a - a_0, a + a_0]$ (see [12]).

References

1. Datko, R.: Not all feedback stabilized hyperbolic systems are robust with respect to small time delays in their feedbacks. SIAM J. Control Optim. 26, 697–713 (1988)
2. Curtain, R., Zwart, H.: An introduction to infinite-dimensional linear systems. Springer, Heidelberg (1995)
3. Fridman, E., Orlov, Y.: Exponential Stability of Linear Distributed Parameter Systems with Time-Varying Delays. Automatica 45(2), 194–201 (2009)
4. Gu, K., Kharitonov, V.L., Chen, J.: Stability of time-delay systems. Birkhauser, Boston (2003)
5. Hale, J., Verduyn-Lunel, S.: Introduction to functional differential equations. Springer, New York (1993)
6. Henry, D.: Geometric theory of semilinear parabolic equations. Lecture Notes in Mathematics. Springer, Berlin (1981)
7. Huang, C., Vanderwalle, S.: An analysis of delay-dependent stability for ordinary and partial differential equations with fixed and distributed delays. SIAM J. Sci. Comput. 25, 1608–1632 (2004)
8. Ikeda, K., Azuma, T., Uchida, K.: Infinite-dimensional LMI approach to analysis and synthesis for linear time-delay systems. Special issue on advances in analysis and control of time-delay systems. Kybernetika 37, 505–520 (2001)
9. Khalil, H.: Nonlinear Systems. Macmillan, New York (1992)
10. Kolmanovskii, V., Myshkis, A.: Applied Theory of functional differential equations. Kluwer, Dordrecht (1999)

11. Krasnoselskii, M., Zabreyko, P., Pustylnik, E., Sobolevski, P.: Integral Operators in Spaces of Summable Functions, Noordhoff, Leyden (1976)
12. Niculescu, S.I.: Delay effects on stability: A Robust Control Approach. LNCIS, p. 269. Springer, London (2001)
13. Richard, J.P.: Time-delay systems: an overview of some recent advances and open problems. Automatica 39, 1667–1694 (2003)
14. Wang, T.: Stability in abstract functional-differential equations. I, II Math. Anal. Appl. 186, 534–558 (1994)
15. Wu, J.: Theory and Applications of Partial Functional Differential Equations. Springer, New York (1996)

Lyapunov Matrices for Neutral Type Time Delay Systems

Gilberto Ochoa[1], Juan E. Velázquez[1], Vladimir L. Kharitonov[2] and Sabine Mondié[1]

[1] Automatic Control Department, CINVESTAV-IPN, D. F., México
gochoa, jvelazquez, smondie@ctrl.cinvestav.mx
[2] Applied Mathematics and Control Process Department, St.-Petersburg State University, St.-Petersburg, Russia khar@apmath.spbu.ru

Summary. In this contribution two procedures for the computation of Lyapunov matrices of neutral type time delay systems are presented. The first one is a semi-analytic method, that consists in solving a two points boundary value problem for a delay free system. The second one is a piece-wise linear approximation of the Lyapunov matrix. An algorithm for the estimation of the error based on Lyapunov-Krasovskii functionals is also proposed.

1 Introduction

The Lyapunov matrix for time delay systems has proved to be useful in the analysis of time delay systems problems such as the determination of robustness bounds, exponential estimates, among many others. In these contributions, a key step is indeed the construction of the Lyapunov matrix: the auxiliary system of delay free matrix equations with a special set of two points boundary conditions satisfied by the Lyapunov matrix revealed in [6] is the starting point of possible construction methods of the Lyapunov matrix [2]. In order to extend such results to neutral type time delay systems, the construction of the Lyapunov matrix of neutral type time delay systems must be addressed. In [4] the construction for the single delay case is studied within a semi-analytic approach.

The aim of this contribution is to present procedures for the computation of the Lyapunov matrix for neutral type time delay systems in the cases of delays that are multiple of a basic one and in the general case.

As the Lyapunov matrix for neutral systems with delays multiple of a basic one satisfy also an auxiliary system of delay free matrix equations with a special set of two points boundary conditions it is possible to extend the semi-analytic approach. When the basic delay is small compared to the other ones the semi-analytic method becomes unpractical. Furthermore, in the general delay case, as no delay free system is available, it does not apply. For these cases we propose a construction based on a piece-wise linear approximation of the Lyapunov matrix.

The contribution is organized as follows. In Section 2 some basic definitions, concepts and previous results about neutral type time delay systems are recalled; a special emphasis is put on the definition and properties of the Lyapunov matrix for neutral time delay systems and on the corresponding Lyapunov-Krasovskii functional. Section 3 is devoted to the main results of this contribution: a semi-analytic method for the construction of the Lyapunov matrix of

J.J. Loiseau et al. (Eds.): Topics in Time Delay Systems, LNCIS 388, pp. 61–71.
springerlink.com

systems with delays multiple of a basic one and a numerical procedure for the computation of a piece-wise linear approximation of Lyapunov matrix for systems with general delays. An algorithm for the validation of the piece-wise linear approximation is presented in Section 4. A number of numerical examples illustrate the proposed construction. Some concluding remarks end this contribution.

2 Preliminaries

In this section some useful results are given. We consider the following neutral type time delay system of the form

$$\sum_{k=0}^{m} B_k \dot{x}(t-h_k) = \sum_{k=0}^{m} A_k x(t-h_k), \quad t \geq 0, \tag{1}$$

where $B_0 = I$, A_k, $B_k \in \mathbb{R}^{n\times n}$, $k = 0, 1, ..., m$ and $0 = h_0 < h_1 < ... < h_m = h$ are the delays.

We denote by $x(t, \varphi)$ the solution of system (1) with the initial function $\varphi \in C([-h, 0], \mathbb{R})$ and by $x_t(\varphi)$ the segment of the solution $x_t(\varphi) \to x(t+\theta, \varphi)$, $\theta \in [-h, 0]$.

Definition 1. *[1] System (1) is said to be exponentially stable if there exist $\sigma > 0$ and $\gamma \geq 1$ such that for every solution $x(t, \varphi)$ the following estimate holds:*

$$\|x(t,\varphi)\| \leq \gamma e^{-\sigma t} \|\varphi\|_h, \quad t \geq 0.$$

Here $\|\varphi\|_h = \max_{\theta\in[-h,0]} \|\varphi(\theta)\|$.

2.1 Fundamental matrix

Let the $n \times n$ matrix $K(t)$ be a solution of the matrix equation

$$\sum_{k=0}^{m} B_k \dot{K}(t-h_k) = \sum_{k=0}^{m} A_k K(t-h_k), \quad t \geq 0,$$

which satisfies the following conditions:

- Initial condition: $K(\theta) = 0$ for $\theta < 0$ and $K(0) = I$.
- Sewing condition: $\sum_{k=0}^{m} B_k K(t-h_k)$ is continuous for $t > 0$.

$K(t)$ is known as the fundamental matrix of system (1).

2.2 Lyapunov-Krasovskii functionals

Let system (1) be exponentially stable. Let a functional of the form

$$w(x_t) = \sum_{k=0}^{m} x^{\mathsf{T}}(t-h_k) W_k x(t-h_k) + \sum_{k=1}^{m} \int_{-h_k}^{0} x^{\mathsf{T}}(t+\theta) W_{m+k} x(t+\theta) d\theta,$$

where $W_k > 0$, $k = 0, 1, ..., 2m$ be given. Then there exists a quadratic functional $v(x_t)$ such that its time derivative along the solutions of system (1) is equal to $-w(x_t)$ for $t \geq 0$. The functional $v(x_t)$ is called of complete type. It is shown in [4] that the functional $v(x_t)$ is of the form

$$v(x_t) = \sum_{j=0}^{m}\sum_{k=0}^{m} x^{\mathsf{T}}(t)B_j^{\mathsf{T}}U(h_j - h_k)B_k x(t) + 2\sum_{j=0}^{m}\sum_{k=1}^{m} x^{\mathsf{T}}(t)B_j^{\mathsf{T}} \times$$
$$\times \int_{-h_k}^{0} U(h_j - h_k - \theta)\left[A_k x(t+\theta) - B_k \dot{x}(t+\theta)\right] d\theta +$$
$$+ \sum_{j=1}^{m}\sum_{k=1}^{m} \int_{-h_j}^{0}\int_{-h_k}^{0} [x^{\mathsf{T}}(t+\theta_1)A_j^{\mathsf{T}} - \dot{x}^{\mathsf{T}}(t+\theta_1)B_j^{\mathsf{T}}]U(\theta_1 - \theta_2 + h_j - h_k) \times$$
$$\times \left[A_k x(t+\theta_2) - B_k \dot{x}(t+\theta_2)\right] d\theta_2 d\theta_1 +$$
$$+ \sum_{k=1}^{m} \int_{-h_k}^{0} \dot{x}(t+\theta)[W_k + (h_k+\theta)W_{m+k}]x(t+\theta)d\theta. \tag{2}$$

Here

$$U(\tau) = \int_{0}^{\infty} K^{\mathsf{T}}(t)WK(t+\tau)dt \tag{3}$$

is called the Lyapunov matrix of system (1) associated to the matrix

$$W = W_0 + \sum_{k=1}^{m}\left[W_k + h_k W_{m+k}\right].$$

2.3 Properties of the Lyapunov matrix

The following properties of the Lyapunov matrix (3) are crucial for its computation:

- Dynamic property

$$\sum_{k=0}^{m} U'(t-h_k)B_k - \sum_{k=0}^{m} U(t-h_k)A_k = 0. \tag{4}$$

- Symmetry property

$$U(-\tau) = U^{\mathsf{T}}(\tau), \quad \tau \geq 0. \tag{5}$$

- Algebraic property

$$-W = \sum_{i=0}^{m}\sum_{j=0}^{m}[B_i^{\mathsf{T}}U(h_i - h_j)A_j + A_j^{\mathsf{T}}U^{\mathsf{T}}(h_i - h_j)B_i]. \tag{6}$$

Unlike the definition for the Lyapunov matrix given in (3) the following definition of the Lyapunov matrix does not require the stability of system (1).

Definition 2. *[4] Given a symmetric matrix W, the Lyapunov matrix for system (1) associated with W is a solution of equation (4) which satisfies the conditions (5)-(6).*

3 Lyapunov matrix

Clearly, the expression (3) cannot be used for the construction of matrix $U(\tau)$, but the properties of the previous section can. In this section, we present two methods for its computation. The first procedure may only be applied to systems with delays multiple of a basic one. It is based on the fact that the Lyapunov matrix satisfies a delay free system of linear differential matrix equations and that it reduces to the solution of two points boundary value problem. The second one is a a piece-wise linear approximation to Lyapunov matrix that is applicable to a more general delays case.

3.1 A semi-analytic construction

We propose a semi-analytic procedure for the computation of the Lyapunov matrix. We consider systems with delays multiple of a basic one. The property (4) takes the form

$$\sum_{k=0}^{m} U'(t-hk)B_k - \sum_{k=0}^{m} U(t-hk)A_k = 0 \tag{7}$$

where h is the basic delay. Let us introduce the following $2m$ auxiliary matrices

$$X_j(\tau) = U(\tau + jh), \quad j = -m, ..., 0, ..., m-1. \tag{8}$$

If we compute the time derivative of the above expressions and we use the fact that $X_i(\tau)$ satisfies (5)-(7), we arrive at the following set of matrix equations

$$\begin{array}{l} \sum\limits_{i=0}^{m} X'_{j-i}(\tau)B_i = \sum\limits_{i=0}^{m} X_{j-i}(\tau)A_i, \quad j = 0, ..., m-1, \\ \sum\limits_{i=0}^{m} B_i^{\mathsf{T}} X'_{i+j}(\tau) = -\sum\limits_{i=0}^{m} A_i^{\mathsf{T}} X_{i+j}(\tau), \quad j = -m, ..., -1. \end{array}$$

The above expressions are rewritten in vector form as

$$M_1 z'(\tau) = M_2 z(\tau) \tag{9}$$

where $z(\tau) = [vec\,(X_{m-1})\,, ..., vec\,(X_{-m})]^{\mathsf{T}}$ (for details on the function $vec(\cdot)$ see [5]) and matrices M_1, M_2 are the resultant matrices of the polynomials

$$P(\lambda) = I + B_1^{\mathsf{T}}\lambda + ... + B_m^{\mathsf{T}}\lambda^m,$$

$$P_1(\lambda) = B_m^{\mathsf{T}} + B_{m-1}^{\mathsf{T}}\lambda + ... + B_1^{\mathsf{T}}\lambda^{m-1} + I\lambda^m,$$
$$Q(\lambda) = I + A_1^{\mathsf{T}}\lambda + ... + A_m^{\mathsf{T}}\lambda^m,$$
$$Q_1(\lambda) = -A_m^{\mathsf{T}} - A_{m-1}^{\mathsf{T}}\lambda - ... - A_1^{\mathsf{T}}\lambda^{m-1} - I\lambda^m,$$

equivalently

$$M_1 = R^{\otimes}(P, P_1), \; M_2 = R^{\otimes}(Q, Q_1),$$

(for more details of the operator $R^{\otimes}(\cdot,\cdot)$ see [3]).

In order to find a particular solution of system (9) one have to specify the initial conditions. In this problem, the initial conditions are not known but they can be determined from the following two boundary value problem. Expression (8) gives the first set of boundary conditions

$$X_{i+1}(0) = X_i(h)\,, \quad i = -m, ..., -1, 0, 1, ...m-2. \tag{10}$$

The remaining set of boundary conditions follows from writing the algebraic property (6) in terms of the auxiliary matrices (8)

$$-W = \sum_{i=0}^{m} B_i^{\mathsf{T}} X_0(0) A_i + \sum_{i=0}^{m} A_i^{\mathsf{T}} X_0(0) B_i +$$

$$\sum_{i=0}^{m-1} \sum_{j=i+1}^{m} \left[B_j^{\mathsf{T}} X_{j-i-1}(h) A_i + B_i^{\mathsf{T}} X_{j-i-1}^{\mathsf{T}}(h) A_j \right]$$

$$+ \sum_{i=0}^{m-1} \sum_{j=i+1}^{m} \left[A_i^{\mathsf{T}} X_{j-i-1}^{\mathsf{T}}(h) B_j + A_j^{\mathsf{T}} X_{j-i-1}(h) B_i \right]. \tag{11}$$

The above procedure can be summarized as follows:

Theorem 1. *Consider a neutral type time delay system with delays multiple of a basic one. Let $U(\tau)$ be the Lyapunov matrix of the neutral type delay system associated to W. Then there exists a solution $X_j(\tau)$, $j = -m, -m+1, ..., -1, 0, 1, ..., m-1$ of the delay free system (9) such that $X_0(\tau) = U(\tau)$, $\tau \in [0, h]$. The solution satisfies the boundary conditions (10)-(11).*

Example 1. Consider the system

$$\dot{x}(t) + B_1 \dot{x}(t-1) = A_0 x(t) + A_1 x(t-1) + A_2 x(t-2),$$

where

$$A_0 = \begin{bmatrix} -1 & 0 \\ 0 & 2 \end{bmatrix}, \quad A_1 = \begin{bmatrix} 0 & 0.7 \\ 0.7 & 0 \end{bmatrix}, \quad A_2 = \begin{bmatrix} -0.5 & 0 \\ 0 & -0.5 \end{bmatrix} \quad B_1 = \begin{bmatrix} 0 & 0.5 \\ -0.3 & 0 \end{bmatrix}.$$

Let us select $W = I$. The Lyapunov matrix associated to matrix W is the solution of the delay free system (9) which satisfies the boundary conditions (10)-(11). The components of the Lyapunov matrix $U(\tau)$ are sketched on Figure 1.

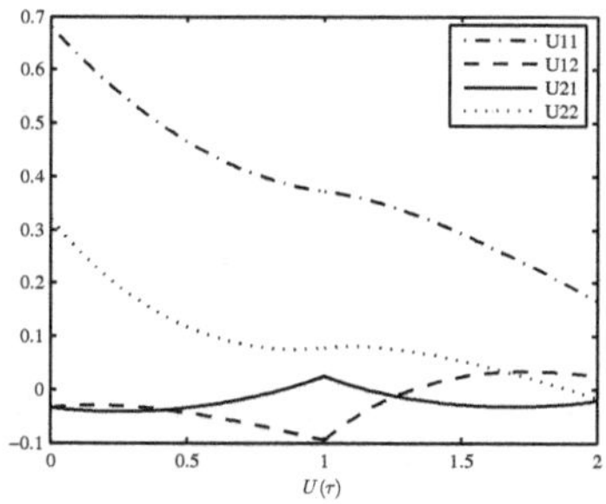

Fig. 1. Components of matrix $U(\tau)$

3.2 A piece-wise linear approximation

In this subsection an algorithm for the computation of a piece-wise linear approximation of the Lyapunov matrix $U(\tau)$ is presented.

Let us go back to the general delay case, where the dynamic equation is of the form (4). We divide first each interval $[h_j, h_{j+1}]$ into η_{j+1}, $j = 0, 1, ..., m-1$ equal segments, where the total number of partitions is $N = \sum_{k=1}^{m} \eta_k$. Denote the points of this partition as τ_j, $j = 0, 1, ..., N$. The length of each segment is defined as $\sigma_j = \tau_{j+1} - \tau_j$, $j = 0, 1, ..., N-1$. Next, introduce the auxiliary matrices U_j for $\tau > 0$ and Ψ_j for $\tau < 0$, $j = 0, 1, ..., N$. Notice that because of (5) $\Psi_j^{\mathsf{T}} = U_j$. Then, a piece-wise linear approximation of the Lyapunov matrix $U(\tau)$ on the interval for $\tau \in [\tau_j, \tau_{j+1}]$ is

$$\tilde{U}(\xi) = \left(1 + \tfrac{\tau_j - \xi}{\sigma_j}\right) U_j + \left(\tfrac{\xi - \tau_j}{\sigma_j}\right) U_{j+1}, \tag{12}$$

and the initial condition on $\tau \in [-\tau_{j+1}, -\tau_j]$ is

$$\Psi(\xi) = \left(1 + \tfrac{\tau_j + \xi}{\sigma_j}\right) \Psi_j + \left(\tfrac{-\xi - \tau_j}{\sigma_j}\right) \Psi_{j+1}. \tag{13}$$

The integral form of (4) appears as

$$\sum_{k=0}^{m} \left[U(\tau - h_k) - U(-h_k)\right] B_k = \sum_{k=0}^{m} \int_{-h_k}^{\tau - h_k} U(\varphi) d\varphi A_k.$$

If we compute the difference between $U(\tau_{j+1})$ and $U(\tau_j)$ we obtain

$$\sum_{k=0}^{m} \left[U(\tau_{j+1} - h_k) - U(\tau_j - h_k)\right] B_k = \sum_{k=0}^{m} \int_{\tau_j - h_k}^{\tau_{j+1} - h_k} U(\varphi) d\varphi A_k. \tag{14}$$

The sign of the argument of the matrix $U(\cdot)$ depends on the quantities $\tau_{j+1} - h_k$ and $\tau_j - h_k$. We express these quantities as

$$\tau_j - h_k = \begin{cases} \tau_i + \theta_i, \theta_i \in [0, \sigma_i) \text{ if } \tau_j - h_k > 0 \\ -\tau_i + \theta_i, \theta_i \in [0, \sigma_{i-1}) \text{ if } \tau_j - h_k < 0 \end{cases}$$

$$\tau_{j+1} - h_k = \begin{cases} \tau_f + \theta_f, \theta_i \in [0, \sigma_f) \text{ if } \tau_{j+1} - h_k > 0 \\ -\tau_f + \theta_f, \theta_i \in [0, \sigma_{f-1}) \text{ if } \tau_{j+1} - h_k < 0 \end{cases}$$

where τ_i and τ_f are the nearest point of the partition at the left. Using the appropriate approximation (12) for the positive case or (13) for the negative case, the following lemma provides an explicit approximation of equation (14).

Lemma 1. *A piece-wise linear approximation of equation (14) can be computed as*

- *if* $\tau_j - h_k \geq 0$ *then* $\tau_{j+1} - h_k > 0$ *and*

$$U(\tau_{j+1} h) = \left(\tfrac{\sigma_f - \theta_f}{\sigma_f}\right) U_f + \left(\tfrac{\theta_f}{\sigma_f}\right) U_{f+1}$$
$$U(\tau_j - h_k) = \left(\tfrac{\sigma_i - \theta_i}{\sigma_i}\right) U_i + \left(\tfrac{\theta_i}{\sigma_i}\right) U_{i+1},$$

with $U(\tau_j - h_k) = U_0$ *if* $\tau_j - h_k = 0$. *The interval on the integral of the right hand side of (14) is also positive and the linear approximation is*

$$\int_{\tau_j-h_k}^{\tau_{j+1}-h_k} U(\varphi)d\varphi = \left[\frac{\sigma_i^2+\theta_i(\theta_i-2\sigma_i)}{2\sigma_i}\right]U_i + \left[\frac{\sigma_i^2-\theta_i^2}{2\sigma_i}\right]U_{i+1} + \left[\frac{\theta_f(2\sigma_f-\theta_f)}{2\sigma_f}\right]U_f$$

$$+\left[\frac{\theta_f^2}{2\sigma_f}\right]U_{f+1} + \sum_{l=0}^{a-1}\left[\frac{\sigma_{l+(i+1)}}{2}U_{l+(i+1)} + \frac{\sigma_{l+(i+1)}}{2}U_{(l+1)+(i+1)}\right],$$

where $\alpha = f - (i+1)$.

- *If* $\tau_{j+1} - h_k \leq 0$ *then* $\tau_j - h_k < 0$ *and*

$$U(\tau_{j+1} - h_k) = \left(\frac{\theta_f}{\sigma_{f-1}}\right)U_{f-1}^{\mathsf{T}} + \left(\frac{\sigma_{f-1}-\theta_f}{\sigma_{f-1}}\right)U_f^{\mathsf{T}}$$
$$U(\tau_j - h_k) = \left(\frac{\theta_i}{\sigma_{i-1}}\right)U_{i-1}^{\mathsf{T}} + \left(\frac{\sigma_{i-1}-\theta_i}{\sigma_{i-1}}\right)U_i^{\mathsf{T}},$$

with $U(\tau_{j+1} - h_k) = U_0$ *if* $\tau_{j+1} - h_k = 0$. *The interval on the integral of the right hand side of (14) is also negative and the linear approximation appears as*

$$\int_{\tau_j-h_k}^{\tau_{j+1}-h_k} U(\varphi)d\varphi = \left[\frac{\sigma_{i-1}^2-\theta_i^2}{2\sigma_{i-1}}\right]U_{i-1}^{\mathsf{T}} + \left[\frac{\sigma_{i-1}^2+\theta_i(\theta_i-2\sigma_{i-1})}{2\sigma_{i-1}}\right]U_i^{\mathsf{T}} +$$

$$+\sum_{l=0}^{a-1}\left[\frac{\sigma_{(i-1)-(l+1)}}{2}U_{(i-1)-(l+1)}^{\mathsf{T}} + \frac{\sigma_{(i-1)-(l+1)}}{2}U_{(i-1)-l}^{\mathsf{T}}\right] +$$

$$+\left[\frac{\theta_f^2}{2\sigma_{f-1}}\right]U_f^{\mathsf{T}} + \left[\frac{\theta_f(2\sigma_{f-1}-\theta_f)}{2\sigma_{f-1}}\right]U_f^{\mathsf{T}}.$$

where $\alpha = (i-1) - f$.

- *If* $\tau_{j+1} - h_k > 0$ *and* $\tau_j - h_k < 0$ *then*

$$\int_{\tau_j-h_k}^{\tau_{j+1}-h_k} U(\varphi)d\varphi = \left[\frac{\sigma_{i-1}^2-\theta_i^2}{2\sigma_{i-1}}\right]U_{i-1}^{\mathsf{T}} + \left[\frac{\sigma_{i-1}^2+\theta_i(\theta_i-2\sigma_{i-1})}{2\sigma_{i-1}}\right]U_i^{\mathsf{T}} +$$

$$\sum_{l=0}^{i-2}\left[\frac{\sigma_{(i-1)-(l+1)}}{2}U_{(i-1)-(l+1)}^{\mathsf{T}} + \frac{\sigma_{(i-1)-(l+1)}}{2}U_{(i-1)-l}^{\mathsf{T}}\right] +$$

$$+\sum_{l=0}^{f-1}\left[\frac{\sigma_l}{2}U_l + \frac{\sigma_l}{2}U_{l+1}\right] + \left[\frac{\theta_f(2\sigma_f-\theta_f)}{2\sigma_f}\right]U_f + \left[\frac{\theta_f^2}{2\sigma_f}\right]U_{f+1}.$$

From the previous lemma it is easy to see that we obtain a set of N matrix equations for $N+1$ unknown matrices U_j, $j = 0, 1, ..., m$. If we add to this set of matrix equations the algebraic property we obtain $N+1$ matrix equations to $N+1$ unknown matrices. When writing equation (6) in terms of the matrices U_j, $j = 0, 1, ..., m$, one can observe that the argument of $U(\cdot)$ depends on $h_i - h_j$ and that either $h_i - h_j < 0$, $h_i - h_j = 0$ or $h_i - h_j > 0$. Then $h_i - h_j$ can be written as $\tau_k + \theta_k = h_i - h_j$ where τ_k is the nearest point at the left and $\theta_k \in [0, \sigma_k)$. Then each term of the algebraic property (6) can be expressed in terms of the matrices U_i, $i = 0, 1, ..., N$ as follows:

- if $h_i - h_j < 0$

$$U(h_i - h_j) = U_{k-1}^{\mathsf{T}}\left(\frac{\theta_k}{\sigma_{k-1}}\right) + U_k^{\mathsf{T}}\left(\frac{\sigma_{k-1}-\theta_k}{\sigma_{k-1}}\right),$$
$$U^{\mathsf{T}}(h_i - h_j) = U_{k-1}\left(\frac{\theta_k}{\sigma_{k-1}}\right) + U_k\left(\frac{\sigma_{k-1}-\theta_k}{\sigma_{k-1}}\right),$$

- if $h_i - h_j = 0$

$$U(h_i - h_j) = U^{\mathsf{T}}(h_i - h_j) = U_0,$$

- if $h_i - h_j > 0$

$$U(h_i - h_j) = U_k\left(\frac{\sigma_k-\theta_k}{\sigma_k}\right) + U_{k+1}\left(\frac{\theta_k}{\sigma_k}\right),$$
$$U^{\mathsf{T}}(h_i - h_j) = U_k^{\mathsf{T}}\left(\frac{\sigma_k-\theta_k}{\sigma_k}\right) + U_{k+1}^{\mathsf{T}}\left(\frac{\theta_k}{\sigma_k}\right).$$

Example 2. To illustrate this numerical procedure, let us consider the system of Example 1 with $\eta_1 = \eta_2 = 5$. The components of the piece-wise linear approximation of the Lyapunov matrix $U(\tau)$ are sketched on Figure 2

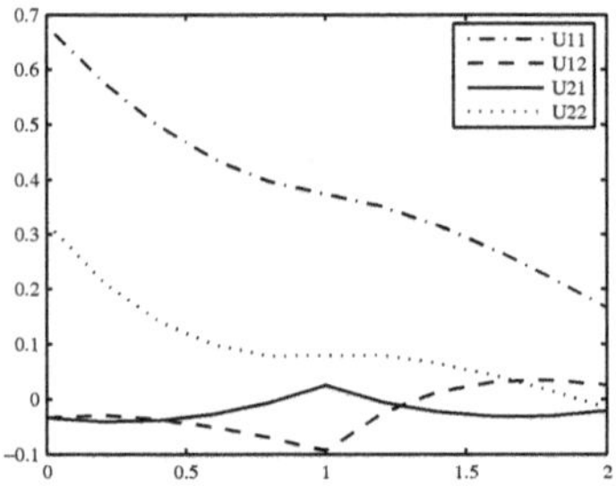

Fig. 2. A piece-wise linear approximation

4 Error estimation

The semi-analytic procedure cannot be applied to systems with delays not multiple of a basic one, thus it is not possible to make a direct comparison between the semi-analytic procedure and the piece-wise linear approximation as in the previous example. In this section we present an algorithm that gives a measure of the quality of the numerical approximation for the case of general delays.

As $U(\tau)$ satisfies the symmetry condition (5), one can set the initial conditions $\Phi(\tau)$, equal to the matrix $U^{\mathsf{T}}(-\tau)$ obtained in the piece-wise linear approximation procedure. It is then possible to compute an approximation of the matrix $U(\tau)$ as the solution of the dynamic condition (4) using the step by step method, see [1]. This approximate Lyapunov matrix is denoted as $\tilde{U}(\tau, \Phi)$.

We introduce the following error matrices

$$\Delta(\tau) = \hat{U}(\tau, \Phi) - \Phi(-\tau) \text{ for } \tau \in [0, h_m], \text{ and } \Delta_0 = W - \hat{W},$$

where

$$-\hat{W} = \sum_{j=0}^{m}\sum_{k=0}^{m}[B_j^{\mathsf{T}}\Phi(h_j - h_k)A_k + A_k^{\mathsf{T}}\Phi^{\mathsf{T}}(h_j - h_k)B_j].$$

Notice that the approximate matrix $\hat{U}(\tau,\Phi)$ satisfies the dynamic property (4) but not necessarily the symmetry property (5) and the algebraic property (6). Let us replace the matrix $U(\tau)$ in the functional (2) by the approximate matrix $\hat{U}(\tau,\Phi)$ and denote the new functional as $\hat{v}(x_t)$. The time derivative of $\hat{v}(x_t)$ along the trajectories of system (1) is equal to

$$\begin{aligned}
\tfrac{d}{dt}\hat{v}(x_t) &= -w(x_t) + \sum_{j=0}^{m}\sum_{k=0}^{m}\dot{x}^{\mathsf{T}}(t)B_j^{\mathsf{T}}\left\{\begin{matrix}-\Delta(h_j - h_k), \text{ if } h_j - h_k \geq 0\\ \Delta^{\mathsf{T}}(h_k - h_j), \text{ if } h_j - h_k < 0\end{matrix}\right\}B_k x(t) +\\
&+ x^{\mathsf{T}}(t)\{\Delta_0 + \sum_{j=0}^{m}\sum_{k=0}^{m}[B_j^{\mathsf{T}}\Delta^{\mathsf{T}}(h_k - h_j)A_k + A_k^{\mathsf{T}}\Delta(h_k - h_j)B_j]\}x(t) +\\
&+ \sum_{j=0}^{m}\sum_{k=1}^{m}[x^{\mathsf{T}}(t)A_j^{\mathsf{T}} - \dot{x}^{\mathsf{T}}(t)B_j^{\mathsf{T}}]\int_{-h_k}^{0}\left\{\begin{matrix}\Delta(h_j - h_k - \theta), \text{ if } h_j - h_k - \theta \geq 0\\ -\Delta^{\mathsf{T}}(h_k - h_j + \theta), \text{ if } h_j - h_k - \theta < 0\end{matrix}\right\}\\
&\times [A_k x(t+\theta) - B_k\dot{x}(t+\theta)]d\theta. \qquad (15)
\end{aligned}$$

We evaluate the quality of the approximate matrix $\hat{U}(\tau,\Phi)$ by comparing the time derivative of the functional $\hat{v}(x_t)$ and $v(x_t)$: the smaller the difference between the time derivatives is, the better the approximation is. To this end we define the quantities

$$\rho = \max_{\tau\in[0,h_m]}\{\|\Delta(\tau)\|\},\ \alpha_i = \|A_i\|, i = 0,1,\ldots,m,$$
$$\beta_i = \|B_i\|, i = 0,1,\ldots,m.$$

We summarize the above error computation procedure as follows:

Theorem 2. *Let a functional (2) be given. The difference between its time derivative along the trajectories of system (1) and the time derivative of functional $\hat{v}(x_t)$, constructed from the Lyapunov matrix $\hat{U}(\tau,\Phi)$ is bounded as follows*

$$\begin{aligned}
\left|\frac{d}{dt}\hat{v}(x_t) - \frac{d}{dt}v(x_t)\right| &\leq \sum_{k=0}^{m}\nu_k\|x(t-h_k)\|^2 + \sum_{k=1}^{m}\nu_{m+k}\int_{-h_k}^{0}\|x(t+\theta)\|^2 d\theta +\\
&+ \sum_{k=0}^{m}v_k\|\dot{x}(t-h_k)\|^2 + \sum_{k=1}^{m}v_{m+k}\int_{-h_k}^{0}\|\dot{x}(t+\theta)\|^2 d\theta. \qquad (16)
\end{aligned}$$

where

$$\begin{aligned}
\nu_0 = \|\Delta_0\| + \rho\sum_{j=0}^{m}&\left[\sum_{k=0}^{m}\beta_j\left(2\alpha_k + \frac{1}{2}\beta_k\sum_{l=1}^{m}(\alpha_l + \beta_l) + \beta_k\alpha_0\right) + \right.\\
&\left. + \frac{1}{2}\sum_{k=1}^{m}(\alpha_j + \alpha_0\beta_j)(\alpha_k + \beta_k)h_k\right]
\end{aligned}$$

and

$$\nu_l = \tfrac{1}{2}\rho\alpha_l \sum_{j=0}^{m} \beta_j \left[\sum_{k=0}^{m} \beta_k + \sum_{k=1}^{m} (\alpha_k + \beta_k) h_k \right]$$
$$\nu_{m+l} = \tfrac{1}{2}\rho\alpha_l \sum_{j=0}^{m} \left[(\alpha_j + \alpha_0\beta_j) + \sum_{k=1}^{m} (\alpha_k + \beta_k) \beta_j \right]$$
$$\upsilon_k = \tfrac{1}{2}\rho\beta_l \sum_{j=0}^{m} \beta_j \left[\sum_{k=0}^{m} \beta_k + \sum_{k=1}^{m} (\alpha_k + \beta_k) h_k \right]$$
$$\upsilon_{m+k} = \tfrac{1}{2}\rho\beta_l \sum_{j=0}^{m} \left[(\alpha_j + \alpha_0\beta_j) + \sum_{k=1}^{m} (\alpha_j + \beta_k) \beta_j \right]$$

Proof. The estimates are obtained by substituting $\dot{x}(t)$ by $\sum_{l=0}^{m} A_l x(t-h_l) - \sum_{l=1}^{m} B_l \dot{x}(t-h_l)$ into (15) followed by straightforward but tedious computations.

Example 3. Once again let us consider the system of Example 1. By the "step by step "method, using $\Phi(-\tau)$ as initial conditions, one can find an approximation of the Lyapunov matrix $U(\tau)$. The components of matrix $\tilde{U}(\tau,\Phi)$ are sketched on Figure 3. Computing the estimates of the error according expression (16) one obtain that

$$\left| \frac{d}{dt}\hat{v}(x_t) - \frac{d}{dt}v(x_t) \right| \leq 0.1124 \left\| x(t) \right\|^2 + 0.0095 \left\| x(t-1) \right\|^2$$

$$+0.01508 \int_{-1}^{0} \left\| x(t+\theta) \right\|^2 d\theta + 0.0068 \left\| x(t-2) \right\|^2 + 0.0107 \int_{-2}^{0} \left\| x(t+\theta) \right\|^2 d\theta +$$

$$+0.0068 \left\| \dot{x}(t-1) \right\|^2 + 0.0107 \int_{-1}^{0} \left\| \dot{x}(t+\theta) \right\|^2 d\theta.$$

where $\alpha_0 = 2$, $\alpha_1 = 0.7$, $\alpha_2 = 0.5$, $\beta_0 = 1$, $\beta_1 = 0.5$, $\beta_2 = 0$, $\rho = 0.0049$ and

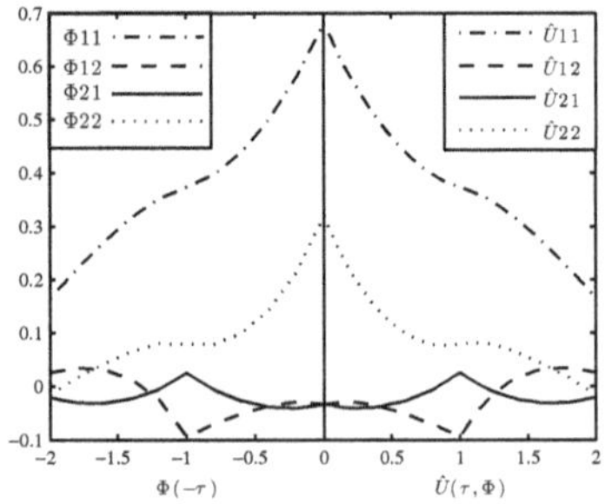

Fig. 3. Components of the matrix $\hat{U}(\tau,\Phi)$

$\|\Delta_0\| = 2.2822 \times 10^{-16}$. As one can see the computed estimates are small enough to consider that matrix $\tilde{U}(\tau,\Phi)$ a good approximation of the Lyapunov matrix $U(\tau)$.

5 Conclusions

In this contribution, the problem of constructing the Lyapunov matrix of neutral type time delay systems is addressed. A semi-analytic approach for systems with delays multiple of a basic one and a piece-wise linear approximation that can be used for the case of general delays are presented. An estimate of the error is also given. The computation of these matrices is a key step for using functionals of complete type in the analysis of time delay systems of neutral type.

References

1. Bellman, R., Cooke, K.L.: Differential Difference Equations. Academic Press, New York (1963)
2. García-Lozano, H., Kharitonov, V.L.: Lyapunov matrices for time delay systems with commensurate delays. In: Proceedings of the 2nd IFAC Symposium on Systems, Structure and Control, Oaxaca, México, pp. 102–106 (2004)
3. Gohberg, I.C., Lerer, L.E.: Resultants of matrix polynomials. Bull. Amer. Math. Soc. 82, 565–567 (1976)
4. Kharitonov, V.L.: Lyapunov functionals and Lyapunov matrices for neutral type time delay systems: a single delay case. International Journal of Control 78, 783–800 (2005)
5. Lancaster, P., Tismenetsky, M.: The theory of matrices. Academic Press, Orlando (1985)
6. Louisell, J.: A matrix method for determining the imaginary axis eigenvalues of a delay system. IEEE Transactions on Automatic Control 48, 2008–2012 (2001)

Stability of Coupled Differential-Difference Equations with Block Diagonal Uncertainty

Yi Liu[1], Hongfei Li[2], and Keqin Gu[1]

[1] Department of Mechanical and Industrial Engineering, Southern Illinois University Edwardsville, Edwardsville, IL 62026-1805, USA.
yliuaa@siue.edu and kgu@siue.edu

[2] Department of Mathematics, Yulin College, Shaanxi Yulin, 719000, China.
lhf8165@163.com

Summary. This article discusses the stability problem of coupled linear differential-difference equations with block-diagonal uncertainty. New stability criteria in the form of linear matrix inequality are derived using the discretized Lyapunov-Krasovskii functional method. A number of examples are presented to illustrate the effectiveness of this method. It is shown that the stability conditions derived can be extended to evaluation of H_∞ performance.

1 Introduction

Coupled differential-difference equations represent a very general class of time-delay systems of practical importance. It is also known as the lossless propagation model, due to the fact that it is the form often obtained after simplifying some hyperbolic partial differential equations describing lossless propagation systems [16] [1] [14]. Indeed, it includes as special cases systems with multiple commensurate delays, neutral time-delay systems, and some singular time-delay systems [8]. A study of its stability problem using complete quadratic Lyapunov-Krasovskii functional is carried out in [8] where the existence of a quadratic Lyapunov-Krasovskii functional is a necessary and sufficient condition for such systems to be asymptotically stable. In [9], the stability of more general coupled differential-functional equations is discussed, and a discretized Lyapunov-Krasovskii functional method is used to study the stability problem of uncertain differential-difference equations. Similar to time-delay systems of retarded type discussed in [6] [7], it is shown in [9] that a discretized Lyapunov-Krasovskii functional method usually approaches the analytical stability limit rather quickly. There does not appear much difficulty in applying other methods based on a complete quadratic form of Lyapunov-Krasovskii functional, such as sum-of-square method in [15], to such systems.

In this chapter, the discretized Lyapunov-Krasovskii method is applied to obtain the LMI stability conditions of coupled differential-difference equations with block diagonal uncertainty. The chapter is organized as follows. Section 2 introduces the system of coupled differential-difference equations and restates the stability criteria developed in [9]. Section 3 presents the stability conditions of such system with block-diagonal uncertainty. It shows that the norm-bounded uncertainty discussed in [13] is included as a special case of this

J.J. Loiseau et al. (Eds.): Topics in Time Delay Systems, LNCIS 388, pp. 73–83.
springerlink.com

setting. Two numerical examples are presented to show the effectiveness of our method. Section 4 discusses how our results can be applied to evaluate H_∞ performance.

The notation in this chapter is rather standard. $\mathbb{R}$ is the set of real numbers, $\mathbb{R}^n$ and $\mathbb{R}^{m\times n}$ represent the set of real n-vectors and m by n matrices, respectively. $\mathbb{R}_+$ and $\bar{\mathbb{R}}_+$ denote the sets of positive and non-negative real numbers, respectively. For a given $r \in \mathbb{R}_+$, which usually corresponds to the delay, $\mathcal{PC}$ is the set of bounded, right continuous, and piecewise continuous $\mathbb{R}^n$-valued functions defined on $[-r, 0)$. For a given function $y(t) \in \mathbb{R}^n$ and $\sigma \in \bar{\mathbb{R}}_+$, we define $y_\sigma \in \mathcal{PC}$ by $y_\sigma(\theta) = y(\sigma+\theta)$, $\theta \in [-r, 0)$. We use $||\cdot||$ to denote 2-norm for vectors and matrices. For $\phi \in \mathcal{PC}$, $||\phi|| \triangleq \sup_{-r\le\theta<0} ||\phi(\theta)||$, and $||(\psi,\phi)|| \triangleq \max\{||\phi||, ||\psi||\}$ for any $\psi \in \mathbb{R}^m$. For an $A \in \mathbb{R}^{m\times n}$, A^T denotes the transpose of A. If $A = A^T$, we use $A > 0$ or $A \ge 0$ to indicate that A is positive definite or positive semi-definite, respectively . Parallel notation is used to denote negative definiteness and semi-definiteness.

2 Problem Setup and Background

Consider the uncertain systems represented by the following coupled differential-difference equations

$$\dot{x}(t) = A(t)x(t) + B(t)y(t-r), \tag{1}$$

$$y(t) = C(t)x(t) + D(t)y(t-r), \tag{2}$$

where $x \in \mathbb{R}^m$, $y \in \mathbb{R}^n$, and $A(t), B(t), C(t), D(t)$ are unknown uncertain time-varying matrices of commensurate dimensions. It is known, however, that they belong to a known compact set Ω, i.e.

$$(A(t), B(t), C(t), D(t)) \in \Omega \text{ for all } t \ge 0. \tag{3}$$

It is shown in [9] that the system is asymptotically stable if the difference equation (2) is input-to-state stable (considering $x(t)$ as the input), and there exists a continuous quadratic Lyapunov-Krasovskii functional

$$V : \mathbb{R} \times \mathbb{R}^m \times \mathcal{PC} \to \mathbb{R},$$

such that

$$\varepsilon||\psi||^2 \le V(t,\psi,\phi) \le M||(\psi,\phi)||^2, \tag{4}$$

$$\dot{V}(t,\psi,\phi) \le -\varepsilon||\psi||^2 \tag{5}$$

for some $\varepsilon > 0$ and $M > 0$.

Furthermore, let V be the following quadratic functional

$$V(\psi,\phi) = \psi^T P\psi + 2\psi^T \int_{-r}^0 Q(\eta)\phi(\eta)d\eta + \int_{-r}^0 \int_{-r}^0 \phi^T(\xi)R(\xi,\eta)\phi(\eta)d\xi d\eta + \int_{-r}^0 \phi(\eta)S(\eta)\phi(\eta)d\eta, \tag{6}$$

where $P = P^T \in \mathbb{R}^{m\times m}$, $Q(\eta) \in \mathbb{R}^{m\times n}$, $R(\xi,\eta) = R^T(\eta,\xi) \in \mathbb{R}^{n\times n}$, $S(\eta) = S^T(\eta) \in \mathbb{R}^{n\times n}$. Choose Q, R and S to be continuous and piecewise linear, the following lemma is shown in [9].

Lemma 1. *The system represented by (1) and (2) is asymptotically stable if there exist matrix* $P = P^T \in \mathbb{R}^{m\times m}$*, and continuous piecewise linear matrices* $Q_i \in \mathbb{R}^{m\times n}$*,* $S_i = S_i^T \in \mathbb{R}^{n\times n}$*, and* $R_{ij} = R_{ji}^T \in \mathbb{R}^{n\times n}$*,* $i = 0, 1, ..., N$*,* $j = 0, 1, ..., N$*, such that*

$$\begin{pmatrix} P & \tilde{Q} \\ \tilde{Q}^T & \tilde{R} + \frac{1}{h}\tilde{S} \end{pmatrix} > 0 \tag{7}$$

and

$$\begin{pmatrix} \Delta & Y^s & Y^a & Z \\ & R_d + \frac{1}{h}S_d & 0 & 0 \\ & & \frac{3}{h}S_d & 0 \\ \textit{Symmetric} & & & S_N \end{pmatrix} > 0 \tag{8}$$

are satisfied for all $(A, B, C, D) \in \Omega$*, where* $h = \frac{r}{N}$ *and*

$$\tilde{Q} = \begin{pmatrix} Q_0 & Q_1 & \dots & Q_N \end{pmatrix}, \tag{9}$$

$$\tilde{S} = \mathit{diag}\begin{pmatrix} S_0 & S_1 & \dots & S_N \end{pmatrix}, \tag{10}$$

$$\tilde{R} = \begin{pmatrix} R_{00} & R_{01} & \dots & R_{0N} \\ R_{10} & R_{11} & \dots & R_{1N} \\ \vdots & \vdots & \ddots & \vdots \\ R_{N0} & R_{N1} & \dots & R_{NN} \end{pmatrix}, \tag{11}$$

$$\Delta = \begin{pmatrix} -A^TP - PA - Q_NC - C^TQ_N^T & -PB + Q_ND + Q_0 \\ -B^TP + D^TQ_N + Q_0 & S_0 \end{pmatrix}, \tag{12}$$

$$Z = \begin{pmatrix} C^TS_N \\ D^TS_N \end{pmatrix}, \tag{13}$$

$$S_d = \mathit{diag}\begin{pmatrix} S_{d1} & S_{d2} & \dots & S_{dN} \end{pmatrix}, \ S_{di} = \frac{1}{h}(S_i - S_{i-1}),$$

$$R_d = \begin{pmatrix} R_{d11} & R_{d12} & \dots & R_{d1N} \\ R_{d21} & R_{d22} & \dots & R_{d2N} \\ \vdots & \vdots & \ddots & \vdots \\ R_{dN1} & R_{dN2} & \dots & R_{dNN} \end{pmatrix}, \ R_{dij} = \frac{1}{h}(R_{ij} - R_{i-1,j-1}),$$

$$Y^s = \begin{pmatrix} Y_{11}^s & Y_{12}^s & \dots & Y_{1N}^s \\ Y_{21}^s & Y_{22}^s & \dots & Y_{2N}^s \end{pmatrix}, \ Y^a = \begin{pmatrix} Y_{11}^a & Y_{12}^a & \dots & Y_{1N}^a \\ Y_{21}^a & Y_{22}^a & \dots & Y_{2N}^a \end{pmatrix},$$

and

$$\begin{aligned} Y_{1i}^s &= -\frac{1}{2}A^T(Q_{i-1} + Q_i) - \frac{1}{2}C^T(R_{i-1,N}^T + R_{i,N}^T) + \frac{1}{h}(Q_i - Q_{i-1}), \\ Y_{2i}^s &= -\frac{1}{2}B^T(Q_{i-1} + Q_i) - \frac{1}{2}D^T(R_{i-1,N}^T + R_{i,N}^T) + \frac{1}{2}(R_{i-1,0}^T + R_{i,0}^T), \\ Y_{1i}^a &= \frac{1}{2}A^T(Q_i - Q_{i-1}) + \frac{1}{2}C^T(R_{i,N}^T - R_{i-1,N}^T), \\ Y_{2i}^a &= \frac{1}{2}B^T(Q_i - Q_{i-1}) + \frac{1}{2}D^T(R_{i,N}^T - R_{i-1,N}^T) - \frac{1}{2}(R_{i,0}^T - R_{i-1,0}^T). \end{aligned}$$

In general, the uncertainty set Ω contains an infinite number of points. This means that one needs to check (8) infinitely many times. A number of special forms of Ω allows this to be reduced to checking a finite number of LMIs. This includes cases when Ω is polytopic or norm-bounded, which is discussed in [9] and [13], respectively. In the subsequent text, we will show that (8) can also be reduced to a finite number of LMIs when the uncertainty is in block-diagonal structure.

3 Systems with Block-Diagonal Uncertainty

3.1 Problem Formulation

Consider the following linear coupled feedback system

$$\dot{x}(t) = A_n x(t) + B_n y(t-r) + E_1 u(t), \tag{14}$$
$$y(t) = C_n x(t) + D_n y(t-r) + E_2 u(t), \tag{15}$$
$$z(t) = G_1 x(t) + G_2 y(t-r) + Lu(t), \tag{16}$$

where $x \in \mathbb{R}^m$, $y \in \mathbb{R}^n$, $u(t) \in \mathbb{R}^p$, $z(t) \in \mathbb{R}^q$, and $L \in \mathbb{R}^{q\times p}$, $E_1 \in \mathbb{R}^{m\times p}$, $E_2 \in \mathbb{R}^{n\times p}$, $G_1 \in \mathbb{R}^{q\times m}$, $G_2 \in \mathbb{R}^{q\times n}$ are known constant real matrices. The system is subject to the uncertain feedback

$$u(t) = \Delta(z(t)), \tag{17}$$

where Δ is a possibly nonlinear causal operator [20] [21] . It is assumed that the characteristics of the uncertainty is such that

$$\frac{1}{2}\int_0^t z^T(\tau)K_z z(\tau)d\tau - \frac{1}{2}\int_0^t u^T(\tau)K_u u(\tau)d\tau \geq 0 \tag{18}$$

for any $t \geq 0$, and

$$(K_u, K_z) \in \mathcal{K}_{pq} \subset \mathbb{R}^{p\times p} \times \mathbb{R}^{q\times q},$$

where $\mathcal{K}_{pq}$ is a known convex set. With the dimensions implied by the context, we omit the subscript of $\mathcal{K}_{pq}$. Typically, this describes block diagonal dynamic or memoryless linear or nonlinear uncertainty.

3.2 Stability Conditions

Based on the quadratic Lyapunov-Krasovskii functional given by (6), we conclude the following.

Proposition 1. *The system described by (14) to (16) subject to (17) is internally stable and input-output stable if there exists a quadratic Lyapunov-Krasovskii functional $V(t, x_t)$ in form of (6) such that for some $\varepsilon > 0$ and arbitrary $x_t \in \mathcal{PC}$,*

$$V(t, x_t, y(t)) \geq \varepsilon \|x(t)\|^2 \tag{19}$$

and its derivative along the solution of (14) satisfies

$$\dot{V}(t, x_t, y(t)) + z^T K_z z - u^T K_u u \leq -\varepsilon(\|x(t)\|^2 + \|u(t)\|^2). \tag{20}$$

Proof. One may follow the same step of the proof for Proposition 8.3 in [7].

Proposition 2. *A Lyapunov-Krasovskii functional $V(t, x_t, y(t))$ satisfies both (19) and (20) if for some $(K_u, K_z) \in \mathcal{K}$, there exist matrices $P = P^T$ and continuous piecewise linear $Q_i = Q_i^T, S_i = S_i^T, i = 0, 1, ...N$; $R_{i,j} = R_{j,i}^T, i = 0, 1...N, j = 0, 1, ...N$ such that*

$$\begin{pmatrix} P & \tilde{Q} \\ \tilde{Q}^T & \tilde{R} + \frac{1}{h}\tilde{S} \end{pmatrix} > 0, \tag{21}$$

and

$$\begin{pmatrix} \Delta & Y^s & Y^a \\ & R_d + \frac{1}{h}S_d & 0 \\ Symmetric & & \frac{3}{h}S_d \end{pmatrix} > 0, \tag{22}$$

where h, $\tilde{Q}$, $\tilde{R}$, $\tilde{S}$, R_d and S_d inherit the expression in lemma 1 and

$$\Delta = \begin{pmatrix} \Delta_{00} & \Delta_{01} & \Delta_{02} \\ & \Delta_{11} & \Delta_{12} \\ Symmetric & & \Delta_{22} \end{pmatrix},$$

$$\begin{aligned}
\Delta_{00} &= K_u - LK_zL - E_2^T S_N E_2, \\
\Delta_{01} &= -(L^T K_z G_1 + E_1^T P + E_2^T Q_N^T + E_2^T S_N C), \\
\Delta_{02} &= -(E_2^T S_N D + L^T K_z G_2), \\
\Delta_{11} &= -(A^T P + PA + Q_N C + C^T Q_N + C^T S_N C + G_1^T K_z G_1), \\
\Delta_{12} &= -(PB + Q_N D - Q_0 + C^T S_N D + G_1^T K_z G_2, \\
\Delta_{22} &= S_0 - D^T S_N D - G_2^T K_z G_2,
\end{aligned} \tag{23}$$

$$Y^s = \begin{pmatrix} Y_{01}^s & Y_{02}^s & \dots & Y_{0N}^s \\ Y_{11}^s & Y_{12}^s & \dots & Y_{1N}^s \\ Y_{21}^s & Y_{22}^s & \dots & Y_{2N}^s \end{pmatrix}, \quad Y^a = \begin{pmatrix} Y_{01}^a & Y_{02}^a & \dots & Y_{0N}^a \\ Y_{11}^a & Y_{12}^a & \dots & Y_{1N}^a \\ Y_{21}^a & Y_{22}^a & \dots & Y_{2N}^a \end{pmatrix},$$

where

$$\begin{aligned}
Y_{0i}^s &= -\frac{1}{2}E_1^T(Q_i + Q_{i-1}) - \frac{1}{2}E_2^T(R_{i,N}^T + R_{i-1,N}), \\
Y_{1i}^s &= -\frac{1}{2}A^T(Q_{i-1} + Q_i) - \frac{1}{2}C^T(R_{i-1,N}^T + R_{i,N}^T) + \frac{1}{h}(Q_i - Q_{i-1}), \\
Y_{2i}^s &= -\frac{1}{2}B^T(Q_{i-1} + Q_i) - \frac{1}{2}D^T(R_{i-1,N}^T + R_{i,N}^T) + \frac{1}{2}(R_{i-1,0}^T + R_{i,0}^T), \\
Y_{0i}^a &= \frac{1}{2}E_1^T(Q_i - Q_{i-1}) + \frac{1}{2}E_2^T(R_{i,N}^T - R_{i-1,N}^T), \\
Y_{1i}^a &= \frac{1}{2}A^T(Q_i - Q_{i-1}) + \frac{1}{2}C^T(R_{i,N}^T - R_{i-1,N}^T), \\
Y_{2i}^a &= \frac{1}{2}B^T(Q_i - Q_{i-1}) + \frac{1}{2}D^T(R_{i,N}^T - R_{i-1,N}^T) - \frac{1}{2}(R_{i,0}^T - R_{i-1,0}^T).
\end{aligned}$$

Proof. After replacing $y(t)$, $z(t)$ by the right-hand side of (15) and (16), this proof can be completed by following the similar procedures of the proof for Proposition 4 in [6].

3.3 Uncertainty Characterization

Quite similar to the discussion in [20] [21], the structure of $\mathcal{K}$ depends on the uncertainty characterization. We briefly restate the conclusion in [20] [21] for the sake of completeness. A very general uncertainty can be modeled as feedback block-diagonal uncertainty described by

$$u = (u_1^T, u_2^T, ..., u_l^T)^T,\ z = (z_1^T, z_2^T, ..., z_l^T)^T,$$

where $u_i^T \in \mathbb{R}^{p_i}$, $z_i^T \in \mathbb{R}^{q_i}$ and

$$u_i = \Delta_i z_i.$$

We will assume Δ_i all have unit gain, otherwise, a standard scaling procedure can be used. Correspondingly, $\mathcal{K}$ consists of all diagonal matrices

$$(\text{diag}(K_{u_1}, K_{u_2}, ..., K_{u_l}), \text{diag}(K_{z_1}, K_{z_2}, ..., K_{z_l})),$$

where

$$(K_{u_i}, K_{z_i}) \in \mathcal{K}_i \subset \mathbb{R}^{p_i \times p_i} \times \mathbb{R}^{q_i \times q_i}.$$

The structure of $\mathcal{K}_i$ depends on the characteristics of the uncertainty, as described in the following.

Case I: Bounded possibly nonlinear uncertainty

It can be characterized as

$$\frac{1}{2}\int_0^t u_i^T(\tau)K_{u_i}u_i(\tau)d\tau \leq \frac{1}{2}\int_0^t z_i^T(\tau)K_{z_i}z_i(\tau)d\tau.$$

Then we can choose

$$\mathcal{K}_i = \{(k_i I_{p_i}, k_i I_{q_i})\,|k_i \in \mathbb{R}_+\}.$$

Case II: Linear repeated blocks

In this case, u_i and z_i are further divided into v_i subvectors

$$u_i = (u_{i1}, u_{i2}, ..., u_{iv_i}),\ z_i = (z_{i1}, z_{i2}, ..., z_{iv_i}),$$

where

$$u_{ij} \in \mathbb{R}^{p_{i0}},\ z_{ij} \in \mathbb{R}^{q_{i0}},\ j = 1, 2, ..., v_i.$$

Each pair of u_{ij} and y_{ij} is repeated by an identical linear uncertain operator

$$u_{ij} = \Delta_{i0} y_{ij},\ j = 1, 2, ..., v_i.$$

Due to linearity

$$\sum_{j=1}^{v_i} \lambda_j u_{ij} = \Delta_{i0} \sum_{j=1}^{v_i} \lambda_j z_{ij},$$

it can be shown that we can choose

$$\mathcal{K}_i = \left\{(K_i \otimes I_{p_{i0}}, K_i \otimes I_{q_{i0}})\,\middle|K_i \in \mathbb{R}^{v_i \times v_i}, K_i > 0\right\},$$

where $\otimes$ represents Kronecker product of matrices.

Let $\mathcal{D}$ be the Banach space consisting of all the operator Δ described above such that (18) is satisfied for some $(K_u, K_z) \in \mathcal{K}$. Define

$$\mathcal{F} = \left\{L_F : \mathbb{R}^q \to \mathbb{R}^p \mid \|F(t)\| \leq 1 \text{ for } F(t) \in \mathbb{R}^{p\times q}\right\},$$

where L_F denotes the left-multiplication transformation by matrix F. Clearly, $\mathcal{F}$ is a subset of $\mathcal{D}$. Bearing this relationship in mind, by eliminating $u(t)$ and $z(t)$ (14) to (15) and setting $L = 0$, it is easy to see the that norm-bounded uncertainty system investigated in [13], i.e.,

$$\dot{x}(t) = (A_n + \Delta A(t))x(t) + (B_n + \Delta B(t))y(t-r), \tag{24}$$

$$y(t) = (C_n + \Delta C(t))x(t) + (D_n + \Delta D(t))y(t-r), \tag{25}$$

with

$$\begin{pmatrix} \Delta A & \Delta B \\ \Delta C & \Delta D \end{pmatrix} = \begin{pmatrix} E_1 \\ E_2 \end{pmatrix} F(t) \begin{pmatrix} G_1 & G_2 \end{pmatrix}, \ \|F(t)\| \leq 1, \tag{26}$$

is a special case of the uncertain feedback system by (14) to (16). Evidently, the latter being input-output stable and internal stable is equivalent to the former being asymptotically stable. The norm-bounded uncertainty can be thus characterized as *bounded, possibly nonlinear, uncertainty.* Consequently, K_u and K_z shall be chosen as to have only one diagonal block. In other words, $\mathcal{K}$ is to be specified by

$$\mathcal{K} = \{(kI_p, kI_q) \mid k \in \mathbb{R}_+\}. \tag{27}$$

Indeed, with K_u and K_z represented by (27), one can verify easily, through Schur's compliment, that the LMI stability conditions shown in Proposition 2 are equivalent to (7) and (17) in [13].

3.4 Numerical Examples

We present here two examples to illustrate the effectiveness of our methods. The generality and advantage of our models can be clearly observed.

Example 1. Consider the following uncertain system

$$\dot{x}(t) = \left(\begin{pmatrix} -2 & 0 \\ 0 & -0.9 \end{pmatrix} + E_a F_1(t) G_a\right) x(t) + \left(\begin{pmatrix} -1 & 0 \\ -1 & -1 \end{pmatrix} + E_b F_2(t) G_b\right) y(t-r),$$

$$y(t) = \left(\begin{pmatrix} 1 & 0 \\ 0 & 1 \end{pmatrix} + E_c F_3(t) G_c\right) x(t) + \left(\begin{pmatrix} 0.2 & 0 \\ 0 & 0.2 \end{pmatrix} + E_d F_4(t) G_d\right) + y(t-r)$$

where

$$E_a = E_b = 0.2I, G_a = G_b = I$$
$$E_c = E_d = 0.1I, G_c = G_d = 0.5I$$

and

$$\|F_i\| \leq 1 \text{ for } i = 1, 2, 3, 4.$$

One shall note that unlike the norm-bounded uncertainty by (26) where the uncertain matrix multiplier F is shared, the uncertainties assoicated with system matrices here are independent of each other. It is not difficult to see that this system is equivalent to the uncertain feed-back system (14) to (16) with

$$E_1 = \begin{pmatrix} E_a & 0 & E_b & 0 \end{pmatrix}, \ E_2 = \begin{pmatrix} 0 & E_c & 0 & E_d \end{pmatrix}$$
$$G_1 = \begin{pmatrix} G_a & G_c & 0 & 0 \end{pmatrix}^T, G_2 = \begin{pmatrix} 0 & 0 & G_b & G_d \end{pmatrix}^T$$

and subject to

$$u(t) = Fz(t)$$

where

$$F = \text{diag}(F_1(t)\ F_2(t)\ F_3(t)\ F_4(t)),\ \|F_i\| \leq 1 \text{ for } i = 1, 2, 3, 4.$$

Here we use a bisection process to find the maximum r, denoted by $r_{\max}$, that can satisfy the condition of Proposition 2 for different N. The numerical results are listed in the following table.

N	1	2	3	4
$r_{\max}$	1.5862	1.5916	1.5919	1.5919

Example 2. Consider the singular system with time-delay

$$\begin{pmatrix} 1 & 0 \\ 0 & 0 \end{pmatrix} \dot{x}(t) = (\begin{pmatrix} 0.5 & 0 \\ 0 & -1 \end{pmatrix} + \triangle A_0)x(t) + (\begin{pmatrix} -1.1 & 1 \\ 0 & 0.5 \end{pmatrix} + \triangle A_1)x(t-r),$$

where

$$\triangle A_0 = \begin{pmatrix} 0.1 \\ 0.1 \end{pmatrix} F(t) \begin{pmatrix} 1 & 0 \end{pmatrix},\ \triangle A_1 = \begin{pmatrix} 0.2 \\ 0 \end{pmatrix} F(t) \begin{pmatrix} 1 & 1 \end{pmatrix},$$

with $|F(t)| \leq 1$. Let

$$z(t) = x_1(t),\ \ y(t) = x(t),$$

the above system of can be expressed as

$$\dot{z}(t) = (0.5 + \triangle A)\, z(t) + (\begin{pmatrix} -1.1 & 1 \end{pmatrix} + \triangle B)\, y(t-r),$$
$$y(t) = \left(\begin{pmatrix} 1 \\ 0 \end{pmatrix} + \triangle C\right) z(t) + \left(\begin{pmatrix} 0 & 0 \\ 0 & 0.5 \end{pmatrix} + \triangle D\right) y(t-r),$$

where the uncertainty matrices $\triangle A, \triangle B, \triangle C, \triangle D$ are described by

$$\begin{pmatrix} \triangle A & \triangle B \\ \triangle C & \triangle D \end{pmatrix} = \begin{pmatrix} 1 & 1 \\ 0 & 0 \\ 1 & 0 \end{pmatrix} \tilde{F}(t) \begin{pmatrix} 0.1 & 0 & 0 \\ 0 & 0.2 & 0.2 \end{pmatrix},$$

with

$$\tilde{F}(t) \in \left\{\text{diag}\left(F(t)\ F(t) \right) | \|F(t)\| \leq 1 \right\}. \tag{28}$$

First, the uncertainty-free case is considered. This is investigated in [17], [5], [18] and [19], and the estimates of maximum delay that the system can tolerate without losing its stability were cited in the following table. The last column gives the estimate obtained by applying Proposition 2 with $N = 2$.

method	[17]	[5]	[18]	[19]	N=2
$r_{\max}$	0.5567	0.8708	0.9680	1.0660	1.1216

It can be seen that our result is the least conservative, and indeed is very close to the analytical solution $r_{\max} = 1.121595$.

For the system with time-varying uncertainty of (28), if we treat the uncertainty as norm-bounded and apply the stability conditions in [13], then we have to consider the uncertainty set as

$$\{\tilde{F}(t) \mid \|\tilde{F}(t)\| \leq 1\}.$$

which, however, is much larger than the true uncertainty set (28). The computational results for the the maximum delays $r_{\max}$ with this enlarged uncertainty set are listed in the following table.

N	1	2	3	4
$r_{\max}$	0.6942	0.6950	0.6951	0.6951

In view of the block-diagonal structure of $\tilde{F}(t)$, however, we may apply the stability conditions of Proposition 2. Indeed, $\tilde{F}(t)$ by (28) can be characterized as the simplest form of *Linear repeated blocks* discussed above, whence $\mathcal{K}$ is chosen as

$$\mathcal{K}=\{(K\otimes I_1, K\otimes I_1)\,|K\in\mathbb{R}^{2\times 2}, K>0\}.$$

The numerical results are estimated in the following table.

N	1	2	3	4
$r_{\max}$	0.7712	0.7931	0.7937	0.7938

The conservatism introduced with the enlarged uncertainty assumption is obvious, compared with the result yields immediately above.

>From another perspective, as $F(t)$ is a scalar function, the uncertainty is actually polytopic with two vertices corresponding to $F(t)$ assuming the values of -1 and 1, hence Lemma 1 can also be used to estimate $r_{\max}$ for stability. This gives the same results as using the block-diagonal uncertainty formulation.

4 Discussion on H$_\infty$ Performance

Under the framework discussed so far, there does not appear much more difficulty to discuss the H_∞ problem of the coupled linear time-delay system under dynamical uncertainty. Let u and z in (14) to (16) be partitioned such that

$$z=\begin{pmatrix}z_1 & z_2\end{pmatrix}^T, u=\begin{pmatrix}u_1 & u_2\end{pmatrix}^T,$$

$$\begin{pmatrix}u_1\\u_2\end{pmatrix}=\begin{pmatrix}\Delta_1 & \\ & \Delta_2\end{pmatrix}\begin{pmatrix}z_1\\z_2\end{pmatrix},$$

where $z_k(t)\in\mathbb{R}^{p_k}$, $u_k(t)\in\mathbb{R}^{q_k}$, $\|\Delta_1\|=\frac{1}{\gamma}$ for some given $\gamma>0$ and $\Delta_2\in\mathcal{D}$. Then the stability problem of the system under the uncertainty is equivalent to the H_∞ problem of system (14) to (16)

$$\int_0^t z_1^T(\tau)z_1(\tau)dt\le\gamma^2\int_0^t u_1^T(\tau)u_1(\tau)dt \tag{29}$$

subject to the feedback uncertainty $u_2=\Delta_2(z_2)$. Without loss of generality, we may assume $\gamma=1$. Choose

$$\mathcal{K}_{pq}=\{(\text{diag}(k_1I_{p_1},K_{u_2}),\text{diag}(k_1I_{q_1},K_{z_2})|\;k_1\in\mathbb{R}_+,(K_{u_2},K_{z_2})\in\mathcal{K}_{p_2q_2}\},$$

where K_{u_2} and K_{z_2} are to be specified by the characteristics of Δ_2, then (29) is satisfied for all $\Delta_2\in\mathcal{D}$ if the conditions in Proposition 2 are satisfied for above defined K_u and K_z. To see this, integrate (20) from 0 to t, considering zero initial conditions, to obtain

$$\begin{aligned}
&\int_0^t z_1^T(\tau)z_1(\tau)dt - \int_0^t u_1^T(\tau)u_1(\tau)dt \\
&\le -\varepsilon \int_0^t (\|x(t)\|^2 + \|u_1(t)\|^2 + \|u_2(t)\|^2)d\tau \\
&- V(t, x_t, y(t)) - \int_0^t \left[z_2^T(\tau)K_z z_2(\tau) - u_2^T(\tau)K_u u_2(\tau)\right] d\tau \\
&\le \int_0^t \left[u_2^T(\tau)K_u u_2(\tau) - z_2^T(\tau)K_z z_2(\tau)\right] d\tau,
\end{aligned}$$

since $\Delta_2 \in \mathcal{D}$ guarantees

$$\int_0^t u_2^T(\tau)K_u u_2(\tau)d\tau \le \int_0^t z_2^T(\tau)K_z z_2(\tau)d\tau,$$

then (29) is satisfied.

5 Conclusions

The stability problem of coupled differential-difference equations with block-diagonal uncertainty is discussed. This model included a number of important classes of systems as special cases. Discretized Lyapunov Method is used and the effectiveness of this method is illustrated by examples. The stability conditions derived is extended to discussion of H_∞ performance.

References

1. Brayton, R.: Nonlinear oscillations in a distributed network. Quart. Appl. Math. 24, 289–301 (1976)
2. Boyd, S., EI, G.L., Feron, E., Balakrishnan, V.: Linear Matrix Inequalities in System and Control Theory. SIAM, Philadelphia (1994)
3. Doyle, J.C.: Analysis of feedback systems with structured uncertainties. Proc. IEE, Part D 129(6), 242–250 (1982)
4. Gahinet, P., Nemirovski, A., Laub, A., Chilali, M.: LMI Control Toolbox for Use with MATLAB, Mathworks, Natick, MA (1995)
5. Gao, H.L., Zhu, S.Q., Cheng, Z.L., Xu, B.G.: Delay-dependent state feedback guaranteed cost control uncertain singular time-delay systems. In: Proceedings of IEEE Conference on Decision and Control, and European Control Conference, pp. 4354–4359 (2005)
6. Gu, K.: A further refinement of discretized Lyapunov functional method for the stability of time-delay systems. Int. J. Control 74(10), 967–976 (2001)
7. Gu, K., Kharitonov, V.L., Chen, J.: Stability of Time-Delay Systems. Birkhäuser, Boston (2003)
8. Gu, K., Niculescu, S.I.: Stability Analysis of Time-Delay Systems: A Lyapunov Approach. In: Loría, A., Lamnabhi-Lagarrigue, F., Panteley, E. (eds.) Advanced Topics in Control Systems Theory, Lecture Notes from FAP 2005. Springer, London (2006)
9. Gu, K., Liu, Y.: Lyapunov-Krasovskii Functional for Coupled Differential-Functional Equations. In: Proc. 46th IEEE Conference on Decision and Control, New Orleans, LA, pp. 2083–2088 (2007)
10. Hale, J.K., Verduyn-Lunel, S.: Introduction to Functional Differential Equations. Springer, New York (1993)

11. Krasovskii, N.N.: Stability of Motion (Russian), Moscow (English translation). Stanford University Press, Stanford (1959)
12. Lee, S., Lee, H.S.: Modeling, Design, and Evaluation of Advanced Teleoperator Control Systems with Short Time Delay. IEEE Transactions on Robotics and Automation 9(5), 607–623 (1993)
13. Liu, Y., Li, H.F., Gu, K.: Stability of Differential-Difference Equations with Norm-Bounded Uncertainty. In: Proc. 7th World Congress on Intelligent Control and Automation, Chongqing, China (2008)
14. Niculescu, S.I.: Delay Effects on Stability—A Robust Control Approach. LNCIS, p. 269. Springer, London (2001)
15. Peet, M., Papachristodoulou, A., Lall, S.: On Positive Forms and the Stability of Linear Time-Delay Systems. In: Proc. 45th Conference on Decision and Control, San Diego, CA, pp. 187–193 (2007)sss
16. Răsvan, V.: Dynamical Systems with lossless propagation and neutral functional differential equations. In: Proc. MTNS 1998, Padoue, pp. 527–531 (1998)
17. Zhong, R.X., Yang, Z.: Delay-dependent robust control of descriptor systems with time delay. Asian Journal of Control 8(1), 36–44 (2006)
18. Fridman, E., Shaked, U.: H_∞ control of linear state-delay descriptor systems: an LMI approach. Linear Algebra and Its Applications 351(1), 271–302 (2002)
19. Wu, Z.G., Zhou, W.N.: Delay-dependent Robust Stabilization for Uncertain Singular Systems with State Delay. Acta Automatica Sinica 33(7), 714–718 (2007)
20. Gu, K.: Stability of Linear Time-Delay Systems with Block-Diagonal Uncertainty. Proc. 1998 American Control Conference 3, 1943–1947 (1998)
21. Han, Q., Gu, K.: On Robust Stability of Time-Delay Systems with Block-Diagonal Uncertainty. Proc. 40th IEEE Conference on Decision and Control 4, 3202–3207 (2001)

On Pole Assignment and Stabilizability of Neutral Type Systems

Rabah Rabah[1], Grigory M. Sklyar[2], and Alexander V. Rezounenko[3]

[1] IRCCyN, École des Mines de Nantes, 4 rue Alfred Kastler, 44307 Nantes, France. rabah@emn.fr
[2] Institute of Mathematics, University of Szczecin, 70451 Szczecin, Wielkopolska 15, Poland. sklar@univ.szczecin.pl
[3] Department of Mechanics and Mathematics, Kharkov University, 4 Svobody sqr., Kharkov, 61077, Ukraine. rezounenko@univer.kharkov.ua

Summary. In this note we present a systematic approach to the stabilizability problem of linear infinite-dimensional dynamical systems whose infinitesimal generator has an infinite number of instable eigenvalues. We are interested in strong non-exponential stabilizability by a linear feed-back control. The study is based on our recent results on the Riesz basis property and a careful selection of the control laws which preserve this property. The investigation may be applied to wave equations and neutral type delay equations.

1 Introduction

For the linear finite-dimensional control system

$$\dot{x} = Ax + Bu, \tag{1}$$

the problem of stabilizability is naturally connected to the possibility to move the eigenvalues of A, which are in the closed right-half plane by a linear feedback $u = Fx$ to eigenvalues of $A + BF$ which are in the open left-half plane, such that the closed-loop system $\dot{x} = (A + BF)x$ becomes asymptotically stable. If we can do that with arbitrary given eigenvalues for A and $A + BF$, we say that the system is completely stabilizable (cf. [27]) or that the pole assignment problem is solvable [26]. The last property is connected with the complete controllability: the system is completely stabilizable if it is completely controllable. When the system is not completely controllable, the problem of stabilizability may be solvable if the unstable modes of the spectrum are controllable. One can then obtain asymptotic stability by feedback, which is in fact also exponential stability. It is well known that the situation in infinite dimensional spaces is much more complicated (see for example [3, 27] and references therein). The possibility to move the spectrum is not so simply connected with the controllability property and the last one also is multiply defined (exact, approximative, spectral). The spectrum itself is not sufficient to describe the asymptotic behavior of the solution of the infinite dimensional system even in Hilbert spaces [25]. Our purpose is to analyze the pole assignment problem in the particular situation of a large class of linear neutral type

J.J. Loiseau et al. (Eds.): Topics in Time Delay Systems, LNCIS 388, pp. 85–93.
springerlink.com

systems. Our approach is based one the infinite dimensional framework. Let us first give some precision about the used notions.

Let us suppose that the system (1) is given in Hilbert spaces X, for the state, and U for the control. There are essentially two notions of stabilizability: exponential and strong (asymptotic, non-exponential).

Definition 1. *The system is exponentially stabilizable if for some linear feedback F the semigroup of the closed loop system $\mathrm{e}^{(A+BF)t}$ verifies:*

$$\exists M_\omega > 1, \quad \exists \omega > 0, \quad \forall x, \quad \|\mathrm{e}^{(A+BF)t}x\| \leq M_\omega \mathrm{e}^{-\omega t}\|x\|.$$

The system is strongly stabilizable if

$$\forall x, \quad \|\mathrm{e}^{(A+BF)t}x\| \to 0, \quad \text{as} \quad t \to \infty.$$

For some particular systems, the two notions are equivalent. For example it is the case for linear parabolic partial differential equations with discrete spectrum or for linear retarded systems. For linear neutral type systems and hyperbolic partial differential equations they are different: such systems may be asymptotically stable but not exponentially stable and then the same situation occurs for stabilizability (see for example [1, 18, 23, 25]). This situation is related to the location of the spectrum near the imaginary axis (see our paper about the stability problem [17]). For the neutral type systems, as for other infinite dimensional systems (for example, hyperbolic partial differential systems), the spectrum may contain an infinite set close to the imaginary axis. In [17, 18] we gave an analysis of this situation on stability conditions. It is shown that even complete information on the location of the spectrum of the operator A (and $A + BF$) does not provide the description of the cases when the system is stable or unstable. As clearly indicated in [18, Theorem 22, p.415], there is an example of two systems (of neutral type) which have the same spectrum in the open left-half plane but one of them is asymptotically stable while the other is unstable. The reason for this is that not only the location of the spectrum (eigenvalues in the case of discrete spectrum), but also the geometric characteristics (the structure of eigenspaces and generalized eigenspaces) are important. In this connection we notice that even for particular systems (except finite-dimensional) the complete description of the stability properties is not available at the present time. In such a situation, the stabilizability problem inherits many of the technical difficulties arisen in the study of the stability.

The second main difficulty of the stabilization problem in infinite dimensional spaces is related to the action of the control, namely the dimension of the control variable and the quality of the feedback. So, it is known [24, 20, 14] that one can essentially change the spectrum of the system by use of a feedback only in the case when it is exactly controllable (this means, in particular, that the image of B is infinite dimensional). For example, it is possible to achieve for any $\mu > \mu_0$:

$$\sigma(A + BF) = \sigma(-A^* - \mu I).$$

When the control space is finite dimensional our possibilities to influence spectrum by feedback is restricted. Only a finite number of eigenvalues may be assigned by a linear feedback (bounded or A-bounded, see [11, 3]). The neutral type systems may have an infinite number of unstable eigenvalues that must be moved from the right half plane by finite dimensional feedback. In the present paper the pole assignment in this case is considered.

After a preliminary section where the infinite dimensional model of the neutral type system is given, we consider in Section 3 the case of the abstract system (1) with a one-dimensional control and with an operator A having a Riesz basis of eigenvectors. The second part of Section 3 is concerned with a more general abstract case, when there is no Riesz basis of eigenvectors. Section 4 is devoted to the case of neutral type system, after that we give some concluding remarks.

2 The neutral type system and the infinite dimensional model

We consider the following neutral type system

$$\dot{z}(t) = A_{-1}\dot{z}(t-1) + \int_{-1}^{0} A_2(\theta)\dot{z}(t+\theta)\mathrm{d}\theta + \int_{-1}^{0} A_3(\theta)z(t+\theta)\mathrm{d}\theta \tag{2}$$

where A_{-1} is a constant $n \times n$-matrix, $\det A_{-1} \neq 0$, A_2, A_3 are $n \times n$-matrices whose elements belong to $L_2(-1,0)$.

In our previous work [18] we analyzed asymptotic stability conditions of the system (2). One of the main point of the cited work is the fact that for (2) it may appear asymptotic non exponential stability (see also [1] for the behavior of solutions of a class of neutral type systems). We gave a detailed analysis of non exponential stability in terms of the spectral properties of the matrix A_{-1}. As a continuation of those results we consider in the present work the control system

$$\dot{z}(t) = A_{-1}\dot{z}(t-1) + \int_{-1}^{0} A_2(\theta)\dot{z}(t+\theta)\mathrm{d}\theta + \int_{-1}^{0} A_3(\theta)z(t+\theta)\mathrm{d}\theta + Bu, \tag{3}$$

where B is a $n \times p$-matrix, and study the property for this system of being asymptotic stable after a choice of a feedback control law. Namely, we say that the system (2) is asymptotically stabilizable if there exists a linear feedback control $u(t) = F(z_t(\cdot)) = F(z(t+\cdot))$ such that the system (2) becomes asymptotically stable.

It is obvious that for linear systems in finite dimensional spaces the linearity of the feedback implies that the control is bounded in every neighborhood of the origin. For infinite dimensional spaces the situation is much more complicated. The boundedness of the feedback law $u = F(z_t(\cdot))$ depends on the topology of the state space.

When the asymptotic stabilizability is achieved by a feedback law which does not change the state space and is bounded with respect to the topology of the state space, then we call it *regular* asymptotic stabilizability. Under our assumption on the state space, namely $H^1([-1,0],\mathbb{C}^n)$, the natural linear feedback is

$$Fz(t+\cdot) = \int_{-1}^{0} F_2(\theta)\dot{z}(t+\theta)\mathrm{d}t + \int_{-1}^{0} F_3(\theta)z(t+\theta)\mathrm{d}t, \tag{4}$$

where $F_2(\cdot), F_3(\cdot) \in L_2(-1,0;\mathbb{C}^n)$.

Several authors (see for example [7, 12, 13, 4] and references therein) use feedback laws which for our system may take the form

$$\sum_{i=1}^{k} F_i\dot{z}(t-h_i) + \int_{-1}^{0} F_2(\theta)\dot{z}(t+\theta)\mathrm{d}t + \int_{-1}^{0} F_3(\theta)z(t+\theta)\mathrm{d}t. \tag{5}$$

This feedback law is not bounded in $H^1([-1,0],\mathbb{C}^n)$ and then stabilizability is not regular. If the original system is not formally stable (see [5]), i.e. the pure neutral part (when

$A_2 = A_3 = 0$) is not stable, the non regular feedback (4) is necessary to stabilize. Later we shall return to this issue from an operator point of view.

Let us consider know the operator model of the system (3) used in [18] (see also [2]):

$$\dot{x} = \mathcal{A}x + \mathcal{B}u, \quad x(t) = \begin{pmatrix} y(t) \\ z_t(\cdot) \end{pmatrix}, \tag{6}$$

where $\mathcal{A}$ is the generator of a C_0-semigroup and is defined by

$$\mathcal{A}x(t) = \mathcal{A}\begin{pmatrix} y(t) \\ z_t(\cdot) \end{pmatrix} = \begin{pmatrix} \int_{-1}^{0} A_2(\theta)\dot{z}_t(\theta)\mathrm{d}\theta + \int_{-1}^{0} A_3(\theta)z_t(\theta)\mathrm{d}\theta \\ \mathrm{d}z_t(\theta)/\mathrm{d}\theta \end{pmatrix}, \tag{7}$$

with the domain

$$\mathcal{D}(\mathcal{A}) = \{(y, z(\cdot)) : z \in H^1([-1,0]; \mathbb{C}^n), y = z(0) - A_{-1}z(-1)\} \subset M_2, \tag{8}$$

where $M_2 \stackrel{\text{def}}{=} \mathbb{C}^n \times L_2(-1,0;\mathbb{C}^n)$. The operator $\mathcal{B} : \mathbb{C}^p \to M_2$ is defined by the $n \times p$-matrix B as follows $\mathcal{B}u \stackrel{\text{def}}{=} \left(\begin{smallmatrix} Bu \\ 0 \end{smallmatrix}\right)$. The relation between the solution of the delay system (3) and the system (6) is $z_t(\theta) = z(t+\theta), \theta \in [-1,0]$. This model was used in particular in [18] for the analysis of the stability of the system (2) and in [15] for the analysis of the controllability problems (see also [2, 10]).

>From the operator point of view, the regular feedback law (4) means a perturbation of the infinitesimal generator $\mathcal{A}$ by the operator $\mathcal{BF}$ which is relatively $\mathcal{A}$-bounded (cf. [8]) and verifies $\mathcal{D}(\mathcal{A}) = \mathcal{D}(\mathcal{A} + \mathcal{BF})$. Such a perturbation does not mean, in general, that $\mathcal{A} + \mathcal{BF}$ is the infinitesimal generator of a C_0-semigroup. However, in our case, this fact is verified directly [18, 15] since after the feedback we get also a neutral type system like (2) with $\mathcal{D}(\mathcal{A}) = \mathcal{D}(\mathcal{A} + \mathcal{BF})$ (see below for more details).

>From a physical point of view, $\mathcal{A}$-boundedness of the stabilizing feedback $\mathcal{F}$ means that the energy added by the feedback remains uniformly bounded in every neighborhood of 0 (see also another point of view in [5]). Hence the problem of *regular* asymptotic stabilizability for the systems (3),(6) is to find a linear relatively $\mathcal{A}$-bounded feedback $u = \mathcal{F}x$ such that the operator $\mathcal{A} + \mathcal{BF}$ generates a C_0-semigroup $\mathrm{e}^{(\mathcal{A}+\mathcal{BF})t}$ with $\mathcal{D}(\mathcal{A} + \mathcal{BF}) = \mathcal{D}(\mathcal{A})$ and for which $\|\mathrm{e}^{(\mathcal{A}+\mathcal{BF})t}x\| \to 0$, as $t \to \infty$ for all $x \in \mathcal{D}(\mathcal{A})$.

3 Main approach to the problem of infinite pole assignment

This section presents the main methodology that we propose to solve the problem of infinite pole assignment. We consider first the case when there exists a Riesz basis of eigenvectors and the after that the case where there is only a basis of invariant subspaces.

3.1 Riesz basis of eigenvectors

This approach has been developed in [21] and then essentially extended to a more general systems as will be described in the next sections.

We consider a system

$$\dot{x} = Ax + Bu, \tag{9}$$

where A generates a contractive semigroup $\{e^{At}\}_{t\geq 0}$ in a Hilbert space H; B is a bounded operator from a Hilbert space U to H. We consider equation (9) under the assumptions:

i) A is an unbounded operator with discrete spectrum consisting of simple eigenvalues $\{\lambda_k\}_{k=1}^{\infty}$,

ii) there exists a constant $C_\sigma \equiv \frac{1}{2} \min_{i \neq j} |\lambda_i - \lambda_j| > 0$, i.e. the spectrum is separated,

iii) the space U is one dimensional, so we associate B with a vector $b \in H$; besides, if $\{\phi_n\}_{n=1}^{\infty}$ is an orthonormal eigenbasis $A\phi_n = \lambda_n \phi_n$, then $b_n = \langle b, \phi_n \rangle \neq 0$, $n \in \mathbb{N}$, i.e. the system is approximatively controllable.

One of the key facts in our approach is the possibility to establish the existence of a special Riesz basis of the state space. We recall the definition.

Definition 2. *A basis $\{\psi_j\}$ of a Hilbert space H is called a Riesz basis if there are an orthonormal basis $\{\phi_j\}$ of H and a linear bounded invertible operator R, such that $R\psi_j = \phi_j$.*

If the operator A has a basis of eigenvectors, it is important also that there is a Riesz basis of eigenvectors of the operator $A + bq^*$, where by q^* we denote the functional defined by $q^*x = \langle x, q \rangle$. It is obvious that the vector q must verify some condition. It will be seen from the next sections that this key property could be extended to a more general situation (e.g. neutral type systems) where not only Riesz basis of eigenvectors, but Riesz basis of invariant subspaces should be investigated.

The following assertion is of prime importance for our considerations since it gives a simple characterization of controls which do not destroy the Riesz basis property.

Theorem 1. *Let $||b|| \cdot ||q|| < C_\sigma/2$, where $C_\sigma \equiv \frac{1}{2} \min |\lambda_i - \lambda_j| > 0$ and the family of eigenvectors $\{\phi_n\}$ of A constitute a Riesz basis of H. Then the eigenvectors ψ_k of the operator $\widetilde{A} \equiv A + bq^*$ constitute a Riesz basis of H as well.*

This result was first proved in [21] for the case of a skew-adjoint operator A. Using the Riesz basis property we have the following main result for the system (9) under the above assumptions i)-iii).

Theorem 2. *[21] Let $\{\widetilde{\lambda}_n\}_{n=1}^{\infty}$ be any set of complex numbers such that*

i) $|\lambda_n - \widetilde{\lambda}_n| < C_\sigma$, $n \in \mathbb{N}$;

ii) $\sum_{n=1}^{\infty} \frac{|\lambda_n - \widetilde{\lambda}_n|^2}{|b_n|^2} < \frac{C_\sigma}{||b||^2}$,

where $C_\sigma, b_n \equiv \langle b, \phi_n \rangle$ and λ_n are as in Theorem 1, the family of eigenvectors $\{\phi_n\}$ of A constitute a Riesz basis of H. Then there exists a unique control $u(x) = q^*x$ such that the spectrum $\sigma(\widetilde{A})$ of the operator $\widetilde{A} = A + bq^*$ is $\{\widetilde{\lambda}_n\}_{n=1}^{\infty}$ and, moreover, the corresponding eigenvectors $\widetilde{A}\psi_n = \widetilde{\lambda}_n \psi_n$, constitute a Riesz basis.

This theorem gives the description how to move the (simple) eigenvalues λ_n by using an one-dimensional bounded control (of the form $u(x) = q^*x$) inside of circles of radii proportional to $r_n \cdot b_n$, where $b_n \equiv \langle b, \phi_n \rangle$ and $\{r_n\}_{n=1}^{\infty} \in \ell^2$.

3.2 Riesz basis of invariant subspaces

In this section we present an abstract approach to the stabilization problem for a general operator model and this approach will be used in the next section to stabilize the neutral type system.

The approach presented here is a generalization of the idea proposed in [21] for the problem of infinite pole assignment for the case of the wave (partial differential) equation. In this case, the operator under consideration is skew-adjoint with a simple spectrum while in the present case it does not satisfy neither first nor second assumption. Nevertheless, the main idea of [21], after necessary improvements, allows us to treat more general case including neutral type operator model.

Here we use the notation $\mathcal{A}$ for an operator satisfying the assumptions given below. As it will be shown in the next section, the operator defined in (7) satisfies these assumptions, so the reader mainly interested in the neutral type system may simply look at $\mathcal{A}$ as at the operator (7).

We denote the points

$$\lambda_m^{(k)} = \ln|\mu_m| + \mathrm{i}(\arg \mu_m + 2\pi k), m = 1,..,\ell; k \in \mathbb{Z}$$

and the circles $L_m^{(k)}(r^{(k)})$ centered at $\lambda_m^{(k)}$ with radii $r^{(k)}$, satisfying

$$\sum_{k\in\mathbb{Z}} (r^{(k)})^2 < \infty. \tag{10}$$

Let H be a complex Hilbert space. We consider an infinitesimal generator $\mathcal{A}$ of a C_0-semigroup in H with domain $\mathcal{D}(\mathcal{A}) \subset H$. We have the following assumptions:

H1) The spectrum of $\mathcal{A}$ consists of the eigenvalues only which are located in the circles $L_m^{(k)}(r^{(k)})$, where radii $r^{(k)}$ satisfy (10). Moreover, there exists N_1 such that for any k, satisfying $|k| \geq N_1$, the total multiplicity of the eigenvalues, contained in the circles $L_m^{(k)}(r^{(k)})$, equals $p_m \in \mathbb{N}$, i.e. the multiplicity is finite and does not depend on k.

We need the spectral projectors

$$P_m^{(k)} = \frac{1}{2\pi\mathrm{i}} \int_{L_m^{(k)}} R(\mathcal{A}, \lambda)\mathrm{d}\lambda \tag{11}$$

to define the subspaces $V_m^{(k)} = P_m^{(k)} H$

H2) There exists a sequence of invariant for operator $\mathcal{A}$ finite-dimensional subspaces which constitute a Riesz basis in H. More precisely, there exists N_0 large enough, such that for any $N \geq N_0$, these subspaces are $\{V_m^{(k)}\}_{\substack{|k|\geq N\\ m=1,..,\ell}}$ and W_N, where the last one is the $2(N+1)n$-dimensional subspace spanned by all eigen- and rootvectors, corresponding to all eigenvalues of $\mathcal{A}$, which are outside of all circles $L_m^{(k)}$, $|k| \geq N, m = 1,..,\ell$.

The scalar product and the norm in which all the finite-dimensional subspaces $V_m^{(k)}$ and W_N are orthogonal and form a Riesz basis of subspaces are denoted by $\langle\cdot,\cdot\rangle_0$ and $\|\cdot\|_0$.

H3) The system is of single input, i.e. the operator $\mathcal{B} : \mathbb{C} \to H$ is the operator of multiplication by $\mathbf{b} \in H$.

The main result of this section is the following Theorem (the proof may be found in [19]).

Theorem 3. *[19]* **(On infinite pole assignment).** *Assume the assumptions H1)-H3) are satisfied. Consider an infinite set of circles $L_m^{(k)}(r^{(k)})$ such that each $L_m^{(k)}(r^{(k)})$ contains only one simple eigenvalue of $\mathcal{A}$, i.e. $p_m = 1$. We denote the set of indexes of these circles by $m \in I$.*

We assume that $\mathbf{b} \in H$ is not orthogonal to eigenvectors φ_m^k, of $\mathcal{A}^*$ for $m \in I$ i.e.

$$\left\langle \mathbf{b}, \varphi_m^k \right\rangle_0 \neq 0 \quad \textit{for all} \quad |k| \geq N, m \in I \tag{12}$$

and

$$\lim_{k\to\infty} k \cdot \left|\left\langle \mathbf{b}, \varphi_m^k \right\rangle_0\right| = c_m, \in \mathbb{R} \quad \textit{for all} \quad m \in I. \tag{13}$$

Then there exists $N_2 \geq N$ *such that for any family of complex numbers* $\widetilde{\lambda}_m^k \in L_m^{(k)}(r^{(k)})$, $m \in I, |k| \geq N_2$ *there exists a linear control* $\mathcal{F} : \mathcal{D}(\mathcal{A}) \to \mathbb{C}$, *such that*

1) the complex numbers $\widetilde{\lambda}_m^k$ *are eigenvalues of the operator* $\mathcal{A} + \mathcal{B}\mathcal{F}$;
2) the operator $\mathcal{B}\mathcal{F} : \mathcal{D}(\mathcal{A}) \to H$ *is relatively* $\mathcal{A}$*-bounded.*

The condition (12) means that the eigenvalues to be changed are controllable, and then may be moved. The condition (13) represents a certain boundedness of the control operator.

The result is that the controllable eigenvalues may be moved arbitrarily in some neighborhoods of the initial eigenvalues.

4 Application to neutral type systems

The main contribution of this paper is that under some controllability conditions on the unstable poles of the system, we can assign arbitrarily the eigenvalues of the closed loop system into circles centered at the unstable eigenvalues of the operator $\mathcal{A}$ with radii r_k such that $\sum r_k^2 < \infty$. This is, in some sense, a generalization of the classical pole assignment problem in finite dimensional space. Precisely we have the following

Theorem 4. *Consider the system (3) under the following assumptions:*

1) All the eigenvalues of the matrix A_{-1} *satisfy* $|\mu| \leq 1$.
2) All the eigenvalues $\mu_j \in \sigma_1 \stackrel{\text{def}}{=} \sigma(A_{-1}) \cap \{z : |z| = 1\}$ *are simple (we denote their index* $j \in I$*).*

Then the system (3) is regularly asymptotic stabilizable if

3) $\operatorname{rank}[\Delta_{\mathcal{A}}(\lambda)\ B] = n$ *for all* $\operatorname{Re}\lambda \geq 0$, *where*

$$\Delta_{\mathcal{A}}(\lambda) = -\lambda I + \lambda \mathrm{e}^{-\lambda} A_{-1} + \lambda \int_{-1}^{0} \mathrm{e}^{\lambda s} A_2(s) \mathrm{d}s + \int_{-1}^{0} \mathrm{e}^{\lambda s} A_3(s) \mathrm{d}s,$$

4) $\operatorname{rank}[\mu I - A_{-1}\ B] = n$ *for all* $|\mu| = 1$.

In fact the controllable eigenvalues may be arbitrarily assigned in some neighborhoods of the initial eigenvalues. Let us also precise that the neutral part of the system (namely here, the matrix A_{-1}) is not modified).

5 Conclusion

Under some controllability condition we obtain that some infinite part of the spectrum of a neutral type system may be moved arbitrarily by a finite dimensional regular feedback. The counterpart is that it may be made only in some neighborhoods of the original eigenvalues. But for neutral type systems it is sufficient to insure asymptotic stability provided that the spectrum close to the imaginary axis is simple.

References

1. Brumley, W.E.: On the asymptotic behavior of solutions of differential-difference equations of neutral type. J. Differential Equations 7, 175–188 (1970)
2. Burns, J.A., Herdman, T.L., Stech, H.W.: Linear functional-differential equations as semigroups on product spaces. SIAM J. Math. Anal. 14(1), 98–116 (1983)
3. Curtain, R.F., Zwart, H.: An introduction to infinite-dimensional linear systems theory. Springer, New York (1995)
4. Dusser, X., Rabah, R.: On exponential stabilizability of linear neutral systems. Math. Probl. Eng. 7(1), 67–86 (2001)
5. Loiseau, J.J., Cardelli, M., Dusser, X.: Neutral-type time-delay systems that are not formally stable are not BIBO stabilizable. Special issue on analysis and design of delay and propagation systems. IMA J. Math. Control Inform. 19(1-2), 217–227 (2002)
6. Hale, J.K., Verduyn Lunel, S.M.: Theory of functional differential equations. Springer, New York (1993)
7. Hale, J.K., Verduyn Lunel, S.M.: Strong stabilization of neutral functional differential equations. IMA Journal of Mathematical Control and Information 19(1-2), 5–23 (2002)
8. Kato, T.: Perturbation theory for linear operators. Springer, Heidelberg (1980)
9. Korobov, V.I., Sklyar, G.M.: Strong stabilizability of contractive systems in Hilbert space. Differentsial'nye Uravn. 20, 1862–1869 (1984)
10. Verduyn Lunel, S.M., Yakubovich, D.V.: A functional model approach to linear neutral functional differential equations. Integral Equa. Oper. Theory 27, 347–378 (1997)
11. Nefedov, S.A., Sholokhovich, F.A.: A criterion for stabilizability of dynamic systems with finite-dimensional input. Differentsial'nye Uravneniya 22(2), 223–228 (1986) (Russian); English translation in the same journal edited by Plenum, New York pp. 163–166
12. O'Connor, D.A., Tarn, T.J.: On stabilization by state feedback for neutral differential equations. IEEE Transactions on Automatic Control 28(5), 615–618 (1983)
13. Pandolfi, L.: Stabilization of neutral functional differential equations. J. Optimization Theory and Appl. 20(2), 191–204 (1976)
14. Rabah, R., Karrakchou, J.: On exact controllablity and complete stabilizability for linear systems in Hilbert spaces. Applied Mathematics Letters 10(1), 35–40 (1997)
15. Rabah, R., Sklyar, G.M.: The analysis of exact controllability of neutral type systems by the moment problem approach. SIAM J. Control and Optimization 46(6), 2148–2181 (2007)
16. Rabah, R., Sklyar, G.M., Rezounenko, A.V.: Generalized Riesz basis property in the analysis of neutral type systems. C. R. Math. Acad. Sci. Paris 337(1), 19–24 (2003)
17. Rabah, R., Sklyar, G.M., Rezounenko, A.V.: On strong stability and stabilizability of systems of neutral type. In: Advances in time-delay systems. LNCSE, vol. 38, pp. 257–268. Springer, Heidelberg (2004)
18. Rabah, R., Sklyar, G.M., Rezounenko, A.V.: Stability analysis of neutral type systems in Hilbert space. J. Differential Equations 214(2), 391–428 (2005)
19. Rabah, R., Sklyar, G.M., Rezounenko, A.V.: On strong regular stabilizability for linear neutral type systems. J. Differential Equations 245(3), 569–593 (2008)
20. Sklyar, G.M.: The problem of the perturbation of an element of a Banach algebra by a right ideal and its application to the question of the stabilization of linear systems in Banach spaces, vol. 230, pp. 32–35. Vestn. Khar'kov. Univ. (1982) (Russian)
21. Sklyar, G., Rezounenko, A.: A theorem on the strong asymptotic stability and determination of stabilizing control. C.R. Acad. Sci. Paris, Ser. I. 333, 807–812 (2001)
22. Sklyar, G.M., Rezounenko, A.V.: Strong asymptotic stability and constructing of stabilizing controls. Mat. Fiz. Anal. Geom. 10(4), 569–582 (2003)

23. Sklyar, G.M., Ya, S.V.: On Asymptotic Stability of Linear Differential Equation in Banach Space. Teor, Funk. Funkt. Analiz. Prilozh. 37, 127–132 (1982)
24. Slemrod, M.: A note on complete controllability and stabilizability for linear control systems in Hilbert space. SIAM J. Control 12, 500–508 (1973)
25. van Neerven, J.: The asymptotic behaviour of semigroups of linear operators. In: Operator Theory: Advances and Applications, p. 88. Birkhäuser, Basel (1996)
26. Wonham, W.M.: Linear multivariable control. A geometric approach, 3rd edn. Springer, New York (1985)
27. Zabczyk, J.: Mathematical Control Theory: an introduction. Birkäuser, Boston (1992)

Part II

Numerical Methods and Algorithms

SOS Methods for Stability Analysis of Neutral Differential Systems

Matthew M. Peet[1], Catherine Bonnet[1], and Hitay Özbay[2]

[1] M. M. Peet and C. Bonnet are with INRIA-Rocquencourt, Domaine de Voluceau, Rocquencourt BP105 78153, Le Chesnay Cedex, France. matthew.peet@inria.fr, catherine.bonnet@inria.fr.

[2] H. Özbay is with the Department of Electrical and Electronics Engineering, Bilkent University, Bilkent, 06800 Ankara, Turkey hitay@bilkent.edu.tr.

Summary. This paper gives a description of how "sum-of-squares" (SOS) techniques can be used to check frequency-domain conditions for the stability of neutral differential systems. For delay-dependent stability, we adapt an approach of Zhang et al. [10] and show how the associated conditions can be expressed as the infeasibility of certain semialgebraic sets. For delay-independent stability, we propose an alternative method of reducing the problem to infeasibility of certain semialgebraic sets. Then, using Positivstellensatz results from semi-algebraic geometry, we convert these infeasibility conditions to feasibility problems using sum-of-squares variables. By bounding the degree of the variables and using the Matlab toolbox SOSTOOLS [7], these conditions can be checked using semidefinite programming

1 Introduction

In this paper, we consider the problem of verification of certain frequency-domain tests for stability of delay systems of the neutral type. We show that several frequency-domain conditions can be reduced to optimization problems on the cone of positive semidefinite matrices. Such an approach is an alternative to the classical graphical tests (e.g. the Nyquist criterion). The motivation for using optimization-based methods is as follows

a) The computational complexity of these methods is well-established. Computational complexity provides a standard benchmark for the difficulty of a given problem. A condition which is expressed as an optimization problem, therefore, will have well-known computational properties.
b) While the accuracy of a result based on graphical methods may be limited by the resolution and range of the plot, a feasible result from an optimization-based method serves as a readily verifiable certificate of stability.

The use of frequency domain criteria for analysis of linear systems has an extensive history and we will not make an attempt to catalogue the list of accomplishments in this field. We do note, however, that while for finite-dimensional systems, time-domain-based LMI methods currently compete successfully with frequency-domain-based graphical criteria, the same can not be said to be true for infinite-dimensional systems.

J.J. Loiseau et al. (Eds.): Topics in Time Delay Systems, LNCIS 388, pp. 97–107.
springerlink.com

In [2], we have already considered some simple delay-independent stability conditions for systems of the neutral type using SOS techniques. Our aim here is to develop more sophisticated methods which will allow us to check the delay-dependant H_∞-stability of neutral systems. The paper is organized as follows. We begin in section 2 by recalling some background on polynomial optimization and "sum-of-squares".

In section 3, we show how a method based on dichotomy arguments which was proposed by Zhang *et al* [10] for the stability analysis of retarded-type delay systems can also be applied to the case of neutral-type systems. The main theorems which enable us to formulate delay-dependant stability in terms of feasibility of semi-algebraic sets and SOS polynomials conditions are given. It is shown that the problem of the appearance of asymptotic chains of roots, common for neutral systems, is solved by the Zhang method of rational over-approximation. In Section 4 we give some numerical examples which show the efficacy of the proposed method. Finally a conclusion is given in section 5.

2 The Positivstellensatz and Sum-of-Squares

A polynomial, p, is said to be *positive* on $G \subset \mathbb{R}^n$ if

$$p(x) \geq 0 \quad \text{for all } x \in G.$$

If G is not mentioned, then it is assumed $G = \mathbb{R}^n$. A *semialgebraic set* is a subset of $\mathbb{R}^n$ defined by polynomials p_i, as

$$G := \{x \in \mathbb{R}^n \ : p_i(x) \geq 0, \, i = 1, \ldots, k\}.$$

Given a polynomial, the question of whether it is positive has been shown to be NP-hard. A polynomial, p, is said to be sum-of-squares (SOS) in variables x, denoted $p \in \Sigma_s[x]$ if there exist a finite number of other polynomials, g_i such that

$$p(x) = \sum_{i=1}^{k} g_i(x)^2.$$

A sum-of-squares polynomial is positive, but a positive polynomial may not be sum-of-squares. A necessary and sufficient condition for the existence of a sum-of-squares representation for a polynomial, p, of degree $2d$ is the existence of a positive semidefinite matrix, Q, such that

$$p(x) = Z(x)^T Q Z(x),$$

where Z is any vector whose elements form a basis for the polynomials of degree d. Positivstellensatz results are "theorems of the alternative" which say that either a semialgebraic set is feasible or there exists a sum-of-squares refutation of feasibility. The Positivstellensatz that we use in this paper is that given by Stengle [9].

Theorem 1 (Stengle). The following are equivalent

1.
$$\left\{x : \begin{matrix} p_i(x) \geq 0 \;\; i = 1, \ldots, k \\ q_j(x) = 0 \;\; j = 1, \ldots, m \end{matrix}\right\} = \emptyset$$

2. There exist $t_i \in \mathbb{R}[x]$, $s_i, r_{ij}, \ldots \in \Sigma_s[x]$ such that

$$-1 = \sum_i q_i t_i + s_0 + \sum_i s_i p_i + \sum_{i \neq j} r_{ij} p_i p_j + \cdots$$

We use $\mathbb{R}[x]$ to denote the real-valued polynomials in variables x. For a given degree bound, the conditions associated with Stengle's Positivstellensatz can be represented as a semidefinite program. Note that, in general, no such upper bound on the degree bound will be known a-priori.

3 Stability of Neutral-Type Systems

In this section, we consider aspects of the stability of general neutral-type systems of the form

$$\dot{x}(t) = \sum_{i=1}^{m} B_i \dot{x}(t - \tau_i) + \sum_{j=0}^{m} A_j x(t - \tau_j) \tag{1}$$

where $A_i, B_j \in \mathbb{R}^{n \times n}$, $\tau_0 = 0$ and $\tau_i \geq 0$.

3.1 Dichotomy Methods

The term *dichotomy* is used to describe methods based on the bisection of the complex plane into right and left half-planes. These methods are based on a continuity argument that states that if a system is stable for one value of a parameter and unstable for another value, then for some intermediate value of the parameter, the system must have a pole on the imaginary axis. For retarded and for neutral-type systems which satisfy certain conditions, this argument is valid due to the following theorem from Datko [3]. Let

$$G_\alpha(s) := \left[s \left(I - \sum_{i=1}^{m} B_i e^{-\alpha s \gamma_i} \right) - \sum_{j=0}^{m} A_j e^{-\alpha s \gamma_j} \right]. \tag{2}$$

Theorem 2 (Datko). Consider G_α as given by Equation (2). If

$$\det \left[I - \sum_{i=1}^{m} B_i e^{-s\gamma_i} \right] = 0$$

has all roots lying in some left half-plane $\operatorname{Re} s \in (-\infty, -\beta_0]$, $\beta_0 > 0$, then

$$\sigma(G_\alpha) := \sup_{\det G_\alpha(s)=0} \operatorname{Re} s$$

is continuous on $\alpha \in [0, \infty)$.

In general, a system of the form 1 will define a transfer function. The poles of this transfer function will be defined by the roots of $\det G(s) := \sum_{i=0}^{n} q_i(s) e^{-\tau_i s}$ where G_α is given by Equation (2). Here the q_i are complex polynomials with $\deg q_0 \geq \deg q_i$ for $i = 1, \ldots, n$. In general, for neutral-type systems, $\deg q_0 = \deg q_i$ for at least one $i \geq 1$.

3.2 Generalization of a method of Zhang et al.

In this subsection, we consider the approach of [10], wherein the exponential term is "covered" by a set of rational transfer functions parameterized by a single parameter. The size of the set, or the range of values of the parameter, is determined by the degree of the rational functions. We begin with the definition of the Padé approximate of e^{-s}.

$$R_m(s) = \frac{P_m(s)}{P_m(-s)}$$

where

$$P_m(s) = \sum_{k=0}^{m} \frac{(2m-k)!m!(-s)^k}{(2m)!k!(m-k)!}.$$

Now define the sets of irrational and rational functions.

$$\Omega_d(\omega, h) := \{e^{-\imath\tau\omega} \ : \tau \in [0, h]\}$$
$$\Omega_m(\omega, h) := \{R_m(\imath\alpha_m\tau\omega) \ : \ \tau \in [0, h]\}$$

Where $\alpha_m := \frac{1}{2\pi}\min\{\omega > 0 \,|\, R_m(\imath\omega) = 1\}$. For example, $\alpha_3 = 1.2329$, $\alpha_4 = 1.0315$, and $\alpha_5 = 1.00363$.

The following is a key result of [10].

Lemma 13 (Zhang et al.). For every integer $m \geq 3$, the following statements hold.

1. All poles of $R_m(s)$ are in the open left half complex plane.
2. $\Omega_d(\omega, h) \subset \Omega_m(\omega, h)$ for any $h \geq 0$ and $\omega \geq 0$.
3. $\lim_{m\to\infty} \alpha_m = 0$

Typically, Lemma 13 is used to prove that a delay-differential system has no poles in the closed right half-plane for an interval of delay of the form $[0, h]$. This is illustrated by the following theorem.

Theorem 3. Suppose G, as given by Equation (2), satisfies the conditions of Theorem 2 and $\det G(s) := \sum_{i=0}^{n} q_i(s)e^{-\tau_i s}$. Let $m \geq 3$, and suppose that

$$\{s \in C \, : \, \sum_{i=0}^{n} q_i(s) = 0, \; \mathrm{Re}\, s \geq 0\} = \emptyset$$

and

$$\{\omega \geq 0, \; \tau_i \in [0, \, h_i\,] \, : \, \sum_{i=0}^{n} q_i(\omega\imath) R_m(\alpha_m\tau_i\omega\imath) = 0\} = \emptyset.$$

Then

$$\{s \in C, \; \tau_i \in [0, \, h_i\,] \, : \, \sum_{i=0}^{n} q_i(s)e^{-\tau_i s} = 0, \; \mathrm{Re}\, s \geq 0\} = \emptyset.$$

Theorem 3 is a trivial generalization of the work of Zhang et al. [10] to neutral-type systems. For retarded-type systems, the work of [10] proposed the construction of a parameter-dependent state-space system. In a later work, [1] proposed the use of a generalized version of the KYP lemma to check the conditions associated with the retarded case through the construction of a perturbed singular system. In this paper, we use a "sum-of-squares" approach based on the application of the Positivstellensatz results described in Section 2.

For convenience, define the following functions.

$$g_r(\omega,\tau) := \mathrm{Re}\Big(\sum_{i=0}^{n} q_i(\omega\imath) P_m(\alpha_m \tau_i \omega\imath) \prod_{\substack{j=0\\ j\neq i}}^{n} P_m(-\alpha_m \tau_j \omega\imath)\Big)$$

$$g_i(\omega,\tau) := \mathrm{Im}\Big(\sum_{i=0}^{n} q_i(\omega\imath) P_m(\alpha_m \tau_i \omega\imath) \prod_{\substack{j=0\\ j\neq i}}^{n} P_m(-\alpha_m \tau_j \omega\imath)\Big)$$

Lemma 14. The following are equivalent

1. The following set is infeasible.

$$\{\omega \geq 0,\ \tau_i \in [0,\ h_i\,] \ :\ \sum_{i=0}^{n} q_i(\omega\imath) R_m(\alpha_m \tau_i \omega\imath) = 0\}$$

2. The following real semi-algebraic set is infeasible.

$$\Big\{\omega, \tau_i \in \mathbb{R} : \omega \geq 0,\ \ \tau(h_i - \tau_i) \geq 0,\ g_r(\omega,\tau_i) = 0,\ g_i(\omega,\tau_i) = 0\Big\}$$

3. There exist polynomials $t_1, t_2 \in \mathbb{R}[\omega,\tau]$ and SOS polynomials $s_i \in \Sigma_s[\omega,\tau]$ such that

$$\begin{aligned} -1 =& s_0 + t_1 g_r + t_2 g_i + \omega s_1 + \tau_0(h_0 - \tau_0)s_2 \\ &+ \omega\tau_0(h_0 - \tau_0)s_3 + \omega\tau_1(h_1 - \tau_1)s_4 + \cdots \end{aligned}$$

Proof. By Lemma 13, all roots of $P_m(-s)$ are in the open left half complex plane, and so Statement 1 is equivalent to infeasibility of the following set.

$$\Big\{\omega \geq 0,\ \tau_i \in [0,\ h_i\,] \ :\ \sum_{i=0}^{n} q_i(\omega\imath) P_m(\alpha_m \tau_i \omega\imath) \prod_{\substack{j=0\\ j\neq i}}^{n} P_m(-\alpha_m \tau_j \omega\imath) = 0\Big\}$$

Now, $\tau \in [0, h]$ is equivalent to $\tau(h - \tau) \geq 0$ and so Statement 1 is equivalent to infeasibility of the set in Statement 2. Furthermore, the real or imaginary part of a complex polynomial is a polynomial in the real and complex parts of the complex argument. Therefore, the set in Statement 2 is real semi-algebraic.

That Statement 2 is equivalent to Statement 3 is an immediate consequence of Theorem 1.

We now give delay-dependent stability conditions for neutral-type systems.

Lemma 15. Define

$$r_m := \inf\left\{\left\|\sum_{i=0}^{n} q_i(\omega\imath) R_m(\alpha_m \tau_i \omega\imath)\right\| \ :\ \omega \geq 0,\ \tau_i \in [0,\ h_i\,]\right\}$$

and

$$r_e := \inf\left\{\left\|\sum_{i=0}^{n} q_i(\omega\imath) e^{-\tau_i \omega\imath}\right\| \ :\ \omega \geq 0,\ \tau_i \in [0,\ h_i\,]\right\}.$$

Then for any $m \geq 3$, $r_e \geq r_m$.

Proof. Suppose there exists an $\omega_w \geq 0$ and $\tau_{i,w} \in [0, h_i]$ such that

$$w = \left\| \sum_{i=0}^{n} q_i(\omega_w \imath) e^{-\tau_{i,w}\omega_w \imath} \right\|$$

Now, by Lemma 13, there exists $\tau'_{i,w} \in [0, h_i]$ such that

$$\left\| \sum_{i=0}^{n} q_i(\omega_w \imath) R_m(\alpha_m \tau'_{i,w} \omega_w \imath) \right\| = \left\| \sum_{i=0}^{n} q_i(\omega_w \imath) e^{-\tau_{i,w}\omega_w \imath} \right\| = w$$

Therefore, $r_m \leq w$ and hence $r_m \leq r_e$.

Theorem 4. Suppose G_τ, given by Equation (2), satisfies the conditions of Theorem 2 and $\det G_\tau(s) := \sum_{i=0}^{n} q_i(s) e^{-\tau_i s}$. Let $m \geq 3$ and suppose that

$$\{s \in C : \sum_{i=0}^{n} q_i(s) = 0, \operatorname{Re} s \geq 0\} = \emptyset$$

and that there exist polynomials $t_1, t_2 \in \mathbb{R}[\omega, \tau]$ and SOS polynomials $s_i \in \Sigma_s[\omega, \tau]$ such that

$$\begin{aligned} -1 =& s_0 + t_1 g_r + t_2 g_i + \omega s_1 + \tau_1(h_1 - \tau_1) s_2 \\ &+ \omega \tau_1 (h_1 - \tau_1) s_3 + \omega \tau_2 (h_2 - \tau_2) s_4 + \cdots \end{aligned}$$

Then the system defined by Equation 1 is H_∞ stable for all $\tau_i \in [0, h_i]$

Proof. If the conditions of the theorem are satisfied, then by Lemma 14

$$H_m(s) := \sum_{i=0}^{n} q_i(s) R_m(\alpha_m \tau_i s)$$

has no roots on the imaginary axis for any $\tau_i \in [0, h_i]$. Therefore, by Theorem 3 $\det G_\tau(s)$ has no roots on the closed right half plane. Therefore $\det G_\tau(s)^{-1}$ is analytic on the closed right half plane. It remains to show that $\det G_\tau(s)^{-1}$ is bounded on the imaginary axis.

Since H_m is a polynomial, then for any $\tau_i \in [0, h_i]$, it has a finite number of roots, none of which are on the imaginary axis. We conclude that for any fixed $\tau_i \in [0, h_i]$,

$$\inf_{\omega \geq 0} \left\| \sum_{i=0}^{n} q_i(\omega \imath) R_m(\alpha_m \tau_i \omega \imath) \right\| > 0$$

Therefore, since the $[0, h_i]$ are compact sets,

$$r_m := \inf \left\{ \left\| \sum_{i=0}^{n} q_i(\omega \imath) R_m(\alpha_m \tau_i \omega \imath) \right\| \ : \omega \geq 0,\ \tau_i \in [0,\ h_i] \right\} > 0.$$

Therefore, by Lemma 15,

$$r_e := \inf \left\{ \left\| \sum_{i=0}^{n} q_i(\omega \imath) e^{-\tau_i \omega \imath} \right\| \ : \omega \geq 0,\ \tau_i \in [0,\ h_i] \right\} \geq r_m > 0.$$

This proves that

$$\|\det G_\tau^{-1}\|_\infty = \sup_{\omega \geq 0} \|\det G_\tau(\imath\omega)\|^{-1} \leq \frac{1}{r_m}$$

and so $\det G_\tau(s)^{-1} \in H_\infty$ for any $\tau_i \in [0, h_i]$.

3.3 Delay-Independent Stability

To check stability independent of delay, the conditions of Theorem 4 may be modified slightly by considering the semialgebraic set

$$\Big\{\omega, \tau_i \in \mathbb{R} : \omega \geq 0,\ \ \tau_i \geq 0,\ g_r(\omega, \tau_i) = 0,\ g_i(\omega, \tau_i) = 0\Big\}.$$

In this section, we propose an alternative approach which eliminates the need for potentially high-order rational approximations. We begin by noting the following proposition which gives conditions that, when combined with the results of Datko, can be used to prove that no poles enter the right half-plane for any value of delay.

Proposition 1. *The following conditions are equivalent*

- *The set* $\{\omega \in \mathbb{R},\, \tau \in \mathbb{R}^k\ :\ \omega \neq 0,\ \tau_k \geq 0\ for\ k = 1, \ldots, n,\ \sum_{k=0}^{n} q_k(\imath\omega)e^{-\imath\omega\tau_k} = 0\}$ *is empty.*
- *The set* $\{\omega \in \mathbb{R},\, \tau \in \mathbb{C}^k\ :\ \omega \neq 0,\ |z_k| = 1,\ \sum_{k=0}^{n} q_k(s)z_k = 0\}$ *is empty.*

Proof. We prove 2) $\Longrightarrow$ 1) by contrapositive. Suppose 1) is false for some ω_0, τ_k. Then let $z_k = e^{-\tau_k\omega_0\imath}$ $\omega = \omega_0$, which is feasible for 2). Therefore, by contrapositive 2) $\Longrightarrow$ 1).

For 1) $\Longrightarrow$ 2), suppose 2) is false. Then there exists $\omega_0 \neq 0$ and $z \in \mathbb{C}^k$ with $|z_k| = 1$ such that $\sum_{k=0}^{n} q_k(\imath\omega_0)z_k = 0$. We can write $z_k = e^{-\imath a_k}$ with $\operatorname{sign} a_k = \operatorname{sign} \omega_0$. Let $\tau_k = \frac{a_k}{\omega_0}$ and $\omega = \omega_0$. Then we get

$$\sum_{k=0}^{n} q_k(\imath\omega)e^{-\imath\omega\tau_k} = \sum_{k=0}^{n} q_k(\imath\omega_0)e^{-\imath\omega_0\tau_k} = \sum_{k=0}^{n} q_k(\imath\omega_0)e^{-\imath a_k} = \sum_{k=0}^{n} q_k(\imath\omega_0)z_k = 0.$$

This contradicts 1) and completes the proof.

Proposition 1 is incomplete in that it does not prove H_∞ or exponential stability. This is because neutral-type delay systems can have infinite roots which may approach the imaginary axis, creating an unbounded high frequency response. This problem was avoided in the previous section by using rational approximations. To deal with this issue, we consider the case of commensurate delays. This is made explicit by the following Lemma.

Lemma 1. *The following conditions are equivalent*

1. *The set* $\{s \in \imath\mathbb{R}^+,\, \tau_k \in \mathbb{R}^+\ :\ \sum_{k=0}^{n} q_k(s)e^{-sk\tau} = 0\}$ *is empty.*
2. *The following semialgebraic set is empty*

$$\Big\{\omega, z_i, z_r \in \mathbb{R}\ :\ 1 - z_i^2 + z_r^2 = 0, \qquad \operatorname{Re}\Big(\sum_{k=0}^{n} q_k(\imath\omega)(z_r + \imath z_i)^k\Big) = 0,$$
$$\operatorname{Im}\Big(\sum_{k=0}^{n} q_k(\imath\omega)(z_r + \imath z_i)^k\Big) = 0\Big\}$$

3. There exist polynomials $t_1, t_2, t_3 \in \mathbb{R}[\omega, z_r, z_i]$ *and* $s \in \Sigma_s$ *such that*

$$\begin{aligned}-1 =& (z_r^2 + z_i^2 - 1)\, t_1 - \operatorname{Re}\Big(\sum_{k=0}^{n} q_k(\imath\omega)(z_r + \imath z_i)^k\Big)\, t_2 \\ &+ \operatorname{Im}\Big(\sum_{k=0}^{n} q_k(\imath\omega)(z_r + \imath z_i)^k\Big)\, t_3 + s. \end{aligned} \tag{3}$$

The proof is an application of Theorem 1. This leads to a simple condition for stability of neutral systems with commensurate delays.

Theorem 5. Suppose G_τ, as given by Equation (2), satisfies the conditions of Theorem 2 and $\det G_\tau(s) := \sum_{k=0}^{n} q_i(s) e^{-k\tau s}$. Suppose that

$$\{s \in C : \sum_{i=0}^{n} q_i(s) = 0,\ \operatorname{Re} s \geq 0\} = \emptyset$$

and that there exist polynomials $t_1, t_2, t_3 \in \mathbb{R}[\omega, z_r, z_i]$ and $s \in \Sigma_s$ such that

$$\begin{aligned}-1 =& (z_r^2 + z_i^2 - 1)\, t_1 - \operatorname{Re}\Big(\sum_{k=0}^{n} q_k(\imath\omega)(z_r + \imath z_i)^k\Big)\, t_2 \\ &+ \operatorname{Im}\Big(\sum_{k=0}^{n} q_k(\imath\omega)(z_r + \imath z_i)^k\Big)\, t_3 + s. \end{aligned} \tag{4}$$

Then the system defined by Equation 1 is both H_∞ and exponentially stable for all $\tau \geq 0$

Proof. The proof is similar to that of Theorem 4. By combining Theorem 2, Lemma 1, and the assumption, G_τ has no roots on the imaginary axis for any positive value of τ. Moreover, for the case of commensurate delays, it can be shown that the conditions of Theorem 2 imply that a lack of roots in the closed right half-plane implies both exponential and H_∞ stability.

3.4 Verifying the Datko Conditions

In this section, we briefly consider the problem of verifying the conditions associated with Theorem 2. In particular, we would like to show that

$$\det\left[I - \sum_{i=1}^{m} B_i e^{-s\gamma_i}\right] = 0$$

has all roots lying in some left half-plane $\operatorname{Re} s \in (-\infty, -\beta_0]$, $\beta_0 > 0$.

A simple sufficient condition, also proposed in [3], is given by the following.

Proposition 16. Suppose $B_i = 0$ for $i = 1, \dots m-1$ and $\det(\lambda I - B_m)$ has all roots in the disc $|\lambda| < 1$. Then the conditions of Theorem 2 are satisfied.

More generally, if the delays are commensurate, then the condition is equivalent to exponential stability of an expanded discrete-time linear system of the general form

$$\begin{bmatrix} x_1(k+1) \\ \vdots \\ x_n(k+1) \end{bmatrix} = \begin{bmatrix} B_1 & \cdots & B_n \\ I & & \\ & I & \end{bmatrix} \begin{bmatrix} x_1(k) \\ \vdots \\ x_n(k) \end{bmatrix}.$$

If the delays are non-commensurate, then an SOS condition can also be given.

Lemma 17. Let $\epsilon > 0$. If

$$\left\{ z_i \in \mathbb{C} : \det\left(I + \sum_{i=1}^{m} B_i z_i \right) = 0, |z_i| \le e^{\epsilon T_i} \right\} = \emptyset,$$

then

$$\det\left[I + \sum_{i=1}^{m} B_i e^{-s\gamma_i} \right] = 0$$

has all roots in the left half-plane $\{\mathrm{Re}\, s \le -\epsilon\}$ for all $\gamma_i \le T_i$.

Proof. Proof by contradiction. Suppose that there exist $(\gamma_1, \cdots, \gamma_n) \in \mathbb{R}^n$ with $\gamma_i \le T_i$ for $i = 1, \cdots, m$ and $s_0 \in \{\mathrm{Re}\, s \ge -\epsilon\}$ such that

$$\det\left(I + \sum_{i=1}^{m} B_i e^{-\gamma_i s_0} \right) = 0.$$

Let $s = s_0$ and $z_i = e^{-\gamma_i s_0}$. Then $s_0 \in \{\mathrm{Re}\, s \ge -\epsilon\}$ and $|z_i| \le e^{\epsilon T_i}$ with

$$\det\left(I + \sum_{i=1}^{n} B_i z_i \right) = 0,$$

which contradicts the statement of the lemma.

The conditions of Lemma 17 can be verified using SOS by application of Theorem 1.

4 Numerical Example

Table 1. Stability Regions

	$m = 3$	$m = 4$	$m = 5$	actual value
maximum τ	1.805	2.157	2.217	2.2255
required degree	14	18	22	

Example 1: In this example, we consider a somewhat arbitrarily chosen example to illustrate the delay-dependent stability condition associated with Theorem4. We use the example chosen by [4], [6], and [5] among many others.

$$\dot{x}(t) - B\dot{x}(t-\tau) = A_0 x(t) + A_1 x(t-\tau)$$

where

$$A_0 = \begin{bmatrix} -0.9 & 0.2 \\ 0.1 & -0.9 \end{bmatrix}, \quad A_1 = \begin{bmatrix} -1.1 & -0.2 \\ -0.1 & -1.1 \end{bmatrix},$$
$$B = \begin{bmatrix} -0.2 & 0 \\ 0.2 & -0.1 \end{bmatrix}.$$

To test stability, we first verify the Datko conditions, which hold by Proposition 16. We now replace the characteristic equation

$$g(s) = \det\left(sI - A_0 - A_1 e^{-\tau s} - Bse^{-\tau s}\right)$$

with the family of rational approximations R_m to get a new characteristic polynomial family

$$H_m(s,\tau) = \det\Big(sP_m(-\alpha_m \tau s)I - A_0 P_m(-\alpha_m \tau s) \\ - A_1 P_m(\alpha_m \tau s) - BsP_m(\alpha_m \tau s)\Big)$$

We then create the real polynomial functions

$$g_i(\omega,\tau) := \mathrm{Re}\, H_m(\imath\omega,\tau)$$

and

$$g_r(\omega,\tau) := \mathrm{Re}\, H_m(\imath\omega,\tau)$$

This can be done automatically in Matlab using the function `cpoly2rpoly` contained in the software package available online at [8].

We now use SOSTools [7] to find polynomials $t_1, t_2 \in \mathbb{R}[\omega,\tau]$ and SOS polynomials $s_i \in \Sigma_s[\omega,\tau]$ for $i = 0,\ldots,3$ such that

$$-1 = s_0 + t_1 g_r + t_2 g_i + \omega s_1 + \tau(h-\tau)s_2 + \omega\tau(h-\tau)s_3$$

By using this method, we are able to prove stability of the system for the values of delay listed in Table 1. This table also lists the degree of the refutation necessary. The accuracy is roughly comparable to what is currently available using existing time-domain methods. Note that the accuracy is restricted only by the value of α_m.

5 Conclusion

We have a proposed a method to check delay-dependant stability of neutral-type delay systems which involves the Padé approximate of e^{-s} and uses sum-of-squares methods to prove the infeasibility of certain semi-algebraic sets. The method is applied to a standard example from the literature.

References

1. Bliman, P.A., Iwasaki, T.: LMI characterisation of robust stability for time-delay systems: singular perturbation approach. In: Proc. IEEE Conference on Decision and Control, San Diego, USA (2006)
2. Bonnet, C., Peet, M.M.: Using the positivstellensatz for stability analysis of neutral delay systems in the Frequency Domain. In: Proc. 7th IFAC Workshop on Time-Delay Systems, Nantes, France (2007)
3. Datko, R.: A procedure for determination of the exponential stability of certain differential-difference equations. Quarterly of Applied Mathematics 36, 279–292 (1978)
4. Fridman, E.: New Lyapunov-Krasovskii functionals for stability of linear retarded and neutral-type systems. Automatica 43, 309–319 (2001)

5. Han, Q.L.: On stability of linear neutral systems with mixed delays: a discretized Lyapunov functional approach. Automatica 41(7), 1209–1218 (2005)
6. He, Y., Wu, M., She, J.H., Liu, G.P.: Delay-dependent robust stability criteria for uncertain neutral systems with mixed delays. Systems & Controls Letters 51(1), 57–65 (2004)
7. Parrilo, P.A.: Web Site for SOSTOOLS (2004), `http://www.cds.caltech.edu/sostools`
8. Peet, M.: Web Site for Matthew M. Peet. Url:http: `http://www-rocq.inria.fr/~peet`
9. Stengle, G.: A nullstellensatz and a positivstellensatz in semialgebraic geometry. Mathematische Annalen 207, 87–97 (1973)
10. Zhang, J., Knospe, C., Tsiotras, P.: Stability of linear time-delay systems: a delay-dependant criterion with a tight conservatism bound. In: Proc. American Control Conference, Anchorage, USA (2002)

Nonlinear Time-Delay Systems: A Polynomial Approach Using Ore Algebras

Miroslav Halás

Institute of Control and Industrial Informatics
Faculty of Electrical Engineering and Information Technology
Slovak University of Technology
Ilkovičova 3, 812 19 Bratislava, Slovakia
miroslav.halas@stuba.sk

Summary. In this work Ore algebra defined over the field of meromorphic functions is introduced to provide a polynomial approach to nonlinear time-delay systems. In comparison to the systems without delays the polynomials related to the nonlinear time-delay systems belong now to the multivariate skew polynomial ring and have the sense of differential time-delay operators. Then the condition for accessibility of nonlinear time-delay systems is presented in terms of common left factors of such polynomials. Transfer functions of nonlinear time-delay systems are introduced as well and their basic properties are shown.

1 Introduction

Algebraic approach of differential forms, originally developed for nonlinear systems without delays, [5] and [1], was recently extended to the case of time-delay systems [13, 14, 15, 18] and was shown to be effective in solving control problems like accessibility and observability, disturbance decoupling, etc.

On the other side, in the case of systems without delays, there exists, in comparison to the machinery of one-forms, an alternative approach in which the system properties are described by skew polynomials from non-commutative polynomial rings. Such polynomials act as differential [19, 20] or shift [11, 12] operators on the differentials of the system inputs and outputs. This approach allows us, for instance, to introduce the accessibility condition expressed in terms of common left factors of skew polynomials derived from the input-output equation. Moreover, the polynomial approach, after defining quotients of skew polynomials, provides also possibility to introduce transfer functions of nonlinear systems, both continuous [6, 7] and discrete-time [8, 9]. Such transfer functions show many properties we expect from transfer functions, like the invariance to state transformations, transfer function algebra and others and was, for instance, already used in [10] to recast and solve the nonlinear model matching problem.

However, a similar polynomial approach is not yet available for nonlinear time-delay systems and in what follows, it is, therefore, extended also to this case.

J.J. Loiseau et al. (Eds.): Topics in Time Delay Systems, LNCIS 388, pp. 109–119.
springerlink.com

2 Algebraic setting

The mathematical setting, to be used in this paper for dealing with nonlinear time-delay systems , was recently introduced in [13, 14, 15, 18] and will be now briefly reviewed. In order to avoid technicalities, we use slightly abbreviated notations and the reader is referred to those works for detailed technical constructions which are not found here.

The nonlinear time-delay systems considered in this paper are objects of the form

$$\begin{aligned} \dot{x}(t) &= f(\{x(t-i), u(t-j); i,j \geq 0\}) \\ y(t) &= g(\{x(t-i), u(t-j); i,j \geq 0\}) \end{aligned} \tag{1}$$

where the entries of f and g are meromorphic functions and $x \in \mathbf{R}^n$, $u \in \mathbf{R}^m$ and $y \in \mathbf{R}^p$ denote state, input and output to the system.

Note that it is not restrictive to assume $i, j \in \mathbf{N}$ since all the delays can be considered as multiples of an elementary delay h [14].

Let $\mathcal{K}$ be the field of meromorphic functions of $\{x(t-i), u^{(k)}(t-j); i,j,k \geq 0\}$ and let $\mathcal{E}$ be the formal vector space over $\mathcal{K}$ given by

$$\mathcal{E} = \operatorname{span}_{\mathcal{K}}\{\mathrm{d}\xi; \xi \in \mathcal{K}\}$$

The delay operator δ is defined on $\mathcal{K}$ and $\mathcal{E}$ as

$$\begin{aligned} \delta(\xi(t)) &= \xi(t-1) \\ \delta(\alpha(t)\mathrm{d}\xi(t)) &= \alpha(t-1)\mathrm{d}\xi(t-1) \end{aligned} \tag{2}$$

for any $\xi(t) \in \mathcal{K}$ and $\alpha(t)\mathrm{d}\xi(t) \in \mathcal{E}$. The delay operator (2) induces the non-commutative polynomial ring $\mathcal{K}[\delta]$ with the multiplication given by the commutation rule

$$\delta a(t) = a(t-1)\delta$$

for any $a(t) \in \mathcal{K}$. The ring $\mathcal{K}[\delta]$ thus represents the ring of linear shift (delay) operators.

Properties of the system (1) can be now analyzed by introducing the machinery of one-forms known from systems without delays [5, 1]. This time, rather than vector spaces we introduce modules over $\mathcal{K}[\delta]$, generally

$$\mathcal{M} = \operatorname{span}_{\mathcal{K}[\delta]}\{\mathrm{d}\xi; \xi \in \mathcal{K}\}$$

Such an approach was shown to be effective in solving a number of control problems, like accessibility and observability, disturbance decoupling, to name a few. See for instance [13, 14, 15, 18].

2.1 Pseudo-linear algebra

To handle different types of linear operators, as for instance differential, shift, difference, q-shift, we, to advantage, introduce univariate skew polynomial rings, or Ore rings [2]. Such structures allow us to handle those types of operators from a uniform standpoint.

Definition 1. *Let K be a field and $\sigma : K \to K$ an automorphism of K. A map $\delta : K \to K$ which satisfies*

$$\begin{aligned} \delta(a+b) &= \delta(a) + \delta(b) \\ \delta(ab) &= \sigma(a)\delta(b) + \delta(a)b \end{aligned} \tag{3}$$

is called a pseudo-derivation (or a σ-derivation).

Definition 2. *The left skew polynomial ring given by σ and δ is the ring $K[x;\sigma,\delta]$ of polynomials in x over K with the usual addition, and the multiplication given by the commutation rule*

$$xa = \sigma(a)x + \delta(a) \tag{4}$$

for any $a \in K$.

Elements of such a ring are called skew polynomials or non-commutative polynomials or Ore polynomials [16, 17]. The commutation rule (4) actually represents the action of the corresponding operator on polynomials.
Any skew polynomial ring $K[x;\sigma,\delta]$ has no zero divisors, that is, it forms non-commutative integral domain. Moreover, it satisfies the left Ore condition (existence of a common left multiple).

Lemma 1 (left Ore condition). *For all non-zero $a, b \in K[x;\sigma,\delta]$, there exist non-zero $a_1, b_1 \in K[x;\sigma,\delta]$ such that $a_1 b = b_1 a$.*

So skew polynomial rings over a field are left Ore rings. Different rings of linear operators can be defined simply by choosing appropriate σ and δ.

Example 1. If K with a derivation $\frac{\mathrm{d}}{\mathrm{d}t}$ is a differential field, then $K[s; 1_K, \frac{\mathrm{d}}{\mathrm{d}t}]$ is the ring of linear ordinary differential operators and we interpret (4) as a rule for differentiation

$$sa(t) = a(t)s + \frac{\mathrm{d}}{\mathrm{d}t}a(t)$$

for any $a(t) \in K$.

Example 2. If K is a difference field and σ over K is the automorphism which takes t to $t-1$, then $K[\delta;\sigma,0]$ is the ring of linear ordinary shift (recurrence) operators and we interpret (4) as a rule for shifting

$$\delta a(t) = a(t-1)\delta$$

for any $a(t) \in K$.

2.2 Ore algebras

However, when one wants to handle more complicated types of operators, as for instance differential time-delay, univariate skew polynomial rings are not enough. In such cases, one has to switch from univariate case to multivariate, or in other words, from Ore rings to Ore algebras. The idea of using Ore algebras in control theory was recently developed in [3].

So, in order to treat the differential time-delay structures, we mix up the constructions of two above examples [3]. This is, in fact, possible due to the following [17].

Lemma 2. *If A is a left Ore ring, so is $A[x;\sigma,\delta]$.*

This allows us to introduce a multivariate skew polynomial ring of the form

$$K[x_1;\sigma_1,\delta_1]...[x_s;\sigma_s,\delta_s]$$

Such a ring is said to be Ore algebra [4] if the σ_i's and δ_j's commute and satisfy $\sigma_i(x_j) = x_j$, $\delta_i(x_j) = 0$.

Note that this does not mean that the elements of K commute with the x_i's.

An illustrative application of Ore algebras in linear time-delay systems can be found in [3] from which we carried over this idea to the nonlinear case.

Naturally, to handle nonlinear time-delay systems we define Ore algebra

$$\mathcal{K}[\delta;\sigma,0][s;1_{\mathcal{K}},\frac{\mathrm{d}}{\mathrm{d}t}]$$

where σ takes t to $t-1$. That is, δ and s stand for delay and, respectively, derivative operator. We will use the abbreviated notation $\mathcal{K}[\delta, s]$.

Note that maps σ and $\frac{\mathrm{d}}{\mathrm{d}t}$ clearly commute, $\delta s = s\delta$, that is, $\sigma(\frac{\mathrm{d}}{\mathrm{d}t}a(t)) = \frac{\mathrm{d}}{\mathrm{d}t}\sigma(a(t))$ for any $a(t) \in \mathcal{K}$, and satisfy $\sigma(s) = s$ and $\frac{\mathrm{d}}{\mathrm{d}t}(\delta) = 0$. So in dealing with polynomials from $\mathcal{K}[\delta, s]$ we have to take into account the following three commutation rules

$$\begin{aligned} sa(t) &= a(t)s + \dot{a}(t) \\ \delta a(t) &= a(t-1)\delta \\ \delta s &= s\delta \end{aligned}$$

We can get a normal form of polynomials by moving δ and s on the right of each summand. For example, $(\delta+1)(a(t)s+1) = \delta a(t)s + a(t)s + \delta + 1 = a(t-1)\delta s + a(t)s + \delta + 1$.

Finally, remark that no additional advantages appear when employing the ring $\mathcal{K}[s;1_{\mathcal{K}},\frac{\mathrm{d}}{\mathrm{d}t}][\delta;\sigma,0]$. As δ and s commute both structures are, in fact, isomorphic.

3 Polynomial description of the input-output equation

Let $\mathcal{K}$ be now the field of meromorphic functions of $\{y^{(l)}(t-i), u^{(k)}(t-j); 0 \le l \le n-1; i,j,k \ge 0\}$.

Consider a nonlinear time-delay single-input single-output control system described by the input-output equation

$$y^{(n)}(t) = \phi(\{y^{(k)}(t-i), u^{(k)}(t-j); 0 \le k \le n-1; i,j \ge 0\}) \tag{5}$$

where $u \in \mathbf{R}$ and $y \in \mathbf{R}$ denote input and output to the system and $\phi \in \mathcal{K}$. The system (5) can be now represented in terms of polynomials in the Ore algebra $\mathcal{K}[\delta, s]$. After differentiating (5) we get

$$\mathrm{d}y^{(n)}(t) - \sum_{\substack{k=0\\ i\ge 0}}^{n-1} \frac{\partial\phi}{\partial y^{(k)}(t-i)}\mathrm{d}y^{(k)}(t-i) = \sum_{\substack{k=0\\ j\ge 0}}^{n-1} \frac{\partial\phi}{\partial u^{(k)}(t-j)}\mathrm{d}u^{(k)}(t-j)$$

which can be rewritten as

$$a(\delta, s)\mathrm{d}y(t) = b(\delta, s)\mathrm{d}u(t) \tag{6}$$

where

$$\begin{aligned} a(\delta, s) &= s^n - \sum_{k=0,i\ge 0}^{n-1} \frac{\partial\phi}{\partial y^{(k)}(t-i)}\delta^i s^k \\ b(\delta, s) &= \sum_{k=0,j\ge 0}^{n-1} \frac{\partial\phi}{\partial u^{(k)}(t-j)}\delta^j s^k \end{aligned}$$

are polynomials in $\mathcal{K}[\delta, s]$.

So the equation (6) represents the system behaviour in the terms of skew polynomials $a(\delta, s), b(\delta, s)$.

Consequently, in case of multi-input multi-output systems we would obtain a polynomial matrix description $A(\delta, s)\mathrm{d}y(t) = B(\delta, s)\mathrm{d}u(t)$.

Example 3. Consider the system $\ddot{y}(t) = \dot{y}(t-1)u(t-1)$. Then

$$\mathrm{d}\ddot{y}(t) - u(t-1)\mathrm{d}\dot{y}(t-1) = \dot{y}(t-1)\mathrm{d}u(t-1)$$

and

$$(s^2 - u(t-1)\delta s)\mathrm{d}y(t) = \dot{y}(t-1)\delta\mathrm{d}u(t)$$

3.1 Accessibility condition

Accessibility of nonlinear systems , based on the concept of autonomous elements, can be dealt with by introducing the accessibility filtration [13, 18]. However, like in the case of the systems without delays [11, 19, 20], we can now express the condition for accessibility of nonlinear time-delay systems in terms of common left factors of skew polynomials derived from the input-output equation (5).

Definition 3. *A one-form $\omega \in \mathrm{span}_{\mathcal{K}[\delta]}\{\mathrm{d}\xi; \xi \in \mathcal{K}\}$ is said to be an autonomous element for the system (5) if there exists an integer ν and coefficients $\alpha_i \in \mathcal{K}[\delta]$ such that*

$$\alpha_0\omega + \cdots \alpha_\nu\omega^{(\nu)} = 0 \tag{7}$$

Theorem 1. *The nonlinear time-delay system (5) has no autonomous elements if and only if the polynomials $a(\delta, s)$ and $b(\delta, s)$ have no common left factors.*

Proof. Neccessity: Suppose that (5) is not accessible; that is, there exists $\omega \in \mathrm{span}_{\mathcal{K}[\delta]}\{\mathrm{d}\xi;$ $\xi \in \mathcal{K}\}$ such that (7) is satisfied. Then ω can be expressed as

$$\omega = \tilde{a}(\delta, s)\mathrm{d}y(t) - \tilde{b}(\delta, s)\mathrm{d}u(t)$$

and (7) as

$$\rho(\delta, s)\omega = \rho(\delta, s)[\tilde{a}(\delta, s)\mathrm{d}y(t) - \tilde{b}(\delta, s)\mathrm{d}u(t)] = 0$$

After matching the latter with (6), rewritten in the form $a(\delta, s)\mathrm{d}y(t) - b(\delta, s)\mathrm{d}u(t) = 0$, we get $a(\delta, s) = \rho(\delta, s)\tilde{a}(\delta, s)$ and $b(\delta, s) = \rho(\delta, s)\tilde{b}(\delta, s)$; that is, $a(\delta, s), b(\delta, s)$ have a common left factor $\rho(\delta, s)$.

Sufficiency: Suppose that $a(\delta, s), b(\delta, s)$ have a (nontrivial) common left factor $\rho(\delta, s)$. Then the equation (6) can be written as

$$\rho(\delta, s)[\tilde{a}(\delta, s)\mathrm{d}y(t) - \tilde{b}(\delta, s)\mathrm{d}u(t)] = 0$$

Let $\omega = \tilde{a}(\delta, s)\mathrm{d}y(t) - \tilde{b}(\delta, s)\mathrm{d}u(t)$. Then we obtain $\rho(\delta, s)\omega = 0$ which implies the existence of an autonomous element.

Clearly, one can conclude that

Theorem 2. *The nonlinear time-delay system (5) is accessible if the polynomials $a(\delta, s)$ and $b(\delta, s)$ have no common left factors.*

Example 4. Consider the system $\ddot{y}(t) = \dot{y}(t-1)u(t) + y(t-1)\dot{u}(t) - \dot{y}(t-1) + y(t-2)u(t-1)$. After differentiating we get

$$\begin{aligned}(s^2 - u(t)s\delta - \dot{u}(t)\delta + s\delta - u(t-1)\delta^2)\mathrm{d}y(t) &= \\ = (\dot{y}(t-1) + y(t-1)s + y(t-2)\delta)\mathrm{d}u(t)\end{aligned}$$

The polynomials have a common left factor.

$$(s+\delta)(s - u(t)\delta)\mathrm{d}y(t) = (s+\delta)y(t-1)\mathrm{d}u(t)$$

The system is not accessible and can be reduced to

$$\begin{aligned}(s - u(t)\delta)\mathrm{d}y(t) &= y(t-1)\mathrm{d}u(t) \\ \mathrm{d}\dot{y}(t) - u(t)\mathrm{d}y(t-1) &= y(t-1)\mathrm{d}u(t)\end{aligned}$$

which after integrating yields $\dot{y}(t) = y(t-1)u(t)$.

4 Transfer functions of nonlinear time-delay systems

The transfer function formalism for nonlinear systems without delays was recently developed in [6, 7, 8, 9].

In what follows we introduce such a formalism for time-delay case. The first step is to construct the quotient field of skew polynomials. In that respect Ore condition plays a key role. Obviously, $\mathcal{K}[\delta, s]$ is still a left Ore ring and the left Ore condition still holds on. Hence, each two elements of $\mathcal{K}[\delta, s]$ have a common left multiple. $\mathcal{K}[\delta, s]$ can be, therefore, embedded to a non-commutative quotient field [16, 17] by defining quotients as

$$\frac{a}{b} = b^{-1} \cdot a$$

where $a, b \in \mathcal{K}[\delta, s]$ and $b \neq 0$. Addition is defined by reducing two quotients to the same denominator

$$\frac{a_1}{b_1} + \frac{a_2}{b_2} = \frac{\beta_2 a_1 + \beta_1 a_2}{\beta_2 b_1} \tag{8}$$

where $\beta_2 b_1 = \beta_1 b_2$ by Ore condition. Multiplication is defined by

$$\frac{a_1}{b_1} \cdot \frac{a_2}{b_2} = \frac{\alpha_1 a_2}{\beta_2 b_1} \tag{9}$$

where $\beta_2 a_1 = \alpha_1 b_2$ again by Ore condition. The resulting quotient field of skew polynomials is denoted by $\mathcal{K}\langle\delta, s\rangle$.

Note that due to the non-commutative multiplication they, of course, differ from the usual rules. In particular, in case of the multiplication (9) we, in general, cannot simply multiply numerators and denominators, nor cancel them in a usual manner. We neither can commute them as the multiplication in $\mathcal{K}\langle\delta, s\rangle$ is non-commutative as well.

Once we have defined a fraction of two skew polynomials we can introduce transfer functions of nonlinear time-delay systems.

Definition 4. *An element $F(\delta, s) \in \mathcal{K}\langle\delta, s\rangle$ such that $\mathrm{d}y(t) = F(\delta, s)\mathrm{d}u(t)$ is said to be a transfer function of the single-input single-output nonlinear time-delay system (1) (respectively (5)).*

In the case of multi-input multi-output system, we think of $F(\delta, s)$ as a matrix with the entries in $\mathcal{K}\langle\delta, s\rangle$ and $F(\delta, s)$ is then referred to as a transfer matrix.

Like in case of the input-output equation (5), where naturally

$$F(\delta, s) = \frac{b(\delta, s)}{a(\delta, s)}$$

also state-space representation (1) can be expressed in terms of skew polynomials. After differentiating (1) we get

$$\begin{aligned} \mathrm{d}\dot{x}(t) &= A\mathrm{d}x(t) + B\mathrm{d}u(t) \\ \mathrm{d}y(t) &= C\mathrm{d}x(t) + D\mathrm{d}u(t) \end{aligned}$$

where A, B, C, D are appropriate matrices with the entries in $\mathcal{K}[\delta]$. Now, we can write

$$\begin{aligned} (sI - A)\mathrm{d}x(t) &= B\mathrm{d}u(t) \\ \mathrm{d}y(t) &= C\mathrm{d}x(t) + D\mathrm{d}u(t) \end{aligned}$$

from which follows

$$F(\delta, s) = C(sI - A)^{-1}B + D \tag{10}$$

Clearly, the elements of $(sI - A)$ are skew polynomials from the ring $\mathcal{K}[\delta, s]$. Hence, inverting $(sI - A)$ requires solving linear equations in non-commutative fields, see [16]. To find the left-hand inverse of $(sI - A)$ one can use Gauss-Jordan elimination considering (8) and (9), see also [6].

Note that the transfer function is defined by employing the standard algebraic formalism of differential forms, following the lines in [5] which introduces the notion of a one-form in a formal and abstract way. In particular, it is not necessary to deal here with the linearization of the system along a trajectory using the Kähler-type differential which leads to a time-varying linear system.

Example 5. Consider the system

$$\begin{aligned} \dot{x}_1(t) &= x_2(t-1) \\ \dot{x}_2(t) &= x_2(t)u(t) \\ y(t) &= x_1(t) \end{aligned}$$

After differentiating we get

$$A = \begin{pmatrix} 0 & \delta \\ 0 & u(t) \end{pmatrix}, \; B = \begin{pmatrix} 0 \\ x_2(t) \end{pmatrix}, \; C = \begin{pmatrix} 1 & 0 \end{pmatrix}$$

Note that

$$(sI - A)^{-1} = \begin{pmatrix} \frac{1}{s} & \frac{\delta}{s^2 - u(t-1)s} \\ 0 & \frac{1}{s - u(t)} \end{pmatrix}$$

Finally

$$\begin{aligned} F(\delta, s) = C(sI - A)^{-1}B &= \frac{\delta}{s^2 - u(t-1)s} \cdot x_2(t) \\ &= \frac{x_2(t-1)\delta}{s^2 - u(t-1)s} \end{aligned}$$

Remark that $x_2(t-1) = \dot{y}(t)$ and

$$\begin{aligned} \mathrm{d}y(t) &= \frac{\dot{y}(t)\delta}{s^2 - u(t-1)s}\mathrm{d}u(t) \\ (s^2 - u(t-1)s)\mathrm{d}y(t) &= \dot{y}(t)\delta\mathrm{d}u(t) \\ \mathrm{d}\ddot{y}(t) - u(t-1)\mathrm{d}\dot{y}(t) &= \dot{y}(t)\mathrm{d}u(t-1) \end{aligned}$$

which after integrating yields $\ddot{y}(t) = \dot{y}(t)u(t-1)$.

4.1 Invariance of the transfer functions

Transfer functions of nonlinear time-delay systems have many properties we expect from transfer functions.

Theorem 3. *Transfer function (10) of the nonlinear time-delay system (1) is invariant with respect to the state transformation* $\xi(t) = \phi(\{x(t-i); i \geq 0\})$.

Proof. For any state transformation $\xi(t) = \phi(\{x(t-i); i \geq 0\})$ one has $\mathrm{d}\xi(t) = T\mathrm{d}x(t)$ where T is an appropriate unimodular (and therefore invertible) matrix from $\mathcal{K}^{n\mathrm{x}n}[\delta]$. In the new coordinates we get

$$\begin{aligned} \mathrm{d}\dot{\xi}(t) &= (TAT^{-1} + \dot{T}T^{-1})\mathrm{d}\xi(t) + TB\mathrm{d}u(t) \\ \mathrm{d}y(t) &= CT^{-1}\mathrm{d}\xi(t) + D\mathrm{d}u(t) \end{aligned}$$

where $\dot{T} = \frac{\mathrm{d}}{\mathrm{d}t}(T)$ with $\frac{\mathrm{d}}{\mathrm{d}t}$ applied pointwise to T. Note that $\frac{\mathrm{d}}{\mathrm{d}t}(\delta) = 0$. Thus, the transfer function reads as

$$\begin{aligned} F(s) &= CT^{-1}(sI - TAT^{-1} - \dot{T}T^{-1})TB + D \\ &= C(T^{-1}sT - A - T^{-1}\dot{T})B + D \end{aligned}$$

After applying the commutation rule $sT = Ts + \dot{T}$, we get

$$F(s) = C(sI - A)^{-1}B + D$$

which completes the proof.

Example 6. Consider the system

$$\begin{aligned} \dot{x}(t) &= -x^2(t)u(t-1) \\ y(t) &= 1/x(t) \end{aligned}$$

which is, in fact, linear. One can easily check that $\dot{y}(t) = u(t-1)$. The corresponding state transformation is $\xi(t) = 1/x(t)$. After differentiating, $A = (-2x(t)u(t-1))$, $B = (-x^2(t)\delta)$, $C = (-1/x^2(t))$, we get

$$\begin{aligned} F(\delta, s) &= C(sI - A)^{-1}B \\ &= -\frac{1}{x^2(t)} \cdot \frac{-x^2(t)\delta}{s + 2x(t)u(t-1)} \\ &= \frac{x^2(t)\delta}{(s + 2x(t)u(t-1))x^2(t)} = \frac{x^2(t)\delta}{x^2(t)s} = \frac{\delta}{s} \end{aligned}$$

which represents the linear system.

4.2 Transfer function algebra

We can also introduce algebra of transfer functions of nonlinear time-delay systems. Each system structure can be divided into three basic types of connections: series, parallel and feedback, see Fig. 1.

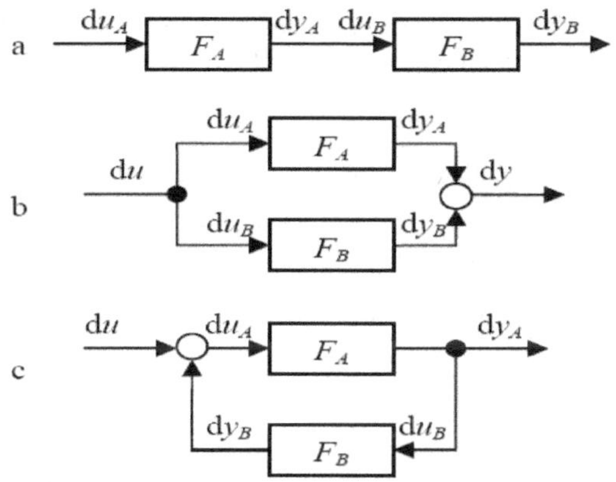

Fig. 1. Series, parallel and feedback connection

For a series connection, it follows that $\mathrm{d}y_B(t) = F_B(\delta, s)\mathrm{d}u_B(t) = F_B(\delta, s)\mathrm{d}y_A(t) = F_B(\delta, s)F_A(\delta, s)\mathrm{d}u_A(t)$. Thus

$$F(\delta, s) = F_B(\delta, s)F_A(\delta, s) \tag{11}$$

For parallel and feedback connection we get

$$\begin{aligned} F(\delta, s) &= F_A(\delta, s) + F_B(\delta, s) \\ F(\delta, s) &= (1 - F_A(\delta, s)F_B(\delta, s))^{-1} \cdot F_A(\delta, s) \end{aligned} \tag{12}$$

Note that due to the non-commutative multiplication the rules (11) and (12) have to be kept exactly as they are.
Following example demonstrates how to handle a series connection of two time-delay systems. It also serves as a motivation why it can be interesting to use the transfer function formalism. This idea was already used for instance in [10] to solve the nonlinear model matching problem.

Example 7. Consider two systems

$$\dot{y}_A(t) = y_A(t)u_A(t-1) \qquad y_B(t) = \ln u_B(t)$$

Transfer functions are following

$$F_A(\delta, s) = \frac{y_A(t)\delta}{s - u_A(t-1)} \qquad F_B(\delta, s) = \frac{1}{u_B(t)}$$

The systems are combined together in a series connection. For the connection A→B, when $u_B(t) = y_A(t)$, the resulting transfer function is

$$\begin{aligned}F(\delta,s) &= F_B(\delta,s)F_A(\delta,s) = \frac{1}{u_B(t)} \cdot \frac{y_A(t)\delta}{s - u_A(t-1)} \\ &= \frac{1}{y_A(t)} \cdot \frac{y_A(t)\delta}{s - u_A(t-1)} \\ &= \frac{y_A(t)\delta}{y_A(t)s + \dot{y}_A(t) - u_A(t-1)y_A(t)} = \frac{\delta}{s}\end{aligned}$$

Hence, the combination A→B is linear from an input-output point of view $\dot{y}_B(t) = u_A(t-1)$. However, when the systems are connected as B→A, that is $u_A(t) = y_B(t)$, the result is different

$$\begin{aligned}F(\delta,s) &= F_A(\delta,s)F_B(\delta,s) = \frac{y_A(t)\delta}{s - u_A(t-1)} \cdot \frac{1}{u_B(t)} \\ &= \frac{y_A(t)\delta}{u_B(t-1)s - u_B(t-1)\ln u_B(t-1)}\end{aligned}$$

This time, it does not yield a linear system.

5 Conclusions

In this paper the polynomial approach to nonlinear time-delay systems was introduced. Unlike in the systems without delays the polynomials belong now to the multivariate skew polynomial ring, called Ore algebra. Polynomials from this ring act as differential time-delay operators on the differentials of the system inputs and outputs. Like in the systems without delays the condition for accessiblity of nonlinear time-delay systems was stated in terms of common left factors of the polynomials derived from the input-output equation. Introduced Ore algebra forms a non-commutative integral domain in which the left Ore condition is satisfied. This allowed us to define quotient field over the Ore algebra. Such quotients were suggested as transfer functions of nonlinear time-delay systems, as they satisfy many properties we expect from transfer functions. Firstly, their invariance to state transformations was shown and then the transfer function algebra was introduced. Mathemathical tools introduced in this paper represent thus an alternative approach to nonlinear time-delay systems and open new possibilities in analysis and feedback design.

References

1. Aranda-Bricaire, E., Kotta, Ü., Moog, C.H.: Linearization of discrete-time systems. SIAM Journal of Control Optimization 34, 1999–2023 (1996)
2. Bronstein, M., Petkovšek, M.: An introduction to pseudo-linear algebra. Theoretical Computer Science 157, 3–33 (1996)
3. Chyzak, F., Quadrat, A., Robertz, D.: Effective algorithms for parametrizing linear control systems over Ore algebras. Applicable Algebra in Engineering, Communications and Computing 16, 319–376 (2005)
4. Chyzak, F., Salvy, B.: Non-commutative elimination in Ore algebras proves multivariate identities. Journal of Symbolic Computation 26, 187–227 (1998)
5. Conte, G., Moog, C.H., Perdon, A.M.: Nonlinear Control Systems: An Algebraic Setting. Springer, London (1999)
6. Halás, M.: An algebraic framework generalizing the concept of transfer functions to nonlinear systems. Automatica 44(5), 1181–1190 (2008)

7. Halás, M., Huba, M.: Symbolic computation for nonlinear systems using quotients over skew polynomial ring. In: Proc. 14th Mediterranean Conference on Control and Automation, Ancona, Italy (2006)
8. Halás, M., Kotta, Ü.: Extension of the concept of transfer function to discrete-time nonlinear control systems. In: Proc. European Control Conference, Kos, Greece (2007)
9. Halás, M., Kotta, Ü.: Transfer Functions of Discrete-time Nonlinear Control Systems. Proc. Estonian Acad. Sci. Phys. Math. 56, 322–335 (2007)
10. Halás, M., Kotta, Ü., Moog, C.H.: Transfer Function Approach to the Model Matching Problem of Nonlinear Systems. In: Proc. 17th IFAC World Congress, Seoul, Korea (2008)
11. Kotta, Ü.: Irreducibility conditions for nonlinear input-output difference equations. In: Proc. 39th IEEE Conf. on Decision and Control, Sydney, Australia (2000)
12. Kotta, Ü., Tõnso, M.: Irreducibility conditions for discrete-time nonlinear multi-input multi-output systems. In: Proc. 6th IFAC Symposium NOLCOS, Stuttgart, Germany (2004)
13. Márquez-Martínez, L.A.: Note sur l'accessibilité des systèmes non linéaires à retards. Comptes Rendus de l'Academie des Sciences Paris 329, 545–550 (1999)
14. Márquez-Martínez, L.A., Moog, C.H., Velasco-Villa, M.: The structure of nonlinear time-delay systems. Kybernetika 36, 53–62 (2000)
15. Moog, C.H., Castro-Linares, R., Velasco-Villa, M., Márquez-Martínez, L.A.: The disturbance decoupling problem for time-delay nonlinear systems. IEEE Transactions on Automatic Control 45, 305–309 (2000)
16. Ore, O.: Linear equations in non-commutative fields. Annals of Mathematics 32, 463–477 (1931)
17. Ore, O.: Theory of non-commutative polynomials. Annals of Mathematics 32, 480–508 (1933)
18. Xia, X., Márquez-Martínez, L.A., Zagalak, P., Moog, C.H.: Analysis of nonlinear time-delay systems using modules over non-commutative rings. Automatica 38, 1549–1555 (2002)
19. Zheng, Y., Willems, J., Zhang, C.: Common factors and controllability of nonlinear systems. In: Proc. 36th IEEE Conf. on Decision and Control, San Diego, California USA (1997)
20. Zheng, Y., Willems, J., Zhang, C.: A polynomial approach to nonlinear system controllability. IEEE Transactions on Automatic Control 46, 1782–1788 (2001)

Stable Periodic Solutions of Time Delay Systems Containing Hysteresis Nonlinearities

Alexander M. Kamachkin[1] and Alexander V. Stepanov[2]

[1] St. Petersburg State University, St. Petersburg, Russia, akamachkin@mail.ru
[2] St. Petersburg State University, St. Petersburg, Russia, stepanov17@yandex.ru

Summary. Hysteresis oscillating systems with time delay effect in feedback loop represent an object for study of the paper. The relay oscillator under consideration consists of linear system (its proper stability is uncertain) in feedback, relay hysteresis and, optionally, of an additive external action which is periodic continuous bounded function of time. Sufficient conditions are obtained for existing of asymptotically orbitally stable periodic modes of unperturbed systems and for stable periodic modes of perturbed systems. Simple analysis of perturbed periodic modes configuration is provided. An account is supported by simple illustrative examples.

1 Introduction

The question on existence of periodic modes in nonlinear control systems and problem of exact construction of such modes are among central problems for automatic control theory. Special difficulties appear when dealing with systems containing the so-called essential nonlinearities, which are nonanalytic functions of phase (for example, relay nonlinearities). In this article, one system with hysteresis nonlinearity will be considered. Nonlinearities of this sort, for example, may describe spatial delay of control mechanism, e.g. autopilot or stabilizer [1]. Furthermore, study of physical processes in systems containing springing or magnetic elements, electrical relays etc. on certain assumptions gives rise to different mathematical models of hysteresis nonlinearity [2, 3]. Numerous works are devoted to the analysis of problem of periodic oscillations presence in such systems [4, 5, 6, 7, 8]. A wide variety of questions concerned to sliding modes in nonlinear systems considered in [9].

In parallel with analysis problem consisting in phase space description when system parameters are fixed there arises the synthesis problem which is to select system parameters in such a way that the control system has to perform in prescribed mode. For example control law synthesis process for one second-order system with non-ideal relay and periodic external action is described in details in [10, 11].

Some results concerning control systems containing hysteresis nonlinearities will be extrapolated below in case of time delay presence in such systems.

J.J. Loiseau et al. (Eds.): Topics in Time Delay Systems, LNCIS 388, pp. 121–132.
springerlink.com

2 Models under consideration

Let us consider following systems

$$\dot{x} = Ax + cu(t-\tau), \tag{1}$$

$$\dot{x} = Ax + c\big(\varphi(t) + u(t-\tau)\big), \tag{2}$$

here $x = x(t) \in E^n,\ \ t \geq t_0 \geq \tau,\ \ A$ is real $n \times n$ matrix, $\ c \in E^n,\ \ \varphi(t)$ is scalar T-periodic continuous function of time describing external disturbance acting upon the system, control statement u defined in the following way:

$$u(t-\tau) = f\left(\sigma(t-\tau)\right), \quad \sigma(t-\tau) = \gamma' x(t-\tau),$$

where $\gamma \in E^n,\ \ \|\gamma\| \neq 0,\ \ f$ describes a nonlinear element of hysteresis sort:

$$f(\sigma) = \begin{cases} m_1, & \sigma < l_2, \\ m_2, & \sigma > l_1, \end{cases} \qquad l_1 < l_2, \quad m_1 < m_2.$$

Vector $x(t), t \in [t_0 - \tau,\ t_0]$, is considered to be known. Quantity $\tau > 0$ describes time delay of actuator or observer.

3 Stationary system

By the analogy with [8] following result may be formulated:

Lemma 1. *If*

$$-\gamma' A^{-1} c m_1 > l_2, \quad -\gamma' A^{-1} c m_2 < l_1,$$

and Re $\lambda < 0$, *for any eigenvalue* λ *of matrix* A, *then system (1) has at least one non-trivial periodic solution.*

Proof. The system under consideration has two formal positions of equilibrium $p_{1,\,2}$ such as

$$\dot{x} = Ap_i + cm_i = 0,$$

that is

$$p_i = -A^{-1} c m_i.$$

Since Re $\lambda < 0$, then there exists the positive constant C such as all motions of the systems tends in time to closed ball $\{\|x\| \leq C\}$. Denote

$$S = \big\{x \in E^n : \ \|x\| \leq C,\ \gamma' x = l_1\big\} \neq \varnothing.$$

It is obvious that any solution of system (1) has infinitely many switching points when

$$\gamma' p_1 > l_2, \quad \gamma' p_2 < l_1,$$

and this system generates mapping of closed set S to itself. The latest inequalities hold true with the assumptions of the lemma. Application of the principle of fixed point completes the proof. ■

Sufficient conditions for orbital stability of periodic solutions of system (1) will be obtained below. Suppose that exists the periodic solution of system (1) having two switching points $s_{1,\,2} \in E^n$ such as

$$\gamma' s_1 = l_1, \quad \gamma' s_2 = l_2.$$

In that case there exists a pair of points $\hat{s}_{1,\,2} \in E^n$ ("virtual" switching points) and real constants $\tau_{1,\,2},\, \tau_i > \tau$ such as

$$\hat{s}_1 = e^{A\tau} s_1 + \int_0^{\tau} e^{A(\tau - s)} c m_2\, ds, \quad s_2 = e^{A(\tau_1 - \tau)} \hat{s}_1 + \int_0^{\tau_1 - \tau} e^{A(\tau_1 - \tau - s)} c m_1\, ds,$$

$$\hat{s}_2 = e^{A\tau} s_2 + \int_0^{\tau} e^{A(\tau - s)} c m_1\, ds, \quad s_1 = e^{A(\tau_2 - \tau)} \hat{s}_2 + \int_0^{\tau_2 - \tau} e^{A(\tau_2 - \tau - s)} c m_2\, ds.$$

Note that hereafter "switching points" are only points lying on switching hyperplane, but in fact switching of control action occurs in virtual switching points.

Let us generalize one of the results cited in [6]. Denote

$$A_1 = \left(E - \frac{(A s_2 + c m_1)\, \gamma'}{\gamma'\, (A s_2 + c m_1)} \right) e^{A\tau_1}, \quad A_2 = \left(E - \frac{(A s_1 + c m_2)\, \gamma'}{\gamma'\, (A s_1 + c m_2)} \right) e^{A\tau_2}.$$

Theorem 1. *If*

$$A = \|A_1\, A_2\| < 1,$$

and

$$\gamma'\, (A s_1 + c m_2) \neq 0, \quad \gamma'\, (A s_2 + c m_1) \neq 0,$$

then concerned periodic mode of system (1) is asymptotically orbitally stable.

Proof. Since

$$s_2 = e^{A\tau_1} s_1 + \int_0^{\tau} e^{A(\tau_1 - s)} c m_2\, ds + \int_{\tau}^{\tau_1} e^{A(\tau_1 - s)} c m_1\, ds,$$

$$s_1 = e^{A\tau_2} s_2 + \int_0^{\tau} e^{A(\tau_2 - s)} c m_1\, ds + \int_{\tau}^{\tau_2} e^{A(\tau_2 - s)} c m_2\, ds,$$

then

$$\frac{\partial s_1}{\partial s_2} = e^{A\tau_2},$$

$$\frac{\partial s_1}{\partial \tau_2} = A\, e^{A\tau_2} s_2 + e^{A(\tau_2 - \tau)} c(m_2 - m_1) + e^{A\tau_2} c m_1 = A s_1 + c m_2.$$

Smilarly,

$$\frac{\partial s_2}{\partial s_1} = e^{A\tau_1}, \quad \frac{\partial s_2}{\partial \tau_1} = A s_2 + c m_1.$$

So, as

$$d s_1 = \frac{\partial s_1}{\partial s_2}\, d s_2 + \frac{\partial s_1}{\partial \tau_2}\, d\tau_2, \quad d s_2 = \frac{\partial s_2}{\partial s_1}\, d s_1 + \frac{\partial s_2}{\partial \tau_1}\, d\tau_1,$$

and

$$d\left(\gamma'\, s_1(s_2,\, \tau_2)\right) = 0, \quad d\left(\gamma'\, s_2(s_1,\, \tau_1)\right) = 0,$$

then

$$d\tau_2 = -\frac{\gamma' e^{A\tau_2}}{\gamma'(As_1 + cm_2)} ds_2, \quad d\tau_1 = -\frac{\gamma' e^{A\tau_1}}{\gamma'(As_2 + cm_1)} ds_1.$$

Hence

$$ds_1 = e^{A\tau_2} ds_2 - \frac{e^{A\tau_2}(As_1 + cm_2)\,\gamma' e^{A\tau_2}}{\gamma'\,(As_1 + cm_2)} ds_2 = A_2\, ds_2,$$

$$ds_2 = e^{A\tau_1} ds_1 - \frac{e^{A\tau_1}(As_2 + cm_1)\,\gamma' e^{A\tau_1}}{\gamma'\,(As_2 + cm_1)} ds_1 = A_1\, ds_1,$$

and

$$ds_2^{(i+1)} = A_1\, A_2\, ds_2^{(i)}, \quad i = 0,\, 1,\, 2,\, \ldots$$

Application of the principle of fixed point completes the proof. ■

Example 1. Consider a system (1), where

$$A = \begin{pmatrix} 0.1 & 0 & 0 \\ 0 & -1 & 10 \\ 0 & -10 & -1 \end{pmatrix}, \quad c = \begin{pmatrix} 1 \\ 1 \\ 1 \end{pmatrix}, \quad \gamma = \begin{pmatrix} -0.4 \\ 0 \\ -0.8 \end{pmatrix},$$

$$l_{1,\,2} = \mp 0.3, \quad m_{1,\,2} = \mp 1, \quad \tau = 0.1.$$

As it may be checked directly, system (1) has three origin-symmetric periodic solutions with two switching points. The half-period τ_1 for each of these solutions may be derived from the equation

$$\gamma'\, s_1(\tau_1) + l = 0,$$

where

$$s_1(\tau_1) = -\left(e^{A\tau_1} + E\right)^{-1}\left(e^{A\tau_1} - 2e^{A(\tau_1 - \tau)} + E\right) A^{-1}\, cm. \tag{3}$$

The latter equation has three solutions corresponding to periodic modes

$$\tau_1^{(1)} = 0.2873..., \ \tau_1^{(2)} = 0.8960..., \ \tau_1^{(3)} = 1.4199...,$$

(the respective switching points are given by (3)). After some intermediate calculation one can verify that

$$A^{(1)} \approx 1.05, \quad A^{(2)} \approx 0.76, \quad A^{(3)} \approx 0.52,$$

so the system under consideration has at least two asymptotically orbitally stable periodic solutions.

4 Perturbed system

Let us colligate some of the results obtained in [10]. Consider a second-order system

$$\ddot{y} + g_1\dot{y} + g_2 y = u(t - \tau) + \varphi(t), \quad t \geq t_0, \tag{4}$$

here $y(t) \in R$ is sought-for time variable, $g_{1,\,2}$ are real constants,

$$u(t - \tau) = f\left(\sigma(t - \tau)\right), \quad \sigma = \alpha_1 y + \alpha_2 \dot{y},$$

$\alpha_{1,\,2}$ are real constants.

In order to deal with system (4) special linear transform may be applied [10, 11].

Let us rewrite system (4) in vector form. Assume

$$z_1 = y, \quad z_2 = \dot{y},$$

in that case

$$\begin{aligned} \dot{z} &= Pz + q\left(\varphi(t) + u(t-\tau)\right), \\ u(t-\tau) &= f\left(\sigma(t-\tau)\right), \quad \sigma = \alpha' z, \end{aligned} \tag{5}$$

where $z' = \begin{pmatrix} z_1 & z_2 \end{pmatrix}$,

$$P = \begin{pmatrix} 0 & 1 \\ -g_2 & -g_1 \end{pmatrix}, \quad q = \begin{pmatrix} 0 \\ 1 \end{pmatrix}, \quad \alpha = \begin{pmatrix} \alpha_1 \\ \alpha_2 \end{pmatrix}.$$

Suppose that characteristic determinant

$$D(s) = \det\left(P - Es\right)$$

has real simple roots $\lambda_{1,\,2}$, and vectors $q,\ Pq$ are linearly independent. In that case system (5) may be reduced to the form (2), where

$$A = \begin{pmatrix} \lambda_1 & 0 \\ 0 & \lambda_2 \end{pmatrix}, \quad c = \begin{pmatrix} 1 \\ 1 \end{pmatrix},$$

by means of nonsingular linear transformation

$$z = Tx, \tag{6}$$

$$x = \begin{pmatrix} x_1 \\ x_2 \end{pmatrix}, \quad T = \begin{pmatrix} \frac{N_1(\lambda_1)}{D'(\lambda_1)} & \frac{N_1(\lambda_2)}{D'(\lambda_2)} \\ \frac{N_2(\lambda_1)}{D'(\lambda_1)} & \frac{N_2(\lambda_2)}{D'(\lambda_2)} \end{pmatrix}.$$

Here

$$D'(\lambda_j) = \frac{d}{ds}\, D(s)\bigg|_{s=\lambda_j}, \quad N_j(s) = \sum_{i=1}^{2} q_i D_{ij}(s),$$

$D_{ij}(s)$ is algebraic supplement for element lying in the intersection of i-th row and j-th column of determinant $D(s)$.

Note that

$$\sigma = \gamma' x, \quad \gamma = T' \alpha.$$

Furthermore, since

$$\gamma_i = -\left(D'\left(\lambda_i\right)\right)^{-1} \sum_{j=1}^{2} \alpha_j\, N_j\left(\lambda_i\right), \quad i = 1, 2.$$

then

$$\begin{aligned} \gamma_1 &= \left(\lambda_1 - \lambda_2\right)^{-1}\left(\alpha_1 + \alpha_2\lambda_1\right), \\ \gamma_2 &= \left(\lambda_2 - \lambda_1\right)^{-1}\left(\alpha_1 + \alpha_2\lambda_2\right). \end{aligned}$$

5 Finding of periodic solutions

Transformation (6) leads to the following system:

$$\begin{cases} \dot{x}_1 = \lambda_1 x_1 + f\left(\sigma(t-\tau)\right) + \varphi(t), \\ \dot{x}_2 = \lambda_2 x_2 + f\left(\sigma(t-\tau)\right) + \varphi(t). \end{cases} \tag{7}$$

If, for example,

$$\alpha_1 = -\lambda_1 \alpha_2,$$

then

$$\gamma_1 = 0, \quad \gamma_2 = \alpha_2, \quad \sigma = \gamma_2 x_2.$$

Function f in that case is independent of variable x_1, and

$$\dot{\sigma} = \lambda_2 \sigma + \gamma_2 \left(f(\gamma_2 \cdot x_2(t-\tau)) + \varphi(t)\right)$$

Solution of the latest equation when $f = m_{1,2}$ has the following form:

$$\sigma\left(t,\, t_0,\, \sigma_{2,\,0},\, m_i\right) = e^{\lambda_2(t-t_0)}\sigma_0 + \gamma_2\, e^{\lambda_2 t} \int_{t_0}^{t} e^{-\lambda_2 s}\Big(m_i + \varphi(s)\Big)\, ds.$$

Let us derive necessary conditions for existing of $(\tau_1 + \tau_2)$-periodic solution of system (7) having two switching points $s_{1,\,2}$.

$$\hat{\sigma}_1 = e^{\lambda_2 \tau}\sigma_1 + \gamma_2\, e^{\lambda_2(t_0+\tau)} \int_{t_0}^{t_0+\tau} e^{-\lambda_2 s}\Big(m_2 + \varphi(s)\Big)\, ds,$$

$$\sigma_2 = e^{\lambda_2(\tau_1-\tau)}\hat{\sigma}_1 + \gamma_2\, e^{\lambda_2(t_0+\tau_1)} \int_{t_0+\tau}^{t_0+\tau_1} e^{-\lambda_2 s}\Big(m_1 + \varphi(s)\Big)\, ds,$$

$$\hat{\sigma}_2 = e^{\lambda_2 \tau}\sigma_2 + \gamma_2\, e^{\lambda_2(t_0+\tau_1+\tau)} \int_{t_0+\tau_1}^{t_0+\tau_1+\tau} e^{-\lambda_2 s}\Big(m_1 + \varphi(s)\Big)\, ds,$$

$$\sigma_1 = e^{\lambda_2(\tau_2-\tau)}\hat{\sigma}_2 + \gamma_2\, e^{\lambda_2(t_0+\tau_1+\tau_2)} \int_{t_0+\tau_1+\tau}^{t_0+\tau_1+\tau_2} e^{-\lambda_2 s}\Big(m_2 + \varphi(s)\Big)\, ds,$$

here $\tau_i > \tau$.

Consequently,

$$\begin{aligned} \sigma_1 = e^{\lambda_2(\tau_2-\tau)} &\left(e^{\lambda_2 \tau}\sigma_2 + \gamma_2\, e^{\lambda_2(t_0+\tau_1+\tau)} \int_{t_0+\tau_1}^{t_0+\tau_1+\tau} e^{-\lambda_2 s}\Big(m_1 + \varphi(s)\Big)\, ds\right) + \\ &+ \gamma_2\, e^{\lambda_2(t_0+\tau_1+\tau_2)} \int_{t_0+\tau_1+\tau}^{t_0+\tau_1+\tau_2} e^{-\lambda_2 s}\Big(m_2 + \varphi(s)\Big)\, ds = \\ &= e^{\lambda_2 \tau_2}\sigma_2 + \gamma_2\, e^{\lambda_2(t_0+\tau_1+\tau_2)} \int_{t_0+\tau_1}^{t_0+\tau_1+\tau} e^{-\lambda_2 s}\Big(m_1 + \varphi(s)\Big)\, ds + \\ &+ \gamma_2\, e^{\lambda_2(t_0+\tau_1+\tau_2)} \int_{t_0+\tau_1+\tau}^{t_0+\tau_1+\tau_2} e^{-\lambda_2 s}\Big(m_2 + \varphi(s)\Big)\, ds = \\ = e^{\lambda_2 \tau_2}\sigma_2 &+ \gamma_2 \lambda_2^{-1} e^{\lambda_2(\tau_2-\tau)}\Big(e^{\lambda_2 \tau} - 1\Big)\, m_1 + \gamma_2 \lambda_2^{-1}\Big(e^{\lambda_2(\tau_2-\tau)} - 1\Big)\, m_2 + \gamma_2 I_2, \end{aligned}$$

$$I_2 = \int_0^{\tau_2} e^{\lambda_2(\tau_2-s)}\varphi(s + t_0 + \tau_1)\, ds.$$

Similarly

$$\sigma_2 = e^{\lambda_2(\tau_1-\tau)}\left(e^{\lambda_2\tau}\sigma_1 + \gamma_2\, e^{\lambda_2(t_0+\tau)}\int_{t_0}^{t_0+\tau} e^{-\lambda_2 s}\Big(m_2+\varphi(s)\Big)\,ds\right)+$$
$$+\gamma_2\, e^{\lambda_2(t_0+\tau_1)}\int_{t_0+\tau}^{t_0+\tau_1} e^{-\lambda_2 s}\Big(m_1+\varphi(s)\Big)\,ds =$$
$$= e^{\lambda_2\tau_1}\sigma_2 + \gamma_2\, e^{\lambda_2(t_0+\tau_1)}\int_{t_0}^{t_0+\tau} e^{-\lambda_2 s}\Big(m_2+\varphi(s)\Big)\,ds+$$
$$+\gamma_2\, e^{\lambda_2(t_0+\tau_1)}\int_{t_0+\tau}^{t_0+\tau_1} e^{-\lambda_2 s}\Big(m_1+\varphi(s)\Big)\,ds =$$
$$= e^{\lambda_2\tau_1}\sigma_1 + \gamma_2\lambda_2^{-1}e^{\lambda_2(\tau_1-\tau)}\left(e^{\lambda_2\tau}-1\right)m_2 + \gamma_2\lambda_2^{-1}\left(e^{\lambda_2(\tau_1-\tau)}-1\right)m_1 + \gamma_2 I_1,$$
$$I_1 = \int_0^{\tau_1} e^{\lambda_2(\tau_1-s)}\varphi(s+t_0)\,ds,$$

So,

$$\sigma_1 = \left(1-e^{\lambda_2(\tau_1+\tau_2)}\right)^{-1}\lambda_2^{-1}\left[e^{\lambda_2(\tau_1+\tau_2)}m_1 + e^{\lambda_2\tau_2}\left(m_2-m_1\right)-m_2\right]+$$
$$+\left(1-e^{\lambda_2(\tau_1+\tau_2)}\right)^{-1}\left[e^{\lambda_2\tau_2}I_2+I_1\right],$$
$$\sigma_2 = \left(1-e^{\lambda_2(\tau_1+\tau_2)}\right)^{-1}\lambda_2^{-1}\left[e^{\lambda_2(\tau_1+\tau_2)}m_2 + e^{\lambda_2\tau_1}\left(m_1-m_2\right)-m_1\right]+$$
$$+\left(1-e^{\lambda_2(\tau_1+\tau_2)}\right)^{-1}\left[e^{\lambda_2\tau_1}I_1+I_2\right].$$

Let $s_{i\,2} = \sigma_i/\gamma_2$, denote

$$s_{1\,1} = \left(1-e^{\lambda_1(\tau_1+\tau_2)}\right)^{-1}\lambda_1^{-1}\left[e^{\lambda_1(\tau_1+\tau_2)}m_1 + e^{\lambda_1\tau_2}\left(m_2-m_1\right)-m_2\right]+$$
$$+\left(1-e^{\lambda_1(\tau_1+\tau_2)}\right)^{-1}\left[e^{\lambda_1\tau_2}I_2+I_1\right],$$
$$s_{2\,1} = \left(1-e^{\lambda_1(\tau_1+\tau_2)}\right)^{-1}\lambda_1^{-1}\left[e^{\lambda_1(\tau_1+\tau_2)}m_2 + e^{\lambda_1\tau_1}\left(m_1-m_2\right)-m_1\right]+$$
$$+\left(1-e^{\lambda_1(\tau_1+\tau_2)}\right)^{-1}\left[e^{\lambda_1\tau_1}I_1+I_2\right],$$

then

$$s_i = \left(s_{i\,1},\, s_{i\,2}\right)',\quad i=1,2.$$

Denote

$$\psi_1(t) = \sigma(t_0+t,\, t_0,\, l_1,\, m_2) - l_2,$$
$$\psi_2(t) = \sigma(t_0+\tau+t,\, t_0+\tau,\, \hat{l}_1,\, m_1) - l_2,$$
$$\psi_3(t) = \sigma(t_0+\tau_1+t,\, t_0+\tau_1,\, l_1,\, m_1) - l_1,$$
$$\psi_4(t) = \sigma(t_0+\tau_1+\tau+t,\, t_0+\tau_1+\tau,\, \hat{l}_2,\, m_2) - l_1,$$

where

$$\hat{l}_1 = \sigma(t_0+\tau, t_0, l_1, m_2),$$
$$\hat{l}_2 = \sigma(t_0+\tau_1+\tau, t_0+\tau_1, l_2, m_1).$$

Following result may be formulated.

Theorem 2. *Consider a system*

$$\begin{cases} \sigma_1 = l_1, \\ \sigma_2 = l_2 \end{cases} \tag{8}$$

with respect to m_1, m_2 *(if equations in (8) are congruent then assume* $m_2 = -m_1$ *and consider only the first equation with respect to* m_1*). Let this system has a solution* $\hat{m}_{1,\,2}$ *such as* $\hat{m}_1 < \hat{m}_2$*. Let for given* $\gamma = \left(0,\, \gamma_2\right)'$ *and* $m_{1,\,2} = \hat{m}_{1,\,2}$ *conditions*

$$\begin{cases} \psi_1(t) < 0, & t \in [0, \tau), \\ \psi_2(t) < 0, & t \in [0, \tau_1 - \tau), \\ \psi_3(t) > 0, & t \in [0, \tau), \\ \psi_4(t) > 0, & t \in [0, \tau_2 - \tau), \end{cases} \tag{9}$$

are satisfied. In that case system (7) has a stable $(\tau_1 + \tau_2)$*-periodic solution with two switching points* s_i*, if* $\lambda_1 < 0$ *and*

$$(\tau_1 + \tau_2)\, T^{-1} \in N.$$

Proof. In order to prove the theorem it is enough to note that under above-listed conditions system (7) settles self-mapping of switching lines $\sigma = l_i$. Moreover, for any $x^{(i)}$ lying on switching line,

$$x_1^{(i+1)} = e^{\lambda_1(\tau_1+\tau_2)} x_1^{(i)} + \Theta, \quad \Theta \in R,$$

and in general case ($\Theta \neq 0$) the latter difference equation has stable solution only if $\lambda_1 < 0$. ■

In order to pass onto variables z_i it is enough to effect linear transform (6).

Note that conditions (9) may be readily verified using mathematical symbolic packages.

Example 2. Consider a system (5), where

$$g_1 = -0.1, \quad g_2 = -0.02, \quad l_{1,\,2} = \mp 1, \quad \varphi(t) = \sin(3t), \quad \tau = 0.5.$$

In that case

$$\lambda_1 = -0.1, \quad \lambda_2 = 0.2,$$

$$T = \begin{pmatrix} -10/3 & 10/3 \\ 1/3 & 2/3 \end{pmatrix}.$$

Let $\gamma_2 = -1,\ t_0 = \pi/3,\ \tau_{1,\,2} = \pi$, then

$$m_1 = -0.74195..., \quad m_2 = -m_1,$$

conditions (9) are satisfied. In z-coordinates,

$$\alpha = \left(T^{-1}\right)' \gamma = \begin{pmatrix} -1/10 \\ -1 \end{pmatrix}.$$

Plot of the periodic solution is presented below (virtual switching points are marked by the stars).

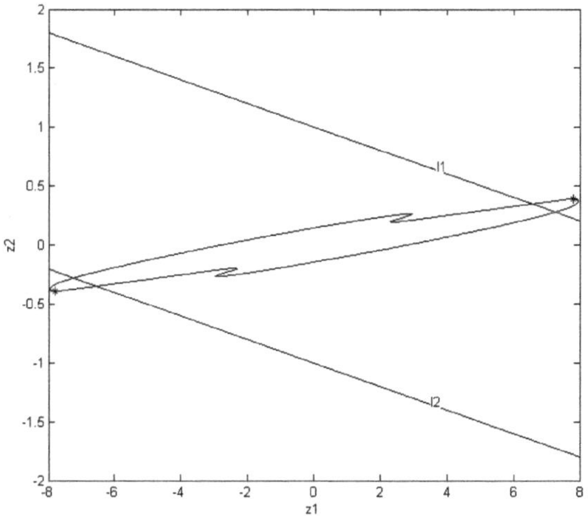

Fig. 1.

Similarly, if $\gamma_2 = -1,\ \tau_1 = 2\pi/3,\ \tau_2 = 4\pi/3$, then

$$m_1 = -1.64206...,\ m_2 = 1.01756...,$$

conditions (9) are satisfied, etc.

Note that here the exact solution of linear system (8) is known.

Example 3. In conclusion of the section let us consider a system (5), where

$$g_1 = 0.8, \quad g_2 = -0.2 \quad (\lambda_1 = -1, \quad \lambda_2 = 0.2),$$
$$\sigma(t) = 0.2y(t) - \dot{y}(t), \quad l_1 = 0, \quad l_2 = 1, \quad m_{1,2} = \mp 2,$$
$$\varphi(t) = 3\sin(\pi t) + 5\sin\left(\frac{3\pi t}{2}\right).$$

Here

$$T = \begin{pmatrix} -5/6 & 5/6 \\ 5/6 & 1/6 \end{pmatrix},$$

and the system can be transformed to x-coordinates and studied as above, but we'll restrict ourselves with numerical experiments for some particular values of parameter τ.

Suppose $t_0 = 0$; $y(0) = \dot{y}(0) = 0$; $u(t) = m_1$, when $t \in [-\tau, 0]$. Plots for some limit cycles in z-coordinates are cited below (switching points are marked by daggers, virtual switching points are marked by the stars).

Note that here the dynamics of the system essentially depends on quantity of delay. Steady-state solutions have different periods and number of switchings. Moreover, for some values of τ and for given initial values solutions of the systems are infinitely move away from origin as the time tends to infinity.

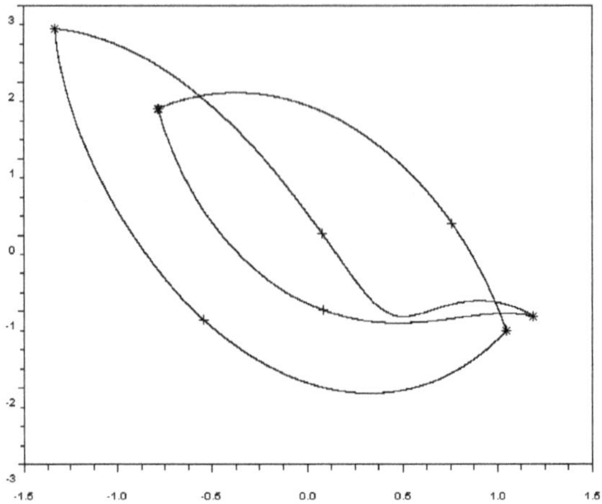

Fig. 2. $\tau = 0.5$

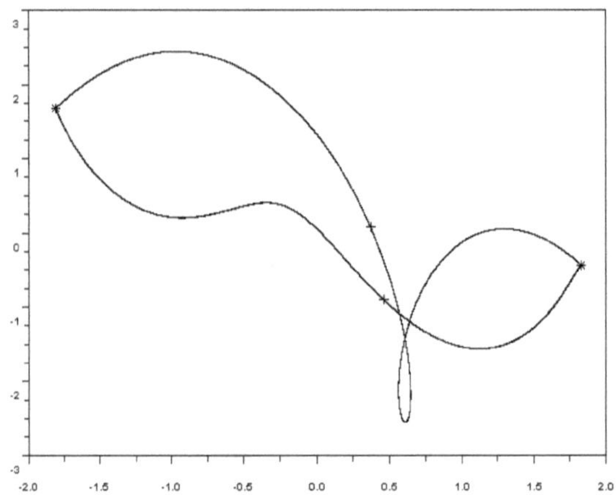

Fig. 3. $\tau = 0.75$

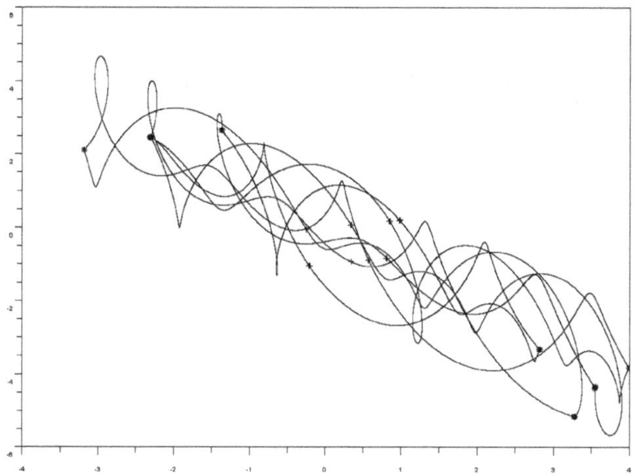

Fig. 4. $\tau = 1.5$

It is evident that here the complexity of analytical computations for exact finding of periodic solutions of the systems increases extensively. So the value of prior and attendant numerical experiments is undoubted (but of course the subject of analytical tracing cannot be wholly covered by computational modeling).

6 Conclusions

The results obtained above suppose further development. For example, calculations carried out in sections 4, 5 may be repeated for systems of dimension higher than two; more complicated nonlinearities may be considered (the same as in [11]). Simplification of conditions (9) when external action $\varphi(t)$ has an appointed form (e.g. linear combination of sinusoids with commensurable periods) represents a special problem.

References

1. Zubov, V.: Theory of optimal control of a vessel and other mobile objects, Sudpromgiz, Leningrad (1966) (in Russian)
2. Krasnoselsky, M., Pokrovsky, A.: Systems with hysteresis. Springer, New York (1989)
3. Visitin, A.: Differential models of hysteresis. Springer, Berlin (1994)
4. Astrom, K.J., Lee, T.H., Tan, K.K., Johansson, K.H.: Recent advances in relay feedback methods-a survey. Proc. of the IEEE. Conf. on Systems, Man and Cybernetics 3, 2616–2621 (1995)
5. Varigonda, S., Georgiou, T.T.: Dynamics of relay relaxation oscillators. IEEE Trans. on Automatic Control 46(1), 65–77 (2001)
6. Zubov, S., Zubov, N.: Mathematical methods of stabilization of dynamic systems. St. Petersburg Univ. Press, St. Petersburg (1996) (in Russian)
7. Bobylev, N.A., Boltyanskii, V.V., Yu, V.S., Kalashnikov, V.V., Kolmanovskii, V.B., Kozyakin, V.S., Kravchenko, A.A., Krasnosel'skii, A.M., Pokrovskii, A.V.: Mathematical Systems Theory, Nauka, Moscow (1986) (in Russian)

8. Kamachkin, A.M.: Existence and uniqueness of periodic solution of the relay system with hysteresis. Differential equations 8(8), 1505–1506 (1972) (in Russian)
9. Utkin, V.: Sliding modes in control and optimization. Springer, London (1992)
10. Kamachkin, A., Shamberov, V.: Finding of periodic solutions in nonlinear dynamic systems. St. Petersburg State University, St. Petersburg (2002) (in Russian)
11. Nelepin, R.: Methods of Nonlinear Vibrations Theory and their Application for Control Systems Investigation. St. Petersburg Univ. Press, St. Petersburg (2002) (in Russian)

SOS for Nonlinear Delayed Models in Biology and Networking

Antonis Papachristodoulou[1] and Matthew M. Peet[2]

[1] Department of Engineering Science, University of Oxford, Parks Road, Oxford, OX1 3PJ, U.K. antonis@eng.ox.ac.uk

[2] INRIA Rocquencourt, Domaine de Voluceau, BP 105, 78153 Le Chesnay Cedex, France. matthew.peet@inria.fr

Summary. In this chapter we illustrate how the Sum of Squares decomposition can be used for understanding the stability properties of models of models of biological and communication networks. The models we consider are in the form of nonlinear delay differential equations with multiple, incommensurate delays. Using the sum of squares approach, appropriate Lyapunov-Krasovskii functionals can be constructed, both for testing delay-independent and delay-dependent stability. We illustrate the methodology using examples from congestion control for the Internet and gene regulatory networks.

1 Introduction

Delay differential equations can be used to model the behavior of dynamical systems in which aftereffect is important. Examples include communication and computer networks (e.g., Internet congestion control), population dynamics of species interaction (e.g., predator-prey models with maturation), biological systems (e.g., gene regulatory networks with auto-regulation) etc. In all these examples, the delays inherent in the systems are important and cannot be ignored in modelling or analysis as this can lead to incorrect or misleading conclusions about the system under study.

A major problem in trying to understand the properties of communication and biological networks is that many times, multiple, incommensurate time-delays are required to accurately represent them. For example, when routers attempt to control Internet congestion by feeding back price signals to the users, there is an intrinsic propagation delay whose size is a function of the distance of the router from the user. Moreover, because users typically use many links in order to transmit their data, the feedback signals will be outdated by multiple incommensurate delays. As another example, consider predator-prey interactions - in this case, multiple time-delays would be needed to adequately model the system, due to the difference in the maturation rates of different species.

Nonlinear delay differential equations with multiple time delays are, however, very difficult to analyze. Even in the case of *linear* system descriptions, it is known that for multiple time-delays lying in the interval $\tau_i \in [0, \overline{\tau_i}]$ the question of stability is NP-hard [1]. Many results that have been developed in the past concentrate on linear systems with multiple

J.J. Loiseau et al. (Eds.): Topics in Time Delay Systems, LNCIS 388, pp. 133–143.
springerlink.com

time-delays and the use of frequency domain tools for performing the stability test [2, 3, 4, 5]. Other results use Lyapunov arguments [6, 7, 8, 9], but these tests are in general more conservative.

For the case of nonlinear time-delay systems, very few approaches have been proposed for algorithmic analysis of systems with multiple, incommensurate time-delays, despite the fact that such models appear naturally in many practical systems. In this chapter, we will show how the sum of squares decomposition of multivariable polynomials can be used to construct appropriate Lyapunov-Krasovskii functionals for polynomial time-delay systems in order to verify stability. This chapter concentrates more on the applications than the theory, for which the reader is referred to [10, 11, 12, 13].

We first present two examples where multiple time-delays appear naturally – communication networks and biology – and in Section 3 briefly outline the tools that can be used to answer the stability question. We return to these examples in Section 4 before concluding the chapter.

1.1 Notation

The notation we will be using is standard. $\mathbb{R}$ denotes the reals and $\mathbb{R}^n$ the $n-$ dimensional Euclidean space. We denote by $\mathbb{R}[x]$ the ring of real polynomials in variables x and by $Z_d[x]$ we denote the vector of monomials in x of degree d or less. We consider delays $\tau_i \in \mathbb{R}_+$, $i = 0, 1, \dots, K$ with $0 = \tau_0 < \tau_1 < \dots < \tau_K = \tau$ and define $H = \{-\tau_0, \dots, -\tau_k\}$ and $H^c = [-\tau, 0] \backslash H$. We will also denote the intervals $I_i = [-\tau_i, -\tau_{i-1}]$. Further, denote $C([-\tau, 0], \mathbb{R}^n)$ the Banach space of continuous functions mapping the interval $[-\tau, 0]$ into $\mathbb{R}^n$ with the norm on C is defined as $\|\phi\| = \sup_{-\tau \le \theta \le 0} |\phi(\theta)|$. Suppose $\sigma \in \mathbb{R}$, $\rho \ge 0$ and $x \in C([\sigma - \tau, \sigma + \rho], \mathbb{R}^n)$; then for any $t \in [\sigma, \sigma + \rho]$, define $x_t \in C$ by $x_t(\theta) = x(t+\theta), \theta \in [-\tau, 0]$. We will also be using symbolic independent variables to reference state and delayed state variables. To facilitate understanding, $\hat{x}_k^{(i)}$ will reference $x_i(t - \tau_k)$ where $l = 1, \dots, K$. We will use $\hat{x}_k$ to denote the vector whose elements are $\hat{x}_k^{(i)}$, $i = 1, \dots, n$. In a similar fashion, $\hat{y}_k^{(i)}$ will be used for $x_i(t + \theta - \tau_k)$ an $\hat{z}_k^{(i)}$ for $x_i(t + \xi - \tau_k)$.

2 The Network Models

2.1 Network Congestion Control for the Internet

Internet congestion control [14] is a distributed algorithm to allocate available bandwidth to competing sources so as to avoid congestion collapse by ensuring that link capacities are not exceeded. The need for congestion control for the Internet emerged in the mid-1980s, when congestion collapse resulted in unreliable file transfer. In 1988 Jacobson [15] proposed an admittedly ingenious scheme for congestion control. The shortcomings of this scheme and its successors such as TCP Reno and Vegas have only recently become apparent: they are not scalable to arbitrary networks with very large capacities and multiple, non-commensurate time-delays. New designs of Active Queue Management (AQM) and/or Transmission Control Protocol (TCP) schemes have been proposed that provide scalable stability in the presence of heterogeneous delays, which can be verified at least for the linearization about a steady-state.

The simplest adequate modeling framework for network congestion control is in the form of nonlinear deterministic delay-differential equations [16, 17, 18]. Some work has been done on the analysis of such systems usually for the single-bottleneck link case, using either Lyapunov-Razumikhin functions or Lyapunov-Krasovskii functionals, IQC methods, passivity etc.

Let us consider a simple network instantiation of what is known as a 'primal' congestion control scheme, shown in Figure 1. It consists of $L = 3$ links, labeled $l = 1, 2, 3$ and $S = 2$ sources, $i = 1, 2$. For this network, we define an $L \times S$ routing matrix R which is 1 if source

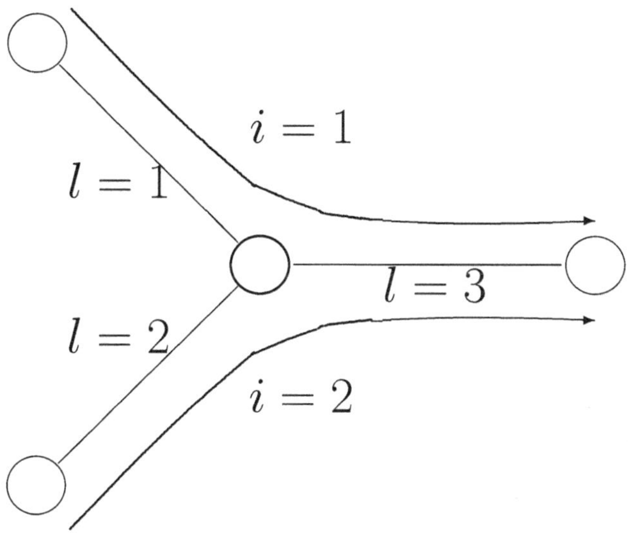

Fig. 1. A network topology under consideration.

i uses link l and 0 otherwise:

$$R = \begin{bmatrix} 1 & 0 \\ 0 & 1 \\ 1 & 1 \end{bmatrix}$$

The architecture of network congestion control is shown in Figure 2. To each source i we associate a transmission rate x_i. All sources whose flow passes through resource l contribute to the *aggregate rate* y_l for resource l, the rates being added with some forward time delay $\tau_{i,l}^f$. Hence we have:

$$y_l(t) = \sum_{i=1}^{S} R_{li} x_i(t - \tau_{i,l}^f) \triangleq r_f(x_i, \tau_{i,l}^f) \tag{1}$$

The resources l react to the aggregate rate y_l by setting congestion information p_l, the price at resource l. This is the Active Queue Management part of the picture. The prices of all the links that source i uses are aggregated to form q_i, the *aggregate price* for source i, again through a delay $\tau_{i,l}^b$:

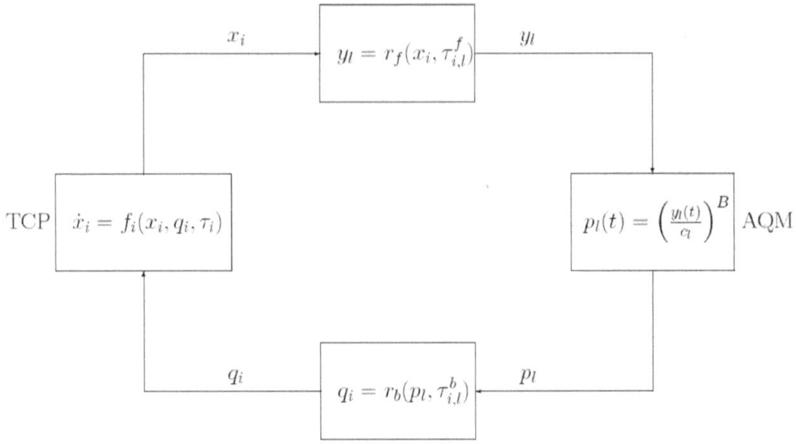

Fig. 2. The internet as an interconnection of sources and links through delays.

$$q_i(t) = \sum_{l=1}^{L} R_{li} p_l(t - \tau_{i,l}^b) \triangleq r_b(p_l, \tau_{i,l}^b) \tag{2}$$

The prices q_i can then be used to set the rate of source i, x_i, which completes the loop. The forward and backward delays can be combined to yield the Round Trip Time (RTT):

$$\tau_i = \tau_{i,l}^f + \tau_{i,l}^b, \quad \forall\, l \tag{3}$$

The capacity of link l is given by c_l. For a general network, the interconnection is shown in Figure 2. In this section we choose the control laws for TCP and AQM as follows:

$$p_l(t) = \left(\frac{y_l(t)}{c_l}\right)^B$$
$$\dot{x}_i(t) = 1 - x_i(t - \tau_i) q_i(t)$$

Here $p_l(t)$ corresponds to the probability that the queue length exceeds B in a $M/M/1$ queue with arrival rate $y_l(t)$ and capacity c_l, and the source law corresponds to a queue length with proportionally fair source dynamics.

In the particular case of the network shown in Figure 1, the closed loop dynamics become:

$$\dot{x}_1 = 1 - x_1(t - \tau_1) \left[\left(\frac{x_1(t - \tau_1)}{c_1} \right)^B + \left(\frac{x_1(t - \tau_1) + x_2(t - \tau_{1,3}^b - \tau_{2,3}^f)}{c_3} \right)^B \right]$$

$$\dot{x}_2 = 1 - x_2(t - \tau_2) \left[\left(\frac{x_2(t - \tau_2)}{c_2} \right)^B + \left(\frac{x_2(t - \tau_2) + x_1(t - \tau_{1,3}^f - \tau_{2,3}^b)}{c_3} \right)^B \right]$$

We will show in the examples section how a model like this can be analyzed.

2.2 Autoregulation in gene regulatory networks

Cells live in complex environments and need to respond to a range of different signals and parameter changes, such as temperature, pressure, the presence of nutrients etc. At the same time, the cell needs to respond to several internal signals. The way cells react to such changes is by producing appropriate proteins that act upon the internal and external environments, aiming to counter their effect.

Upon an environmental change, special proteins, called *transcription factors* become activated and bind to DNA to regulate the rate at which target genes are read. This is a two-step process: first, genes are read (transcribed) into mRNA, which is then translated into protein. This means, of course, that there are two time-delays from the time RNA polymerase and transcription factors bind to DNA and the time the protein is produced: translation cannot happen before transcription has been completed.

Another important aspect is that the resulting proteins can act on the environment but may also be transcription factors for other genes, in a feedback fashion. In particular, transcription factors can act as *activators*, in which case they increase the transcription rate of a gene, or *repressors*, in which case they reduce it. The strength of the effect of a transcription factor on the transcription rate of its target gene is described by an *input function*:

- In the case of activation, denoted by $X \to Y$, i.e., where the production of protein Y is controlled by a transcription factor X that acts as an activator, the number of molecules of Y is a function of the concentration of X in its active form, X^*. This is a monotonically increasing, S-shaped function, which is usually captured using a so-called Hill function:

$$f(x^*) = \frac{\beta_X (x^*)^n}{K^n + (x^*)^n}$$

  where K is the activation coefficient, n is the Hill coefficient and β is the maximal expression level.
- In the case of repression, denoted by $X \dashv Y$, the Hill function is a decreasing S-shaped function of the form

$$f(x^*) = \frac{\beta_X K^n}{K^n + (x^*)^n}$$

  The latter means that the production of protein Y is controlled by X which acts as an inhibitor.

The simplest gene regulatory network we can consider is a negative auto-regulation motif, in which the gene product acts as a transcription factor which represses its own production. We can model this as follows:

$$\begin{aligned}\dot{x}_1 &= -\lambda_1 x_1 + c_1 x_2(t - \tau_1)\\ \dot{x}_2 &= -\lambda_2 x_2 + \frac{c_2}{1 + (x_1(t - \tau_2))^n}\end{aligned}$$

where x_1 denotes the concentration of the protein while x_2 denotes the concentration of the mRNA produced by transcription of the gene encoding the protein. In the above model, λ_1 and λ_2 are natural degradation/dilution rates, while c_1 and c_2 are translation and transcription rates respectively. Also, τ_1 is the time required for the mRNA to be produced and transported to the ribosomes in order to initiate the transcription process, while τ_2 is the time-delay required before the production of the protein can actually regulate its own production. The parameter n is a Hill coefficient. We will analyze this model in Section 4.

In the next section, we discuss briefly the tools that we will develop for analyzing nonlinear time-delay systems with multiple, incommensurate time-delays.

3 Methodology

Consider a time-delay system with multiple delays of the form

$$\dot{x}(t) = f(x(t), x(t-\tau_1), \dots, x(t-\tau_K)) \tag{4}$$

where $f : \mathbb{R}^{n\times(K+1)} \to \mathbb{R}^n$ with $f(0,\dots,0) = 0$ is such that a unique solution to the above delay-differential equation exists from an appropriate initial condition close to 0. In the sequel, we assume that the vector field $f \in \mathbb{R}[\hat{x}_0, \hat{x}_1, \dots, \hat{x}_k]$. One type of Lyapunov functional that can be used to verify the delay-dependent (DD) stability of the zero steady-state takes the form

$$V(\phi) = \int_{-\tau}^{0} v_1(\phi(0), \phi(\theta), \theta) d\theta + \int_{-\tau}^{0}\int_{-\tau}^{0} v_2(\phi(\xi), \phi(\theta), \theta, \xi)\, d\theta\, d\xi. \tag{5}$$

This functional has a derivative of the general form

$$\begin{aligned}\dot{V}(\phi) = &\int_{-\tau}^{0} v_3(\phi(0), \phi(-\tau_1), \dots, \phi(-\tau_K), \phi(\theta), \theta) d\theta \\ &- \int_{-\tau}^{0}\int_{-\tau}^{0} \left(\frac{\partial v_2}{\partial \theta} + \frac{\partial v_2}{\partial \xi}\right) d\theta\, d\xi, \end{aligned} \tag{6}$$

where the map v_1, v_2 to v_3 will be presented later on.

In this chapter we will consider Lyapunov-Krasovskii functionals for studying the delay-independent and delay-dependent stability of the zero equilibrium of (4). This approach will transform these notions of stability into polynomial non-negativity conditions which are in general difficult to test. A sufficient condition for polynomial non-negativity which is worst-case polynomial-time verifiable is the existence of a sum of squares (SOS) decomposition. More details on positive polynomials and the SOS decomposition can be found in [19, 20, 21, 22]. Software, SOSTOOLS to support the theory is also available [23].

Denote by Σ the SOS cone and by Σ_d the subset of Σ of polynomials of degree d of less. First note that if $a \in \mathbb{R}[x]$ is an SOS, then it is globally nonnegative. In order to ensure that it is positive *definite* and radially unbounded we can use a polynomial 'shaping' function $\varphi(x)$. For example, given a polynomial $a(x)$ of degree $2d$, if we let

$$\varphi(x) = \sum_{i=1}^{n}\sum_{j=1}^{d} \epsilon_{ij} x_i^{2j}, \quad \sum_{j=1}^{d} \epsilon_{ij} \geq \gamma \quad \forall\, i = 1, \dots, n, \tag{7}$$

with γ a positive number, and $\epsilon_{ij} \geq 0$ for all i and j, then the condition $a(x) - \varphi(x) \in \Sigma$ guarantees the positive definiteness of $a(x)$, i.e., $a(x) > 0, x \neq 0$. Moreover, $a(x)$ is radially unbounded.

Another issue is related to testing non-negativity of a polynomial over a bounded domain instead of globally. Conditional satisfiability conditions can be tested using a generalization to the S-procedure, which is based on Putinar's representation [24] in Real Algebraic Geometry.

Given $p \in \mathbb{R}[x]$, suppose we want to ensure that $p(x) > 0$ on the set $D = \{x \in \mathbb{R}^n : g_i(x) \leq 0, i = 1, \dots, N_1\}$. Then one can search for Lagrange-type multipliers $\lambda_i \in \Sigma_k$ so that $p(x) + \sum_{i=1}^{N_1} \lambda_i(x) g_i(x) \in \Sigma$. Searching for $\lambda_i(x)$ of a fixed degree k so that the above expression is SOS is a semidefinite programme. Note that if p and g_i are quadratic forms and λ_i are constants, the above test is indeed the S-procedure, which can fail if $i \geq 2$. However, it has been shown in [24] that if D is compact and another mild condition holds on the $g_i(x)$ (the highest degree homogeneous parts of the g_i's have no common zeros in $\mathbb{R}^n$ except at 0), then there is a k for which the above test will succeed – it is indeed a necessary and sufficient condition. Other tests can also be formulated, which may have different properties. See e.g. [25].

3.1 Independent of Delay stability, Multiple-Delay case

The equilibrium of a time-delay system is said to be Independent Of Delay (IOD) stable if it is stable for all positive values of the time-delays. For simplicity, here we only develop the case in which there is a single steady-state and we seek to show global stability. Simple sufficient IOD stability conditions for global stability can be found in the following proposition:

Proposition 1. *Consider the system given by (4) with* 0 *a steady state. Suppose there exist functions* $v_0, v_{1i} : \mathbb{R}^n \to \mathbb{R}$ *for* $i = 1, \ldots, K$, *a positive definite radially unbounded function* $\varphi : \mathbb{R}^n \to \mathbb{R}$ *and a non-negative function* $\psi : \mathbb{R}^n \to \mathbb{R}$ *such that for all* $\hat{x}_0, \ldots, \hat{x}_K, \hat{y}_0 \in \mathbb{R}^n$:

1) $v_0(\hat{x}_0) - \varphi(\hat{x}_0) \geq 0$,
2) $v_{1i}(\hat{y}_0) \geq 0$ *for all* $i = 1, \ldots, K$,
3) $\nabla_{\hat{x}_0} v_0(\hat{x}_0)^T f(\hat{x}_0, \hat{x}_1, \ldots, \hat{x}_K) + \sum_{i=1}^K (v_{1i}(\hat{x}_0) - v_{1i}(\hat{x}_i)) + \psi(\hat{x}_0) \leq 0$.

Then the steady-state is globally stable. If $\psi(\hat{x}_0) > 0$, *then the steady-state is globally asymptotically stable.*

Proof. Consider the following Lyapunov-Krasovskii functional:

$$V(\phi) = v_0(\phi(0)) + \sum_{i=1}^{K} \int_{-\tau_i}^{0} v_{1i}(\phi(\theta))d\theta. \tag{8}$$

The first two conditions guarantee that this functional is positive definite and radially unbounded, by construction of φ. Differentiating $V(\phi)$ along the system trajectories we get:

$$\dot{V} = \nabla_{\hat{x}_0} v_0(\hat{x}_0)^T f(\hat{x}_0, \hat{x}_1, \ldots, \hat{x}_K) + \sum_{i=1}^{K} (v_{1i}(\hat{x}_0) - v_{1i}(\hat{x}_i))$$

which is non-positive by condition 3) in the above proposition. Therefore, the steady-state is globally stable using the Lyapunov Krasovskii Theorem [7]. Moreover, if $\psi(\hat{x}_0) > 0$ then we have $\dot{V} < 0$ and therefore the steady-state is globally asymptotically stable.

The cases of local stability, robustness analysis etc., can be dealt with in a unified manner. See [10] for more details.

Delay-Dependent, Multiple-Delay Case

The Lyapunov functions that one needs to use in order to show delay-dependent stability in the presence of multiple, incommensurate delays have piecewise continuous kernels, and take the form (5). Sufficient conditions for DD stability can then be written as follows:

Proposition 2. *Consider the system given by (4) with* 0 *a steady state and* f *a polynomial vector field. Given a vector of monomials* $Z_d[\hat{y}_0, \theta]$ *consider* $Q \succeq 0$ *and let* Q_{ij} *be the* i,j*th block of* Q *and*

$$v_{2ij}(\hat{y}_0, \hat{z}_0, \theta, \xi) = Z_d^T[\hat{y}_0, \theta] Q_{ij} Z_d[\hat{z}_0, \xi], \text{ for } i = 1, \ldots, K, \quad j = 1, \ldots, K.$$

Furthermore, suppose there exist polynomials $v_{1i}(\hat{x}_0, \hat{y}_0, \theta)$, $r_{1i}(\hat{x}_0, \theta)$ *and* $r_{2i}(\hat{x}_0, \hat{x}_1, \ldots, \hat{x}_K, \theta)$ *for* $i = 1, \ldots, K$, *a positive definite radially unbounded polynomial function* $\varphi(\hat{x}_0) > 0$ *and a non-negative polynomial function* $\psi(\hat{x}_0)$ *such that for all* $\hat{y}_0, \hat{x}_0, \ldots, \hat{x}_K \in \mathbb{R}^n$:

1) $v_{1i}(\hat{x}_0, \hat{y}_0, \theta) + r_{1i}(\hat{x}_0, \theta) - \varphi(\hat{x}_0) \geq 0$, *for all* $\theta \in [-\tau_i, -\tau_{i-1}]$ *for each* $i = 1, \ldots, K$.

2)
$$\nabla_{\hat{x}_0} v_{1i}(\hat{x}_0, \hat{y}_0, \theta)^T f(\hat{x}_0, \hat{x}_1, \ldots, \hat{x}_K) - \frac{\partial v_{1i}}{\partial \theta}(\hat{x}_0, \hat{y}_0, \theta)$$
$$+ \frac{1}{\tau_i - \tau_{i-1}} v_{1i}(\hat{x}_0, \hat{x}_{i-1}, -\tau_{i-1})$$
$$- \frac{1}{\tau_i - \tau_{i-1}} v_{1i}(\hat{x}_0, \hat{x}_i, -\tau_i) + 2\sum_{j=1}^{K} v_{2ij}(\hat{x}_{j-1}, \hat{y}_0, -\tau_{j-1}, \theta)$$
$$-2\sum_{j=1}^{K} v_{2ij}(\hat{x}_j, \hat{y}_0, -\tau_j, \theta) + r_{2i}(\hat{x}_0, \hat{x}_1, \ldots, \hat{x}_K, \theta) + \psi(\hat{x}_0) \leq 0$$
for all $\theta \in [-\tau_i, -\tau_{i-1}]$ *and for each* $i = 1, \ldots, K$,

3) $\sum_{i=1}^{K} \int_{I_i} r_{1i}(\hat{x}_0, \theta) d\theta = 0,,$

4) $\sum_{i=1}^{K} \int_{I_i} r_{2i}(\hat{x}_0, \hat{x}_1, \ldots, \hat{x}_K, \theta) d\theta = 0.$

5) There exist $R \succeq 0$ *and let* R_{ij}*, the i,jth block of* R *satisfy*

$$\frac{\partial v_{2ij}}{\partial \theta}(\hat{y}_0, \hat{z}_0, \theta, \xi) + \frac{\partial v_{2ij}}{\partial \xi}(\hat{y}_0, \hat{z}_0, \theta, \xi) = Z_d[\hat{y}_0, \theta]^T R_{ij} Z_d[\hat{z}_0, \xi],$$
$$i, j = 1, \ldots, K.$$

Then the steady-state is globally stable. If $\psi(\hat{x}_0) > 0$*, then the steady-state is globally asymptotically stable.*

Proof. The proof of this proposition is based on ensuring that the following functional is a Lyapunov-Krasovksii functional:

$$V(\phi) = \sum_{i=1}^{K} \int_{I_i} v_{1i}(\phi(0), \phi(\theta), \theta) d\theta + \sum_{i=1}^{K} \sum_{j=1}^{K} \int_{I_i} \int_{I_j} v_{2ij}(\phi(\theta), \phi(\xi), \theta, \xi) d\theta d\xi. \tag{9}$$

Conditions (1) and (3) in the proposition ensure that $V(\phi) > 0$, so the first Lyapunov condition is satisfied and moreover V is radially unbounded.

The derivative of this functional along the trajectories of (4) is:

$$\dot{V}(\phi) = \sum_{i=1}^{K} \int_{I_i} \begin{pmatrix} \nabla_{\phi(0)} v_{1i}(\phi(0), \phi(\theta), \theta)^T f(\phi(0), \phi(-\tau_1), \ldots, \phi(-\tau_K)) \\ -\frac{\partial v_{1i}}{\partial \theta}(\phi(0), \phi(\theta), \theta) + \frac{1}{\tau_i - \tau_{i-1}} v_{1i}(\phi(0), \phi(-\tau_{i-1}), -\tau_{i-1}) \\ -\frac{1}{\tau_i - \tau_{i-1}} v_{1i}(\phi(0), \phi(-\tau_i), -\tau_i) \\ +2\sum_{j=1}^{K} v_{2ij}(\phi(-\tau_{j-1}), \phi(\theta), -\tau_{j-1}, \theta) \\ -2\sum_{j=1}^{K} v_{2ij}(\phi(-\tau_j), \phi(\xi), -\tau_j, \theta) \end{pmatrix} d\theta$$
$$- \sum_{i=1}^{K} \sum_{j=i}^{K} \int_{I_i} \int_{I_j} \left(\frac{\partial v_{2ij}}{\partial \theta}(\phi(\theta), \phi(\xi), \theta, \xi) + \frac{\partial v_{2ij}}{\partial \xi}(\phi(\theta), \phi(\xi), \theta, \xi) \right) d\theta d\xi$$

The non-positivity of this time derivative is ensured by condition (2) when condition (4) is taken into account, as well as the decomposition given by (5). Therefore $V(\phi)$ is a Lyapunov-Krasovskii functional that proves global stability of the steady-state. If $\psi(\phi) > 0$, then $\dot{V}(\phi) < 0$ and the steady-state is globally asymptotically stable.

It is sometimes beneficial to consider other criteria for the stability analysis, such as combining all the above derivative conditions into one in more variables. For example, we can consider other Lyapunov functions, e.g., of the form:

$$V(x_t) = V_0(x(t)) + \sum_{i=1}^{K} \int_{-\tau_i}^{0} V_{1i}(\theta_i, x(t), x(t+\theta_i))d\theta_i$$
$$+ \sum_{i=1}^{K} \int_{-\tau_i}^{-\tau_{i-1}} \int_{t+\theta_i}^{t} V_{2i}(x(\zeta))d\zeta d\theta_i \tag{10}$$

Positivity tests similar to the ones shown earlier can be given in this case too.

We now turn to the examples we have considered in Section 2.

4 Examples

4.1 Network Congestion Control for the Internet

Consider the example developed in Section 2.1. Let $B = 2$, $c_1 = c_2 = 3$, and $c_3 = 1$. Then the steady state of this system is $(x_1^*, x_2^*) = (0.6242, 0.6242)$. We consider delay sizes such that $\tau_{1,3}^b = 63$ms, $\tau_{1,3}^f = 93$ms, $\tau_{2,3}^f = 49$ms, $\tau_{2,3}^b = 77$ms. The system equations about the new steady-state become:

$$\dot{x}_1' = 1 - [x_1'(t-0.154) + x_1^*]$$
$$\left[\left(\frac{x_1'(t-0.154)+x_1^*}{3}\right)^2 + \left(x_1'(t-0.154) + x_2'(t-0.14) + x_1^* + x_2^*\right)^2\right]$$
$$\dot{x}_2' = 1 - [x_2'(t-0.126) + x_2^*]$$
$$\left[\left(\frac{x_2'(t-0.126)+x_2^*}{3}\right)^2 + \left(x_2'(t-0.126) + x_1'(t-0.14) + x_1^* + x_2^*\right)^2\right]$$

where $x_i(t) = x_i'(t) + x^*$. The linearization of this system about the steady-state is stable and a Lyapunov function of the form (9) can be constructed. The same Lyapunov function can be constructed in a region of the state-space satisfying $\|x_{1_t}'\| \leq 0.8x_1^*$ and $\|x_{2_t}'\| \leq 0.8x_2^*$, thus showing that the steady-state is locally stable. The Lyapunov functional has the following form: all v_{1i} are 4th order, quadratic in $(\phi(0), \phi(\theta))$ and at most second order in θ and Z_d to construct v_{2ij} are third order: at most first order in $\phi(\theta)$ and second order in θ.

4.2 Stability of a negative gene autoregulation model with delays

Consider now the model developed in Section 2.2. We choose the following parameters: $\lambda_1 = \lambda_2 = 0.03$, $c_1 = c_2 = 1$ and investigate the stability bounds for the delays τ_i. The model we consider is very similar to the one developed in [26].

When the Hill coefficient is equal to 1, i.e., $n = 1$, the system is delay-independent stable and this can be verified through the construction of a Lyapunov Krasovskii functional. In this case we need to perform a local stability analysis as it does not make sense to consider negative initial conditions for concentration profiles. The resulting Lyapunov functional is quartic in all variables and is of the form (8).

5 Conclusions

In this chapter we have studied application examples from Biology and Communication networks and shown how the Sum of Squares decomposition of multivariable polynomials can be used to analyze the system behaviour.

References

1. Toker, O., Ozbay, H.: Complexity issues in robust stability of linear delay-differential systems. Math. Control, Signals, Syst. 9, 386–400 (1996)
2. Breda, D., Maset, S., Vermiglio, R.: Pseudospectral differencing methods for characteristic roots of delay differential equations. SIAM Journal on Scientific Computing 27(2), 482–495 (2005)
3. Ergenc, A.F., Olgac, N., Fazelinia, H.: Extended kronecker summation for cluster treatment of LTI systems with multiple delays. SIAM Journal on Control and Optimization 46(1), 143–155 (2007)
4. Fazelinia, H., Sipahi, R., Olgac, N.: Stability robustness analysis of multiple time-delayed systems using "building block" concept. IEEE Transactions on Automatic Control 52(5), 799–810 (2007)
5. Gu, K., Niculescu, S.I., Chen, J.: On stability crossing curves for general systems with two delays. Journal of Mathematical Analysis and Applications 311, 231–253 (2005)
6. Cao, Y.Y., Sun, Y.X., Cheng, C.: Delay-dependent robust stabilization of uncertain systems with multiple state delays. IEEE Transactions on Automatic Control 43(11), 1608–1612 (1998)
7. Hale, J.K., Verduyn Lunel, S.M.: Introduction to Functional Differential Equations. In: Applied Mathematical Sciences, p. 99. Springer, New York (1993)
8. Mehdi, D., Boukas, E.K., Liu, Z.K.: Dynamical systems with multiple time-varying delays: Stability and stabilizability. Journal of Optimization Theory and Applications 113(3), 537–565 (2002)
9. Xu, B., Liu, Y.H.: Delay dependent/delay independent stability of linear systems with multiple time-varying delays. IEEE Transactions on Automatic Control 48(4), 697–701 (2003)
10. Papachristodoulou, A.: Analysis of nonlinear time delay systems using the sum of squares decomposition. In: Proceedings of the American Control Conference, Boston, US (2004)
11. Peet, M., Lall, S.: Constructing Lyapunov functions for nonlinear delay-differential equations using semidefinite programming. In: Proceedings of IFAC Symp. NOLCOS (2004)
12. Papachristodoulou, A.: Scalable Analysis of Nonlinear Systems Using Convex Optimization. PhD thesis, California Institute of Technology, Pasadena, CA (2005)
13. Peet, M.M.: Stability and Control of Functional Differential Equations. PhD thesis, Stanford University, Palo Alto, CA (2006)
14. Srikant, R.: The Mathematics of Internet Congestion Control. Birkhäuser, Boston (2003)
15. Jacobson, V.: Congestion avoidance and control. In: Proceedings of ACM SIGCOMM 1988 (1988), `http://ftp://ftp.ee.lbl.gov/papers/congavoid.ps.Z`
16. Shakkottai, S., Srikant, R.: How good are deterministic fluid models of internet congestion control. In: Proc. of IEEE Infocom (2002)
17. Mathis, M., Semske, J., Mahdavi, J., Ott, T.: The macroscopic behavior of the TCP congestion avoidance algorithm. Computer Communication Review (1997)

18. Padhye, J., Firoiu, V., Towsley, D., Kurose, J.: Modeling TCP throughput: A simple model and its empirical validation. In: Proc. ACM Sigcomm (2001)
19. Parrilo, P.A.: Structured Semidefinite Programs and Semialgebraic Geometry Methods in Robustness and Optimization. Ph.D thesis, California Institute of Technology, Pasadena, CA (2000)
20. Lasserre, J.B.: Global optimization with polynomials and the problem of moments. SIAM Journal of Optimization 11, 796–817 (2001)
21. Nesterov, Y.: Squared functional systems and optimization problems. In: High Performance Optimization Methods, pp. 405–439. Kluwer Academic Publishers, Dordrecht (2000)
22. Henrion, D., Garulli, A. (eds.): Positive Polynomials in Control. Springer, Heidelberg (2005)
23. Prajna, S., Papachristodoulou, A., Parrilo, P.A.: SOSTOOLS – Sum of Squares Optimization Toolbox, User's Guide (2002), `http://www.cds.caltech.edu/sostools`
24. Putinar, M.: Positive polynomials on compact semialgebraic sets. Indiana University Mathematics Journal 42(3), 969–984 (1993)
25. Bochnak, J., Coste, M., Roy, M.F.: Real Algebraic Geometry. Springer, Heidelberg (1998)
26. Bernard, S., Čajavec, B., Pujo-Menjouet, L., Mackey, M.C., Herzel, H.: Modelling transcriptional feedback loops: the role of Gro/TLE1 in Hes1 oscillations. Phil. Trans. R. Soc. A 364, 1155–1170 (2006)

TRACE-DDE: a Tool for Robust Analysis and Characteristic Equations for Delay Differential Equations

Dimitri Breda[1], Stefano Maset[2], and Rossana Vermiglio[3]

[1] Dipartimento di Matematica e Informatica, Università degli Studi di Udine, via delle Scienze, 206, I-33100 Udine, Italy dimitri.breda@dimi.uniud.it
[2] Dipartimento di Matematica e Informatica, Università degli Studi di Trieste, via Valerio 12, I-34127 Trieste, Italy maset@units.it
[3] Dipartimento di Matematica e Informatica, Università degli Studi di Udine, via delle Scienze, 206, I-33100 Udine, Italy rossana.vermiglio@dimi.uniud.it

Summary. In the recent years the authors developed numerical schemes to detect the stability properties of different classes of systems involving delayed terms. The base of all methods is the use of pseudospectral differentiation techniques in order to get numerical approximations of the relevant characteristic eigenvalues. This chapter is aimed to present the freely available Matlab package TRACE-DDE devoted to the computation of characteristic roots and stability charts of linear autonomous systems of delay differential equations with discrete and distributed delays and to resume the main features of the underlying pseudospectral approach.

1 Introduction

It is nowadays widely recognized that many real phenomena in physics, engineering, chemistry, biology, economics etc. are better modeled and simulated if time delays are taken into consideration. Delay systems are fundamental in control theory, where the effects of delays on stability are a crucial problem [19], [21], [18], but key applications can be found also in machining tool where the role of parameters such as spindle speed and feed are stability determining [23], [16]. Moreover, time delays are concerned in other fields involving different models whose stability characteristics are important, e.g. age-structured population dynamics [4], neutral and advanced-retarded functional differential equations [9] and partial differential equations with delays [10].

All delay systems share the common feature of being influenced, in their present evolution, by information on their past history. Much of their interest is concerned with the asymptotic stability analysis of the linear (or linearized) case. The lack of good estimates of the parameter values involved in system models (e.g. delays, but not only) leads to develop suitable criteria to determine not only whether a nominal system is stable or not, but an entire stability region of parameters due to this uncertainty (robust stability). In particular, when we deal with two variable parameters, we talk about *stability charts*.

J.J. Loiseau et al. (Eds.): Topics in Time Delay Systems, LNCIS 388, pp. 145–155.
springerlink.com

In order to tackle the two above questions (nominal and robust stability), several techniques have been proposed in the more or less recent literature. In the exhaustive monograph [19] many graphical and analytical tests to this aim are reported, but in general they work only for restricted classes of delay differential equations (DDEs), e.g. constant delays, single-delay dynamics, commensurate delays and so on. Here we consider a system of m linear DDEs with multiple discrete and distributed delays:

$$y'(t) = L_0 y(t) + \sum_{l=1}^{k} L_l y(t-\tau_l) + \int_{-\tau}^{0} M(\theta) y(t+\theta) d\theta, \; t \geq 0, \tag{1}$$

where $L_0, L_1, \ldots, L_k \in \mathbb{C}^{m \times m}$, $0 = \tau_0 < \tau_1 < \ldots < \tau_k = \tau$ and $M : [-\tau, 0] \to \mathbb{C}^{m \times m}$ is a piecewise smooth function. Indeed in the implementation of the method we write the distributed term as

$$\int_{-\tau}^{0} M(\theta) y(t+\theta) d\theta = \sum_{l=1}^{k_d} \int_{-\tau_l}^{-\tau_{l-1}} M_l(\theta) y(t+\theta) d\theta, \; t \geq 0,$$

where M_l, $l = 1, \ldots, k_d$, is smooth (without loss of generality one can assume $k_d = k$). However, here we restrict to the standard case (1) with smooth M.

In the last few years we devoted our research activity to the numerical computation of the characteristic roots of (1), i.e. the (infinitely many) roots of $\det(\Delta(\lambda)) = 0$ where

$$\Delta(\lambda) = \lambda I - \sum_{l=0}^{k} L_l e^{-\lambda \tau_l} - \int_{-\tau}^{0} M(\theta) e^{\lambda\theta} d\theta, \; \lambda \in \mathbb{C},$$

since it is well known that the zero solution of (1) is asymptotically stable if and only if these roots have strictly negative real part ([11]).

Several numerical approaches for characteristic roots computation have been proposed, which are based on the discretization of either the solution operator associated to (1) or the infinitesimal generator of the solution operator semigroup. We briefly recall that the solution operator $T(t)$, $t \geq 0$, associated to (1) is defined by $T(t)\varphi = y_t$, $\varphi \in X$, where $X = C([-\tau, 0], \mathbb{C}^m)$ is endowed with the maximum norm, y_t is the *state* of the system, i.e. the function $y_t(\theta) = y(t+\theta)$, $\theta \in [-\tau, 0]$, and y is the solution of (1) with initial data $y_0 = \varphi \in X$ at $t = 0$. The family $\{T(t)\}_{t \geq 0}$ is a C_0-semigroup with infinitesimal generator $\mathcal{A} : D(\mathcal{A}) \subseteq X \to X$ given by

$$\mathcal{A}\varphi = \varphi', \; \varphi \in D(\mathcal{A}), \tag{2}$$

with domain

$$D(\mathcal{A}) = \left\{ \varphi \in X \; : \; \varphi' \in X, \; \varphi'(0) = \sum_{l=0}^{k} L_l \varphi(-\tau_l) + \int_{-\tau}^{0} M(\theta)\varphi(\theta) d\theta \right\}. \tag{3}$$

So (1) can be restated as the following abstract Cauchy problem ([11]) on the state space X:

$$\begin{cases} \dfrac{du}{dt}(t) = \mathcal{A}u(t), \; t > 0, \\ u(0) = \varphi \in X, \end{cases}$$

whose solution is $u(t) = y_t$ whenever $\varphi \in D(\mathcal{A})$. The two following important results ([11], [15])

1. $\det(\Delta(\lambda)) = 0 \Leftrightarrow \forall t > 0,\ \lambda = \frac{1}{t} \ln \mu$ for some $\mu \in \sigma(T(t)) \setminus \{0\}$;
2. $\det(\Delta(\lambda)) = 0 \Leftrightarrow \lambda \in \sigma(\mathcal{A})$;

where $\sigma(\cdot)$ denotes the spectrum, suggest the idea to turn the characteristic roots approximation problem into a corresponding eigenvalue problem for suitable matrix discretization of either $T(t)$ (*solution operator* approach) or $\mathcal{A}$ (*infinitesimal generator* approach).

Engelborghs and Roose proposed in [13] the solution operator approach via linear multistep (LMS) time integration of (1) without distributed delay term. Their method computes approximations to the roots from a large, standard and sparse eigenvalue problem and it is implemented in the Matlab package DDE-BIFTOOL for DDEs bifurcation analysis [12]. The distributed delay case is considered in [17] by using LMS methods and in [3] by using Runge-Kutta (RK) methods. The complete development of the infinitesimal generator approach first appeared in [5] and [2] where a matrix approximation $\mathcal{A}_N$ of $\mathcal{A}$ is obtained discretizing the derivative in (2) by RK or LMS method, respectively, ending up with a large and sparse eigenvalue problem as in [13]. Finally, the infinitesimal generator approach via pseudospectral differencing methods has been proposed in [8]. The technique is based on the exact differentiation of interpolants at selected sets of $N+1$ nodes. Although the resulting differentiation matrix is non-sparse, the advantage of the well-known "spectral accuracy" (see [24] and the bibliography therein) allows very accurate approximations with small matrix dimension. This behavior represents in fact, for sufficiently small tolerance, the outstanding advantage of this method compared to the above mentioned schemes. Among those methods for stability detection not based on the semigroup structure of the problem it is worthy to mention the Cluster Treatment of Characteristic Roots (CTCR) by Olgac and Sipahi (see e.g. [20]) and the Quasi Polynomial mapping based Root-finder (QPmR) by Vyhlídal and Zítek (see [25]).

The chapter is organized as follows. In Section 2 we present the Graphic User Interface (GUI) Matlab package TRACE-DDE, acronym for Tool for Robust Analysis and Characteristic Equations for Delay Differential Equations. The software allows an efficient and reliable determination of the characteristic roots and the stability chart of systems of DDEs like (1). Its usage is explained by following a tutorial example while the underlying methods are recalled in the forthcoming sections. In particular, in Section 3 we shortly recover from [8] the basic facts about the numerical computation of characteristic roots of (1) through the pseudospectral discretization of the associated infinitesimal generator. Then, in Section 4, we recall the main lines of a recent work of the authors on the approximation of level curves of surfaces ([6],[7]). The chapter is closed with several experimental results showing features, appearance and performances of the new tool.

2 The graphic user interface

TRACE-DDE is a Matlab package equipped with a GUI devoted to the robust analysis and the computation of the characteristic roots of delay systems entering the class (1). The software is publicly available (`http://users.dimi.uniud.it/~dimitri.breda/software.html`) and its core is made of the two algorithms briefly described in the forthcoming Sections 3 and 4.

In this section we limit ourselves to a brief description through the computation of the roots and of a stability chart for the system [14]

$$y'(t) = L_0 y(t) + L_1 y(t-a) + \int_{-b}^{0} M(\theta) y(t+\theta) d\theta \tag{4}$$

with coefficients matrices

$$L_0 = \begin{pmatrix} -3 & 1 \\ -24.646 & -35.430 \end{pmatrix}, L_1 = \begin{pmatrix} 1 & 0 \\ 2.35553 & 2.00365 \end{pmatrix}, M = \begin{pmatrix} 1 & -\theta \\ 1 & -\theta \end{pmatrix},$$

where a and b are uncertain parameters both with nominal value 1.

Once the source package has been downloaded, thanks to the graphical suite, the user is required only to type `tracedde` at the Matlab prompt and then to follow the self-contained messages through the forthcoming windows. The first couple of windows (Figure 1) help the user in choosing the type of computation (i.e. characteristic roots or stability chart of a given system) and in inserting the preliminary data such as m, k and k_d as given in (1). In particular, the menu "LOAD" in the second window allows to upload existing data saved in previous sessions.

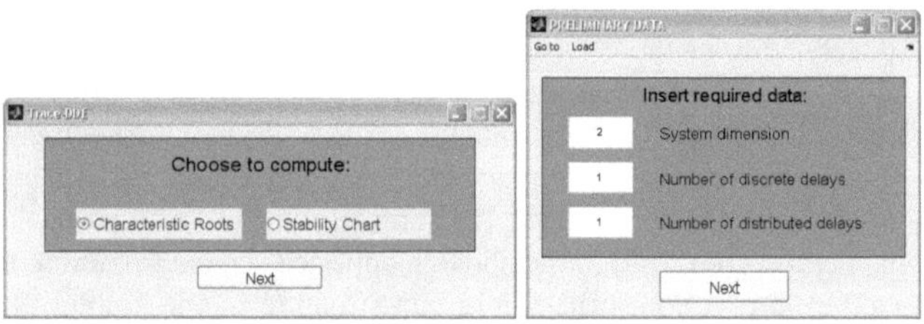

Fig. 1. Starting windows: problem choice (left) and system data (right).

Supposing to start with the roots computation (Section 3), the window in Figure 2, left, appears and the user is asked to fill the boxes with the required data by using the standard Matlab syntax for matrices (this is the only Matlab requirement). The notation follows the representation of the chosen DDE appearing at the bottom. The maximum number of roots $m(N + 1)$ (with N the discretization parameter, see Section 3) is determined automatically since $N = 20$ is set by default. Obviously, the users requiring more roots or more accuracy may change this value in the source code. Once the computation is performed, the result window (Figure 2, right) is produced and several options (such as, print, plot, etc.) are available.

In order to switch to the stability chart computation (Section 4), one can run `tracedde` again or go directly to the appropriate window (Figure 3, left) by selecting the relevant choice in the "GO TO" menu in Figure 2, left. With this second choice all data appear automatically and the user is required only to choose the uncertain parameters using variables (a and b for (4)) and insert their ranges in the allowed boxes. Then the option to plot the stability boundaries (Figure 3, right, observe the triangulation structure along the boundary) or surface is provided.

We are not going to give here a full description of all the features provided in TRACE-DDE. The motivation is twofold: on the one hand, the GUI structure should be self-contained, on the other hand we want to focus on the general purpose of the package. Nevertheless, we would like to mention, among others, the presence of error-alerting messages, data saving and loading facilities and zooming of stability chart zones. Other details can be found in the relevant web site. Anyway, let us invite possible users to report their feedbacks to the authors.

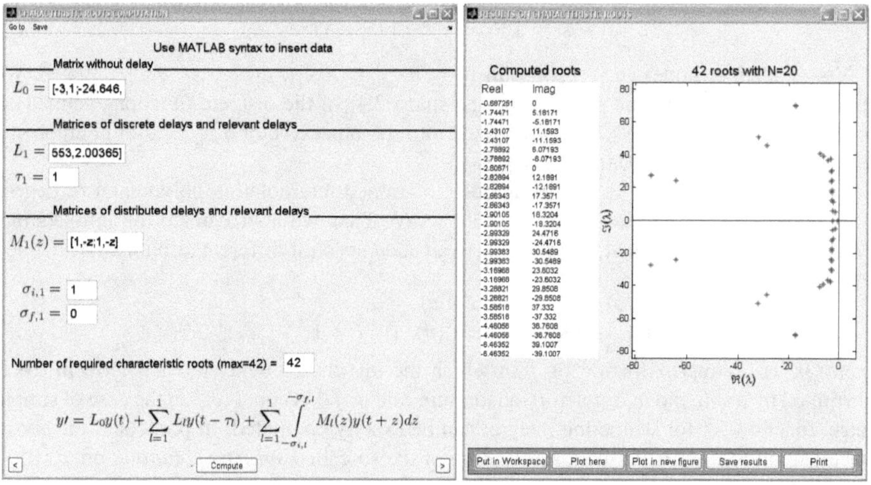

Fig. 2. Characteristic roots windows: data (left) and results (right).

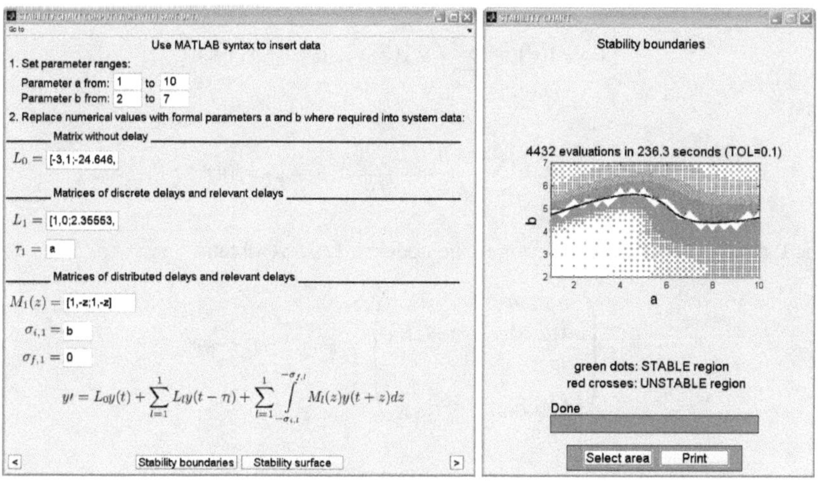

Fig. 3. Stability charts windows: data (left) and results (right).

3 The pseudospectral approach

For the sake of simplicity, let us rewrite the system of DDEs (1) as $y'(t) = f(y_t)$ where $f : X \to \mathbb{C}^m$ is defined by

$$f(\varphi) = \sum_{l=0}^{k} L_l \varphi(-\tau_l) + \int_{-\tau}^{0} M(\theta)\varphi(\theta)d\theta, \; \varphi \in X. \tag{5}$$

For a given N, N positive integer, let us consider the mesh

$$\Omega_N = \{\theta_{N,i},\ i = 0, 1, \ldots, N\}$$

of $N+1$ distinct nodes in $[-\tau, 0]$ with $0 = \theta_{N,0} > \theta_{N,1} > \ldots > \theta_{N,N} \geq -\tau$. We replace the continuous state space X by the space X_N of the discrete functions defined on the mesh Ω_N, i.e. any $\varphi \in X$ is discretized into the block-vector $x \in X_N$ of components $x_i = \varphi(\theta_{N,i}) \in \mathbb{C}^m$, $i = 0, 1, \ldots, N$.

Let now $\mathcal{L}_N x$, $x \in X_N$, be the unique $\mathbb{C}^m$-valued interpolating polynomial of degree $\leq N$ with $(\mathcal{L}_N x)(\theta_{N,i}) = x_i$, $i = 0, 1, \ldots, N$. Thus we approximate the infinitesimal generator $\mathcal{A}$ by the matrix $\mathcal{A}_N : X_N \to X_N$, called spectral differentiation matrix, defined as follows:

$$\begin{cases} (\mathcal{A}_N x)_0 = f_N(\mathcal{L}_N x), \\ (\mathcal{A}_N x)_i = (\mathcal{L}_N x)'(\theta_{N,i}),\ i = 1, \ldots, N, \end{cases} \tag{6}$$

where f_N is an approximation of f in which the distributed delay integral term in (5) is substituted by a suitable interpolatory quadrature rule or f_N is equal to f in the case of simple matrix function M for which the integral can be exactly computed. In particular, the above discretization follows from the discretization of the so-called *splicing condition* on $\varphi'(0)$ in (3) for what concerns the first block-row, while the remaining part is due to the derivative action (2) of the infinitesimal generator.

By using the Lagrange representation of $\mathcal{L}_N x$, i.e.

$$(\mathcal{L}_N x)(\theta) = \sum_{j=0}^{N} \ell_{N,j}(\theta) x_j,\ \theta \in [-\tau, 0],$$

where

$$\ell_{N,j}(\theta) = \sum_{\substack{i=0 \\ i \neq j}}^{N} \frac{\theta - \theta_{N,i}}{\theta_{N,j} - \theta_{N,i}},\ \theta \in [-\tau, 0],$$

are the Lagrange coefficients relevant to the nodes in Ω_N, we obtain

$$\mathcal{A}_N = \begin{pmatrix} a_0 & a_1 & \cdots & a_N \\ d_{10} & d_{11} & \cdots & d_{1N} \\ \vdots & \vdots & \ddots & \vdots \\ d_{N0} & d_{N1} & \cdots & d_{NN} \end{pmatrix} \in \mathbb{C}^{m(N+1) \times m(N+1)}$$

with

$$\begin{cases} a_j = f_N\left(\ell_{N,j}(\cdot) I_m\right),\ j = 0, 1, \ldots, N, \\ d_{ij} = \ell'_{N,j}(\theta_{N,i}) I_m,\ i = 1, \ldots, N,\ j = 0, 1, \ldots, N, \end{cases}$$

where, by definition, $f_N\left(\ell_{N,j}(\cdot) I_m\right) = \left(f_N\left(\ell_{N,j}(\cdot) e^{(i)}\right)\right)_{i=1}^{m}$ with $e^{(i)}$'s the canonical vectors in $\mathbb{R}^m$. Explicit expressions of the d_{ij}'s for particular choices of Ω_N (e.g. Chebyshev extremal points) can be found in [24, §6].

Therefore the original infinite-dimensional problem of the numerical computation of the characteristic roots of (1) can be turned into the finite-dimensional eigenvalue problem for the matrix $\mathcal{A}_N$, i.e. eigenvalues of $\mathcal{A}_N$ directly approximate the characteristic roots. If one wonders how much accurate are these approximations, the following theorem, whose proof is detailed in [8] through a complete convergence analysis, should serve as an answer.

Theorem 1. *Assume to use the Chebyshev extremal nodes, i.e.*

$$\theta_{N,i} = \frac{\tau}{2}\left(\cos\left(i\frac{\pi}{N}\right) - 1\right),\ i = 0, 1, \ldots, N,$$

and that the function f_N *in (6) satisfies* $\sup_{N\in\mathbb{N}} \|f_N\| < +\infty$. *Let* $\lambda^* \in \mathbb{C}$ *be a characteristic root of (1) with multiplicity* ν. *Then, for sufficiently large* N, $\mathcal{A}_N$ *has exactly* ν *eigenvalues* λ_i, $i = 1, \ldots, \nu$ *(counted with their multiplicity), such that*

$$\max_{i=1,\ldots,\nu} |\lambda^* - \lambda_i| \leq \left[C_2 \left(\varepsilon_N + \frac{1}{\sqrt{N}} \left(\frac{C_1}{N} \right)^N \right) \right]^{1/\nu} \tag{7}$$

holds where C_1 *and* C_2 *are constants independent of* N *and* ε_N *is the error due to the approximate computation of the distributed term.*

It is important to point out that the spectral accuracy (i.e. $O(N^{-N})$) of the method is preserved in the behavior of the final error (7), if the distributed term in (6) can be exactly computed, i.e. $\varepsilon_N = 0$. On the other hand, we also maintain the spectral accuracy if we make the further assumptions that the function M in (1) is C^∞ with equilimited derivatives and that a interpolatory quadrature rule based on Chebyshev nodes as given in Theorem 1 is used for the distributed term (Clenshaw-Curtis quadrature). Finally, we observe that if the distributed term is approximated by some software within a tolerance TOL, then the error bound in Theorem 1 is still valid with $\varepsilon_N = \text{TOL}$ and thus the convergence follows spectral accuracy down to this value.

As to conclude the section, we give an explicit expression of the discretization matrix for the single constant delay scalar equation

$$y'(t) = ay(t) + by(t - \tau), \ a, b \in \mathbb{C}.$$

In this case, it is easy to see that $\mathcal{A}_N \in \mathbb{C}^{(N+1)\times(N+1)}$ is given by

$$\mathcal{A}_N = \begin{pmatrix} a & 0 & \cdots & 0 & b \\ \ell'_{N,0}(\theta_{N,1}) & \ell'_{N,1}(\theta_{N,1}) & \cdots & \ell'_{N,N-1}(\theta_{N,1}) & \ell'_{N,N}(\theta_{N,1}) \\ \vdots & \vdots & \ddots & \vdots & \\ \ell'_{N,0}(\theta_{N,N}) & \ell'_{N,1}(\theta_{N,N}) & \cdots & \ell'_{N,N-1}(\theta_{N,N}) & \ell'_{N,N}(\theta_{N,N}) \end{pmatrix}$$

whenever $\theta_{N,N} = -\tau$.

4 Detection of stability boundaries

In this section we briefly describe an adaptive strategy which can be employed in order to detect the stability boundaries of a system of DDEs in the plane of two uncertain parameters. The problem can be viewed as a particular instance of the more general question of computing the level curves of a surface function $z = f(x, y)$ where, possibly, f does not have an explicit form but rather it can be evaluated for any choice of x and y in a given rectangular region of the (x, y)-plane. In fact, the complete stability map in the parameters plane is the set of level curves $f(x, y) = 0$ where f is the function giving the real part of the rightmost eigenvalue governing the system dynamics. Hence, in our case, f corresponds to an eigenvalue problem, possibly of large dimension (e.g. space-discretized partial differential equations), and its computation at one point (x, y) could be heavy.

Here we describe only the underlying idea of the algorithm while we refer the interested reader to [6] and [7] for a fully detailed version of the methodology which is implemented in TRACE-DDE.

A first rude (but simple) technique is used in the Matlab's `contour` plot. It is based on a uniform grid dividing the plane region into rectangular cells at whose vertices f is evaluated. If the signs at the vertices of a cell are not equal, a segment of level curve crosses the cell. Crossing points are approximated by linear interpolation and joint together they reproduce the searched boundary. The final accuracy depends only on the grid size. This tool inevitably calculates many values of f which are useless and in order to reduce the amount of computation an adaptive refinement is proposed.

In alternative to the obvious rectangular refinement (each cell is divided into four rectangles by the mid points), we consider a triangular refinement by which each cell is divided at the first step into four triangles by the cell center and any of these latter into two triangles in all the forthcoming steps by the height foot. Triangulation is more efficient in reducing the overall number of evaluation points. In fact, every triangular refinement requires one new evaluation (the cell center or the height foot) while the rectangular one requires five new evaluations (the cell center plus the mid point of each edge). Moreover it can be accompanied with a couple of other tricks (described in [7]) to further reduce the computational cost while preserving the desired accuracy.

5 A test set

In this section we collect a set of results which we believe it might be useful for comparison purposes. Analytical knowledge of exact characteristic roots and stability boundaries is, in fact, restricted to rather exceptional cases. To this aim we cite [15]: "*For the autonomous matrix equation...*" $x'(t) = Ax(t) + Bx(t-r)$ "*...the exact region of stability as an explicit function of A, B and r is not known and probably will never be known. The reason is simple to understand because the characteristic equation is so complicated. It is therefore worthwhile to obtain methods for determining approximations to the region of stability.*" Hence we report in the sequel a list of case-studies with relevant plots of the rightmost part of their (approximated) spectrum, a stability chart and the value of the righmost root accurate to machine precision. All results are obtained by using the core algorithms in TRACE-DDE and concern the following systems and equations:

Example 1. [13]

$$y'(t) = ay(t) + by(t-1).$$

Example 2. [20]

$$y'(t) = \begin{pmatrix} -6.45 & -12.1 \\ 1.5 & -0.45 \end{pmatrix} y(t) + \begin{pmatrix} -6 & 0 \\ 1 & 0 \end{pmatrix} y(t-\tau_1) + \begin{pmatrix} 0 & 4 \\ 0 & -2 \end{pmatrix} y(t-\tau_2)$$

Example 3. [14]

$$y'(t) = L_0 y(t) + L_1 y(t-1) + \int_{-\tau_1}^{-0.1} M_1 y(t+\theta)d\theta + \int_{-\tau_2}^{-0.5} M_2 y(t+\theta)d\theta$$

with coefficients matrices

$$L_0 = \begin{pmatrix} -3 & 1 \\ -24.646 & -35.430 \end{pmatrix}, L_1 = \begin{pmatrix} 1 & 0 \\ 2.35553 & 2.00365 \end{pmatrix},$$

$$M_1 = \begin{pmatrix} 2 & 2.5 \\ 0 & -0.5 \end{pmatrix}, M_2 = \begin{pmatrix} -1 & 0 \\ 0 & -1 \end{pmatrix}.$$

Example 4. [1]

$$y'(t) = L_0 y(t) + L_1(y(t-\tau_1) + y(t-\tau_2)) + \\ + L_2(y(t-2\tau_1) + y(t-2\tau_2)) + L_3 y(t-\tau_1-\tau_2)$$

whose coefficients matrices $L_l \in \mathbb{C}^{8\times 8}$, $l = 0, 1, 2, 3$, can be found in [1].

For every example we report in Table 1 the parameters nominal values, the corresponding rightmost root (or imaginary couple) λ_r, the tolerance at which this latter is approximated (which determines the number of significant digits) and the discretization index N at which such an accuracy is met first. Almost full machine precision is performed for Examples 2 and 3 with very low N. As for Example 1, the accuracy is halved since $\lambda_r = 1$ is a double root as it can be easily verified through the characteristic equation (as stated in Theorem 1). As for the last Example 4, the final accuracy level is rather low due to ill-conditioning of the coefficients matrices: this may serve as an interesting case-study to test a root-finder performances on dramatically ill-posed problems.

Table 1. Nominal values and righmost roots.

example	nominal values	N	λ_r	TOL
1	$a = 2, b = e$	10	0.9999999 ± 0.0000002i	10^{-7}
2	$\tau_1 = 0.2, \tau_2 = 0.3$	14	$-1.3352684760607 \pm 9.1194833474024$i	10^{-13}
3	$\tau_1 = 0.3, \tau_2 = 1$	13	-1.24623812459204	10^{-14}
4	$\tau_1 = \tau_2 = 5 \times 10^{-4}$	17	-22.616 ± 3617.030i	10^{-3}

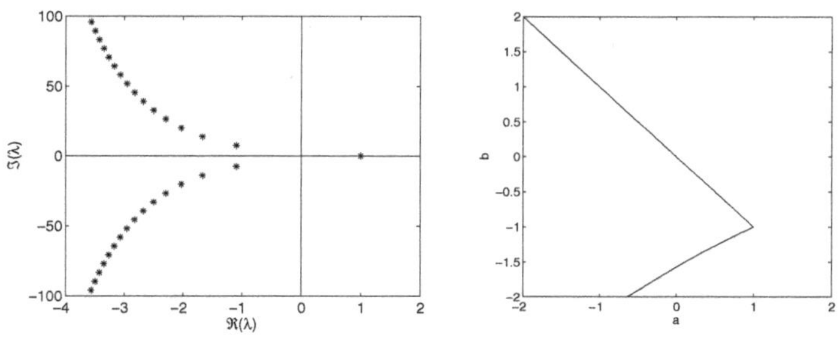

Fig. 4. Spectrum and stability chart for Example 1.

6 Acknowledgments

The authors are grateful to Daniele Sechi who built the first GUI version of TRACE-DDE for Macintosh platforms as part of his B.Sc. thesis in computer science at University of Udine [22].

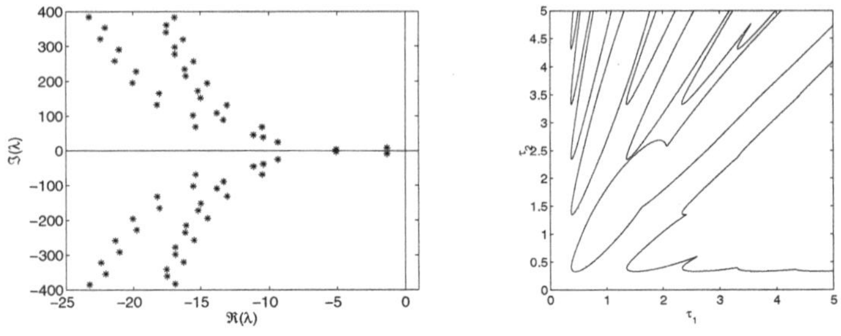

Fig. 5. Spectrum and stability chart for Example 2.

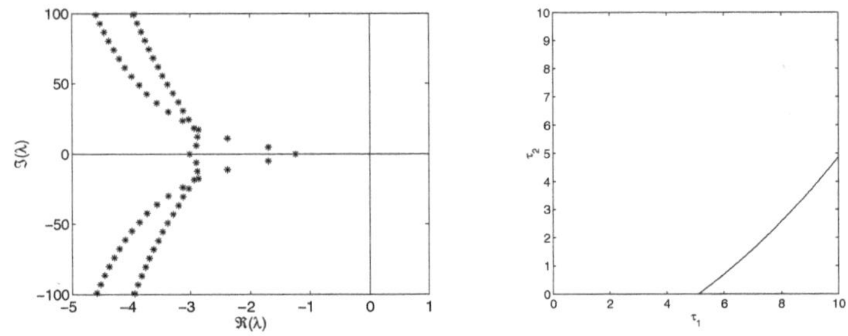

Fig. 6. Spectrum and stability chart for Example 3.

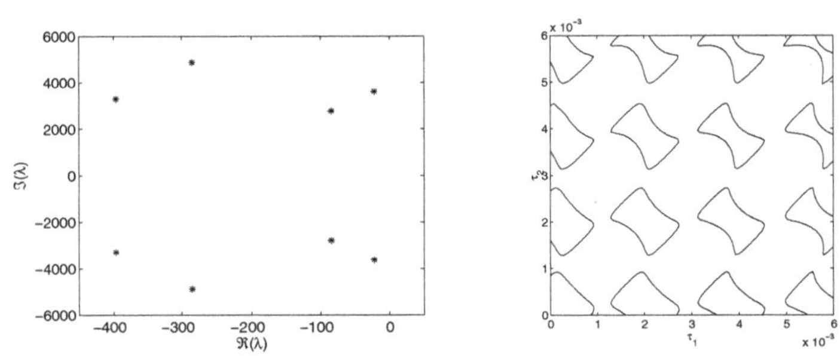

Fig. 7. Spectrum and stability chart for Example 4.

References

1. Altintas, Y., Engin, S., Budak, E.: Analytical stability prediction and design of variable pitch cutters. ASME J. Manuf. Sci. E.-T. 121, 173–178 (1999)
2. Breda, D.: The infinitesimal generator approach for the computation of characteristic roots for delay differential equations using BDF methods. Technical Report RR17/2002, Department of Mathematics and Computer Science, University of Udine (2002)
3. Breda, D.: Solution operator approximation for characteristic roots of delay differential equations. Appl. Numer. Math. 56(3-4), 305–317 (2006)

4. Breda, D., Iannelli, M., Maset, S., Vermiglio, R.: Stability analysis of the Gurtin-MacCamy model. J. Numer. Anal. 46(2), 980–995 (2008)
5. Breda, D., Maset, S., Vermiglio, R.: Computing the characteristic roots for delay differential equations. IMA J. Numer. Anal. 24(1), 1–19 (2004)
6. Breda, D., Maset, S., Vermiglio, R.: Efficient computation of stability charts for linear time delay systems. In: Proc. ASME-IDETC/CIE 2005, Long Beach, USA (2005)
7. Breda, D., Maset, S., Vermiglio, R.: An adaptive algorithm for efficient computation of level curves of surfaces. To appear in Numer Algorithms (2009) (preprint)
8. Breda, D., Maset, S., Vermiglio, R.: Pseudospectral differencing methods for characteristic roots of delay differential equations. SIAM J. Sci. Comput. 27(2), 482–495 (2005)
9. Breda, D., Maset, S., Vermiglio, R.: Pseudospectral approximation of eigenvalues of derivative operators with non-local boundary conditions. Appl. Numer. Math. 56(3-4), 318–331 (2006)
10. Breda, D., Maset, S., Vermiglio, R.: Numerical approximation of characteristic values of partial retarded functional differential equations. Numer. Math. (2009), doi:10.1007/s00211-009-0233-7
11. Diekmann, O., van Gils, S.A., Verduyn Lunel, S.M., Walther, H.O.: Delay Equations - Functional, Complex and Nonlinear Analysis. In: AMS series, vol. 110. Springer, New York (1995)
12. Engelborghs, K., Luzyanina, T., Roose, D.: Numerical bifurcation analysis of delay differential equations using DDE-BIFTOOL. ACM T. Math. Software 28(1), 1–21 (2002)
13. Engelborghs, K., Roose, D.: On stability of LMS methods and characteristic roots of delay differential equations. SIAM J. Numer. Anal. 40(2), 629–650 (2002)
14. Fattouh, A., Sename, O., Dion, J.M.: h_∞ controller and observer design for linear systems with point and distributed delays. In: Proc. 2nd IFAC workshop on Linear Time Delay Systems, Ancona, Italy (2000)
15. Hale, J.K., Verduyn Lunel, S.M.: Introduction to functional differential equations. AMS series, vol. 99. Springer, New York (1993)
16. Insperger, T., Stépán, G.: Updated semi-discretization method for periodic delay-differential equations with discrete delay. Int. J. Numer. Meth. Eng. 61, 117–141 (2004)
17. Luzyanina, T., Engelborghs, K., Roose, D.: Computing stability of differential equations with bounded distributed delays. Numer. Algorithms 34(1), 41–66 (2003)
18. Michiels, W., Niculescu, S.I.: Stability and stabilization of time-delay systems. An eigenvalue based approach. In: Advances in Design and Control, vol. 12. SIAM, Philadelphia (2007)
19. Niculescu, S.-I.: Delay Effects on Stability: A Robust Control Approach. In: TLNCIS Monograph, vol. 269. Springer, London (2001)
20. Olgac, N., Sipahi, R.: An exact method for the stability analysis of time delayed LTI systems. IEEE T. Automat. Contr. 47(5), 793–797 (2002)
21. Richard, J.-P.: Time-delay systems: an overview of some recent advances and open problems. Automatica 39, 1667–1694 (2003)
22. Sechi, D.: Sviluppo di interfaccia grafica per lo studio della stabilità di sistemi differenziali con ritardo (in italian). Master's thesis, University of Udine (2005)
23. Stépán, G.: Retarded dynamical systems. Longman, Harlow (1989)
24. Trefethen, L.N.: Spectral methods in MATLAB. Software - Environment - Tools series. SIAM, Philadelphia (2000)
25. Vyhlídal, T., Zítek, P.: Mapping the spectrum of a retarded time-delay system utilizing root distribution features. In: Proc. 6th IFAC workshop on Linear Time Delay Systems, L'Aquila, Italy (2006)

Analysis of Numerical Integration for Time Delay Systems

Alfredo Bellen[1] and Stefano Maset[2]

[1] Universitá di Trieste. Dipartimento di Matematica e Informatica bellen@units.it
[2] Universitá di Trieste. Dipartimento di Matematica e Informatica maset@units.it

Summary. The paper provides an original presentation of the numerical integration of general Retarded Functional Differential Equations as a sequence of approximations of the states of the system. The global error analysis is developed for one-step methods and order conditions are provided for a suitable generalization of explicit Runge-Kutta methods.

1 Introduction

Several mathematical models of evolution systems arising in many fields such as Astronomy, Biology, Economics, Engineering, Medicine, etc. are described by differential equations involving arguments which refer to the past. Such systems are often called Time Delay Systems (TDS), mostly in Control Theory. There is a huge literature in Control Theory concerning stability analysis of TDS (see, for example, the surveys [6], [2] and [3]) and, in this context, simulating such systems by a numerical integration often provides the sole approach for studying issues like global stability.

For a comprehensive presentation of the issues related to the numerical analysis of equations involving discrete delays we refer the interested reader to [1]. Here, we develop an original presentation of numerical methods for the more general class of the Retarded Functional Differential Equations (RFDE) which include both distributed and discrete delays. Moreover, we take the view that numerical methods provide dynamical systems approximating the dynamical system to be simulated.

Let us then consider the Cauchy problem , or Initial Value Problem (IVP), for the RFDE

$$y'(t) = F(t, y_t) \tag{1}$$

where $y(t) \in \mathbb{R}^d$, $F : \mathbb{R} \times \mathcal{C} \longrightarrow \mathbb{R}^d$, $\mathcal{C}$ is the space of continuous functions from $[-r, 0]$, with $r > 0$, to $\mathbb{R}^d$ called the *state space*, and $y_t \in \mathcal{C}$ is the function

$$y_t(\theta) = y(t+\theta), \quad \theta \in [-r, 0], \tag{2}$$

called the *state*, or the *solution segment*, at t. The space $\mathcal{C}$ is equipped with the norm

$$\|\phi\| = \max_{\theta \in [-r,0]} |\phi(\theta)|. \tag{3}$$

where $|\cdot|$ denote a norm on $\mathbb{R}^d$.

J.J. Loiseau et al. (Eds.): Topics in Time Delay Systems, LNCIS 388, pp. 157–166.
springerlink.com

In the following, we assume that F is continuous and globally Lipschitz continuous with respect to second argument.

Special instances of (1) are the equations with *discrete delay*, or *Delay Differential Equations* (DDEs),

$$y'(t) = f\left(t, y(t), y(t-\tau_1), \cdots y(t-\tau_s)\right), \tag{4}$$

where the functional F requires the computation of the state y_t on a discrete set of values, and equations with *distributed delay*, or *Delay-Integro-Differential Equations* (DIDEs)

$$y'(t) = f\left(t, y(t), \int_{-\tau}^{0} k\left(t, \theta, y(t+\theta)\right) d\theta\right), \tag{5}$$

where F requires the computation of the state on a continuous set of values collected by an integral. In both cases, the delays τ_i and τ are in $[-r, 0]$ and may depend on t. Remark that equations with *state dependent* delay, such as

$$y'(t) = f\left(t, y(t), y(t - \tau(t, y(t)))\right), \tag{6}$$

are not included in this mathematical framework since they must be treated in a more restricted state space (see [4]).

In the IVP for RFDEs (1), we need to assign an initial point $t_0 \in \mathbb{R}$ and an initial state $\phi_0 \in \mathcal{C}$ for which it takes the form

$$\begin{cases} y'(t) = F(t, y_t), & t \geq 0, \\ y(t_0 + \theta) = \phi_0(\theta), & \theta \in [-r, 0]. \end{cases} \tag{7}$$

It is well-known that, under the assumption of continuity and Lipschitz continuity with respect to the second argument of the functional F, for any $(t_0, \phi_0) \in \mathbb{R} \times \mathcal{C}$ there exists a unique solution of (7), which will be denoted by $y\left(\cdot; t_0, \phi_0\right)$ and defined on $[-r, +\infty)$.

The *solution operator* for the RFDE (1) is the map T associating to an initial point $\sigma \in \mathbb{R}$, an initial state $\varphi \in \mathcal{C}$ and a stepsize $h \geq 0$, the state

$$T(\sigma, \varphi, h) = y\left(\cdot; \sigma, \varphi\right)_{\sigma+h} \in \mathcal{C}$$

i.e.

$$T(\sigma, \varphi, h)\,(\theta) = y\left(\sigma + h + \theta; \sigma, \varphi\right), \quad \theta \in [-r, 0].$$

A *numerical method for RFDEs* provides, for any equation (1), a map $\widetilde{T}$ which approximates the solution operator T. More precisely, $\widetilde{T}$ associates to an initial point $\sigma \in \mathbb{R}$, an initial state $\varphi \in \mathcal{C}$ and a stepsize $h \geq 0$, the state $\widetilde{T}\left(\sigma, \varphi, h\right) \in \mathcal{C}$ which approximates $T\left(\sigma, \varphi, h\right)$.

When the method is applied for the step-by-step computation of the solution y of the IVP (7) on the window $[t_0, t_f]$, $t_f > t_0$, with mesh

$$\Delta = \{t_n : n = 0, 1, 2, ..., N_\Delta\}, \ t_0 < t_1 < t_2 < ... < t_{N_\Delta} = t_f, \tag{8}$$

it yields the finite sequence $\{\phi_n\}_{n=1,2,\ldots,N_\Delta}$, where ϕ_n is an approximation of y_{t_n}, given by

$$\phi_{n+1} = \widetilde{T}\left(t_n, \phi_n, h_{n+1}\right), \quad n = 0, 1, 2, ..., N_\Delta - 1, \tag{9}$$

with $h_{n+1} = t_{n+1} - t_n$. We will denote by h_Δ the maximum stepsize in the mesh Δ, i.e.

$$h_\Delta = \max_{n=1,2,\ldots,N_\Delta} h_n.$$

We deal with numerical methods such that

$$\widetilde{T}(\sigma, \varphi, h) = \eta\left(\cdot; \sigma, \varphi, h\right)_{\sigma+h},$$

i.e.

$$\widetilde{T}(\sigma, \varphi, h)\,(\theta) = \eta\left(\sigma + h + \theta; \sigma, \varphi, h\right), \quad \theta \in [-r, 0],$$

where $\eta\left(\cdot; \sigma, \varphi, h\right)$ is an approximation of $y\left(\cdot; \sigma, \varphi\right)$ given by

$$\eta\left(\sigma + \alpha h; \sigma, \varphi, h\right) = \varphi\left(0\right) + h\Phi\left(\alpha; \sigma, \varphi, h\right), \quad \alpha \in [0, 1], \tag{10}$$

$$\eta\left(\sigma + \theta; \sigma, \varphi, h\right) = \varphi\left(\theta\right), \quad \theta \in [-r, 0], \tag{11}$$

and $\Phi\left(\cdot; \sigma, \varphi, h\right) : [0, 1] \longrightarrow \mathbb{R}^d$ is a suitable continuous *incremental function* satisfying $\Phi\left(0; \sigma, \varphi, h\right) = 0$. The error of the approximation $\eta\left(\cdot; \sigma, \varphi, h\right)$ is given by the function

$$E\left(\cdot; \sigma, \varphi, h\right) := \eta\left(\cdot; \sigma, \varphi, h\right) - y\left(\cdot; \sigma, \varphi, h\right).$$

2 Uniform, discrete and convergence orders

Now, we are in condition to introduce the concepts of uniform order, discrete order and convergence order of a numerical method. To this aim, we consider problems (7) such that:

- F is of class C^l with respect to the second argument for a sufficiently large positive integer l: that is it has (Frechet-)derivatives $F^{(k)} : \mathbb{R} \times \mathcal{C} \longrightarrow \mathcal{L}^k\left(\mathcal{C}, \mathbb{R}^d\right)$, $k = 1, 2, \ldots, l$, which are bounded and continuous with respect to the second argument;
- the solution $y = y\left(\cdot; t_0, \phi_0\right)$ is piecewise of class C^m for a sufficiently large positive integer m: that is there exist a finite sequence $\xi_0, \xi_1, \ldots, \xi_I$ with $t_0 = \xi_0 < \xi_1 < \cdots < \xi_I$, called *discontinuity points*, such that y has continuous derivatives $y^{(k)}$, $k = 1, 2, \ldots, m$, on the intervals $[\xi_i, \xi_{i+1}]$, $i = 0, 1, \ldots, I - 1$, and also on the interval $[\xi_I, +\infty)$.

Definition 1. *Let p be a positive integer. A numerical method for RFDEs is of* numerics ! uniform order *(*discrete order*) p if there exist positive integers l and m such that for any problem (7), where F is of class C^l with respect to the second argument and y is piecewise of class C^m, and for any $t_f > t_0$ there exist constants $H > 0$ and $C > 0$ such that for any $t \in [t_0, t_f)$ and $h \in [0, H)$ we have*

$$\max_{\alpha \in [0,1]} \left|E\left(t + \alpha h; t, y_t, h\right)\right| \leq C h^{p+1} \quad \left(\left|E\left(t + h; t, y_t, h\right)\right| \leq C h^{p+1}\right).$$

where $(t, t + h)$ does not contain discontinuity points of y.

Definition 2. *Let r be a positive integer. A numerical method for RFDEs is of* convergence order *r if there exist positive integers l and m such that for any problem (7), where F is of class C^l with respect to the second argument and y is piecewise of class C^m, and $t_f > t_0$ we have*

$$\max_{n=1,2,\ldots,N_{\Delta_k}} \left\|\phi_n - y_{t_n}\right\| = \mathcal{O}\left(h^r_{\Delta_k}\right), \quad k \longrightarrow +\infty,$$

for any sequence $\{\Delta_k\}_k$ of meshes on $[t_0, t_f]$ such that

$$h_{\Delta_k} \longrightarrow 0, \quad k \longrightarrow +\infty,$$

and, for any k, Δ_k includes all the discontinuity points of y in $[t_0, t_f]$.

The relationship between the convergence order and the uniform and discrete orders of the method is given by the following theorem.

Theorem 1. *Let us consider a numerical method for RFDEs such that there exists a constant* $M \geq 0$ *for which*

$$|\Phi(\alpha;\sigma,\varphi,h) - \Phi(\alpha;\sigma,\psi,h)| \leq M \|\varphi - \psi\|$$

holds for all $\alpha \in [0,1]$, $\sigma \in \mathbb{R}$, $\varphi, \psi \in C$, *and* $h \geq 0$.

If the method has uniform order q *and discrete order* p, *then it has convergence order* $\min\{q+1, p\}$.

Proof. Let us assume that the numerical method is applied for the computation of the solution y of (7) on the interval $[t_0, t_f]$ with mesh

$$\Delta = \{t_n : n = 0, 1, 2, ..., N_\Delta\}$$

and let

$$e_n = \phi_n - y_{t_n}, \quad n = 0, 1, 2, \ldots, N_\Delta,$$

be the error of the method at the mesh point t_n.

For $n = 0, 1, \cdots, N_\Delta - 1$, we have

$$\begin{aligned} e_{n+1} &= \widetilde{T}(t_n, \phi_n, h_{n+1}) - T(t_n, y_{t_n}, h_{n+1}) \\ &= \widetilde{T}(t_n, \phi_n, h_{n+1}) - \widetilde{T}(t_n, y_{t_n}, h_{n+1}) \\ &\quad + \widetilde{T}(t_n, y_{t_n}, h_{n+1}) - T(t_n, y_{t_n}, h_{n+1}), \end{aligned}$$

where

$$\begin{aligned} &\widetilde{T}(t_n, \phi_n, h_{n+1})(\theta) - \widetilde{T}(t_n, y_{t_n}, h_{n+1})(\theta) \\ &= \begin{cases} e_n(0) + h_{n+1}\left[\Phi\left(1 + \frac{\theta}{h_{n+1}}; t_n, \phi_n, h_{n+1}\right)\right. \\ \left. -\Phi\left(1 + \frac{\theta}{h_{n+1}}; t_n, y_{t_n}, h_{n+1}\right)\right] \quad \text{if} - h_{n+1} \leq \theta \leq 0 \\ \\ e_n(h_{n+1} + \theta) \quad \text{if } \theta \leq -h_{n+1}, \end{cases} \end{aligned}$$

and

$$\begin{aligned} &\widetilde{T}(t_n, y_{t_n}, h_{n+1})(\theta) - T(t_n, y_{t_n}, h_{n+1})(\theta) \\ &= \begin{cases} E(t_n + h_{n+1} + \theta; t_n, y_{t_n}, h_{n+1}) \quad \text{if} - h_{n+1} \leq \theta \leq 0 \\ \\ 0 \quad \text{if } \theta \leq -h_{n+1}. \end{cases} \end{aligned}$$

Therefore,

$$\begin{bmatrix} |e_{n+1}(0)| \\ \|e_{n+1}\| \end{bmatrix} \leq \begin{bmatrix} 1 & h_{n+1}M \\ 1 & h_{n+1}M \end{bmatrix} \begin{bmatrix} |e_n(0)| \\ \|e_n\| \end{bmatrix} + \begin{bmatrix} \mathcal{E}_0(n+1;\Delta) \\ \mathcal{E}(n+1;\Delta) \end{bmatrix},$$

where

$$\mathcal{E}_0(n+1;\Delta) := |E(t_n + h_{n+1}; t_n, y_{t_n}, h_{n+1})|$$

and

$$\mathcal{E}(n+1;\Delta) := \max_{\alpha \in [0,1]} \|E(t_n + \alpha h_{n+1}; t_n, y_{t_n}, h_{n+1})\|.$$

Then, for $n = 1, 2, \cdots, N_\Delta$,

$$\begin{aligned}
\begin{bmatrix} |e_n(0)| \\ \|e_n\| \end{bmatrix} &\leq \sum_{i=1}^{n} \left(\prod_{j=i+1}^{n} \begin{bmatrix} 1 & h_j M \\ 1 & h_j M \end{bmatrix} \right) \begin{bmatrix} \mathcal{E}_0(i;\Delta) \\ \mathcal{E}(i;\Delta) \end{bmatrix} \\
&= \begin{bmatrix} \mathcal{E}_0(n;\Delta) \\ \mathcal{E}(n;\Delta) \end{bmatrix} + \sum_{i=1}^{n-1} \left(\prod_{j=i+2}^{n} \begin{bmatrix} 1 & h_j M \\ 1 & h_j M \end{bmatrix} \right) \\
&\quad \cdot \begin{bmatrix} \mathcal{E}_0(i;\Delta) + h_{i+1} M \mathcal{E}(i;\Delta) \\ \mathcal{E}_0(i;\Delta) + h_{i+1} M \mathcal{E}(i;\Delta) \end{bmatrix}
\end{aligned}$$

and so

$$\begin{aligned}
\|e_n\| &= \max\{|e_n(0)|, \|e_n\|\} \\
&\leq \max\{\mathcal{E}_0(n;\Delta), \mathcal{E}(n;\Delta)\} + \sum_{i=1}^{n-1} \left(\prod_{j=i+2}^{n} (1 + h_j M) \right) \\
&\quad \cdot (\mathcal{E}_0(i;\Delta) + h_{i+1} M \mathcal{E}(i;\Delta)) \\
&\leq \max\left\{ h_n \frac{\mathcal{E}_0(n;\Delta)}{h_n}, \mathcal{E}(n;\Delta) \right\} + \sum_{i=1}^{n-1} e^{(t_n - t_i)M} h_i \cdot e^{-h_i M} \frac{\mathcal{E}_0(i,\Delta)}{h_i} \\
&\quad + \sum_{i=1}^{n-1} e^{(t_n - t_{i+i})M} h_{i+1} \cdot \mathcal{E}(i,\Delta) \\
&\leq \max\left\{ h_n \frac{\mathcal{E}_0(n;\Delta)}{h_n}, \mathcal{E}(n;\Delta) \right\} \\
&\quad + \frac{e^{(t_n - t_0)M} - e^{h_n M}}{M} \cdot \max_{i=1,\dots,n-1} e^{-h_i M} \frac{\mathcal{E}_0(i,\Delta)}{h_i} \\
&\quad + \frac{e^{(t_n - t_1)M} - 1}{M} \cdot \max_{i=1,\dots,n-1} \mathcal{E}(i,\Delta).
\end{aligned}$$

If the method has uniform order p and discrete order q, then there exist positive integers l and m and constants $H > 0$ and $C > 0$ such that

$$\max_{i=1,\dots,N_\Delta} \mathcal{E}(i,\Delta) \leq C h_\Delta^{q+1}$$

$$\max_{i=1,\dots,N_\Delta} \frac{\mathcal{E}_0(i,\Delta)}{h_i} \leq C h_\Delta^{p}$$

and then

$$\max_{n=1,\dots,N_\Delta} \|e_n\| \leq \max\left\{ C h_\Delta^{p+1}, C h_\Delta^{q+1} \right\} + \frac{e^{(t_f - t_0)M} - 1}{M} \left(C h_\Delta^{p} + C h_\Delta^{q+1} \right),$$

whenever F is of class $\mathcal{C}^l$, y is of piecewise class $\mathcal{C}^m$, the mesh Δ contains all the discontinuity points of y in $[t_0, t_f]$ and $h_\Delta < H$. We conclude that the method has convegence order $\min\{q+1, p\}$.

3 Continuous Runge-Kutta methods for RFDEs

In this section we present and analyze particular methods (10)-(11) of Runge-Kutta type for RFDEs that will be called Functional-Continuous-Runge-Kutta (FCRK) methods .

Definition 3. *Let ν be a positive integer. A ν-stage FCRK method is a triple*

$$(A(\cdot), b(\cdot), c)$$

where

- $A(\cdot)$ *is an* $\mathbb{R}^{\nu\times\nu}-$*valued polynomial function such that* $A(0) = 0$,
- $b(\cdot)$ *is an* $\mathbb{R}^{\nu}-$*valued polynomial function such that* $b(0) = 0$ *and*
- $c \in \mathbb{R}^{\nu}$ *with* $0 \le c_i \le 1$, $i = 1, ..., \nu$.

Such a method provides the approximation

$$\eta(\sigma + \alpha h; \sigma, \varphi, h) = \varphi(0) + h \sum_{i=1}^{\nu} b_i(\alpha) K_i, \quad \alpha \in [0,1], \tag{12}$$

$$\eta(\sigma + \theta; \sigma, \varphi, h) = \varphi(\theta), \quad \theta \in [-r, 0], \tag{13}$$

where

$$K_i = F\left(\sigma + c_i h, Y^i_{\sigma + c_i h}\right), \; i = 1, ..., \nu \tag{14}$$

and $Y^i : [\sigma - r, \sigma + h] \to \mathbb{R}^d$ is a *stage function* given by

$$Y^i(\sigma + \alpha h) = \varphi(0) + h \sum_{j=1}^{\nu} a_{ij}(\alpha) K_j, \; \alpha \in [0,1], \tag{15}$$

$$Y^i(\sigma + \theta) = \varphi(\theta), \; \theta \in [-r, 0]. \tag{16}$$

The conditions $A(0) = 0$ and $b(0) = 0$ guarantee that the functions η and Y^i, $i = 1, ..., \nu$, are continuous.

In case of an ordinary differential equation, i.e. $r = 0$ in (1), the method coincides with the $\nu-$stage (classical) RK method (A, b, c), where the coefficients a_{ij}, $i, j = 1, ...\nu$, are the values $a_{ij}(c_i)$ and the weights b_i are the values $b_i(1)$. In fact, it provides the map

$$\widetilde{T}(\sigma, \varphi, h) = \eta(\sigma, \varphi, h)(\sigma + h) = \varphi(0) + h \sum_{i=1}^{\nu} b_i(1) K_i$$

where

$$K_i = F\left(\sigma + c_i h, Y^i(\sigma + c_i h)\right), \quad i = 1, ..., \nu,$$

and

$$Y^i(\sigma + c_i h) = \varphi(0) + h \sum_{j=1}^{\nu} a_{ij}(c_i) K_j.$$

For a DDE (4) the equations (14) reads

$$K_i = f\left(\sigma + c_i h, Z^i_0, Z^i_1, \ldots, Z^i_s\right)$$

for $i = 1, ..., \nu$, where the stage values $Z^i_0 \in \mathbb{R}^d$ are given by

$$Z^i_0 = Y^i(\sigma + c_i h) = \varphi(0) + h \sum_{j=1}^{\nu} a_{ij}(c_i) K_j \tag{17}$$

and the stage values $Z^i_k \in \mathbb{R}^d$, $k = 1, \ldots, s$, are given by

$$Z_k^i = Y^i(\sigma + c_i h - \tau_k(\sigma + c_i h))$$
$$= \begin{cases} \varphi(0) + h \sum\limits_{j=1}^{\nu} a_{ij} \left(c_i - \frac{\tau_k(\sigma + c_i h)}{h} \right) K_j & \text{if } c_i h \geq \tau_k(\sigma + c_i h) \\ \varphi(c_i h - \tau_k(\sigma + c_i h)) & \text{if } c_i h \leq \tau_k(\sigma + c_i h) \end{cases}$$

If, for any $i = 1, \ldots, \nu$ and $k = 1, \ldots, s$, we have $c_i h \leq \tau_k(\sigma + c_i h)$, then the values Z_k^i do not depend on the functions $a_{ij}(\cdot)$ which are then involved only in (17) by the discrete values $a_{ij}(c_i)$. Note that such condition is guaranteed if

$$\tau_i(t) \geq \tau_0 > 0, \quad i = 1, \ldots, s, \tag{18}$$

holds for t in a right neighborhood $[\sigma, \sigma + h_0]$ of σ and $h \leq \min\{\tau_0, h_0\}$.

In other words, whenever the non-vanishing delay condition (18) holds and the stepsize h is small enough, the equation may be locally treated as an ODE, and the FCRK method reduces, de facto, to the continuous RK method $(A, b(\cdot), c)$.

On the other hand, for a DIDE (5) the equations (14) reads

$$K_i = f\left(\sigma + c_i h, Z_0^i, Z_1^i\right) \tag{19}$$

for $i = 1, ..., \nu$, where Z_0^i is given by (17) and Z_1^i is given by

$$Z_1^i = \int_{-\tau(\sigma + c_i h)}^{-c_i h} k(\sigma + c_i h, \theta, \varphi(c_i h + \theta)) d\theta$$
$$+ h \int_0^{c_i} k\left(\sigma + c_i h, -h(c_i - \alpha), \varphi(0) + h \sum_{j=1}^{s} a_{ij}(\alpha) K_j\right) d\alpha.$$

Here, unlike for the DDE (4), the value Z_1^i depends on the functions $a_{ij}(\cdot)$ for any choice of the stepsize h.

A FCRK method is called *explicit* if $a_{ij}(\cdot) = 0$ for $j \geq i$. A method is called *implicit* if it is not explicit. For explicit methods the values K_i, $i = 1, ..., \nu$, can be recursively computed as follows:

- $K_1 = F\left(\sigma + c_1 h, Y^1_{\sigma + c_1 h}\right)$ where

$$Y^i(\sigma + \alpha h) = \varphi(0), \ \alpha \in [0, 1],$$
$$Y^i(\sigma + \theta) = \varphi(\theta), \ \theta \in [-r, 0]$$

- $K_i = F\left(\sigma + c_i h, Y^i_{\sigma + c_i h}\right)$, $i = 2, ..., \nu$, where

$$Y^i(\sigma + \alpha h) = \varphi(0) + h \sum_{j=1}^{i-1} a_{ij}(\alpha) K_j, \ \alpha \in [0, 1],$$
$$Y^i(\sigma + \theta) = \varphi(\theta), \ \theta \in [-r, 0].$$

The computation of the values K_i requires ν evaluations of the functional F.

On the contrary, for implicit methods, the values K_i, $i = 1, ..., \nu$, are implicitly defined by (14)-(15)-(16) and they are the solution of a non-linear algebraic system of dimension $d\nu$.

A particular class of implicit FCRK methods arises from polynomial collocation. In this approach, given the distinct nodes c_i, $i = 1, ..., \nu$, in $[0,1]$, the approximation $\eta = \eta(\cdot;\sigma,\phi,h)$ on $[\sigma,\sigma+h]$ is given by a $\nu-$degree polynomial such that

$$\eta'(\sigma + c_i h) = F(\sigma + c_i h, \eta_{\sigma+c_i h}), \; i = 1, ..., \nu,$$
$$\eta(\sigma) = \varphi(0).$$

This corresponds to the FCRK method

$$b_i(\alpha) = a_{ij}(\alpha) = \int_0^\alpha \ell_i(s)ds, \quad \alpha \in [0,1], \quad i,j = 1,\dots,\nu \tag{20}$$

where ℓ_i are the Lagrange coefficients relevant the nodes c_i. In this case, the stage functions Y^i in (15)-(16) coincide with the approximation η in (12)-(13) and the method is included in the frame of the so called *standard approach* developed in [1]. In such an approach, which is based on a Continuous RK method $(A, b(\cdot), c)$ the functional F is split as $F(t, y_t) = f(t, y(t), y_t)$ and, contrary to the FCRK methods, the values K^i in (12) are given just by

$$K_i = f\left(\sigma + c_i h, Y^i, \eta_{\sigma+c_i h}\right), \; i = 1, ..., \nu$$

where

$$Y^i = \varphi(0) + h\sum_{j=1}^{\nu} a_{ij}K_j \, .$$

It is worth remarking that the two approaches differ to each other if the equation does not locally reduce to an ODE, as in DDE when $c_i h \geq \tau_k(\sigma + c_i h)$ for some i, k and h as well as in DIDE. In such cases the standard approach turns out to be implicit even if the underlying Continuous RK method is explicit whereas the FCRK method preserves the explicit/implicit character for any equation and for any step-size h. This makes the method competitive with the standard approach for non-stiff DDEs.

By extending the tableau notation of a classical RK method for ODEs, we denote a FCRK $(A(\cdot), b(\cdot), c)$ by

$$\begin{array}{c|c} c & A(\alpha) \\ \hline & b(\alpha) \end{array} \, .$$

Particular explicit methods are given by the tableaux

$$\begin{array}{c|c} 0 & 0 \\ \hline & \alpha \end{array} \tag{21}$$

and

$$\begin{array}{c|cc} 0 & 0 & 0 \\ 1 & \alpha & 0 \\ \hline & \alpha - \frac{1}{2}\alpha^2 & \frac{1}{2}\alpha^2 \end{array} \tag{22}$$

which are referred to as the *explicit Euler method* and the *Heun method* for RFDEs, respectively, as named by Cryer and Tavernini in the seminal paper [5]. They coincide with the classical explicit Euler and Heun methods when applied to ODEs.

An example of implicit method is

$$\begin{array}{c|c} 1 & \alpha \\ \hline & \alpha \end{array} \, . \tag{23}$$

which can be referred to as the *implicit Euler method* for RFDEs. Remark that it corresponds to the collocation method on the single node $c_1 = 1$.

4 Order conditions

In this section we present the conditions to be imposed to the coefficients of a FCRK in order to obtain a uniform, discrete or convergence order one and two. For proofs and more details we refer to [7].

We assume that the FCRK method satisfies the conditions

$$\sum_{j=1}^{\nu} a_{ij}(\alpha) = \alpha, \quad \alpha \in [0, c_i], \quad i = 1, ..., \nu.$$

which are the analogous of the known conditions

$$\sum_{j=1}^{\nu} a_{ij} = c_i, \quad i = 1, \ldots, \nu,$$

for classical RK methods. As a consequence, $c_1 = 0$ for explicit methods.

The condition for the uniform order one is

$$\sum_{i=1}^{\nu} b_i(\alpha) = \alpha, \quad \alpha \in [0, 1], \tag{24}$$

whereas for the discrete order one the condition is

$$\sum_{i=1}^{\nu} b_i(1) = 1. \tag{25}$$

In view of the convergence Theorem (1), the sole condition (25) also guarantees the convergence order one. Both explicit and implicit Euler methods (21) and (23) have uniform and convergence orders one.

The conditions to be added for the uniform and discrete order two are

$$\sum_{i=1}^{\nu} b_i(\alpha) c_i = \frac{\alpha^2}{2}, \ \alpha \in [0, 1], \tag{26}$$

and

$$\sum_{i=1}^{\nu} b_i(1) c_i = \frac{1}{2}, \tag{27}$$

respectively. Again, according to the Theorem (1), the convergence order two is obtained whenever (24) and (27) hold. The two-stage explicit methods of uniform order two take the form

$$\begin{array}{c|cc} 0 & 0 & 0 \\ c_2 & \alpha & 0 \\ \hline & \alpha - b_2(\alpha) & b_2(\alpha) \end{array}, \tag{28}$$

where $0 < c_2 \leq 1$ and

$$b_2(\alpha) = \frac{\alpha^2}{2c_2}, \quad \alpha \in [0, 1].$$

The foregoing Heun method (22) corresponds to the choice $c_2 = 1$. On the other hand, the convergence order two for methods (28) can be obtained by imposing the weaker condition

$$b_2(1) = \frac{1}{2c_2}.$$

We finish this section by remarking that, in the class of one stage implicit methods, the uniform order two cannot be obtained whereas the convergence order two is performed by the method

$$\begin{array}{c|c} \frac{1}{2} & \alpha \\ \hline & \alpha \end{array}$$

which coincides with the *implicit midpoint rule* when applied to ODEs.

References

1. Bellen, A., Zennaro, M.: Numerical methods for delay differential equations. Oxford University Press, Oxford (2003)
2. Niculescu, S.I.: Delay Effects on Stability, A Robust Control Approach. Springer, Heidelberg (2001)
3. Michiels, W., Niculescu, S.I.: Stability and Stabilization of Time-Delay Systems. An eigenvalue-based approach. In: Advances in Design and Control. SIAM, Philadelphia (2007)
4. Hartung, F., Krisztin, T., Walther, H.O., Wu, J.: Functional Differential Equations with State-Dependent Delays: Theory and Applications. In: Canada, A., Drabeck, P., Fonda, A. (eds.) Handbook of Differential Equations, vol. 3. Elsevier B.V., Amsterdam (2006)
5. Cryer, C., Tavernini, L.: The numerical solution of Volterra functional differential equations by Euler's method. SIAM J. Numer. Anal. 9, 105–129 (1972)
6. Richard, J.P.: Time-delay systems: an overview of some recent advances and open problems. Automatica 39, 1667–1964 (2003)
7. Maset, S., Torelli, L., Vermiglio, R.: Runge-Kutta methods for retarded functional differential equations. Mathematical Models and Methods in Applied Sciences 8, 1203–1251 (2005)

On Algebraic Simplifications of Linear Functional Systems

Thomas Cluzeau[1] and Alban Quadrat[2]

[1] University of Limoges; CNRS; XLIM UMR 6172, DMI; ENSIL, 16 Rue d'Atlantis, Parc Ester Technopole, 87068 Limoges Cedex, France cluzeau@ensil.unilim.fr

[2] INRIA Sophia Antipolis, APICS Project, 2004 Route des Lucioles, BP 93, 06902 Sophia Antipolis Cedex, France, Alban.Quadrat@sophia.inria.fr

Summary. In this paper, we show how to conjointly use module theory and constructive homological algebra to obtain general conditions for a matrix R of functional operators (e.g., differential/shift/time-delay operators) to be equivalent to a block-triangular or block-diagonal matrix $\overline{R}$ (i.e., conditions for the existence of unimodular matrices V and W satisfying that $\overline{R} = V\,R\,W$). These results allow us to simplify the study of many linear functional systems – particularly differential time-delay systems – appearing in control theory and mathematical physics.

Introduction

Given a linear functional system coming from mathematical physics, applied mathematics, engineering sciences or control theory, it is often interesting to simplify the equations of the system before studying its structural properties and using numerical analysis methods.

In the case of linear ordinary differential systems of the form $\dot{x} = A(t)\,x$, where A is a $n \times n$ matrix with entries in the field $\mathbb{Q}(t)$ of rational functions, i.e., $A \in \mathbb{Q}(t)^{n\times n}$, the computer algebra community has developed algorithms for *reducing* (resp., *decomposing*) such systems. Here, reducing (resp., decomposing) the system means finding an invertible change of variables $x = P(t)\,y$ such that the new differential system for y writes $\dot{y} = B(t)\,y$, where $B(t)$ has a block-triangular (resp., a block-diagonal) form. One kind of algorithms to perform this task is based on an object called the *eigenring* of the system $\dot{x} = A(t)\,x$. The eigenring is the ring of matrices $P \in \mathbb{Q}(t)^{n\times n}$ which commute with the derivation $\nabla = \frac{d}{dt}\,I_n - A(t)$, namely, $P\,\nabla = \nabla\,P$, which is equivalent to $\dot{P}(t) = A(t)\,P(t) - P(t)\,A(t)$. If one can find an *idempotent* P in the eigenring, namely, $P^2 = P$, then one can decompose the system $\dot{x} = A(t)\,x$. For more details, we refer the reader to [1, 16, 19] and references therein.

The eigenring approach has been extended to linear (q-)difference systems (see, e.g., [1]) and more recently to algebraic integrable connections (see [16, 17] and references therein). This approach is restricted to the so-called class of *D-finite* linear functional systems ([2]) since it is based on the property that the latter systems are finite-dimensional vector spaces over the base field k. Unfortunately, this approach cannot be extended to the main classes of linear functional systems (e.g., ordinary/partial differential equations, time-delay equations,

J.J. Loiseau et al. (Eds.): Topics in Time Delay Systems, LNCIS 388, pp. 167–178.
springerlink.com

difference equations) appearing in applied mathematics, mathematical physics, engineering sciences and control theory as they are generally not D-finite.

If we consider linear systems of ordinary differential/difference equations with constant coefficients, the *Smith normal form* can be used to determine an equivalent system defined by decoupled scalar equations: here again, this technique simplifies the study of the original system. The influence of the Smith normal form has been particularly evident in the control theory community where it has played an important role in the so-called *polynomial approach* pioneered by Rosenbrock, Kučera, Kailath and others. For more details, we refer to [10, 12, 18] and references therein. A generalization of the Smith normal form to (left/right) principal ideal domains exists and is usually called the *Jacobson normal form* ([8, 15]). However, once again, Smith and Jacobson normal forms do not exist for the main classes of linear functional systems coming from applied mathematics, mathematical physics, engineering sciences or control theory as, for instance, they do not exist for linear systems of partial differential equations or differential time-delay equations.

Starting from these observations and trying to fill the gap, in [5, 6], we have studied the factorization and decomposition problems for a larger class of linear functional systems (determined, over-determined, under-determined) within a *constructive homological algebra approach*. Using the concept of Ore algebras of functional operators (e.g., ordinary/partial differential operators, shift operators, time delay-operators), we have developed in [5, 6] algorithms for computing morphisms between finitely presented left modules over an Ore algebra and idempotent endomorphisms of a finitely presented left module over an Ore algebra. Then, we have given three theorems (recalled hereafter) concerning the factorization, reduction and decomposition of linear functional systems. Note that, as in the eigenring approach, idempotents are crucial for the decomposition problem. Finally, these results naturally find applications in control theory (i.e., study of structural properties, study of the Monge problem and its applications to optimal control, decoupling problems), engineering sciences (e.g., algebraic pre-conditioner to numerical analysis methods) and mathematical physics (e.g., search for quadratic conservation laws). For some applications, see [5, 6].

The purpose of this paper is to give a survey on the main theoretical results developed in [5, 6] based on a constructive homological algebra approach. See [7] for an implementation of those results in the OREMORPHISMS package.

1 Notations and Problems

In this paper, we denote by D a non-commutative ring, $R \in D^{q\times p}$ a $q \times p$ matrix with entries in D and $\mathrm{GL}_p(D)$ the *general linear group* of units of the ring $D^{p\times p}$, i.e., $\mathrm{GL}_p(D) = \{U \in D^{p\times p} \mid \exists\, V \in D^{p\times p} : U\,V = V\,U = I_p\}$. Given $R \in D^{q\times p}$, the latter results are related to the following three problems:

1. *Factorization problem*: Find two matrices $L \in D^{q\times r}$ and $S \in D^{r\times p}$ such that we have the factorization $R = L\,S$.
2. *Reduction problem*: Find two matrices $U \in \mathrm{GL}_p(D)$ and $V \in \mathrm{GL}_q(D)$ such that the matrix $\overline{R} = V\,R\,U^{-1} \in D^{q\times p}$ has a block-triangular form.
3. *Decomposition problem*: Find two matrices $U \in \mathrm{GL}_p(D)$ and $V \in \mathrm{GL}_q(D)$ such that the matrix $\overline{R} = V\,R\,U^{-1} \in D^{q\times p}$ has a block-diagonal form.

In what follows, we shall focus on a particular type of non-commutative polynomial rings called *Ore algebras* of functional operators (e.g., differential operators, time-delay operators, shift operators) over which the existence of *Gröbner bases* is ensured for any *admissible term order*. For a precise definition of an Ore algebra, we refer the reader to [2, 3, 4, 15] or [6, Definition 2.1] and to [2, 13] and the references therein for the concepts of Gröbner bases and admissible term orders. To achieve these conditions, which allow us to constructively work over the ring D, it is sufficient to assume that the Ore algebra D further satisfies the technical hypotheses of [6, Proposition 2.1]. In practice, these conditions are not too restrictive since they are satisfied for the common Ore algebras that we encounter in many examples coming from applied mathematics, mathematical physics, engineering sciences or control theory as, for instance, rings of differential operators, of differential time-delay operators or of shift operators. The reader, unfamiliar with non-commutative polynomial rings, can replace everywhere the Ore algebra D by the commutative polynomial ring $k[x_1, \ldots, x_n]$ with coefficients in a field k (e.g., $k = \mathbb{Q}$).

2 Morphisms and Galois transformations

We recall that the definition of a *left D-module* M is the same as the one of a k-vector space but where the field k is replaced by a ring D and the coefficients of D act on the left on the elements of M, namely, for all $m_1, m_2 \in M$ and for all $a_1, a_2 \in D$, we have $a_1 m_1 + a_2 m_2 \in M$ (see, e.g., [17]).

Let D be a non-commutative Ore algebra of functional operators, $\mathcal{F}$ a left D-module and $R \in D^{q\times p}$ a matrix. Then, the *abelian group*, namely, the $\mathbb{Z}$-module, defined by $\ker_{\mathcal{F}}(R.) = \{\eta \in \mathcal{F}^p \mid R\eta = 0\}$ is called a *linear functional system* or a *behaviour* and $\mathcal{F}$ the *signal space* (see, e.g., [3]).

Within the *algebraic analysis* approach to mathematical systems theory ([11, 14]), the *finitely presented left D-module* $M = D^{1\times p}/(D^{1\times q}\, R)$ is associated with the linear system $\ker_{\mathcal{F}}(R.)$, where $D^{1\times p}$ denotes the left D-module formed by row vectors of length p with entries in D.

Let us introduce a few definitions (see, e.g., [17]).

Definition 1. Let M and M' be left D-modules. We call *left D-homomorphism* or simply *D-morphism* $f : M \longrightarrow M'$, a left D-linear application from M to M', i.e.:

$$\forall\, m_1,\, m_2 \in M,\ \forall\, a_1,\, a_2 \in D: \ f(a_1\, m_1 + a_2\, m_2) = a_1\, f(m_1) + a_2\, f(m_2).$$

We denote by $\hom_D(M, M')$ the abelian group of D-morphisms from M to M'. If $M' = M$, then $\hom_D(M, M)$ is also denoted by $\mathrm{end}_D(M)$ and an element of $\mathrm{end}_D(M)$ is called a *D-endomorphism* of M.

If D is a commutative ring and M and M' are two D-modules, then the abelian group $\hom_D(M, M')$ inherits a D-module structure. However, if D is a non-commutative ring and M and M' are two left D-modules, then $\hom_D(M, M')$ has only an abelian group structure. If D is a k-algebra, where k is a field included in the center $Z(D) = \{a \in D \mid \forall\, b \in D,\, a\, b = b\, a\}$ of the ring D, then $\hom_D(M, M')$ can be endowed with a k-vector space structure. We note that $\mathrm{end}_D(M)$ is a non-commutative ring for the addition and the composition of endomorphisms. The unity of $\mathrm{end}_D(M)$ is id_M. It is called the *D-endomorphism ring* of M. For more details, see, e.g., [17].

Let us now describe the left D-module M in terms of generators and relations. We denote by $\{e_j\}_{j=1,\dots,p}$ (resp., $\{f_i\}_{i=1,\dots,q}$) the standard basis of the *free* left D-module $D^{1\times p}$ (resp., $D^{1\times q}$), namely, e_j (resp., f_i) is the row vector of $D^{1\times p}$ (resp., $D^{1\times q}$) defined by 1 at the jth (resp., i^{th}) position and 0 elsewhere. Moreover, let $\pi : D^{1\times p} \longrightarrow M$ be the D-morphism sending an element $\lambda \in D^{1\times p}$ to its residue class $\pi(\lambda)$ in the quotient left D-module $M = D^{1\times p}/(D^{1\times q}\,R)$. We have $f_i\,R \in (D^{1\times q}\,R)$, for $i = 1,\dots,q$, a fact implying that $\pi(f_i\,R) = 0$ and, by D-linearity, we then get:

$$\forall\, i = 1,\dots,q, \quad \pi(f_i\,R) = \pi\left(\sum_{j=1}^{p} R_{ij}\,e_j\right) = \sum_{j=1}^{p} R_{ij}\,\pi(e_j) = 0. \tag{1}$$

Moreover, using the fact that π is surjective, for all $m \in M$, there exists an element $a = (a_1,\dots,a_p) \in D^{1\times p}$ such that $m = \pi(a) = \sum_{j=1}^{p} a_j\,\pi(e_j)$, proving that the left D-module M is finitely generated by $\{\pi(e_j)\}_{j=1,\dots,p}$ and its generators $\pi(e_j)$'s satisfy the relations (1) and their left D-linear combinations.

The following result explains why the left D-module $M = D^{1\times p}/(D^{1\times q}\,R)$ plays an important role in mathematical systems theory.

Theorem 1 ([11, 14]). *Let D be a non-commutative ring, $R \in D^{q\times p}$ a matrix, $\mathcal{F}$ a left D-module, $M = D^{1\times p}/(D^{1\times q}\,R)$ the left D-module finitely presented by R and $\ker_{\mathcal{F}}(R.) = \{\eta \in \mathcal{F}^p \mid R\,\eta = 0\}$ the corresponding linear functional system. Then, the morphism of abelian groups*

$$\psi : \ker_{\mathcal{F}}(R.) \longrightarrow \hom_D(M,\mathcal{F}), \quad \eta \longmapsto \psi(\eta)$$

– where $f = \psi(\eta)$ is the D-morphism defined by $f(\pi(e_i)) = \eta_i$, for $i = 1,\dots,p$ and $\eta = (\eta_1,\dots,\eta_p)^T$ – is an isomorphism, i.e., we have:

$$\ker_{\mathcal{F}}(R.) \cong \hom_D(M,\mathcal{F}).$$

Theorem 1 shows that the linear functional system $\ker_{\mathcal{F}}(R.)$ only depends on the finitely presented left D-module M and on the left D-module $\mathcal{F}$ ([11]).

The main tool that we use to handle the three previously stated problems is the study of D-morphisms between finitely presented left D-modules. The next proposition, classical in homological algebra (see, e.g., [17]), will play a fundamental role in what follows.

Proposition 1 ([6]). *Let $R \in D^{q\times p}$ and $R' \in D^{q'\times p'}$ be two matrices with entries in a ring D, $M = D^{1\times p}/(D^{1\times q}\,R)$ and $M' = D^{1\times p'}/(D^{1\times q'}\,R')$ two left D-modules respectively finitely presented by R and R' and $\pi : D^{1\times p} \longrightarrow M$ and $\pi' : D^{1\times p'} \longrightarrow M'$ the two canonical projections. A morphism $f \in \hom_D(M,M')$ is then defined by*

$$\forall\, m = \pi(\lambda) \in M, \ \lambda \in D^{1\times p} : \ f(m) = \pi'(\lambda\,P),$$

where $P \in D^{p\times p'}$ is a matrix such that there exists $Q \in D^{q\times q'}$ satisfying:

$$R\,P = Q\,R'. \tag{2}$$

The pair of matrices (P, Q) is defined up to a homotopy equivalence, *namely, if we denote by $R_2' \in D^{r' \times q'}$ a matrix satisfying*

$$\ker_D(.R') \triangleq \{\mu \in D^{1\times q'} \mid \mu\, R' = 0\} = D^{1\times r'}\, R_2',$$

then, for all $Z \in D^{p\times q'}$ and $Z' \in D^{q\times r'}$, the matrices defined by

$$\begin{cases} \overline{P} = P + Z\, R', \\ \overline{Q} = Q + R\, Z + Z'\, R_2', \end{cases} \tag{3}$$

satisfy the relation $R\,\overline{P} = \overline{Q}\, R'$ and define the same D-morphism f.

Proposition 1 can easily be understood if we first note that, contrary to k-vector spaces, a left D-module M is generally not free, i.e., M does not admit a basis. Hence, a D-linear application from a left D-module M to a left D-module M' is generally not defined by the matrix obtained by sending the elements of a basis of M to elements of M'. However, as we have previously seen, $\{\pi(e_j)\}_{j=1,\ldots,p}$ (resp., $\{\pi'(e_k')\}_{k=1,\ldots,p'}$) forms a family of generators of M (resp., M') and a D-morphism $f \in \hom_D(M, M')$ sends the generators of M to certain elements of M', i.e., we have $f(\pi(e_j)) = \sum_{k=1}^{p'} P_{jk}\, \pi'(e_k')$, $j = 1, \ldots, p$, where the P_{jk}'s are elements of D which must satisfy the relations coming from the fact that $f(0) = 0$, i.e., f must send the relations $\sum_{j=1}^{p} R_{ij}\, \pi(e_j) = 0$, $i = 1, \ldots, q$, between the generators $\pi(e_j)$'s of M to 0, i.e., for $i = 1, \ldots, q$, by D-linearity, we have:

$$\begin{aligned} f\left(\sum_{j=1}^{p} R_{ij}\, \pi(e_j)\right) &= \sum_{j=1}^{p} R_{ij}\, f(\pi(e_j)) = \sum_{j=1}^{p} R_{ij} \left(\sum_{k=1}^{p'} P_{jk}\, \pi'(e_k')\right) \\ &= \pi'\left(\sum_{k=1}^{p'} \left(\sum_{j=1}^{p} R_{ij}\, P_{jk}\right) e_k'\right) = 0, \end{aligned}$$

and thus, $\left(\sum_{j=1}^{p} R_{ij}\, P_{j1}, \ldots, \sum_{j=1}^{p} R_{ij}\, P_{jp'}\right) \in D^{1\times q'}\, R'$, i.e., there exists a row vector $Q_i \in D^{1\times q'}$ such that $\left(\sum_{j=1}^{p} R_{ij}\, P_{j1}, \ldots, \sum_{j=1}^{p} R_{ij}\, P_{jp'}\right) = Q_i\, R'$. Hence, we obtain $R\, P = Q\, R'$, where $Q = (Q_1^T \ldots Q_q^T)^T \in D^{q\times q'}$.

We can easily check that the P_{jk}'s are not uniquely defined because, if we also have $f(\pi(e_j)) = \sum_{k=1}^{p'} \overline{P}_{jk}\, \pi'(e_k')$, then we get

$$\forall\, j = 1, \ldots, p, \quad \sum_{k=1}^{p'} (\overline{P}_{jk} - P_{jk})\, \pi'(e_k') = 0 = \pi'\left(\sum_{k=1}^{p'} (\overline{P}_{jk} - P_{jk})\, e_k'\right),$$

and thus, the row vector $\overline{P}_{j\bullet} - P_{j\bullet} = (\overline{P}_{j1} - P_{j1}, \ldots, \overline{P}_{jp'} - P_{jp'})$ belongs to $D^{1\times q'}\, R'$, i.e., there exists $Z_j \in D^{1\times q'}$ satisfying $\overline{P}_{j\bullet} - P_{j\bullet} = Z_j\, R'$. Hence, we obtain $\overline{P} - P = Z\, R'$, where $Z = (Z_1^T \ldots Z_p^T)^T \in D^{p\times q'}$. Finally, if we denote by $R_2' \in D^{r'\times q'}$ a matrix generating $\ker_D(.R') = \{\lambda \in D^{1\times q'} \mid \lambda\, R' = 0\}$, i.e. $\ker_D(.R') = D^{1\times r'}\, R_2'$, and $Z' \in D^{q\times r'}$ is any arbitrary matrix, then we get

$$R\,\overline{P} = R\, P + R\, Z\, R' = Q\, R' + R\, Z\, R' = (Q + R\, Z)\, R' = (Q + R\, Z + Z'\, R_2')\, R',$$

which proves that we have $R\,\overline{P} = \overline{Q}\, R'$, with the notation $\overline{Q} = Q + R\, Z + Z'\, R_2'$.

One of the main interests of computing $\hom_D(M, M')$ is the following corollary of Proposition 1 which is a straightforward consequence of (2).

Corollary 1 ([6]). *Let $\mathcal{F}$ be a left D-module, $R \in D^{q \times p}$, $R' \in D^{q' \times p'}$ and the corresponding two linear functional systems:*

$$\ker_{\mathcal{F}}(R.) = \{\eta \in \mathcal{F}^p \mid R\,\eta = 0\}, \quad \ker_{\mathcal{F}}(R'.) = \{\eta' \in \mathcal{F}^{p'} \mid R'\,\eta' = 0\}.$$

Then, an element $f \in \hom_D(M, M')$ defined by $P \in D^{p \times p'}$ and $Q \in D^{q \times q'}$ satisfying (2) induces the following morphism of abelian groups:

$$\begin{aligned} f^\star : \ker_{\mathcal{F}}(R'.) &\longrightarrow \ker_{\mathcal{F}}(R.) \\ \eta' &\longmapsto \eta = P\,\eta'. \end{aligned}$$

By Corollary 1, we obtain that the abelian group $\hom_D(M, M')$ defines transformations sending the solutions of the second system to those of the first one, i.e., defines a *morphism of behaviours*. In particular, if $M' = M$, then the elements of the D-endomorphism ring $\mathrm{end}_D(M)$ of M define internal transformations of $\ker_{\mathcal{F}}(R.)$, i.e., *Galois-like transformations*. We refer the reader to [6] for applications of the Galois transformations to the computation of quadratic conservation laws of linear systems of partial differential equations.

Moreover, in the next section, we shall see that the endomorphism ring $\mathrm{end}_D(M)$ allows us to factorize, reduce and decompose the left D-module M and its corresponding linear functional system $\ker_{\mathcal{F}}(R.)$.

3 Computations of D-morphisms

Let first study $\hom_D(M, M')$ when D is a commutative ring. As, in this particular case, $\hom_D(M, M')$ is a D-module, we can characterize this D-module by means of generators and relations.

Definition 2. The *Kronecker product* of $E \in D^{q \times p}$ and $F \in D^{r \times s}$ is the matrix defined by:

$$E \otimes F = \begin{pmatrix} E_{11}\,F & \ldots & E_{1p}\,F \\ \vdots & & \vdots \\ E_{q1}\,F & \ldots & E_{qp}\,F \end{pmatrix} \in D^{(q\,r) \times (p\,s)}.$$

Using the Kronecker product, we have the simple but useful lemma.

Lemma 1. *Let D be a commutative ring and $U \in D^{a \times b}$, $V \in D^{b \times c}$ and $W \in D^{c \times d}$. Then, we have*

$$\mathrm{row}(U\,V\,W) = \mathrm{row}(V)\,(U^T \otimes W),$$

with the notation $\mathrm{row}(V) = (V_{1\bullet} \;\ldots\; V_{b\bullet})$, where $V_{i\bullet}$ denotes the ith row of the matrix V.

With the notations of Proposition 1, we get:

$$\begin{cases} \mathrm{row}(R\,P) = \mathrm{row}(R\,P\,I_{p'}) = \mathrm{row}(P)\,(R^T \otimes I_{p'}), \\ \mathrm{row}(Q\,R') = \mathrm{row}(I_q\,Q\,R') = \mathrm{row}(Q)\,(I_q \otimes R'). \end{cases}$$

Hence, the computation of the matrices $P \in D^{p \times p'}$ and $Q \in D^{q \times q'}$ satisfying (2) is reduced to the computation of a set of generators of the D-module:

$$\ker_D\left(.\begin{pmatrix} R^T\otimes I_{p'} \\ -I_q\otimes R' \end{pmatrix}\right) = \left\{\lambda\in D^{1\times(p\,p'+q\,q')}\mid \lambda\begin{pmatrix} R^T\otimes I_{p'} \\ -I_q\otimes R' \end{pmatrix}=0\right\}. \quad (4)$$

If D is a *noetherian ring*, namely, if every ideal of D is finitely generated as a D-module, then the previous D-module is finitely generated and even finitely presented (see, e.g., [17]). In particular, when D is a commutative polynomial ring over a *computational field* k (e.g., $k=\mathbb{Q}$, $\mathbb{F}_p$) (see, e.g., [2, 13]), the kernel defined by (4) can be obtained by means of a Gröbner basis computation. For more details, see, e.g., [3, 4, 6, 13] and the references therein.

The complete algorithm for computing generators of $\hom_D(M,M')$ based on this method is given in [6, Algorithm 2.1]. The relations between these generators can be obtained using the method described in [6, Remark 2.3].

Example 1. We consider a fluid in a tank satisfying Saint-Venant's equations and subjected to a one-dimensional horizontal move ([9]):

$$\begin{cases} y_1(t-2\,h)+y_2(t)-2\,\dot{u}(t-h)=0, \\ y_1(t)+y_2(t-2\,h)-2\,\dot{u}(t-h)=0. \end{cases} \quad (5)$$

Let $D=\mathbb{Q}\,[\partial,\delta]$ be the commutative polynomial ring of differential time-delay operators with rational constant coefficients (i.e., $\partial f(t)=\dot{f}(t)$, $\delta f(t)=f(t-h)$), the system matrix of (5) defined by

$$R=\begin{pmatrix} \delta^2 & 1 & -2\,\partial\,\delta \\ 1 & \delta^2 & -2\,\partial\,\delta \end{pmatrix}\in D^{2\times 3}. \quad (6)$$

and the corresponding finitely presented D-module $M=D^{1\times 3}/(D^{1\times 2}\,R)$.

Applying the previous algorithm to R, we obtain that $\mathrm{end}_D(M)$ is generated by the D-endomorphisms f_{e_1}, f_{e_2}, f_{e_3} and f_{e_4} defined by $f_\alpha(\pi(\lambda))=\pi(\lambda\,P_\alpha)$, for all $\lambda\in D^{1\times 3}$, where

$$P_\alpha=\begin{pmatrix} \alpha_1 & \alpha_2 & 2\,\alpha_3\,\partial\,\delta \\ \alpha_2+2\,\alpha_4\,\partial & \alpha_1-2\,\alpha_4\,\partial & 2\,\alpha_3\,\partial\,\delta \\ \alpha_4\,\delta & -\alpha_4\,\delta & \alpha_1+\alpha_2+\alpha_3\,(\delta^2+1) \end{pmatrix},$$

$$Q_\alpha=\begin{pmatrix} \alpha_1-2\,\alpha_4\,\partial & \alpha_2+2\,\alpha_4\,\partial \\ \alpha_2 & \alpha_1 \end{pmatrix},$$

$\alpha=(\alpha_1,\alpha_2,\alpha_3,\alpha_4)\in D^{1\times 4}$ and $\{e_j\}_{j=1,\ldots,4}$ denotes the standard basis of $D^{1\times 4}$. We can check that the generators $\{f_{e_i}\}_{i=1,\ldots,4}$ of the D-module $\mathrm{end}_D(M)$ satisfy the following D-linear relations:

$$(\delta^2-1)\,f_{e_4}=0,\quad \delta^2\,f_{e_1}+f_{e_2}-f_{e_3}=0,\quad f_{e_1}+\delta^2\,f_{e_2}-f_{e_3}=0. \quad (7)$$

Let us consider the D-module $\mathcal{F}=C^\infty(\mathbb{R})$ and the linear system

$$\ker_{\mathcal{F}}(R.)=\{\eta=(y_1,y_2,u)^T\in\mathcal{F}^3\mid R\,\eta=0\}$$

defined by (5). Then, every $f_\alpha\in\mathrm{end}_D(M)$ defines the D-morphism

$$\begin{aligned} f_\alpha^\star:\ker_{\mathcal{F}}(R.) &\longrightarrow \ker_{\mathcal{F}}(R.), \\ \eta &\longmapsto P_\alpha\,\eta, \end{aligned}$$

i.e., defines a Galois transformation of the linear system $\ker_{\mathcal{F}}(R.)$.

A complete description of the non-commutative ring $\mathrm{end}_D(M)$ is given by the knowledge of the expressions of the compositions $f_{e_i} \circ f_{e_j}$ in the family of generators $\{f_{e_k}\}_{k=1,\ldots,4}$ for $i, j = 1, \ldots, 4$:

$$\begin{cases} f_{e_1} \circ f_{e_i} = f_{e_i} \circ f_{e_1} = f_{e_i}, \\ f_{e_2} \circ f_{e_2} = f_{e_1}, \\ f_{e_2} \circ f_{e_3} = f_{e_3} \circ f_{e_2} = f_{e_3}, \\ f_{e_2} \circ f_{e_4} = 2\,\partial\, f_{e_1} - 2\,\partial\, f_{e_2} + f_{e_4}, \\ f_{e_4} \circ f_{e_2} = -f_{e_4}, \end{cases} \quad \begin{cases} f_{e_3} \circ f_{e_3} = (\delta^2 + 1)\, f_{e_3}, \\ f_{e_3} \circ f_{e_4} = 2\,\partial\, f_{e_1} - 2\,\partial\, f_{e_2} + 2\, f_{e_4}, \\ f_{e_4} \circ f_{e_3} = 0, \\ f_{e_4} \circ f_{e_4} = -2\,\partial\, f_{e_4}. \end{cases} \tag{8}$$

In other words, we get the following *multiplication table* where $f_{e_c} \circ f_{e_r}$ means that we compose an element in the first column with an element in the first row:

$f_{e_c} \circ f_{e_r}$	f_{e_1}	f_{e_2}	f_{e_3}	f_{e_4}
f_{e_1}	f_{e_1}	f_{e_2}	f_{e_3}	f_{e_4}
f_{e_2}	f_{e_2}	f_{e_1}	f_{e_3}	$2\,\partial\, f_{e_1} - 2\,\partial\, f_{e_2} + f_{e_4}$
f_{e_3}	f_{e_3}	f_{e_3}	$(\delta^2 + 1)\, f_{e_3}$	$2\,\partial\, f_{e_1} - 2\,\partial\, f_{e_2} + 2\, f_{e_4}$
f_{e_4}	f_{e_4}	$-f_{e_4}$	0	$-2\,\partial\, f_{e_4}$

If we denote by $D < f_{e_1}, f_{e_2}, f_{e_3}, f_{e_4} >$ the *free D-algebra* (i.e., the non-commutative polynomial ring with coefficients in D) defined by the f_{e_i}'s and

$$I =< (\delta^2 - 1)\, f_{e_4}, \delta^2\, f_{e_1} + f_{e_2} - f_{e_3}, f_{e_1} + \delta^2\, f_{e_2} - f_{e_3}, \ldots, f_{e_4} \circ f_{e_4} + 2\,\partial\, f_{e_4} >$$

the two-sided ideal of $D < f_{e_1}, f_{e_2}, f_{e_3}, f_{e_4} >$ generated by the polynomials obtained from the identities (7) and (8), we then get that:

$$\mathrm{end}_D(M) = D < f_{e_1}, f_{e_2}, f_{e_3}, f_{e_4} > /I.$$

The previous quotient of a non-commutative polynomial algebra can then be studied by means of non-commutative Gröbner bases ([13])

When D is a non-commutative k-algebra, the computation of the k-vector space $\hom_D(M, M')$ is more complicated and we can generally only compute the morphisms of $\hom_D(M, M')$ which are defined by means of a matrix P with a fixed total order in the functional operators ∂_i and a fixed degree in x_i for the numerators and denominators of the polynomial/rational coefficients. As this issue is not crucial for the rest of the paper, we shall not detail that point here and we refer to [6, Algorithm 2.2] for a complete algorithm computing morphisms in the non-commutative case. Note that we can easily study $\hom_D(M, M')$ when it is a finite-dimensional k-vector space and a k-basis of $\hom_D(M, M')$ is known (see [6, 16, 17] and references therein). In particular, it is the case when M and M' are two left *holonomic* modules over the ring $A_n(k)$ of differential operators with coefficients in $k[x_1, \ldots, x_n]$ or *D-finite* linear functional systems ([2, 16, 17]) such as, for instance, $B_n(k)$-*finite* linear systems, where $B_n(k)$ denotes the ring of differential operators with coefficients in $k(x_1, \ldots, x_n)$ (see [6, 16, 17] and references therein).

4 Factorization, reduction and decomposition problems

If $f \in \hom_D(M, M')$, then we can define the following left D-modules:

$$\begin{cases} \ker f = \{m \in M \mid f(m) = 0\}, \\ \operatorname{im} f = \{m' \in M' \mid \exists\, m \in M : \; m' = f(m)\}, \end{cases} \qquad \begin{cases} \operatorname{coim} f = M/\ker f, \\ \operatorname{coker} f = M'/\operatorname{im} f. \end{cases}$$

We now give three main theorems of [6]. Each of them corresponds to one of the three problems given in Section 1.

Theorem 2 ([6] - Factorization problem). *Let us consider the matrix $R \in D^{q\times p}$, the finitely presented left D-module $M = D^{1\times p}/(D^{1\times q}\, R)$ and $f \in \operatorname{end}_D(M)$ defined by $P \in D^{p\times p}$ and $Q \in D^{q\times q}$ satisfying $R\,P = Q\,R$. Let $S \in D^{r\times p}$ and $T \in D^{r\times q}$ be two matrices such that:*

$$\ker_D\left(.\begin{pmatrix} P \\ R \end{pmatrix}\right) = D^{1\times r}\,(S \quad -T).$$

Then, we have the following inclusion of left D-modules $D^{1\times q}\, R \subseteq D^{1\times r}\, S$, which proves the existence of a factorization of the matrix R of the form

$$R = L\,S, \tag{9}$$

for a certain matrix $L \in D^{q\times r}$. In particular, if $\mathcal{F}$ is a left D-module, from (9), we then have $\ker_{\mathcal{F}}(S.) \subseteq \ker_{\mathcal{F}}(R.)$.

Moreover, we have:

1. $\ker f = (D^{1\times r}\, S)/(D^{1\times q}\, R)$. *Hence, f is injective iff there exists a matrix $F \in D^{r\times q}$ such that $S = F\,R$.*
2. $\operatorname{coim} f = D^{1\times p}/(D^{1\times r}\, S)$. *Hence, $f = 0$ iff the matrix S admits a left-inverse $X \in D^{p\times r}$, namely, $X\,S = I_p$.*
3. $\operatorname{im} f = (D^{1\times (p+q)}\,(P^T \quad R^T)^T)/(D^{1\times q}\, R)$.
4. $\operatorname{coker} f = D^{1\times p}/(D^{1\times (p+q)}\,(P^T \quad R^T)^T)$. *Hence, f is surjective iff the matrix $(P^T \quad R^T)^T$ admits a left-inverse $(X_1 \quad X_2) \in D^{p\times (p+q)}$, i.e., we have:*

$$X_1\,P + X_2\,R = I_p.$$

We refer the reader to [3] for constructive algorithms which factorize and compute left-inverses over certain classes of Ore algebras and their implementations in the library ORE-MODULES ([4]).

The second result gives conditions for the existence of a reduction of R.

Theorem 3 ([6] - Reduction problem). *Let us consider the matrix $R \in D^{q\times p}$, the finitely presented left D-module $M = D^{1\times p}/(D^{1\times q}\, R)$ and $f \in \operatorname{end}_D(M)$ defined by $P \in D^{p\times p}$ and $Q \in D^{q\times q}$ satisfying $R\,P = Q\,R$. If the left D-modules* $\ker_D(.P)$, $\operatorname{coim}_D(.P)$, $\ker_D(.Q)$ *and* $\operatorname{coim}_D(.Q)$ *are respectively free of rank m, $p - m$, l, $q - l$, then there exist matrices $U_1 \in D^{m\times p}$, $U_2 \in D^{(p-m)\times p}$, $V_1 \in D^{l\times q}$ and $V_2 \in D^{(q-l)\times q}$ such that $U = (U_1^T \quad U_2^T)^T \in \mathrm{GL}_p(D)$, $V = (V_1^T \quad V_2^T)^T \in \mathrm{GL}_q(D)$ and*

$$\overline{R} = V\,R\,U^{-1} = \begin{pmatrix} V_1\,R\,W_1 & 0 \\ V_2\,R\,W_1 & V_2\,R\,W_2 \end{pmatrix} \in D^{q\times p},$$

where $U^{-1} = (W_1 \quad W_2)$, $W_1 \in D^{p\times m}$ and $W_2 \in D^{p\times (p-m)}$.

The full row rank matrix U_1 (resp., U_2, V_1, V_2) defines a basis of the free left D-module $\ker_D(.P)$ *(resp.,* $\operatorname{coim}_D(.P)$, $\ker_D(.Q)$, $\operatorname{coim}_D(.Q)$*), i.e.,*

$$\begin{cases} \ker_D(.P) = D^{1\times m}\, U_1, \\ \operatorname{coim}_D(.P) = \kappa(D^{1\times (p-m)}\, U_2), \\ \ker_D(.Q) = D^{1\times l}\, V_1, \\ \operatorname{coim}_D(.Q) = \rho(D^{1\times (q-l)}\, V_2), \end{cases}$$

where $\kappa : D^{1\times p} \longrightarrow \operatorname{coim}_D(.P) = D^{1\times p}/\ker_D(.P)$ *denotes the canonical projection and similarly with* $\rho : D^{1\times q} \longrightarrow \operatorname{coim}_D(.Q) = D^{1\times q}/\ker_D(.Q)$.

The last result gives conditions for the existence of a decomposition of R.

Theorem 4 ([6] - Decomposition problem). *Let us consider the matrix* $R \in D^{q\times p}$, *the finitely presented left D-module* $M = D^{1\times p}/(D^{1\times q}\, R)$ *and an idempotent endomorphism* f *of* M *defined by two idempotent matrices* $P \in D^{p\times p}$ *and* $Q \in D^{q\times q}$, *namely, they satisfy* $R\,P = Q\,R$, $P^2 = P$ *and* $Q^2 = Q$. *If the left D-modules* $\ker_D(.P)$, $\operatorname{im}_D(.P) = \ker_D(.(I_p - P))$, $\ker_D(.Q)$ *and* $\operatorname{im}_D(.Q) = \ker_D(.(I_q - Q))$ *are respectively free of rank* m, $p-m$, l, $q-l$, *then there exist four matrices* $U_1 \in D^{m\times p}$, $U_2 \in D^{(p-m)\times p}$, $V_1 \in D^{l\times q}$ *and* $V_2 \in D^{(q-l)\times q}$ *such that*

$$U = (U_1^T \quad U_2^T)^T \in \mathrm{GL}_p(D), \quad V = (V_1^T \quad V_2^T)^T \in \mathrm{GL}_q(D),$$

and

$$\overline{R} = V\,R\,U^{-1} = \begin{pmatrix} V_1\,R\,W_1 & 0 \\ 0 & V_2\,R\,W_2 \end{pmatrix} \in D^{q\times p},$$

where $U^{-1} = (W_1 \quad W_2)$, $W_1 \in D^{p\times m}$ *and* $W_2 \in D^{p\times (p-m)}$.

The full row rank matrix U_1 *(resp.,* U_2, V_1, V_2*) defines a basis of the free left D-module* $\ker_D(.P)$, *(resp.,* $\operatorname{im}_D(.P)$, $\ker_D(.Q)$, $\operatorname{im}_D(.Q)$*) of rank* m *(resp.,* $p-m$, l, $q-l$*). In other words, we have:*

$$\begin{cases} \ker_D(.P) = D^{1\times m}\, U_1, \\ \operatorname{im}_D(.P) = D^{1\times (p-m)}\, U_2, \\ \ker_D(.Q) = D^{1\times l}\, V_1, \\ \operatorname{im}_D(.Q) = D^{1\times (q-l)}\, V_2. \end{cases}$$

Finally, if $\mathcal{F}$ *is a left D-module, then we obtain the following decomposition of the linear functional system* $\ker_{\mathcal{F}}(R.)$:

$$\ker_{\mathcal{F}}(R.) \cong \ker_{\mathcal{F}}(V_1\,R\,W_1.) \oplus \ker_{\mathcal{F}}(V_2\,R\,W_2.).$$

Example 2. If we consider again Example 1, then we can check that the D-endomorphism $f = \frac{1}{2}\,(f_{e_1} + f_{e_2})$ is defined by the following matrices:

$$P = \frac{1}{2}\begin{pmatrix} 1 & 1 & 0 \\ 1 & 1 & 0 \\ 0 & 0 & 2 \end{pmatrix}, \quad Q = \frac{1}{2}\begin{pmatrix} 1 & 1 \\ 1 & 1 \end{pmatrix}.$$

The corresponding factorization of the matrix R defined by (6) is given by:

$$R = L\,S, \quad L = \begin{pmatrix} \delta^2 & 1 \\ 1 & 1 \end{pmatrix}, \quad S = \begin{pmatrix} 1 & -1 & 0 \\ 0 & \delta^2 + 1 & -2\,\partial\,\delta \end{pmatrix}.$$

Moreover, as the entries of P and Q belong to the field $\mathbb{Q}$, using linear algebraic techniques, we can easily compute bases of the free D-modules $\ker_D(.P)$, $\mathrm{coim}_D(.P)$, $\ker_D(.Q)$ and $\mathrm{coim}_D(.Q)$ and we get:

$$U_1 = (1 \ \ -1 \ \ 0), \quad U_2 = \begin{pmatrix} 0 & 1 & 0 \\ 0 & 0 & 1 \end{pmatrix}, \quad V_1 = (1 \ \ -1), \quad V_2 = (0 \ \ 1).$$

Forming $U = (U_1^T \ \ U_2^T)^T \in \mathrm{GL}_3(D)$ and $V = (V_1^T \ \ V_2^T)^T \in \mathrm{GL}_2(D)$, we obtain the following reduction of the matrix R:

$$\overline{R} = V\,R\,U^{-1} = \begin{pmatrix} \delta^2 - 1 & 0 & 0 \\ 1 & \delta^2 + 1 & -2\,\partial\,\delta \end{pmatrix}.$$

Using (8), we can easily check that f is an idempotent of $\mathrm{end}_D(M)$ defined by two idempotent matrices P and Q, i.e., $P^2 = P$ and $Q^2 = Q$. By means of linear algebra, we can compute bases of the free D-modules $\mathrm{im}_D(.P) = \ker_D(.(I_3 - P))$ and $\mathrm{im}_D(.Q) = \ker_D(.(I_2 - Q))$ and we obtain:

$$X_2 = \begin{pmatrix} 1 & 1 & 0 \\ 0 & 0 & 1 \end{pmatrix}, \quad Y_2 = (1 \ \ 1).$$

Forming $X = (U_1^T \ \ X_2^T)^T \in \mathrm{GL}_3(D)$ and $Y = (V_1^T \ \ Y_2^T)^T \in \mathrm{GL}_2(D)$, we finally obtain the following decomposition of the matrix R:

$$\overline{\overline{R}} = Y\,R\,X^{-1} = \begin{pmatrix} \delta^2 - 1 & 0 & 0 \\ 0 & \delta^2 + 1 & -4\,\partial\,\delta \end{pmatrix}.$$

We refer the reader to [6, 7] for more results and examples coming from control theory, mathematical physics and engineering sciences.

Conclusion

In this paper, we have recalled the main theoretical results of [5, 6] on the factorization, reduction and decomposition problems of linear functional systems. The theorems given in Section 4 have been implemented in the OREMORPHISMS package ([7]) and can be used to decompose the main linear differential time-delay systems appearing in the literature of control theory ([6, 7]). However, theoretical difficult questions are still open for getting a general algorithmic test which checks whether or not a finitely presented left D-module M is (completely) decomposable or simple. The algebraic structure of the endomorphism ring $\mathrm{end}_D(M)$ needs to be investigated more accurately. In particular, the study of the *regular elements* of $\mathrm{end}_D(M)$ (namely, $f \in \mathrm{end}_D(M)$ such that there exists $g \in \mathrm{end}_D(M)$ satisfying $f \circ g \circ f = f$) and the idempotents of $\mathrm{end}_D(M)$ will be continued in the future.

References

1. Barkatou, M.A.: Factoring systems of linear functional systems using eigenrings. In: Kotsireas, I., Zima, E. (eds.) Proceedings of the Waterloo Workshop, Computer Algebra 2006, Latest Advances in Symbolic Algorithms, Ontario, Canada, April 10-12, pp. 22–42. World Scientific, Singapore (2007)

2. Chyzak, F., Salvy, B.: Non-commutative elimination in Ore algebras proves multivariate identities. J. Symbolic Comput. 26, 187–227 (1998)
3. Chyzak, F., Quadrat, A., Robertz, D.: Effective algorithms for parametrizing linear control systems over Ore algebras. Appl. Algebra Engrg. Comm. Comput. 16, 319–376 (2005)
4. Chyzak, F., Quadrat, A., Robertz, D.: OREMODULES: A symbolic package for the study of multidimensional linear systems. In: Chiasson, J., Loiseau, J.J. (eds.) Applications of Time-Delay Systems. LNCIS, vol. 352, pp. 233–264. Springer, Heidelberg (2007)
5. Cluzeau, T., Quadrat, A.: Using morphism computations for factoring and decomposing general linear functional systems. In: Proc. Mathematical Theory of Networks and Systems (MTNS), Kyoto, Japan (2006)
6. Cluzeau, T., Quadrat, A.: Factoring and decomposing a class of linear functional systems. Linear Algebra Appl. 428, 324–381 (2008)
7. Cluzeau, T., Quadrat, A.: OREMORPHISMS: A homological algebraic package for factoring, reducing and decomposing linear functional systems. In: Loiseau, J.J., Michiels, W., Niculescu, S.I., Sipahi, R. (eds.) Topics in Time-Delay Systems: Analysis, Algorithms and Control. LNCIS, pp. 179–194. Springer, Heidelberg (2009)
8. Cohn, P.M.: Free rings and their Relations. LMS Monographs 19, 2nd edn. Academic Press, London (1985)
9. Dubois, F., Petit, N., Rouchon, P.: Motion planning and nonlinear simulations for a tank containing a fluid. In: Proc. European Control Conference (ECC), Karlsruhe, Germany (1999)
10. Kailath, T.: Linear Systems. Prentice-Hall, Englewood Cliffs (1980)
11. Kashiwara, M.: Algebraic Study of Systems of Partial Differential Equations. Master Thesis, Tokyo Univ. Mémoire de la Société Mathématiques de France 63 (1995) (English translation) (1970)
12. Kučera V.: Discrete Linear Control. Wiley, Chichester (1979)
13. Levandovskyy, V.: Non-commutative Computer Algebra for polynomial algebras: Gröbner bases, applications and implementation. PhD Thesis, University of Kaiserslautern, Germany (2005)
14. Malgrange, B.: Systèmes à coefficients constants. Séminaire Bourbaki, 1–11 (1962/1963)
15. McConnell, J.C., Robson, J.C.: Noncommutative Noetherian Rings. American Mathematical Society, Providence (2000)
16. van der Put, M., Singer, M.F.: Galois theory of linear differential equations. Grundlehren der mathematischen Wissenschaften, vol. 328. Springer, Heidelberg (2003)
17. Rotman, J.J.: An Introduction to Homological Algebra. Academic Press, London (1979)
18. Rosenbrock, H.H.: State Space and Multivariable Theory. Wiley, Chichester (1970)
19. Singer, M.F.: Testing reducibility of linear differential operators: a group theoretic perspective. Appl. Algebra Engrg. Comm. Comput. 7, 77–104 (1996)
20. Wu, M.: On Solutions of Linear Functional Systems and Factorization of Modules over Laurent-Ore Algebras. PhD thesis of the Chinese Academy of Sciences (China) and the University of Nice-Sophia Antipolis, France (2005)

OreMorphisms: A Homological Algebraic Package for Factoring, Reducing and Decomposing Linear Functional Systems

Thomas Cluzeau[1] and Alban Quadrat[2]

[1] University of Limoges; CNRS; XLIM UMR 6172, DMI; ENSIL, 16 Rue d'Atlantis, Parc Ester Technopole, 87068 Limoges Cedex, France `cluzeau@ensil.unilim.fr`
[2] INRIA Sophia Antipolis, APICS Project, 2004 Route des Lucioles, BP 93, 06902 Sophia Antipolis Cedex, France `Alban.Quadrat@sophia.inria.fr`

Summary. The purpose of this paper is to demonstrate the symbolic package OreMorphisms which is dedicated to the implementation of different algorithms and heuristic methods for the study of the factorization, reduction and decomposition problems of general linear functional systems (e.g., systems of partial differential or difference equations, differential time-delay systems). In particular, we explicitly show how to decompose a differential time-delay system (a string with an interior mass [15]) formed by 4 equations in 6 unknowns and prove that it is equivalent to a simple equation in 3 unknowns. We finally give a list of reductions of classical systems of differential time-delay equations and partial differential equations coming from control theory and mathematical physics.

Introduction

In [6], we have recalled the main theoretical results of [5] on the factorization, reduction and decomposition problems for general linear functional systems obtained within a constructive homological algebra approach. The purpose of this paper is to demonstrate the Maple package OreMorphisms which is dedicated to the implementation of those results.

The OreMorphisms package focuses on the following problems:

- Compute D-morphisms between two finitely presented left D-modules over certain classes of Ore algebras D, i.e., the ones implemented in the package *Ore_algebra* available in the current Maple releases.
- Compute idempotents of the endomorphism ring $\mathrm{end}_D(M)$ of a finitely presented left D-module M (i.e., $f \in \mathrm{end}_D(M)$, $f^2 = f$) and, among the latter, those further defined by idempotent matrices P and Q, i.e., $P^2 = P$ and $Q^2 = Q$.
- Compute presentations of the kernel, image, cokernel, and coimage of a given morphism. Test whether or not a given morphism is injective, surjective or defines a D-isomorphism.
- Compute factorizations, reductions and decompositions of linear functional systems.

The package OreMorphisms is based on the Maple library OreModules ([2, 4]) devoted to the symbolic study of multidimensional systems. The OreModules library and its

J.J. Loiseau et al. (Eds.): Topics in Time Delay Systems, LNCIS 388, pp. 179–194.
springerlink.com

subpackage OREMORPHISMS are both freely available. For more details, see [2] for OREMODULES and [7] for OREMORPHISMS.

1 Computational issues and OREMORPHISMS functions

Hereafter, we use the notations of [6]. In particular, D denotes a non-commutative polynomial ring and $R \in D^{q\times p}$.

To implement our algorithms for factoring, reducing and decomposing linear functional systems, we mainly need to be able to constructively perform the two following tasks:

1. Compute a finite number of pairs (P, Q) of matrices which define endomorphisms of the left D-module $M = D^{1\times p}/(D^{1\times q}\, R)$, i.e., satisfy the relation $R\,P = Q\,R$. From those pairs, compute those that define idempotent D-endomorphisms (namely, $f \in \mathrm{end}_D(M)$ satisfying $f^2 = f$) defined by idempotent matrices P and Q, i.e., $P^2 = P$, $Q^2 = Q$.
2. Decide whether a left D-module is free and, if so, compute a basis of it.

In the package OREMORPHISMS, we have implemented algorithms handling the first point. In certain cases, the computation of idempotent D-endomorphisms defined by idempotent matrices P and Q reduces to calculating the solutions of the algebraic Riccati equation (see [5, Lemma 4.4])

$$\Lambda\, R \Lambda + (P - I_p)\,\Lambda + \Lambda\, Q + Z = 0, \tag{1}$$

where $P^2 = P + Z\,R$, $Q^2 = Q + R\,Z$, $R\,P = Q\,R$, $Z \in D^{p\times q}$ and R has a *full row rank*, namely, $\ker_D(.R) \triangleq \{\lambda \in D^{1\times q} \mid \lambda\, R = 0\} = 0$. Then, the matrices $\overline{P} = P + \Lambda\, R$ and $\overline{Q} = Q + R\,\Lambda$ satisfy $\overline{P}^2 = \overline{P}$, $\overline{Q}^2 = \overline{Q}$ and $R\,\overline{P} = \overline{Q}\,R$. Hence, we have also implemented an algorithm computing solutions of (1).

For the second point, testing whether or not a given finitely presented left D-module over an Ore algebra D is free has been constructively studied in [2, 4] and algorithmic and heuristic methods have been implemented in OREMODULES. Moreover, the (non-trivial) problem of computing bases of free left modules over certain classes of Ore algebras has been made algorithmic:

- When D is the ring of ordinary differential or shift operators with rational coefficients, this can be achieved by means of *Jacobson normal form* computations (see, e.g., [9]). Jacobson normal forms have been implemented by G. Culianez for certain Ore algebras in the package JACOBSON ([9]) of the library OREMODULES.
- When D is a commutative Ore algebra (e.g., differential time-delay operators with constant coefficients), we can use constructive versions of the *Quillen-Suslin theorem* of Serre's conjecture (see, e.g., [11]). A constructive algorithm has recently been implemented by A. Fabiańska (Aachen University) in the Maple package QUILLENSUSLIN ([10, 11]).
- When D is the *Weyl algebra* $A_n(\mathbb{Q})$ (resp., $B_n(\mathbb{Q})$) of differential operators with polynomial (resp., rational) coefficients, a recent algorithm for computing bases of free left D-modules has been developed in [18] based on *Stafford's theorems*. This algorithm is implemented in the package STAFFORD of the OREMODULES library ([2]).

A list of OREMORPHISMS functions is given in Table 1. We use the notation A for the ring of functional operators as D is protected in Maple. The suffix "ConstCoeff" (resp., "Rat") distinguishes the procedures which deal with constant (resp., rational) coefficients from those dealing with polynomial coefficients (no suffix means that the procedures handle the polynomial coefficients case). In this table, $R \in A^{q\times p}$ and $R' \in A^{q'\times p'}$ denote two matrices with coefficients in an Ore algebra A handled in the Maple package *Ore_algebra*. We denote $M = A^{1\times p}/(A^{1\times q}\, R)$ and $M' = A^{1\times p'}/(A^{1\times q'}\, R')$ the two associated finitely presented left A-modules.

2 A worked example using the package OREMORPHISMS

We consider the model of a string with an interior mass studied in [15]

$$\begin{cases} \phi_1(t) + \psi_1(t) - \phi_2(t) - \psi_2(t) = 0, \\ \dot{\phi}_1(t) + \dot{\psi}_1(t) + \eta_1\, \phi_1(t) - \eta_1\, \psi_1(t) - \eta_2\, \phi_2(t) + \eta_2\, \psi_2(t) = 0, \\ \phi_1(t - 2\, h_1) + \psi_1(t) - u(t - h_1) = 0, \\ \phi_2(t) + \psi_2(t - 2\, h_2) - v(t - h_2) = 0, \end{cases} \tag{2}$$

where η_1, η_2 are constant parameters and $h_1, h_2 \in \mathbb{R}_+$ are such that $\mathbb{Q}\, h_1 + \mathbb{Q}\, h_2$ is a 2-dimensional $\mathbb{Q}$-vector space. Let us denote by $A = \mathbb{Q}(\eta_1, \eta_2)\, [d, \sigma_1, \sigma_2]$ the commutative polynomial algebra of differential incommensurable time-delay operators in d, σ_1 and σ_2, where:

$$d\, f(t) = \dot{f}(t), \quad \sigma_1 f(t) = f(t - h_1), \quad \sigma_2 f(t) = f(t - h_2).$$

The system matrix $R \in A^{4\times 6}$ of (2) is defined by:

```
> with(OreModules):
> with(OreMorphisms): with(linalg):

> A:=DefineOreAlgebra(diff=[d,t],dual_shift=[sigma[1],
> x[1]], dual_shift=[sigma[2],x[2]],polynom=[t,x[1],x[2]],
> comm=[eta[1],eta[2]]):

> R:=matrix(4,6,[1,1,-1,-1,0,0,d+eta[1],d-eta[1],-eta[2],
> eta[2],0,0, sigma[1]^2,1,0,0,-sigma[1],0,0,0,1,
> sigma[2]^2,0,-sigma[2]]);
```

$$R := \begin{bmatrix} 1 & 1 & -1 & -1 & 0 & 0 \\ d + \eta_1 & d - \eta_1 & -\eta_2 & \eta_2 & 0 & 0 \\ {\sigma_1}^2 & 1 & 0 & 0 & -\sigma_1 & 0 \\ 0 & 0 & 1 & {\sigma_2}^2 & 0 & -\sigma_2 \end{bmatrix}$$

2.1 Factorization problem

We show how to use OREMORPHISMS for computing a factorization of R of the form $R = L\, S$. We first need to compute the endomorphism ring $\mathrm{end}_A(M)$ of the A-module $M = A^{1\times 6}/(A^{1\times 4}\, R)$ finitely presented by the matrix R.

```
> Endo:=MorphismsConstCoeff(R,R,A):
```

Then, we choose a particular morphism f by selecting the first element P_1 of $Endo[1]$ and compute a matrix Q_1 satisfying $R\, P_1 = Q_1\, R$. The latter operation can be performed by means of the *Factorize* procedure of OREMODULES.

Table 1. List of the main functions of the package OREMORPHISMS

`Morphisms(ConstCoeff,Rat)`	Compute a finite family of matrices $P \in A^{p \times p'}$ which define elements of $\hom_A(M, M')$, i.e., such that there exist matrices $Q \in A^{q \times q'}$ satisfying the relation: $$R\,P = Q\,R'$$
`Idempotents(ConstCoeff,Rat)`	Compute a finite family of matrices $P \in A^{p \times p}$ defining idempotent elements of $\mathrm{end}_A(M)$, i.e., such that there exist three matrices $Q \in A^{q \times q}$, $Z \in A^{p \times q}$ and $Z' \in A^{q \times r}$ satisfying the relations $$R\,P = Q\,R,\; P^2 = P + Z\,R,$$ $$Q^2 = Q + R\,Z + Z'\,R_2,$$ where $\ker_A(.R) = A^{1 \times r}\,R_2$
`IdempotentsMat(ConstCoeff,Rat)`	Compute a finite family of idempotent matrices $P \in A^{p \times p}$ defining idempotent elements of $\mathrm{end}_A(M)$, i.e., such that there exist matrices $Q \in A^{q \times q}$ satisfying the relations $$R\,P = Q\,R,\; P^2 = P,\; Q^2 = Q,$$ where R has full row rank
`Riccati(ConstCoeff,Rat)`	Find a finite family of solutions $\Lambda \in A^{p \times q}$ of the algebraic Riccati equation $$\Lambda\,R\,\Lambda + (P - I_p)\,\Lambda + \Lambda\,Q + Z = 0,$$ where the pair (P, Q) defines an idempotent element of $\mathrm{end}_A(M)$, i.e., satisfies $$R\,P = Q\,R, P^2 = P + Z\,R, Q^2 = Q + R\,Z,$$ with $Z \in A^{p \times q}$ (R has full row rank)
`KerMorphism(Rat)`	Compute the kernel of $f \in \hom_A(M, M')$, i.e., compute two matrices $S \in A^{r \times p}$ and $X \in A^{s \times r}$ such that: $$\ker f = (A^{1 \times r}\,S)/(A^{1 \times q}\,R) \cong A^{1 \times r}/(A^{1 \times s}\,X)$$
`ImMorphism(Rat)`	Compute the image of $f \in \hom_A(M, M')$ defined by a pair of matrices (P, Q), i.e., $$\mathrm{im}\, f = (A^{1 \times (p+q')}\,(P^T \quad R'^T)^T)/(A^{1 \times q'}\,R'),$$ by reducing the rows of $(P^T \quad R'^T)^T$ modulo the left A-module $A^{1 \times q'}\,R'$
`CoimMorphism(Rat)`	Compute the coimage of $f \in \hom_A(M, M')$, i.e., compute a matrix $S \in A^{r \times p}$ such that: $$\mathrm{coim}\, f = A^{1 \times p}/(A^{1 \times r}\,S)$$

CokerMorphism(Rat)	Compute the cokernel of $f \in \hom_A(M, M')$ defined by a pair of matrices (P, Q), i.e., $\operatorname{coker} f = A^{1\times p'}/(A^{1\times(p+q')}\,(P^T \quad R'^T)^T)$
TestInj(Rat)	Test whether or not a given element of $\hom_A(M, M')$ is injective
TestSurj(Rat)	Test whether or not a given element of $\hom_A(M, M')$ is surjective
TestIso(Rat)	Test whether or not a given element of $\hom_A(M, M')$ is an A-isomorphism
HeuristicReduction(Rat)	Compute a reduction of the matrix R, i.e., compute an equivalent matrix with a block-triangular form. The heuristic part corresponds to the computation of bases of the different free left A-modules
HeuristicDecomposition(Rat)	Compute a decomposition of the matrix R, i.e., compute an equivalent matrix with a block-diagonal form. The heuristic part corresponds to the computation of bases of the different free left A-modules

```
> P[1]:=Endo[1,1]; Q[1]:=Factorize(Mult(R,P[1],A),R,A);
```

$$P_1 := \begin{bmatrix} 0 & 0 & \eta_2\,\sigma_2 & \eta_2\,\sigma_2 & 0 & 0 \\ 0 & 0 & 0 & 0 & 0 & 0 \\ 0 & \sigma_2\,\eta_1 & \eta_2\,\sigma_2 & 0 & 0 & 0 \\ 0 & -\sigma_2\,\eta_1 & 0 & \eta_2\,\sigma_2 & 0 & 0 \\ 0 & 0 & \eta_2\,\sigma_2\,\sigma_1 & \eta_2\,\sigma_2\,\sigma_1 & 0 & 0 \\ 0 & \eta_1 - {\sigma_2}^2\eta_1 & 0 & 0 & 0 & \eta_2\,\sigma_2 \end{bmatrix}$$

$$Q_1 := \begin{bmatrix} 0 & 0 & 0 & 0 \\ -\eta_2\,\sigma_2\,\eta_1 - \eta_2\,\sigma_2\,d & \eta_2\,\sigma_2 & 0 & 0 \\ 0 & 0 & 0 & 0 \\ 0 & 0 & 0 & \eta_2\,\sigma_2 \end{bmatrix}$$

By [6, Theorem 2], the matrix S that we are searching for is the one defining the coimage of the endomorphism f of M defined by the previous matrices P_1 and Q_1. So, we compute it using the *CoimMorphism* procedure.

```
> S:=CoimMorphism(R,R,P[1],Q[1],A)[1];
```

$$S := \begin{bmatrix} 1 & 0 & -1 & -1 & 0 & 0 \\ 0 & 1 & 0 & 0 & 0 & 0 \\ 0 & 0 & \sigma_1 & \sigma_1 & -1 & 0 \\ 0 & 0 & 1 & {\sigma_2}^2 & 0 & -\sigma_2 \\ 0 & 0 & -d+\eta_2-\eta_1 & -d-\eta_2-\eta_1 & 0 & 0 \end{bmatrix}$$

The matrix L such that $R = L\,S$ can be obtained by right factoring R by S.

```
> L:=Factorize(R,S,A);
```

$$L := \begin{bmatrix} 1 & 1 & 0 & 0 & 0 & 0 \\ d+\eta_1 & d-\eta_1 & 0 & 1 & 0 & 0 \\ {\sigma_1}^2 & 1 & \sigma_1 & 0 & 0 & 0 \\ 0 & 0 & 0 & 0 & 1 & 0 \end{bmatrix}$$

We note that choosing another endomorphism of M, i.e., another element of $Endo[1]$, would lead to another factorization of the matrix R.

2.2 Reduction problem

We use the package OREMORPHISMS to reduce the matrix R, i.e., to find an equivalent matrix with a block-triangular form. By [6, Theorem 3], this can be done using an endomorphism of M defined by a pair of matrices P and Q provided that the A-modules $\ker_A(.P)$, $\mathrm{coim}_A(.P)$, $\ker_A(.Q)$ and $\mathrm{coim}_A(.Q)$ are free. We use the library OREMODULES to check that these properties are fulfilled and use a heuristic method to compute bases of those free A-modules. We then form the matrices U and V as defined in [6, Theorem 3]. We note that we generally need to use the package QUILLENSUSLIN to compute bases of free modules over a commutative polynomial ring.

```
> U1:=SyzygyModule(P[1],A): EU:=Exti(Involution(U1,A),A,1):
> U2:=LeftInverse(EU[3],A): U:=stackmatrix(U1,U2);
> V1:=SyzygyModule(Q[1],A): EV:=Exti(Involution(V1,A),A,1):
> V2:=LeftInverse(EV[3],A): V:=stackmatrix(V1,V2);
```

$$U := \begin{bmatrix} 1 & 0 & -1 & -1 & 0 & 0 \\ 0 & 1 & 0 & 0 & 0 & 0 \\ 0 & 0 & \sigma_1 & \sigma_1 & -1 & 0 \\ 0 & 0 & -1 & -1 & 0 & 0 \\ 0 & 0 & 1 & 0 & 0 & 0 \\ 0 & 0 & 0 & 0 & 0 & 1 \end{bmatrix} \qquad V := \begin{bmatrix} 1 & 0 & 0 & 0 \\ 0 & 0 & 1 & 0 \\ 0 & 1 & 0 & 0 \\ 0 & 0 & 0 & 1 \end{bmatrix}$$

Then, we can compute the reduction $V\,R\,U^{-1}$ of the matrix R:

```
> R_red:=Mult(V,R,LeftInverse(U,A),A);
```

$$R\_red := \begin{bmatrix} 1 & 1 & 0 & 0 & 0 & 0 \\ {\sigma_1}^2 & 1 & \sigma_1 & 0 & 0 & 0 \\ d+\eta_1 & d-\eta_1 & 0 & -\eta_1-\eta_2-d & -2\,\eta_2 & 0 \\ 0 & 0 & 0 & -{\sigma_2}^2 & 1-{\sigma_2}^2 & -\sigma_2 \end{bmatrix}$$

This reduction can be obtained using the *HeuristicReduction* procedure.

```
> HeuristicReduction(R,P[1],A)[1];
```

$$\begin{bmatrix} 1 & 1 & 0 & 0 & 0 & 0 \\ {\sigma_1}^2 & 1 & \sigma_1 & 0 & 0 & 0 \\ d+\eta_1 & d-\eta_1 & 0 & -\eta_1-\eta_2-d & -2\,\eta_2 & 0 \\ 0 & 0 & 0 & -{\sigma_2}^2 & 1-{\sigma_2}^2 & -\sigma_2 \end{bmatrix}$$

2.3 Decomposition problem

We now show how to use the package OREMORPHISMS to decompose the differential time-delay linear system (2), i.e., to find an equivalent system defined by a block-diagonal matrix. To achieve this decomposition, we first need to compute idempotent endomorphisms of M that are defined by idempotent matrices P and Q i.e., $R\,P = Q\,R$, $P^2 = P$ and $Q^2 = Q$. A way to do that is to use the procedure *IdempotentsMatConstCoeff* of OREMORPHISMS. We need to specify the total order in d, σ_1 and σ_2 of the idempotent matrix P, a piece of information which is specified by the fourth entry of the procedure. We first start by searching for idempotents of M defined by constant matrices.

```
> Idem_order0:=IdempotentsMatConstCoeff(R,Endo[1],A,0)[1];
```

$$Idem\_order0 := \left[\begin{bmatrix} 0&0&0&0&0&0 \\ 0&0&0&0&0&0 \\ 0&0&0&0&0&0 \\ 0&0&0&0&0&0 \\ 0&0&0&0&0&0 \\ 0&0&0&0&0&0 \end{bmatrix}, \begin{bmatrix} 0&-1&1&1&0&0 \\ 0&1&0&0&0&0 \\ 0&0&1&0&0&0 \\ 0&0&0&1&0&0 \\ 0&0&0&0&1&0 \\ 0&0&0&0&0&1 \end{bmatrix} \right]$$

We choose the non-trivial idempotent, i.e., the second entry of $Idem\_order0$:

```
> P[2]:=Idem_order0[2]; Q[2]:=Factorize(Mult(R,P[2],A),R,A);
```

$$P_2 := \begin{bmatrix} 0&-1&1&1&0&0 \\ 0&1&0&0&0&0 \\ 0&0&1&0&0&0 \\ 0&0&0&1&0&0 \\ 0&0&0&0&1&0 \\ 0&0&0&0&0&1 \end{bmatrix} \qquad Q_2 := \begin{bmatrix} 0 & 0&0&0 \\ -d-\eta_1 & 1&0&0 \\ -{\sigma_1}^2 & 0&1&0 \\ 0 & 0&0&1 \end{bmatrix}$$

The fact that $P_2^2 = P_2$ and $Q_2^2 = Q_2$ imply that the A-modules $\ker_A(.P_2)$, $\ker_A(.Q_2)$, $\mathrm{im}_A(.P_2) = \ker_A(.(I_6 - P_2))$ and $\mathrm{im}_A(.Q_2) = \ker_A(.(I_4 - Q_2))$ are projective, and thus, free by the Quillen-Suslin theorem. We need to compute bases of those free A-modules. We then form the matrices U and V as explained in [6, Theorem 4].

```
> U1:=SyzygyModule(P[2],A):
> U2:=SyzygyModule(evalm(1-P[2]),A):
> U:=stackmatrix(U1,U2);
> V1:=SyzygyModule(Q[2],A):
> V2:=SyzygyModule(evalm(1-Q[2]),A):
> V:=stackmatrix(V1,submatrix(V2,[1, 2, 4],1..4));
```

$$U := \begin{bmatrix} 1 & 1 & -1 & -1 & 0 & 0 \\ 0 & 1 & 0 & 0 & 0 & 0 \\ 0 & 0 & 1 & 0 & 0 & 0 \\ 0 & 0 & 0 & 1 & 0 & 0 \\ 0 & 0 & 0 & 0 & 1 & 0 \\ 0 & 0 & 0 & 0 & 0 & 1 \end{bmatrix} \qquad V := \begin{bmatrix} 1 & 0 & 0 & 0 \\ d + \eta_1 & -1 & 0 & 0 \\ {\sigma_1}^2 & 0 & -1 & 0 \\ 0 & 0 & 0 & 1 \end{bmatrix}$$

Now, we can compute the corresponding decomposition $V\,R\,U^{-1}$ of R:

```
> R_dec:=Mult(V,R,LeftInverse(U,A),A);
```

$$R\_dec := \begin{bmatrix} 1 & 0 & 0 & 0 & 0 & 0 \\ 0 & 2\eta_1 & -d + \eta_2 - \eta_1 & -d - \eta_2 - \eta_1 & 0 & 0 \\ 0 & {\sigma_1}^2 - 1 & -{\sigma_1}^2 & -{\sigma_1}^2 & \sigma_1 & 0 \\ 0 & 0 & 1 & {\sigma_2}^2 & 0 & -\sigma_2 \end{bmatrix}$$

We can now try to decompose the second diagonal block matrix S of *R_dec*:

```
> S:=submatrix(R_dec,2..4,2..6):
```

We apply the same technique as above: compute the endomorphism ring of the A-module $N = A^{1\times 5}/(A^{1\times 3}\, S)$ finitely presented by S, find one idempotent defined by idempotent matrices, compute bases of the free A-modules defined by their kernels and images, form the corresponding unimodular matrices and deduce the decomposition of S.

```
> Endo1:=MorphismsConstCoeff(S,S,A):
> Idem1_order0:=IdempotentsMatConstCoeff(S,Endo1[1],A,0)[1];
```

$$Idem1\_order0 := \left[\begin{bmatrix} 0&0&0&0&0 \\ 0&0&0&0&0 \\ 0&0&0&0&0 \\ 0&0&0&0&0 \\ 0&0&0&0&0 \end{bmatrix}, \begin{bmatrix} 1&0&0&0&0 \\ 0&1&0&0&0 \\ 0&0&1&0&0 \\ 0&0&0&1&0 \\ 0&0&0&0&1 \end{bmatrix} \right]$$

We do not obtain a non-trivial idempotent of order 0 by means of the *IdempotentsMatConstCoeff* procedure. Hence, we can try another technique which searches for idempotents which are obtained by homotopies from the trivial idempotent id_N defined by $P_3 = I_5$ and $Q_3 = I_3$, i.e., $S\,P_3 = Q_3\,S$.

```
> P[3]:=diag(1$5):Q[3]:=diag(1$3):Z[3]:=matrix(5,3,[0$15]):
```

We then need to solve the algebraic Riccati equation $\Lambda\,S\,\Lambda + \Lambda = 0$:

```
> Mu:=RiccatiConstCoeff(S,P[3],Q[3],Z[3],A,0,alpha):
```

We choose one solution Λ_1 of the previous algebraic Riccati equation:

```
> Lambda[1]:=subs({b321=0,b521=0},Mu[1,2]);
```

$$\Lambda_1 := \begin{bmatrix} 0 & 1 & 0 \\ 0 & 1 & 0 \\ 0 & 0 & 0 \\ 0 & 0 & 0 \\ 0 & 0 & 0 \end{bmatrix}$$

We get a non-trivial idempotent defined by the following matrices P_4 and Q_4:

```
> P[4]:=simplify(evalm(P[3]+Mult(Lambda[1],S,A)));
> Q[4]:=simplify(evalm(Q[3]+Mult(S,Lambda[1],A)));
```

$$P_4 := \begin{bmatrix} \sigma_1{}^2 & -\sigma_1{}^2 & -\sigma_1{}^2\,\sigma_1 & 0 \\ \sigma_1{}^2 - 1 & -\sigma_1{}^2 + 1 & -\sigma_1{}^2\,\sigma_1 & 0 \\ 0 & 0 & 1 & 0\;0 \\ 0 & 0 & 0 & 1\;0 \\ 0 & 0 & 0 & 0\;1 \end{bmatrix} \quad Q_4 := \begin{bmatrix} 1 & \eta_1 - d + \eta_2 & 0 \\ 0 & 0 & 0 \\ 0 & 1 & 1 \end{bmatrix}$$

We now compute bases of the free A-modules $\ker_A(.P_4)$, $\ker_A(.Q_4)$,, $\mathrm{im}_A(.P_4) = \ker_A(.(I_5 - P_4))$, and $\mathrm{im}_A(.Q_4) = \ker_A(.(I_3 - Q_4))$ and we get the following two unimodular matrices X and Y:

```
> X1:=SyzygyModule(P[4],A):
> X2:=SyzygyModule(evalm(1-P[4]),A):
> X:=stackmatrix(X1,X2);
> Y1:=SyzygyModule(Q[4],A):
> Y2:=SyzygyModule(evalm(1-Q[4]),A):
> Y:=stackmatrix(Y1,Y2);
```

$$X := \begin{bmatrix} \sigma_1{}^2 - 1 & -\sigma_1{}^2 & -\sigma_1{}^2\,\sigma_1 & 0 \\ 1 & -1 & 0 & 0\;0 \\ 0 & 0 & 1 & 0\;0 \\ 0 & 0 & 0 & 1\;0 \\ 0 & 0 & 0 & 0\;1 \end{bmatrix} \quad Y := \begin{bmatrix} 0 & 1 & 0 \\ 1 & 0 & d - \eta_2 - \eta_1 \\ 0 & 1 & 1 \end{bmatrix}$$

Then, we obtain the following decomposition $Y\,S\,X^{-1}$ of the matrix S:

```
> S_dec:=Mult(Y,S,LeftInverse(X,A),A);
```

$$S\_dec :=$$

$$\begin{bmatrix} 1 & 0 & 0 & 0 & 0 \\ 0 & 2\,\eta_1 & -d-\eta_2-\eta_1+{\sigma_2}^2\,d-\eta_2\,{\sigma_2}^2-{\sigma_2}^2\,\eta_1 & 0 & (\eta_1-d+\eta_2)\,\sigma_2 \\ 0 & {\sigma_1}^2-1 & -{\sigma_1}^2+{\sigma_2}^2 & \sigma_1 & -\sigma_2 \end{bmatrix}$$

We continue by considering the second diagonal block matrix T of *S_dec*:

```
> T:=submatrix(S_dec,2..3,2..5):
```

We apply the same technique as above:

```
> P[5]:=diag(1$4): Q[5]:=diag(1$2):
> Z[5]:=matrix(4,2,[0$8]):
```

We compute the solutions of the Riccati equation $\Lambda\,T\,\Lambda+\Lambda=0$:

```
> Mu1:=RiccatiConstCoeff(T,P[5],Q[5],Z[5],A,0,alpha):
```

We choose one solution Λ_2 of the previous algebraic Riccati equation:

```
> Lambda[2]:=subs({b311=0},Mu1[1,1]);
```

$$\Lambda_2 := \begin{bmatrix} -1/(2\,\eta_1) & 0 \\ 0 & 0 \\ 0 & 0 \\ 0 & 0 \end{bmatrix}$$

Hence, we get an idempotent of the endomorphism ring of the A-module finitely presented by T defined by the following matrices P_6 and Q_6:

```
> P[6]:=simplify(evalm(P[5]+Mult(Lambda[2],T,A)));
> Q[6]:=simplify(evalm(Q[5]+Mult(T,Lambda[2],A)));
```

$$P_6 = \begin{bmatrix} 0 & 1/2\,\frac{\eta_1+\eta_2+d-{\sigma_2}^2\,d+\eta_2\,{\sigma_2}^2+{\sigma_2}^2\,\eta_1}{\eta_1} & 0 & -1/2\,\frac{(\eta_1-d+\eta_2)\sigma_2}{\eta_1} \\ 0 & 1 & 0 & 0 \\ 0 & 0 & 1 & 0 \\ 0 & 0 & 0 & 1 \end{bmatrix}$$

$$Q_6 = \begin{bmatrix} 0 & 0 \\ -1/2\,\frac{{\sigma_1}^2-1}{\eta_1} & 1 \end{bmatrix}$$

We now compute bases of the free A-modules $\ker_A(.P_6)$, $\ker_A(.Q_6)$, $\mathrm{im}_A(.P_6) = \ker_A(.(I_4-P_6))$ and $\mathrm{im}_A(.Q_6) = \ker_A(.(I_2-Q_6))$ and we obtain the following unimodular matrices G and H:

```
> G1:=SyzygyModule(P[6],A):
> G2:=SyzygyModule(evalm(1-P[6]),A):
> G:=stackmatrix(G1,G2);
> H1:=SyzygyModule(Q[6],A):
> H2:=SyzygyModule(evalm(1-Q[6]),A):
> H:=stackmatrix(H1,H2);
```

$$G := \begin{bmatrix} 2\,\eta_1 & -d-\eta_2-\eta_1+{\sigma_2}^2\,d-\eta_2\,{\sigma_2}^2-{\sigma_2}^2\eta_1 & 0 & \sigma_2\,\eta_1-\sigma_2\,d+\eta_2\,\sigma_2 \\ 0 & 1 & 0 & 0 \\ 0 & 0 & 1 & 0 \\ 0 & 0 & 0 & 1 \end{bmatrix}$$

$$H := \begin{bmatrix} 1 & 0 \\ {\sigma_1}^2-1 & -2\,\eta_1 \end{bmatrix}$$

Then, we obtain the following decomposition $H\,T\,G^{-1}$ of the matrix T:

```
> T_dec:=Mult(H,T,LeftInverse(G,A),A);
```

$$T\_dec :=$$

$$\begin{bmatrix} 1 & 0 & 0 & 0 \\ 0 & ((-\eta_1+d-\eta_2)\,{\sigma_2}^2+\eta_1-d-\eta_2)\,{\sigma_1}^2+(-d-\eta_1+\eta_2)\,{\sigma_2}^2+d+\eta_2+\eta_1 & -2\,\eta_1\,\sigma_1 & (\eta_1-d+\eta_2)\,\sigma_2\,{\sigma_1}^2+(d-\eta_2+\eta_1)\,\sigma_2 \end{bmatrix}$$

>From the previous three invertible transformations, we can deduce the unimodular matrices that perform all this decomposition process in one step:

```
> W[1]:=Mult(diag(1,1,G),diag(1,X),U,A):
> W[2]:=Mult(diag(1,1,H),diag(1,Y),V,A):
```

The system matrix R is equivalent to the matrix $L = W_2\,R\,W_1^{-1}$.

```
> L:=Mult(W[2],R,LeftInverse(W[1],A),A)):
```

The matrix L then has the form

```
> ShapeOfMatrix(L);
```

$$\begin{bmatrix} 1\,0\,0\,0\,0\,0 \\ 0\,1\,0\,0\,0\,0 \\ 0\,0\,1\,0\,0\,0 \\ 0\,0\,0\,\star\,\star\,\star \end{bmatrix}$$

where the stars $\star$ denote non-trivial elements of A respectively defined by:

```
> collect(L[4,4],{d,sigma[1],sigma[2]});
```

$$((-\eta_1+d-\eta_2)\,{\sigma_2}^2+\eta_1-d-\eta_2)\,{\sigma_1}^2+(-d-\eta_1+\eta_2)\,{\sigma_2}^2+d+\eta_2+\eta_1$$

```
> collect(L[4,5],{d,sigma[1],sigma[2]});
```

$$-2\,\eta_1\,\sigma_1$$

```
> collect(L[4,6],{d,sigma[1],sigma[2]});
```

$$(\eta_1-d+\eta_2)\,\sigma_2\,{\sigma_1}^2+(d-\eta_2+\eta_1)\,\sigma_2$$

The entries of the last row of L can be reduced by means of elementary column operations. Hence, if we consider the following unimodular matrix

$$J :=$$

$$\begin{bmatrix} 1 & 1 & -1 & -1 & 0 & 0 \\ 0 & {\sigma_1}^2-1 & -{\sigma_1}^2 & -{\sigma_1}^2 & \sigma_1 & 0 \\ 0 & 2\,\eta_1 & -2\,\eta_1 & -\eta_1-\eta_2-d+{\sigma_2}^2\,d-{\sigma_2}^2\,\eta_1-\eta_2\,{\sigma_2}^2 & 0 & -(d-\eta_1-\eta_2)\,\sigma_2 \\ 0 & 0 & 0 & 1-{\sigma_2}^2 & 0 & \sigma_2 \\ 0 & 0 & 0 & \sigma_1\,({\sigma_2}^2\,d-{\sigma_2}^2\,\eta_1-\eta_2\,{\sigma_2}^2-d-\eta_2+\eta_1) & -2\,\eta_1 & -\sigma_2\,\sigma_1\,(d-\eta_1-\eta_2) \\ 0 & 0 & 0 & 2\,\sigma_2\,\eta_2 & 0 & -2\,\eta_2 \end{bmatrix}$$

obtained from W_1 by means of elementary operations (see [7]), we finally get the following simpler decomposition $W_2\,R\,J^{-1}$ of R:

```
> R_final:=Mult(W[2],R,LeftInverse(J,A),A);
```

$$R\_final = \begin{bmatrix} 1\,0\,0 & 0 & 0 & 0 \\ 0\,1\,0 & 0 & 0 & 0 \\ 0\,0\,1 & 0 & 0 & 0 \\ 0\,0\,0 & d+\eta_1+\eta_2 & \sigma_1 & \sigma_2 \end{bmatrix}$$

Hence, the differential time-delay system (2) formed by 4 equations in 6 unknowns is equivalent to the following sole equation in 3 unknowns:

$$\dot{x}_1(t) + (\eta_1 + \eta_2)\,x_1(t) + x_2(t-h_1) + x_3(t-h_2) = 0. \tag{3}$$

Using the simple form of (3), we can easily study its structural properties (e.g., controllability, parametrizability, flatness, π-freeness, stability, stabilizability), and thus, those of (2). In particular, we obtain that (3), and thus, (2) is controllable, parametrizable, σ_1-free and σ_2-free (see [3, 15] for the corresponding definitions). The parametrizations of (2) can directly be obtained from the ones of (3) by means of the matrix J^{-1} ([3, 4, 15, 17]). System (3) admits an unstable pole at $-(\eta_1+\eta_2)$, where the η_i's are two positive parameters of (2). Its stabilizability can be studied using, e.g., [17, Proposition 3.8].

3 Others systems decomposed with OREMORPHISMS

In Table 2, we gather a list of different kinds of systems appearing in control theory, mathematical physics and engineering sciences that we have decomposed using OREMORPHISMS. We give the system matrix R, the unimodular matrices U and V and the decomposed equivalent matrix $\overline{R} = V\,R\,U^{-1}$. For more examples, we refer the reader to the OREMORPHISMS web pages ([7]).

Conclusion

In this paper, we have demonstrated the homological algebraic package OREMORPHISMS dedicated to the factorization, reduction and decomposition problems of linear functional systems. The increasing role of homological algebra in mathematical systems theory, mathematical physics and other fields has recently motivated the development of packages based on

Table 2. List of examples computed with OREMORPHISMS

$\overline{R} = V\,R\,U^{-1}$	U & V
[16] $V \begin{pmatrix} \partial & -\partial\,\delta^2 & \alpha\,\partial^2\,\delta \\ \partial\,\delta^2 & -\partial & \alpha\,\partial^2\,\delta \end{pmatrix} U^{-1}$ $=$ $\begin{pmatrix} \partial\,(1-\delta)\,(1+\delta) & 0 & 0 \\ 0 & \partial\,(\delta^2+1) & 2\,\alpha\,\partial^2\,\delta \end{pmatrix}$	$U = \begin{pmatrix} 1 & 1 & 0 \\ 1 & -1 & 0 \\ 0 & 0 & 1 \end{pmatrix}$ $V = \begin{pmatrix} 1 & -1 \\ 1 & 1 \end{pmatrix}$
[16] $V \begin{pmatrix} \partial & -\partial\,\delta^2 & \alpha\,\partial^2\,\delta \\ \partial\,\delta^2 & -\partial & \alpha\,\partial^2\,\delta \end{pmatrix} U^{-1}$ $=$ $\begin{pmatrix} \partial & 0 & 0 \\ 0 & \partial\,(\delta^4-1) & \alpha\,\partial^2\,(\delta^3-\delta) \end{pmatrix}$	$U = \begin{pmatrix} \delta^2 & -1 & \alpha\,\partial\,\delta \\ 1 & 0 & 0 \\ 0 & 0 & 1 \end{pmatrix}$ $V = \begin{pmatrix} 0 & 1 \\ -1 & \delta^2 \end{pmatrix}$
$V \begin{pmatrix} \partial & -\partial\,\delta & -1 \\ 2\,\partial\,\delta & -\partial\,(\delta^2+1) & 0 \end{pmatrix} U^{-1}$ $=$ $\begin{pmatrix} \partial & 0 & 0 \\ 0 & 1 & 0 \end{pmatrix}$ [14]	$U = \begin{pmatrix} -2\,\delta & \delta^2+1 & 0 \\ 2\,\partial\,(1-\delta^2) & \partial\,\delta\,(\delta^2-1) & -2 \\ -1 & \delta/2 & 0 \end{pmatrix}$ $V = \begin{pmatrix} 0 & -1 \\ 2 & -\delta \end{pmatrix}$
$\begin{pmatrix} \partial + \frac{1}{2\,\theta} & 0 & -1 & -1 \\ 0 & \partial + \frac{1}{\theta} & -\frac{d_1}{V_0}\,\delta & -\frac{d_2}{V_0}\,\delta \end{pmatrix}$ $=$ $V^{-1} \begin{pmatrix} \partial + \frac{1}{\theta} & -\frac{1}{V_0}\,\delta & 0 & 0 \\ 0 & 0 & 1 & 0 \end{pmatrix} U$ [12]	$U = \begin{pmatrix} 0 & 1 & 0 & 0 \\ 0 & 0 & d_1 & d_2 \\ \partial + \frac{1}{2\,\theta} & 0 & -1 & -1 \\ 1 & 0 & 0 & 0 \end{pmatrix}$ $V = \begin{pmatrix} 0 & 1 \\ 1 & 0 \end{pmatrix}$
$\begin{pmatrix} \partial + a & -k\,a\,\delta & 0 & 0 \\ 0 & \partial & -1 & 0 \\ 0 & \omega^2 & \partial + 2\,\zeta\,\omega & -\omega^2 \end{pmatrix}$ $= V^{-1} \begin{pmatrix} \partial + a & -a\,k\,\omega^2\,\delta & 0 & 0 \\ 0 & 0 & 1 & 0 \\ 0 & 0 & 0 & 1 \end{pmatrix} U$ [13]	$U =$ $\begin{pmatrix} \omega^2 & \partial & -1 & 0 \\ 0 & 1 & 0 & 0 \\ \omega^2\,(\partial + a) & -\omega^2\,(k\,a\,\delta+1) & -(\partial + 2\,\omega\,\zeta\,) & \omega^2 \\ 0 & \partial & -1 & 0 \end{pmatrix}$ $V = \begin{pmatrix} \omega^2 & \partial + a & 0 \\ \omega^2 & 0 & -1 \\ 0 & 1 & 0 \end{pmatrix}$

$\overline{R} = V\,R\,U^{-1}$	$U \,\&\, V$
$V \begin{pmatrix} \partial_x & a\,\partial_t \\ \partial_t & b\,\partial_x \end{pmatrix} U^{-1}$ $= \begin{pmatrix} \partial_t - 2\,\alpha\,b\,\partial_x & 0 \\ 0 & \partial_t + 2\,\alpha\,b\,\partial_x \end{pmatrix}$ [8]	$U = \frac{1}{2\,\alpha} \begin{pmatrix} -2\,\alpha & 1 \\ 2\,\alpha & 1 \end{pmatrix}$ $V = \begin{pmatrix} 2\,b\,\alpha & -1 \\ 2\,b\,\alpha & 1 \end{pmatrix}$ $4\,a\,b\,\alpha^2 - 1 = 0$
$V \begin{pmatrix} \partial_x & L\,\partial_t + R \\ C\,\partial_t + G & \partial_x \end{pmatrix} U^{-1}$ $=$ $\begin{pmatrix} 1 & 0 \\ 0 & (R + L\,\partial_t)\,(G + C\,\partial_t) - \partial_x^2 \end{pmatrix}$ [8]	$U =$ $\begin{pmatrix} C\,(\partial_x - \alpha\,\partial_t) - \alpha\,G & C\,(L\,\partial_t + R) - \alpha\,\partial_x \\ C & -\alpha \end{pmatrix}$ $V =$ $\begin{pmatrix} C & -\alpha \\ -C\,(\partial_x + \alpha\,\partial_t) - \alpha\,G & \alpha\,\partial_x + C\,(L\,\partial_t + R) \end{pmatrix}$ $\alpha^2 - L\,C = 0$
$V \begin{pmatrix} u\,\rho\,\partial_x & c^2\,\partial_x & 0 \\ 0 & c^2\,\partial_y & u\,\rho\,\partial_x \\ \rho\,\partial_x & u\,\partial_x & \rho\,\partial_y \end{pmatrix} U^{-1}$ $=$ $\begin{pmatrix} \partial_x - 2\,\alpha\,c\,\partial_y & 0 & 0 \\ 0 & \partial_x + 2\,\alpha\,c\,\partial_y & 0 \\ 0 & 0 & \partial_x \end{pmatrix}$ [8]	$U = \begin{pmatrix} 0 & 2\,\alpha\,c\,(c^2 - u^2) & u\,\rho \\ 0 & 2\,\alpha\,c\,(c^2 - u^2) & -u\,\rho \\ u\,\rho & c^2 & 0 \end{pmatrix}$ $V = \begin{pmatrix} 2\,\alpha\,c & 1 & -2\,\alpha\,c\,u \\ 2\,\alpha\,c & -1 & -2\,\alpha\,c\,u \\ 1 & 0 & 0 \end{pmatrix}$ $1 + 4\,(c^2 - u^2)\,\alpha^2 = 0$
$\begin{pmatrix} \partial_t & 0 & -i\,\partial_3 & -(i\,\partial_1 + \partial_2) \\ 0 & \partial_t & -i\,\partial_1 + \partial_2 & i\,\partial_3 \\ i\,\partial_3 & i\,\partial_1 + \partial_2 & -\partial_t & 0 \\ i\,\partial_1 - \partial_2 & -i\,\partial_3 & 0 & -\partial_t \end{pmatrix}$ $= V^{-1}$ $\begin{pmatrix} i\,\partial_3 - \partial_t & -i\,\partial_1 - \partial_2 & 0 & 0 \\ i\,\partial_1 - \partial_2 & i\,\partial_3 + \partial_t & 0 & 0 \\ 0 & 0 & i\,\partial_3 + \partial_t & -i\,\partial_1 - \partial_2 \\ 0 & 0 & -i\,\partial_1 + \partial_2 & -i\,\partial_3 + \partial_t \end{pmatrix}$ U [8]	$U = \begin{pmatrix} 1 & 0 & 1 & 0 \\ 0 & -1 & 0 & -1 \\ -1 & 0 & 1 & 0 \\ 0 & 1 & 0 & -1 \end{pmatrix}$ $V = -\begin{pmatrix} 1 & 0 & -1 & 0 \\ 0 & 1 & 0 & -1 \\ 1 & 0 & 1 & 0 \\ 0 & -1 & 0 & -1 \end{pmatrix}$

more and more powerful homological algebraic techniques as, for instance, OREMODULES ([2, 4]), OREMORPHISMS and HOMALG ([1]). We are convinced that this phenomenon is a precursory sign of a new era where computer algebra and symbolic computation will play the equivalent role for pure mathematics as the one played by numerical analysis in applied mathematics and engineering sciences.

References

1. Barakat, M., Robertz, D.: An abstract package for homological algebra. J. Algebra Appl. 7, 299–317 (2008), `http://wwwb.math.rwth-aachen.de/homalg/`
2. OREMODULES project, `http://wwwb.math.rwth-aachen.de/OreModules`
3. Chyzak, F., Quadrat, A., Robertz, D.: Effective algorithms for parametrizing linear control systems over Ore algebras. Appl. Algebra Engrg. Comm. Comput. 16, 319–376 (2005)
4. Chyzak, F., Quadrat, A., Robertz, D.: OREMODULES: A symbolic package for the study of multidimensional linear systems. In: Chiasson, J., Loiseau, J.J. (eds.) Applications of Time-Delay Systems. LNCIS, vol. 352, pp. 233–264. Springer, Heidelberg (2007)
5. Cluzeau, T., Quadrat, A.: Factoring and decomposing a class of linear functional systems. Linear Algebra Appl. 428, 324–381 (2008)
6. Cluzeau, T., Quadrat, A.: On algebraic simplifications of linear functional systems. In: Loiseau, J.J., Michiels, W., Niculescu, S.I., Sipahi, R. (eds.) Topics in Time-Delay Systems: Analysis, Algorithms and Control. LNCIS, pp. 167–178. Springer, Heidelberg (2009)
7. Cluzeau, T., Quadrat, A.: OREMORPHISMS project,
`http://www.ensil.unilim.fr/~cluzeau/`,
`http://www-sop.inria.fr/personnel/Alban.Quadrat/index.html`
8. Courant, R., Hilbert, D.: Methods of Mathematical Physics. Wiley Classics Library, Wiley (1989)
9. Culianez, G.: Formes de Hermite et de Jacobson: Implémentations et applications. Internship with Quadrat A, INRIA Sophia Antipolis (2005)
10. Fabiańska, A.: QUILLENSUSLIN project,
`http://wwwb.math.rwth-aachen.de/QuillenSuslin`
11. Fabiańska, A., Quadrat, A.: Applications of the Quillen-Suslin theorem in multidimensional systems theory. In: Park, H., Regensburger, G. (eds.) Gröbner Bases in Control Theory and Signal Processing. Radon Series on Computation and Applied Mathematics, vol. 3, pp. 23–106. de Gruyter publisher (2007)
12. Kwakernaak, H., Sivan, R.: Linear Optimal Control Systems. Wiley, Chichester (1972)
13. Manitius, A.: Feedback controllers for a wind tunnel model involving a delay: analytical design and numerical simulations. IEEE Trans. Autom. Contr. 29, 1058–1068 (1984)
14. Mounier, H., Rudolph, J., Petitot, M., Fliess, M.: A flexible rod as a linear delay system. In: Proc. European Control Conference (ECC), Rome (1995)
15. Mounier, H., Rudolph, J., Fliess, M., Rouchon, P.: Tracking control of a vibrating string with an interior mass viewed as delay system. ESAIM Control Optim. Calc. Var. 3, 315–321 (1998)
16. Petit, N., Rouchon, P.: Dynamics and solutions to some control problems for water-tank systems. IEEE Trans. Automatic Control 47, 595–609 (2002)

17. Quadrat, A.: The fractional representation approach to synthesis problems: an algebraic analysis viewpoint. Part II: Internal stabilization, SIAM J. Control & Optimization 42, 300–320 (2003)
18. Quadrat, A., Robertz, D.: Computation of bases of free modules over the Weyl algebras. Journal Symbolic Comput. 42, 1113–1141 (2007)

Part III

Controlling and Observing Delay Systems

Integral Action Controllers for Systems with Time Delays*

Hitay Özbay[1] and A. Nazli Gündeş[2]

[1] Dept. Electrical & Electronics Eng., Bilkent Univ., Ankara, 06800 Turkey, hitay@bilkent.edu.tr
[2] Dept. Electrical & Computer Eng., Univ. of California, Davis, CA 95616, U.S.A., angundes@ucdavis.edu

Summary. Consider a stabilizing controller C_1 for a given plant P. If C_1 and P do not have any zeros at the origin, then one can use a cascade connected PI (proportional plus integral) controller C_{pi} with C_1 and keep the feedback system stable. In this work we examine the allowable range of the integral action gain in C_{pi}, and discuss how C_1 should be chosen to maximize this range for systems with time delays.

1 Introduction

In the design of feedback controllers it is often desirable to use an integrator to be able to track constant reference signals. For example, internal model principle says that the controller must include a copy of the reference signal (or disturbance) generator in order to have a robust tracking (or disturbance rejection), see e.g. [2, 4, 7]. Typically, the reference generator $G_r(s)$ is an unstable system: an integrator (resp. oscillator) if the reference is a constant (resp. a sinusoidal signal). One way to achieve robust asymptotic tracking (or disturbance rejection) is to append G_r to the plant P and then design a controller C_o for the combined "plant" G_rP. Thus $C = C_oG_r$ is a stabilizing controller for P and it achieves the performance objectives, see e.g. [1, 16] for more details.

In this chapter we consider the dual problem: first design a stabilizing controller for the plant, then append a PI term to this controller. A similar problem has been discussed in [3] for finite dimensional systems. Briefly, the problem we deal with can be stated as follows: let C_1 be a stabilizing controller for a time delay system P, and append (in the form of a cascade connection) $C_{pi}(s) = \frac{(s+k_i)}{s}$ to C_1. Note that the proportional gain of the PI controller is set to unity; this is without loss of generality since a non-unity gain can be absorbed into C_1. Assume that P and C_1 do not have any zeros at the origin. Then, there exists k_i such that the feedback system is stable. We examine the range of allowable k_i, and discuss the problem of designing an optimal C_1 so that this range is maximized.

We should indicate that rather than the cascade PI-controller connection to be discussed here, a two-stage parallel connection of controllers is also possible. For example, as before, let C_1 be a a stabilizing feedback controller for a given plant P. If the PI part of the controller,

* This work was supported in part by TÜBİTAK (grant no. EEEAG-105E156).

J.J. Loiseau et al. (Eds.): Topics in Time Delay Systems, LNCIS 388, pp. 197–207.
springerlink.com

C_{pi}, is a stabilizing controller for the new plant $P(I+C_1P)^{-1}$, then the parallel connection of the controller, $C_{pi}+C_1$, is a stabilizing controller for the original plant P, see [9, 14, 16]. One can study the problem choosing the best C_1 so that the allowable range of k_i is maximum. We leave this problem aside, because the techniques to be used in such a study would be similar to the approach taken in this chapter for the cascade connection of the controllers.

This chapter is organized as follows: stability of the feedback system under cascade connection of the PI controller is investigated in Section 2. Design of C_1, maximizing the allowable range of the integral gain, is discussed in Section 3. Concluding remarks are made in Section 4.

Notation used here is standard. In particular, the norm sign $\|\cdot\|$ stands for the $\mathcal{H}_\infty$ norm $\|\cdot\|_\infty$ whenever the argument is in $\mathcal{H}_\infty$.

2 Feedback System Stability Under Cascade Connection of the PI Controller

Consider the feedback system shown in Figure 1 with an r input r output plant whose $r\times r$ transfer matrix is $P(s)$. The $r\times r$ controller transfer matrix is $C(s)$. Assume that P is full rank. The feedback system is said to be stable if $C(1+PC)^{-1}, PC(1+CP)^{-1}, C(I+PC)^{-1}P, P(1+CP)^{-1}$ are in $\mathcal{H}_\infty^{r\times r}$. In this case, we say $C\in\mathcal{S}(P)$, where $\mathcal{S}(P)$ is the set of all controllers stabilizing the feedback system with plant P.

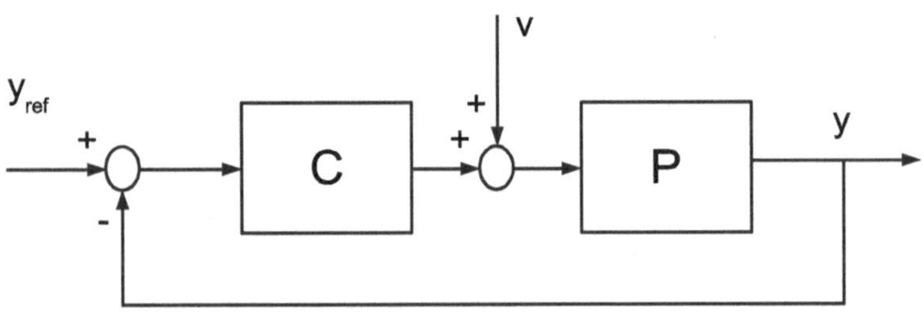

Fig. 1. Feedback System

Let C_1 be in $\mathcal{S}(P)$ and consider the cascade connection $C=C_1C_x$ for some C_x. The result stated below as Theorem 1 addresses the following question: Is the closed-loop system still stable if $C=C_1C_x$, i.e. do we have $C\in\mathcal{S}(P)$?

Theorem 1. *Let P be a given $r\times r$ plant and let $C_1\in\mathcal{S}(P)$. Assume that P and C_1 are full rank and define the complementary sensitivity function for the feedback system with $C=C_1$ as $T_1:=PC_1(I+PC_1)^{-1}$. Then, we have the following two results:*

a) *If $C_y:=C_x-I$ stabilizes $T_1\in\mathcal{H}_\infty^{r\times r}$, then*

$$C=C_1C_x\in\mathcal{S}(P)\,. \tag{1}$$

b) *Let P and C_1 have no transmission-zeros at the origin. Choose any $\hat{K}_P$, $\hat{K}_D\in\mathbb{R}^{r\times r}$, and $\tau\in\mathbb{R}_+$. Define $\Psi\in\mathcal{H}_\infty^{r\times r}$ as*

$$\Psi(s) = \frac{T_1(s)T_1(0)^{-1} - I}{s} + T_1(s)\,(\hat{K}_P + \frac{\hat{K}_D\, s}{\tau s + 1}). \tag{2}$$

Then for $\rho \in \mathbb{R}_+$ satisfying

$$\rho < \|\,\Psi\,\|^{-1} =: \psi^{-1} \tag{3}$$

the controller $C = C_1 C_{pid} \in \mathcal{S}(P)$, where C_{pid} is a PID-controller given by $C_{pid} = I + \hat{C}_{pid}$, where

$$\hat{C}_{pid} = \rho\left(\hat{K}_P + \frac{T_1(0)^{-1}}{s} + \frac{\hat{K}_D\, s}{\tau s + 1}\right). \tag{4}$$

For $\hat{K}_D = 0$, (4) becomes a PI-controller. ∎

Proof of Theorem 1 is given in the Appendix. By part (a) of this theorem the stabilizing controller $C = C_1 C_x$ gives rise to the following complementary sensitivity $T_x = PC(I + PC)^{-1}$:

$$\begin{aligned} T_x &= PC_1(I + C_x PC_1)^{-1}C_x = PC_1(I + PC_1 + C_y PC_1)^{-1}C_x \\ &= T_1(I + C_y T_1)^{-1}(I + C_y) = (I + T_1(I - C_x))^{-1}T_1 C_x. \end{aligned} \tag{5}$$

If $C_x = C_{pid}$ as in (4) of Theorem 1-(b), then T_x in (5) becomes

$$T = (I + T_1\hat{C}_{pid})^{-1}T_1 C_{pid}\,, \tag{6}$$

which can be expressed as

$$\begin{aligned} T &= (\frac{s}{s+\rho}I + T_1(s)\hat{C}_{pid}(s)\frac{s}{s+\rho})^{-1}T_1(s)(\frac{s}{s+\rho}I + \frac{s}{s+\rho}\hat{C}_{pid}) \\ &= (I + \frac{\rho s}{s+\rho}\,\Psi(s)\,)^{-1}T_1(s)\frac{\rho s}{s+\rho}(I + \rho\, s(\hat{K}_P + \frac{T_1(0)^{-1}}{s} + \frac{\hat{K}_D s}{\tau s + 1})). \end{aligned} \tag{7}$$

Therefore, $T(0) = I$ and

$$\|T\| \leq \|(I + \frac{\rho s}{s+\rho}\Psi)^{-1}\| \cdot \|T_1(s)\frac{s}{s+\rho}(I + \hat{C}_{pid})\|\,.$$

Writing $(I + \frac{\rho s}{s+\rho}\,\Psi\,)^{-1} = I - (I + \frac{\rho s}{s+\rho}\Psi)^{-1}\frac{\rho s}{s+\rho}\Psi$, we obtain

$$\|(I + \frac{\rho s}{s+\rho}\,\Psi(s)\,)^{-1}\| \leq 1 + \rho\psi\,\|\,(I + \frac{\rho s}{s+\rho}\Psi)^{-1}\|$$

and hence, $\|\,(I + \frac{\rho s}{s+\rho}\,\Psi(s)\,)^{-1}\| \leq (1 - \rho\,\psi)^{-1}$, and

$$\|\,T\,\| \leq \frac{1}{1-\rho\psi}\,\|\,T_1(s)\,\frac{s}{s+\rho}(I + \hat{C}_{pid}(s)\,)\|. \tag{8}$$

Now suppose that in the PID-controller $\hat{C}_{pid}$ we choose $\hat{K}_P = 0$, $\hat{K}_D = 0$. Then by (4), the PI-controller is

$$C_{pi}(s) = I + \frac{\rho\, T_1(0)^{-1}}{s}\,,$$

where $\rho \in \mathbb{R}_+$ satisfies (3), i.e.,

$$\rho < \|\,\frac{T_1(s)T_1(0)^{-1} - I}{s}\,\|^{-1} =: \psi_o^{-1}\,. \tag{9}$$

In this case, the upper-bound on $\| T \|$ given in (8) becomes

$$\|T\| \leq \frac{1}{1-\rho\psi_o} \|T_1 \, T_1(0)^{-1}\| \cdot \|\frac{sT_1(0)+\rho I}{s+\rho}\|.$$

In particular, if $T_1(0) = I$, i.e. C_1 and/or P contain a pole at $s = 0$, then

$$\|T\| \leq \frac{1}{1-\rho\psi_o} \|T_1\|.$$

>From the above discussion we see that if $\rho \, \psi_o \ll 1$ then the upper bound of $\|T\|$ is close to $\|T_1\|$.

3 Design of C_1 Maximizing the Integral Action Gain

In this section we discuss the design of C_1 for the largest allowable range of ρ, (9), for a class of single input single output (SISO) plants with time delays. In this case, from (9) we see that C_1 should be designed to minimize

$$\psi = \|\frac{T_1(s)T_1^{-1}(0) - 1}{s}\|_\infty, \tag{10}$$

where $T_1 = PC_1(1 + PC_1)^{-1}$ and $C_1 \in \mathcal{S}(P)$.

Solution of this problem will be obtained below in two steps: (i) first we solve the problem for stable plants, then (ii) we extend this solution to cover unstable plants case. In both steps we begin with inner-outer and coprime factorizations of given P, then we solve an $\mathcal{H}_\infty$ optimization problem. Inner-outer factorizations require finding $\mathbb{C}_+$ roots of a quasi-polynomial, for which several algorithms exist by now, see e.g. [5, 12, 17] and their references. Using these algorithms and the methods developed for the $\mathcal{H}_\infty$ control of general infinite dimensional systems, (see e.g. [6] and [8]) we can solve the problem in step (i) for a large class of time delay systems. We will see that the extension (ii) to unstable plants, with finitely many poles in $\mathbb{C}_+$, involves a parameterization of all suboptimal solutions of the problem in (i), and the use of Nevanlinna-Pick interpolation. The mathematical tools for these problems can be found in [6, 11, 15, 18].

3.1 Stable plants

In this section we consider stable SISO plants whose inner-outer factorizations are in the form $P = P_i P_o$ where P_i is inner (all-pass) with $P_i(0) = 1$, and P_o is outer (minimum-phase).

Example. Consider the plant with input/output delay, $h > 0$, and internal delays

$$P(s) = \left(\frac{e^{-hs}}{s+2}\right) \frac{(s+1)+2(s-1)e^{-2s}}{(s+3)+e^{-3s}}. \tag{11}$$

Then, the following is an inner-outer factorization:

$$P_i(s) = -e^{-hs} \frac{(s+1)+2(s-1)e^{-2s}}{2(s+1)+(s-1)e^{-2s}}$$

$$P_o(s) = \frac{-1}{s+2} \frac{2(s+1)+(s-1)e^{-2s}}{(s+3)+e^{-3s}}$$

Clearly, $P_i(0) = 1$, the poles and zeros of P_i are symmetric around the Im-axis, and P_o contains no poles or zeros in the right half plane. ■

The set of all stabilizing controllers is parameterized as

$$\mathcal{S}(P) = \{Q/(1 - PQ) \ : \ Q \in \mathcal{H}_\infty \text{ and } PQ \neq 1\}.$$

Therefore, C_1 must be in the form $C_1 = Q_1/(1 - PQ_1)$, where $Q_1 \in \mathcal{H}_\infty$ is free. Let

$$Q_1(s) = P_o^{-1}(s)\, \frac{Q_i(s)}{(1 + \varepsilon s)^\ell} \tag{12}$$

where $Q_i \in \mathcal{H}_\infty$ is the free parameter, $\varepsilon > 0$, and ℓ is the relative degree of P_o. Then we have

$$T_1(s) = P(s)Q_1(s) = P_i(s)\, \frac{Q_i(s)}{(1 + \varepsilon s)^\ell}.$$

Hence the problem of maximizing the allowable range of ρ reduces to finding

$$\begin{aligned} \psi_o &= \inf_{Q_i \in \mathcal{H}_\infty} \|(\frac{P_i(s)}{(1 + \varepsilon s)^\ell} Q_i(s) Q_i^{-1}(0) - 1)/s\|_\infty \\ &= \|(\frac{P_i(s)}{(1 + \varepsilon s)^\ell} Q_{i,\mathrm{opt}}(s) Q_{i,\mathrm{opt}}^{-1}(0) - 1)/s\|_\infty \end{aligned} \tag{13}$$

and the corresponding optimal $Q_{i,\mathrm{opt}} \in \mathcal{H}_\infty$ solving this problem. Note that optimal solution is not unique: if $Q_{i,o}$ is a solution of (13), then so is $KQ_{i,o}$, for any non-zero constant K. Therefore, we define the normalized free parameter $\widetilde{Q}_i(s) = Q_i(s)Q_i^{-1}(0)$, and try to find optimal $\widetilde{Q}_i$ in the problem (14) defined below. First let $\mathcal{H}_\infty^o = \{\widetilde{Q}_i \in \mathcal{H}_\infty \ : \ \widetilde{Q}_i(0) = 1\}$. Note that

$$\begin{aligned} \psi_o &= \inf_{\widetilde{Q}_i \in \mathcal{H}_\infty^o} \|(\frac{P_i(s)}{(1 + \varepsilon s)^\ell} \widetilde{Q}_i(s) - 1)/s\|_\infty \\ &\geq \inf_{\widetilde{Q}_i \in \mathcal{H}_\infty} \|(\frac{P_i(s)}{(1 + \varepsilon s)^\ell} \widetilde{Q}_i(s) - 1)/s\|_\infty =: \widetilde{\psi}_o. \end{aligned}$$

But the optimal solution of the problem defining $\widetilde{\psi}_o$ must lie in $\mathcal{H}_\infty^o$, because $(\frac{P_i(s)}{(1+\varepsilon s)^\ell}\widetilde{Q}_i(s) - 1)/s$ is in $\mathcal{H}_\infty$ only if $\widetilde{Q}_i(0) = 1$. Therefore, $\widetilde{\psi}_o = \psi_o$ and $\widetilde{Q}_{i,\mathrm{opt}}$ is the optimal solution of

$$\psi_o = \inf_{\widetilde{Q}_i \in \mathcal{H}_\infty} \|(\frac{P_i(s)}{(1 + \varepsilon s)^\ell} \widetilde{Q}_i(s) - 1)/s\|_\infty. \tag{14}$$

The problem (14) is a one-block $\mathcal{H}_\infty$ optimization problem, which can be seen as equivalent to a weighted sensitivity minimization for a stable plant with the sensitivity weigh being an integrator. For a general inner function P_i, the problem (14) can be solved using the techniques developed for the $\mathcal{H}_\infty$ control of infinite dimensional systems, see e.g. [6, 10, 13] and their references. It turns out that the optimal $\widetilde{Q}_i$ is in the form

$$\widetilde{Q}_{i,\mathrm{opt}}(s) = \frac{(1 + \varepsilon s)^\ell}{(1 + \delta s)^{(\ell+1)}} \left(\frac{1 + \psi_o^2 s^2}{P_i(s) + \psi_o s} \right) \tag{15}$$

where $\delta \to 0$ and ψ_o is the largest value of $\psi > 0$ for which we have

$$P_i(j/\psi) = -j. \tag{16}$$

Choosing Q_1 as in (12) with $Q_i = K_1 \widetilde{Q}_{i,\mathrm{opt}}$, for an arbitrary $K_1 \neq 0$, and defining the controller $C_1 = Q_1/(1 - PQ_1)$, we obtain

$$C_1(s) = \frac{K_1 P_o^{-1}(s)}{\frac{(1+\delta s)^{\ell+1}(\psi_o s + P_i(s))}{1+\psi_o^2 s^2} - K_1 P_i(s)} . \tag{17}$$

$\rho = k_i\, K_1 < \psi_o^{-1}$. Hence, depending on the gain K_1 used in C_1 we get an allowable range for k_i,

$$|k_i| < \psi_o^{-1}/|K_1| .$$

Note that ψ_o is invariant and completely determined by the inner part $P_i(s)$ of the plant.

Another interesting problem in this context is to investigate PD (proportional plus derivative) type of $\widetilde{Q}_i(s) = (1 + k_d\, s)$ in (14). More precisely,

$$\psi_{pd} := \inf_{k_d \in \mathbb{R}} \|(\frac{P_i(s)}{(1+\varepsilon s)^\ell}(1 + k_d\, s) - 1)/s\|_\infty = \inf_{k_d \in \mathbb{R}} \|\frac{f(s) - 1}{s} + k_d f(s)\|_\infty . \tag{18}$$

where $f(s) = \frac{P_i(s)}{(1+\varepsilon s)^\ell}$. The function f is in $\mathcal{H}_\infty$ and $f(0) = 1$. This problem has been studied in the context of resilient PD controller design in [14] and a closed form expression is obtained for the optimal k_d.

Example (Revisited) For the example given in (11), the equation (16) can be written as

$$e^{-jh/\psi}\, m(j/\psi) \left(\frac{1 - m(-j/\psi)/2}{1 - m(j/\psi)/2} \right) = -j$$

where $m(s) = e^{-2s}\,\frac{(1-s)}{(1+s)}$. Since $m(s)$ is inner we have $m(j/\psi) = e^{-j\theta_m}$, where $\theta_m = 2/\psi + 2\tan^{-1}(1/\psi)$. We also have $\dfrac{1 - m(-j/\psi)/2}{1 - m(j/\psi)/2} = e^{-j\theta}$,
where $\theta = 2\tan^{-1}\left(\frac{\sin(\theta_m)}{2-\cos(\theta_m)}\right)$. Therefore ψ_o^{-1} is the smallest x satisfying

$$(1 + \frac{h}{2})x + \tan^{-1}(x) + \tan^{-1} \Omega(x) = \frac{\pi}{4} \tag{19}$$

where $\Omega(x) = \dfrac{\sin(2(x + \tan^{-1}(x)))}{2 - \cos(2(x + \tan^{-1}(x)))}$.

We should also note that if we change the inner part of the plant to an input/output delay, $P_i(s) = e^{-hs}$ (i.e. consider $m(s) = 1$), then from (16) we get $\psi_o^{-1} = \pi/2h$, which is precisely the gain margin of the feedback system whose open loop transfer function is e^{-hs}/s.

When we consider a PD type of $\widetilde{Q}_i(s) = (1 + k_d\, s)$, the solution of (18) gives optimal ψ_{pd}^{opt} and the corresponding k_d^{opt}. Figure 2 illustrates how the optimal ψ_o, ψ_{pd}^{opt} and k_d^{opt} vary with h. Note from this figure that the use of PD term does not lead to significant performance degradation (reduction in the largest allowable k_i range) compared to the use of optimal $\widetilde{Q}_i$ of (15).

3.2 Extension to Unstable Plants

Now consider unstable SISO plants factorized as

$$P(s) = \frac{N(s)}{D_i(s)} , \quad N(s) = N_i(s) N_o(s) \tag{20}$$

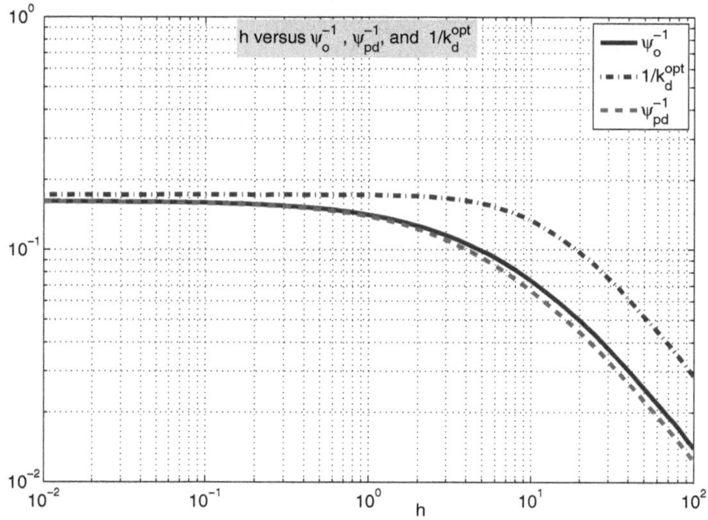

Fig. 2. h versus ψ_o^{-1}, ψ_{pd}^{-1} and $1/k_d^{\rm opt}$.

where $N_o(s)$ is outer, $N_i(s)$ and $D_i(s)$ are inner with D_i being a finite Blaschke product (i.e. the plant has finitely many poles in $\mathbb{C}_+$, and it has no poles on the Im-axis). As before, we will assume that $N_i(0) = 1$. For this type of plants $C_1 \in \mathcal{S}(P)$ if and only if

$$C_1(s) = \frac{X(s) + D_i(s)Q_1(s)}{Y(s) - N(s)Q_1(s)} \quad \text{for some } Q_1 \in \mathcal{H}_\infty \tag{21}$$

where $X, Y \in \mathcal{H}_\infty$ satisfy

$$Y(s) = \frac{1 - N(s)X(s)}{D_i(s)} . \tag{22}$$

Let $p_1, \ldots, p_n$ be the zeros of $D_i(s)$, i.e. poles of $P(s)$ in $\mathbb{C}_+$, and for simplicity of the exposition assume that they are distinct. Then, $Y \in \mathcal{H}_\infty$ if and only if the function $X \in \mathcal{H}_\infty$ satisfy $X(p_i) = 1/N(p_i)$, $i = 1, \ldots, n$. If we use C_1 in the form of (21) as the initial stabilizing controller for the plant P, then

$$T_1(s) = N(s)(X(s) + D_i(s)Q_1(s)).$$

Therefore $\psi = \|(T_1(s)T_1^{-1}(0) - 1)/s\|_\infty$ is obtained as

$$\psi = \|\frac{N(s)N(0)^{-1}Q_{1X}(s)Q_{1X}(0)^{-1} - 1}{s}\|_\infty, \tag{23}$$

where $Q_{1X}(s) = (X(s) + D_i(s)Q_1(s))$. Thus the optimal ψ_o is the smallest ψ over $Q_{1X}(s) = (X(s) + D_i(s)Q_1(s))$ for $Q_1 \in \mathcal{H}_\infty$. Define

$$Q_{1X}(s) =: \frac{N_o^{-1}(s)}{(1+\varepsilon s)^\ell} Q_X(s) \ \text{ where } \varepsilon \searrow 0,$$

and ℓ is the relative degree of $N_o(s)$. Then, we have an invertible relation between the free parameters Q_{1X} and Q_X in $\mathcal{H}_\infty$. Note that the problem (23) is exactly in the form (13) except that $Q_{1X}(s)$ is restricted to have $Q_{1X}(p_i) = X(p_i) = 1/N(p_i)$, whereas in (13) there is no such restriction on the free parameter $Q_i \in \mathcal{H}_\infty$. In summary, we have the following

$$\psi(Q_X) := \|s^{-1}(\frac{N_i(s)}{(1+\varepsilon s)^\ell}\frac{Q_X(s)}{Q_X(0)} - 1)\|_\infty \tag{24}$$

$$\psi_o = \inf\{\, \psi(Q_X) \;:\; Q_X \in \mathcal{H}_\infty \text{ and } Q_X(p_i) = \frac{(1+\varepsilon p_i)^\ell}{N_i(p_i)}\;, i = 1\ldots, n\}. \tag{25}$$

As in Section 3.1, we will be restricting ourselves to $Q_X \in \mathcal{H}_\infty$ such that $Q_X(0) = 1$, because $\psi(KQ_X) = \psi(Q_X)$ for any non-zero K. Thus, in the unstable plants case the problem is modified to finding

$$\psi_o = \inf_{\widetilde{Q}_X} \left\| s^{-1}\left(\frac{N_i(s)}{(1+\varepsilon s)^\ell}\,\widetilde{Q}_X(s) - 1\right)\right\|_\infty$$

$$\text{subject to } \widetilde{Q}_X \in \mathcal{H}_\infty \text{and } \widetilde{Q}_X(p_i) = \frac{(1+\varepsilon p_i)^\ell}{N_i(p_i)}\;, i = 1\ldots, n.$$

For a given $\gamma > \psi_o$, the set of all $\widetilde{Q}_X \in \mathcal{H}_\infty$ satisfying

$$\left\| s^{-1}\left(\frac{N_i(s)}{(1+\varepsilon s)^\ell}\,\widetilde{Q}_X(s) - 1\right)\right\|_\infty \le \gamma \tag{26}$$

can be characterized as

$$\mathcal{Q}_\gamma = \left\{\widetilde{Q}_X(s) = \frac{F_1(s) + F_2(s)U(s)}{F_3(s) + F_4(s)U(s)} \;:\; U \in \mathcal{H}_\infty\,,\; \|U\|_\infty \le 1\right\} \tag{27}$$

where $F_1, \ldots, F_4$ are computed explicitly from the problem data, see e.g. [6]. Therefore, the problem at hand can be transformed to finding the smallest γ for which there exists $U \in \mathcal{H}_\infty$, $\|U\|_\infty \le 1$ such that

$$\frac{F_1(p_i) + F_2(p_i)U(p_i)}{F_3(p_i) + F_4(p_i)U(p_i)} = \frac{(1+\varepsilon p_i)^\ell}{N_i(p_i)} =: \alpha_i\;, \tag{28}$$

for $i = 1\ldots, n$. This leads to a set of interpolation conditions on U

$$U(p_i) = \frac{\alpha_i\, F_3(p_i) - F_1(p_i)}{F_2(p_i) - \alpha_i\, F_4(p_i)} =: \beta_i \tag{29}$$

for $i = 1\ldots, n$. For each fixed γ we can find β_i using for example [6]. Now we need to check whether there exists $U \in \mathcal{H}_\infty$ with $\|U\|_\infty \le 1$ such that $U(p_i) = \beta_i$. This is a Nevanlinna-Pick interpolation problem and it can be solved from the given problem data $\{(p_1\ldots, p_n), (\beta_1, \ldots, \beta_n)\}$, see e.g. [6, 11, 18].

In summary, for unstable plants the problem is solved in two steps:

1. Given $\gamma > \psi_o$, solve the suboptimal version (26) of the problem (13) studied in Section 3.1; characterize all suboptimal solutions in the form (27), i.e. find F_1, F_2, F_3, F_4.
2. Given $p_1, \ldots, p_n$, determine $\beta_1, \ldots, \beta_n$ from the first step. Use this data to check if the Nevanlinna-Pick interpolation problem has a feasible solution. If yes decrease γ, if no increase γ, and repeat Steps 1 and 2; using a bisection in this iteration find the optimal γ_o. For $\gamma = \gamma_o + \epsilon$, where $\epsilon > 0$, the Nevanlinna-Pick problem gives a solution U, which in turn gives our suboptimal Q_X, from which we get Q_{1X} and hence C_1.

Example. Let $h > 0$, and consider $P(s) = \frac{e^{-hs}}{s+2}\left(\frac{s+1+2(s-1)e^{-2s}}{s+3-5e^{-0.5s}}\right)$. This plant is unstable with single pole $p_1 = 0.6367$ in $\mathbb{C}_+$. Therefore, its factorization can be done as (20) where

$$D_i(s) = \frac{s - 0.6367}{s + 0.6367}$$
$$N_i(s) = e^{-hs}\frac{(s+1) + 2(s-1)e^{-2s}}{(1-s)e^{-2s} - 2(s+1)}$$
$$N_o(s) = \frac{(1-s)e^{-2s} - 2(s+1)}{(s+2)(s+3-5e^{-0.5s})}\, D_i(s).$$

For $h = 3$ we have $\psi_o = 8.6744$. This gives $\alpha_1 = N_i(p_1)^{-1} = -14.945$. Applying the procedure described above we find $F_1, \ldots, F_4$ for each fixed $\gamma > \psi_o$, and compute β_1 defined by (29). Since we have single interpolation condition, the solution of Nevanlinna-Pick problem is rather trivial: it is solvable if and only if $|\beta_1| \leq 1$, and as a solution we can take $U(s) = \beta_1$. By using a bi-section search we find that smallest $\gamma > \psi_o$ leading to $|\beta_1| \leq 1$ is $\gamma_o = 13.4485$, which leads to $\beta_1 = -1$, see Figure 3. Thus if we choose $U(s) = -1$, we get

$$\widetilde{Q}_X(s) = \frac{F_1(s) - F_2(s)}{F_3(s) - F_4(s)},$$

where $F_1, \ldots, F_4$ are computed from the solution of the suboptimal one-block $\mathcal{H}_\infty$ problem with $\gamma = 13.45 > \gamma_o$.

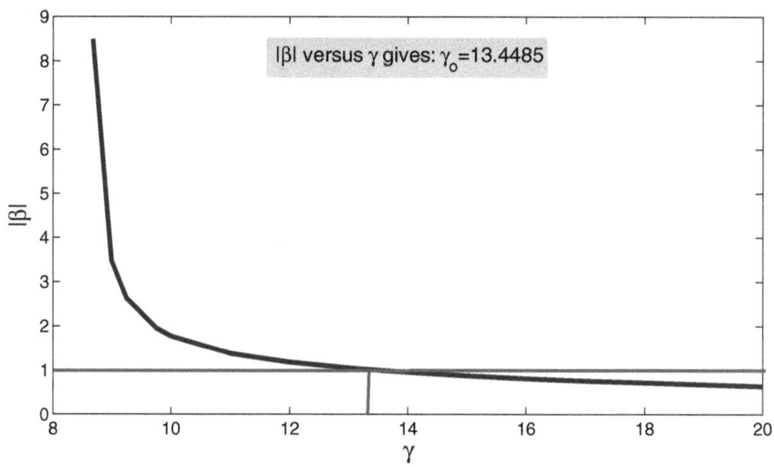

Fig. 3. $|\beta|$ versus γ.

4 Conclusions

A sufficient condition is derived for $C = C_1 C_{pi}$, cascade connection of a PI controller C_{pi}, and an initial stabilizing controller C_1, to stabilize a given plant P. Design of C_1 for the largest allowable range of the integral action gain interval is investigated for stable plants, including systems with internal and input-output delays. We used parametrization of all stabilizing controller to characterize C_1. Then we have seen that the problem at hand reduces to a weighted sensitivity minimization for a stable plant whose inner part is infinite dimensional and the weight is an integrator. When we consider a PD-like $\tilde{Q}_i(s)$ in the parametrization of C_1, the problem becomes finding optimal k_d in (18), which is solved in [14].

For unstable plants the problem of finding the largest allowable range of the integral action gain is solved in two steps. First the a suboptimal one-block problem is solved and in the second step a Nevanlinna-Pick interpolation problem is solved.

We should also point out that the result stated in Section 2 is a sufficient condition. Therefore the largest allowable integral action gain found in Section 3 is within the set of allowable gains characterized by this sufficient condition, which may be conservative. It would be interesting to investigate the level of conservatism in this approach. We leave this open problem to a future study.

APPENDIX

Proof of Theorem 1: **a)** Let $P = \tilde{Y}^{-1}\tilde{X}$ be a left-coprime-factorization (LCF) of P and let $N_1 D_1^{-1}$ be a right-coprime-factorization (RCF) of C_1. Since C_1 stabilizes P, $M_1 := \tilde{Y}D_1 + \tilde{X}N_1$ is unimodular in $\mathcal{H}_\infty^{r\times r}$. With $C_1 \in \mathcal{S}(P)$, we have $Q_1 := C_1(I+PC_1)^{-1} \in \mathcal{H}_\infty^{r\times r}$ and $T_1 := PC_1(I+PC_1)^{-1} = PQ_1 \in \mathcal{H}_\infty^{r\times r}$. Now $C_y = I - C_x$ stabilizes T_1 if and only if $C_y(I+T_1C_y)^{-1} \in \mathcal{H}_\infty^{r\times r}$, which implies $(I+T_1C_y)^{-1} \in \mathcal{H}_\infty^{r\times r}$. Define $D_c := (I+TC_y)^{-1}D_1$, $N_c = N_1 + Q_1C_yD_c$; then N_c, $D_c \in \mathcal{H}_\infty^{r\times r}$. Write $C = C_1C_x$ as $C = C_1 + C_1C_y = N_cD_c^{-1}$. Then $\tilde{Y}D_c + \tilde{X}N_c = \tilde{Y}D_c + \tilde{X}[N_1 + Q_1C_yD_c] = \tilde{Y}D_c + \tilde{X}N_1 + \tilde{Y}PQ_1C_yD_c = \tilde{Y}(I+T_1C_y)D_c + \tilde{X}N_1 = \tilde{Y}(I+T_1C_y)(I+TC_y)^{-1}D_1 + \tilde{X}N_1 = M_1$ is unimodular and hence, $C = C_1C_x \in \mathcal{S}(P)$.
b) Let $P = XY^{-1}$ be an RCF and $C_1 = \tilde{D}_1^{-1}\tilde{N}_1$ be an LCF. Then $C_1 \in \mathcal{S}(P)$ if and only if $\tilde{M}_1 := \tilde{D}_1 Y + \tilde{N}_1 X$ is unimodular; hence, $\det \tilde{M}_1(0) \neq 0$. Since $P, C_1 \in \mathcal{S}(P)$ do not have transmission-zeros at $s = 0$, $\det X(0) \neq 0$ and $\det \tilde{N}_1(0) \neq 0$. Since $\det T_1(0) = \det X(0)\tilde{M}_1(0)\tilde{N}_1(0) \neq 0$, we conclude that $T_1 \in \mathcal{H}_\infty^{r\times r}$ does not have transmission-zeros at $s = 0$. It follows from [9], Proposition 2, that the PID-controller $\hat{C}_{pid}$ in (4) stabilizes $T_1 \in \mathcal{H}_\infty^{r\times r}$. Therefore, by (1), $C_1C_x \in \mathcal{S}(P)$, where $C_x = I + \hat{C}_{pid}$.

References

1. Bélanger, P.R.: Control engineering: A modern approach. Saunders College Pub., Philadelphia (1995)
2. Davison, E.J.: The robust control of a servomechanism problem for linear time-invariant multivariable systems. IEEE Transactions on Automatic Control 21, 25–34 (1976)
3. Davison, E.J., Özgüner, Ü.: Expanding system problem. Systems and Control Letters 1, 255–260 (1982)
4. Desoer, C.A., Wang, Y.T.: On the minimum order of a robust servocompensator. IEEE Transactions on Automatic Control 23, 70–73 (1978)

5. Engelborghs, K., Luzyanina, T., Roose, D.: Numerical bifurcation analysis of delay differential equations using DDE-BIFTOOL. ACM Transactions on Mathematical Software 28, 1–21 (2002)
6. Foias, C., Özbay, H., Tannenbaum, A.: Robust control of infinite dimensional systems: Frequency domain methods. LNCIS, vol. 209. Springer, London (1996)
7. Francis, B.A., Wonham, W.M.: The internal model principle of control theory. Automatica 12, 457–465 (1976)
8. Gümüşsoy, S., Özbay, H.: Remarks on H^∞ controller design for SISO plants with time delays. In: Proc. 5th IFAC Symposium on Robust Control Design, Toulouse, France (2006)
9. Gündeş, A.N., Özbay, H., Özgüler, A.B.: PID controller synthesis for a class of unstable MIMO plants with I/O delays. Automatica 43, 135–142 (2007)
10. Kashima, K., Özbay, H., Yamamoto, Y.: A Hamiltonian-based solution to the mixed sensitivity optimization problem for stable pseudorational plants. Systems & Control Letters 54, 1063–1068 (2005)
11. Krein, M.G., Nudel'man, A.A.: The Markov moment problem and extremal problems. Translations of Mathematical Monographs 50 (1977)
12. Olgaç, N., Sipahi, R.: A practical method for analyzing the stability of neutral type LTI-time delayed systems. Automatica 40, 847–853 (2004)
13. Özbay, H.: A simpler formula for the singular values of a certain Hankel operator. Systems & Control Letters 15, 381–390 (1990)
14. Özbay, H., Gündeş, A.N.: Resilient PI and PD controller designs for a class of unstable plants with I/O delays. Applied and Computational Mathematics 6, 18–26 (2007)
15. Toker, O., Özbay, H.: H^∞ optimal and suboptimal controllers for infinite dimensional SISO plants. IEEE Transactions on Automatic Control 40, 751–755 (1995)
16. Vidyasagar, M.: Control system synthesis: A factorization approach. MIT Press, Cambridge (1985)
17. Vyhlidal, T., Zitek, P.: Mapping the spectrum of a retarded time-delay system utilizing root distribution features. In: Proc. IFAC Workshop on Time-Delay Systems, L'Aquila, Italy (2006)
18. Zeren, M., Özbay, H.: Comments on Solutions to combined sensitivity and complementary sensitivity problem in control systems. IEEE Transactions on Automatic Control 43, 724 (1998)

Stabilization of Neutral Time-Delay Systems

Michaël Di Loreto[1], Catherine Bonnet[2] and Jean Jacques Loiseau[3]

[1] Laboratoire Ampère, UMR CNRS 5005, INSA-Lyon, F-69621, France
michael.di-loreto@insa-lyon.fr

[2] INRIA Rocquencourt, Domaine de Voluceau, B.P. 105, 78153 Le Chesnay Cedex, France
catherine.bonnet@inria.fr

[3] IRCCyN, UMR CNRS 6597, 44321 Nantes Cedex 03, France
jean-jacques.loiseau@irccyn.ec-nantes.fr

Summary. This paper addresses the stabilization problem for neutral time-delay systems. Starting from a transfer approach, we develop a general algebraic framework for the study of the stabilization. The stability considered in this paper is the exponential one. We introduce a class of generalized operators, and we show the existence of a coprime factorization over this class. The main result of this paper concerns strictly proper systems that are not formally stable. It is shown that any stabilizing compensator, which is a real rational fraction in s and $e^{-\theta s}$, is necessarily not proper.

1 Introduction

In this note, we consider neutral type time-delay systems of the general form

$$\dot{x}(t) = \sum_{i=1}^{\kappa} E_i \dot{x}(t - i\theta) + \sum_{i=0}^{\kappa} A_i x(t - i\theta) + \sum_{i=0}^{\kappa} B_i u(t - i\theta), \tag{1}$$

$$y(t) = \sum_{i=0}^{\kappa} C_i x(t - i\theta) + \sum_{i=0}^{\kappa} D_i u(t - i\theta), \tag{2}$$

where $x(t) \in \mathbb{R}^n$ is the instantaneous state, $u(t) \in \mathbb{R}^m$ is the control input, $y(t) \in \mathbb{R}^p$ is the output. The matrices A_i, B_i, E_i, C_i, D_i are real with adequate dimensions, and $E_i \neq 0$ for some $i \in \{1, \ldots, \kappa\}$. All delays are commensurate to $\theta > 0$.

The neutral type systems are a recurrent topic of control research. Their stability analysis is widely the most treated topic in the literature. Since the earlier work of [1], research has been oriented to various approaches. We can cite the algebraic approach [11], [5], [7], where sufficient conditions are established for the stability of (1)-(2), or the functional approach [10]. In general, the study of the characteristic equation of (1)-(2) is not sufficient to conclude on the stability of the system. This equation has an infinite number of roots. The particularity of neutral systems is that these roots admit asymptotic vertical chains in the complex plane [1]. This fact is at the core of various pathologies when an asymptotic chain coincides with the imaginary axis. Among these pathologies, the loss of equivalence between various stabilities

J.J. Loiseau et al. (Eds.): Topics in Time Delay Systems, LNCIS 388, pp. 209–219.
springerlink.com

is the principal one. A neutral system can be exponentially or asymptotically unstable while it is BIBO or L_2-stable, as well as it can be asymptotically stable and exponentially unstable. To learn more on this subject, we refer to [6], [8], [9], [12], and [15].

To get one step insight, let us recall some basic facts on stability of neutral type time-delay systems. According to [5], the system (1)-(2) is said to be formally stable if

$$\mathrm{Rank}(I - \hat{E}(\mathrm{e}^{-\theta s})) = n, \ \ \forall s \in \mathbb{C}, \ \mathrm{Re}(s) \geq 0, \tag{3}$$

where

$$\hat{E}(\mathrm{e}^{-\theta s}) = \sum_{i=1}^{\kappa} E_i \mathrm{e}^{-i\theta s}. \tag{4}$$

The open-loop characteristic equation associated to (1)-(2) is defined by

$$\Delta(s, \mathrm{e}^{-\theta s}) = 0, \tag{5}$$

where

$$\Delta(s, \mathrm{e}^{-\theta s}) = \det\left(s(I - \hat{E}(\mathrm{e}^{-\theta s})) - \hat{A}(\mathrm{e}^{-\theta s})\right),$$

and

$$\hat{A}(\mathrm{e}^{-\theta s}) = \sum_{i=0}^{\kappa} A_i \mathrm{e}^{-i\theta s}. \tag{6}$$

Developing the determinant in (5), $\Delta(s, \mathrm{e}^{-\theta s})$ is of the general form

$$\Delta(s, \mathrm{e}^{-\theta s}) = \sum_{i=0}^{q} \sum_{k=0}^{r} \Delta_{ik} s^i \mathrm{e}^{-k\theta s}, \tag{7}$$

with $\Delta_{ik} \in \mathbb{R}$, for $i = 0$ to q and $k = 0$ to r, for some integers $q, r \in \mathbb{N}$. Hence $\Delta(s, \mathrm{e}^{-\theta s})$ is a quasipolynomial defined over $\mathbb{R}[s, \mathrm{e}^{-\theta s}]$, the ring of real polynomials in the variables s and $\mathrm{e}^{-\theta s}$. The leading coefficient in s of $\Delta(s, \mathrm{e}^{-\theta s})$ is actually a real polynomial in $\mathrm{e}^{-\theta s}$. It is called the principal part of $\Delta(s, \mathrm{e}^{-\theta s})$, and it is denoted by $\mathrm{pp}(\Delta)(\mathrm{e}^{-\theta s})$, *i.e.*

$$\mathrm{pp}(\Delta)(\mathrm{e}^{-\theta s}) = \sum_{k=0}^{r} \Delta_{qk} \mathrm{e}^{-k\theta s}. \tag{8}$$

The coefficient Δ_{q0} is called the principal term of $\Delta(s, \mathrm{e}^{-\theta s})$, and we say that $\Delta(s, \mathrm{e}^{-\theta s})$ has a principal term if $\Delta_{q0} \neq 0$. One says that $\Delta(s, \mathrm{e}^{-\theta s})$ is monic (in s), if $\mathrm{pp}(\Delta)(\mathrm{e}^{-\theta s})$ in (8) is a unit over $\mathbb{R}$, in other words if $\mathrm{pp}(\Delta)(\mathrm{e}^{-\theta s}) = 1$. Note that a monic quasipolynomial has a principal term.

In this paper, we focus on exponential stability. For simplicity, we will just call it stability. This stability notion is related to the location of the roots of (5) in the complex plane. This property is often taken as a definition.

Definition 1. *[13] The system (1)-(2) is said to be stable if there exists $\delta < 0$ such that*

$$\Delta(s, \mathrm{e}^{-\theta s}) \neq 0, \ \forall s \in \mathbb{C}, \ \mathrm{Re}(s) \geq \delta. \tag{9}$$

If the system is formally stable, (5) has only a finite number of zeros in the closed right half complex plane [13]. If the quasipolynomial $\Delta(s, \mathrm{e}^{-\theta s})$ has no principal term, or if its principal part is not stable, (5) has an infinite number of unstable zeros. According to (3), the system (1)-(2) is formally stable if and only if the principal part of $\Delta(s, \mathrm{e}^{-\theta s})$, $\mathrm{pp}(\Delta)(\mathrm{e}^{-\theta s})$, is a stable polynomial in $\mathrm{e}^{-\theta s}$, *i.e.* there exists $\mu < 0$ such that

$$\mathrm{pp}(\Delta)(\mathrm{e}^{-\theta s}) \neq 0, \; \forall s \in \mathbb{C}, \; \mathrm{Re}(s) \geq \mu.$$

>From these basic definitions and properties, we are now in position to develop a mathematical algebraic framework for the analysis of neutral type time-delay systems. This framework is detailed in the next section.

2 Background

A distributed delay is an input-output system defined by a convolution of the form

$$y(t) = \int_{\theta_1}^{\theta_2} f(\tau) u(t-\tau)\, \mathrm{d}\tau, \tag{10}$$

where the kernel f and the input u are assumed to be Laplace transformable in the sense of distributions, and $0 \leq \theta_1 < \theta_2$. The Laplace transform of the kernel $f(t)$ is an entire function. For the case study of time-delay systems with commensurate delays, we focus our attention on real rational fractions in s and $\mathrm{e}^{-\theta s}$. This fraction field is denoted $\mathbb{R}(s, \mathrm{e}^{-\theta s})$. The interest to use distributed delays for the stabilization problem was early pointed out by various authors. Consequently, the ring $\mathcal{G}$ of distributed delays with Laplace transform in $\mathbb{R}(s, \mathrm{e}^{-\theta s})$ was characterized in [3]. It is shown that any element $g(s, \mathrm{e}^{-\theta s}) \in \mathcal{G}$ can be put in the form

$$g(s, \mathrm{e}^{-\theta s}) = \frac{n(s, \mathrm{e}^{-\theta s})}{d(s)}, \tag{11}$$

where $n(s, \mathrm{e}^{-\theta s}) \in \mathbb{R}[s, \mathrm{e}^{-\theta s}]$, $d(s) \in \mathbb{R}[s]$, and $g(s, \mathrm{e}^{-\theta s})$ is an entire rational fraction, which is strictly proper with respect to s.

>From this characterization, the authors consider linear operators that include a finite number of derivatives, pointwise delays, and distributed delays. In other words, they define the set

$$\mathcal{E} = (\mathbb{R}[\mathrm{e}^{-\theta s}] + \mathcal{G})[s]. \tag{12}$$

Any element $p(s, \mathrm{e}^{-\theta s}) \in \mathcal{E}$ is called a pseudopolynomial. It is shown that $\mathcal{E}$ is a Bézout ring.

From (11), any pseudopolynomial $p(s, \mathrm{e}^{-\theta s}) \in \mathcal{E}$ is an entire function, and can be written

$$p(s, \mathrm{e}^{-\theta s}) = \frac{n(s, \mathrm{e}^{-\theta s})}{d(s)} = \frac{\sum_{i=0}^{r} n_i(\mathrm{e}^{-\theta s}) s^i}{\sum_{i=0}^{t} d_i s^i}, \tag{13}$$

where $n_i(\mathrm{e}^{-\theta s}) \in \mathbb{R}[\mathrm{e}^{-\theta s}]$ for $i = 0$ to r, and $d_i \in \mathbb{R}$ for $i = 0$ to t. The degree of $p(s, \mathrm{e}^{-\theta s})$ is defined by $\deg(p) = \deg_s(n) - \deg_s(d)$, which is equal to $(r-t)$ if $n_r \neq 0$ and $d_t \neq 0$. The definitions of monic pseudopolynomial or principal part are natural extensions of the equivalent definitions for quasipolynomials in $\mathbb{R}[s, \mathrm{e}^{-\theta s}]$.

Since $\mathcal{E}$ is a Bézout ring, the notions of divisor, multiple, and coprimeness are well-defined. Under the assumption that the pseudopolynomials p_1 and p_2 are coprime, there exist two pseudopolynomials x_1 and x_2 such that

$$p_1 x_1 + p_2 x_2 = 1. \tag{14}$$

Over $\mathcal{E}$, the division of two pseudopolynomials makes sense if the divisor is monic. When this divisor is not monic, but just has a principal term, this division yields to a quotient and a remainder that are not pseudopolynomials. To analyze more particularly this division property, let us remark that any element $t(s, \mathrm{e}^{-\theta s}) \in \mathbb{R}(s, \mathrm{e}^{-\theta s})$ can be put in the form

$$t(s, \mathrm{e}^{-\theta s}) = \frac{n_t(s, \mathrm{e}^{-\theta s})}{d_t(s, \mathrm{e}^{-\theta s})}, \tag{15}$$

where $n_t(s, \mathrm{e}^{-\theta s})$ and $d_t(s, \mathrm{e}^{-\theta s})$ are two coprime pseudopolynomials. With the notation in (15), we obtain the following result.

Lemma 1. *Let x and y be two pseudopolynomials, where y has a principal term. There exist q and r over $\mathbb{R}(s, \mathrm{e}^{-\theta s})$, with d_q and d_r two pseudopolynomials with principal terms,* $\deg(r) < \deg(y)$, *such that*

$$x = qy + r. \tag{16}$$

Proof. According to (13), we decompose x and y in the form

$$x = \frac{n_x}{d_x}, \quad y = \frac{n_y}{d_y},$$

and we denote $\tilde{x} = n_x d_y$, $\tilde{y} = n_y d_x$. If $\deg(\tilde{x}) < \deg(\tilde{y})$, take $q = 0$ and $r = x$ in (16), and the solution follows. Now, if $\deg(\tilde{x}) \geq \deg(\tilde{y})$, we construct iteratively the sequences r_i and q_i of elements in $\mathbb{R}(s, \mathrm{e}^{-\theta s})$ by $r_0 = \tilde{x}$, and

$$\begin{aligned} q_{i+1} &= \tfrac{\mathrm{pp}(r_i)}{\mathrm{pp}(\tilde{y})} s^{\deg(r_i) - \deg(\tilde{y})} \\ r_{i+1} &= r_i - \tfrac{\mathrm{pp}(r_i)}{\mathrm{pp}(\tilde{y})} s^{\deg(r_i) - \deg(\tilde{y})} \tilde{y}. \end{aligned} \tag{17}$$

Such sequences satisfy $r_i = q_{i+1}\tilde{y} + r_{i+1}$, and $\deg(r_i) < \deg(r_{i+1})$. Since the degrees of the sequence r_i are decreasing, there exists $a \in \mathbb{N}$ such that $\deg(r_a) < \deg(\tilde{y})$. The element r_a satisfies

$$\tilde{x} = \left(\sum_{i=0}^{a-1} q_{i+1} \right) \tilde{y} + r_a. \tag{18}$$

This equality yields $x = qy + r$, where

$$q = \sum_{i=0}^{a-1} q_{i+1}, \quad r = \frac{r_a}{d_x d_y}. \tag{19}$$

By assumption, y has a principal part. This implies that $\tilde{y}$ has itself a principal part. Thus it follows by (17) and (19) that q and r have denominators, d_q and d_r respectively, with principal terms. □

Two remarks can be made on this lemma. The first one is related to the principal term property. If the divisor y has no principal term, it follows from the above lemma that d_q and d_r have no principal terms. Nevertheless, there exists $\nu \in \mathbb{N}^*$ such that

$$\mathrm{e}^{-\nu\theta s} x = qy + r, \tag{20}$$

where q and r are in $\mathbb{R}(s, \mathrm{e}^{-\theta s})$, d_q and d_r have principal terms, and $\deg(r) < \deg(y)$. Hence Lemma 1 can be applied to $\mathrm{e}^{-\nu\theta s} x$ and y. The second remark concerns the non uniqueness of the quotient q and the remainder r in (16). Indeed, for any pseudopolynomial $w \in \mathcal{E}$ such that $\deg(w) < 0$, we have

$$\begin{aligned} x &= qy + r \\ &= (q - w)y + r + wy = q'y + r', \end{aligned}$$

where $q' = q - w$ and $r' = r + wy$ satisfy the degree condition $\deg(r') < \deg(y)$. Consequently, any pair (q', r'), parameterized by $w \in \mathcal{E}$ with strictly negative degree, is another solution of (16).

For the stability analysis of neutral type systems, the stability definition of a pseudopolynomial plays a central role. Accordingly with Definition 1, we say that a pseudopolynomial $p(s, \mathrm{e}^{-\theta s})$ is stable if there exists $\delta < 0$ such that

$$p(s, \mathrm{e}^{-\theta s}) \neq 0, \quad \forall s \in \mathbb{C}, \ \mathrm{Re}(s) \geq \delta.$$

The subring of $\mathcal{E}$ of all stable pseudopolynomials is denoted $\mathcal{E}_s$. One remarks that given a stable pseudopolynomial y, $\mathrm{pp}(y)$ is a stable polynomial in the variable $\mathrm{e}^{-\theta s}$, that is there exists $\mu < 0$ such that

$$\mathrm{pp}(y)(\mathrm{e}^{-\theta s}) \neq 0, \ \forall s \in \mathbb{C}, \ \mathrm{Re}(s) \geq \mu. \tag{21}$$

Under this assumption, it is immediate to verify that the elements q and r in Lemma 1 are by construction stable fractions, in the sense that d_q and d_r are stable pseudopolynomials.

With this general background, we can develop adapted tools for the stabilization problem, based on the attractive framework of proper and stable fractions ring discussed in [16] for linear finite dimensional systems. A first step of such a generalization applied to retarded type time-delay systems was made in [4]. In the next section, we introduce this generalization.

3 Proper and stable factorization

The ring $\mathcal{P}_{\mathcal{E}}$ of proper and stable fractions over $\mathcal{E}$ is defined as the set

$$\left\{ \frac{p}{q} : \ p \in \mathcal{E}, \ q \in \mathcal{E}_s, \ \deg(p) \leq \deg(q) \right\}. \tag{22}$$

The set $\mathcal{P}_{\mathcal{E}}$ is a subring of $\mathbb{R}(s, \mathrm{e}^{-\theta s})$ for the usual additive and multiplicative laws. More generally, an element $t = \frac{p}{q}$ in $\mathbb{R}(s, \mathrm{e}^{-\theta s})$ is said to be proper (respectively strictly proper) if

1. $\deg(p) \leq \deg(q)$ (respectively $\deg(p) < \deg(q)$),
2. q has a principal term.

The properness of $t(s, \mathrm{e}^{-\theta s})$ means that this plant can be realized through a neutral type time-delay system of the form (1)-(2), using a finite bank of integrators and pointwise commensurate delays.

Any transfer function $t(s, \mathrm{e}^{-\theta s}) \in \mathbb{R}(s, \mathrm{e}^{-\theta s})$ admits a proper and stable factorization. Indeed, $t(s, \mathrm{e}^{-\theta s})$ writes as a fraction of two pseudopolynomials, say $n(s, \mathrm{e}^{-\theta s})$ and $d(s, \mathrm{e}^{-\theta s})$. Let $\eta = \max(\deg(n), \deg(d))$, and take any stable pseudopolynomial $q(s, \mathrm{e}^{-\theta s}) \in \mathcal{E}_s$, with $\deg(q) \leq \eta$. Hence,

$$\tilde{n} = \frac{n}{q}, \quad \tilde{d} = \frac{d}{q}$$

is a proper and stable factorization of $t(s, \mathrm{e}^{-\theta s})$, since $\tilde{n}$, $\tilde{d} \in \mathcal{P}_{\mathcal{E}}$, and $t = \frac{\tilde{n}}{\tilde{d}}$. The main interrogation arises for the existence of a coprime proper and stable factorization. Among all factorizations over $\mathcal{P}_{\mathcal{E}}$, we wish to find two elements $\tilde{n}$ and $\tilde{d}$ in $\mathcal{P}_{\mathcal{E}}$ such that $t = \tilde{n}\tilde{d}^{-1}$, where $\tilde{n}$ and $\tilde{d}$ are coprime. The coprimeness of two elements in $\mathcal{P}_{\mathcal{E}}$ has to be understood as the absence of non trivial common factors. The answer of this question is divided into two steps. We distinguish the cases of formally stable and unstable neutral type time-delay systems. In the first case, the existence of a coprime factorization is guaranteed and constructive. For this part, the result is a direct consequence of classical works [17], [4], [14]. However, for formally unstable systems, we will show that such a coprime factorization can possibly not exist.

Theorem 1. *Any transfer function $t(s, \mathrm{e}^{-\theta s}) \in \mathbb{R}(s, \mathrm{e}^{-\theta s})$, which has a denominator with a stable principal part, admits a coprime factorization over $\mathcal{P}_{\mathcal{E}}$.*

Proof. Let $t(s, \mathrm{e}^{-\theta s}) \in \mathbb{R}(s, \mathrm{e}^{-\theta s})$. Write this element as $t = pq^{-1}$, where p and q are two coprime pseudopolynomials. There exist x_1 and x_2 over $\mathcal{E}$ such that $qx_1 + px_2 = 1$. Now, let φ be any stable pseudopolynomial (and consequently with principal term) such that

$$m = \max(\deg(q\varphi), \deg(p\varphi))$$

be non negative. We have

$$q\varphi x_1 + p\varphi x_2 = \varphi. \tag{23}$$

Consider an arbitrary stable polynomial ϕ with degree m. >From the above equality, it follows

$$q\varphi x_1\phi^2 + p\varphi x_2\phi^2 = \varphi\phi^2. \tag{24}$$

Dividing $x_2\phi^2$ by $q\varphi$, there exist, from Lemma 1, γ_1 and γ_2 in $\mathbb{R}(s, \mathrm{e}^{-\theta s})$, with denominators which have stable principal parts, such that

$$x_2\phi^2 = q\varphi\gamma_1 + \gamma_2, \tag{25}$$

and $\deg(\gamma_2) < m$. Thus, combining (24) and (25) gives us

$$\frac{q\varphi}{\phi}\frac{x_1\phi^2 + p\varphi\gamma_1}{\phi\varphi} + \frac{p\varphi}{\phi}\frac{\gamma_2}{\phi\varphi} = 1. \tag{26}$$

Equation (26) is of the form $q_c y_c + p_c x_c = 1$, where all elements are in $\mathcal{P}_{\mathcal{E}}$, and it can be verified that p_c/q_c is a coprime factorization over $\mathcal{P}_{\mathcal{E}}$ of t. □

>From this result, a fraction in $\mathbb{R}(s, \mathrm{e}^{-\theta s})$ with a denominator having a stable principal part admits a proper and stable factorization, which satisfies a Bézout identity. This result is well known [17]. This assertion fails in the case of fractions whose principal part of its denominator is not stable. Indeed, consider the general case of neutral systems that are formally stable or unstable, and take again the proof of Theorem 1. We see that, under the condition that d_t has a principal term, γ_1 and γ_2 are in $\mathbb{R}(s, \mathrm{e}^{-\theta s})$, and their denominators have principal terms. Consequently, we obtain a Bézout like form $q_c x_c + p_c y_c = 1$, where q_c and p_c are elements in $\mathcal{P}_{\mathcal{E}}$ satisfying

$$t(s, \mathrm{e}^{-\theta s}) = \frac{p_c(s, \mathrm{e}^{-\theta s})}{q_c(s, \mathrm{e}^{-\theta s})},$$

but where y_c and x_c are in $\mathbb{R}(s, \mathrm{e}^{-\theta s})$, with denominators having principal terms. In general, the elements x_c and y_c are not defined over $\mathcal{P}_{\mathcal{E}}$, since they can be unstable.
It is worth noting that the elements p_c and q_c in $\mathcal{P}_{\mathcal{E}}$ also satisfy the identity $q_c b_c + p_c a_c = 1$, where a_c and b_c are stable fractions over $\mathbb{R}(s, \mathrm{e}^{-\theta s})$ with monic denominators, that are in general not proper. Indeed, from the proof of Theorem 1, (23) writes

$$q\varphi x_1 + p\varphi x_2 = \varphi,$$

where φ is any stable pseudopolynomial such that $m = \max(\deg(q\varphi), \deg(p\varphi))$ is non negative. Consequently, for any stable pseudopolynomial ϕ with degree greater than m, we get

$$\frac{q\varphi}{\phi}\frac{\phi x_1}{\varphi} + \frac{p\varphi}{\phi}\frac{\phi x_2}{\varphi} = 1. \tag{27}$$

This identity is of the form $q_c b_c + p_c a_c = 1$, where q_c and p_c are by construction in $\mathcal{P}_{\mathcal{E}}$, while a_c and b_c are stable fractions in $\mathbb{R}(s, \mathrm{e}^{-\theta s})$, with monic denominators but not necessarily proper. We will denote by $\mathcal{S}_{\mathcal{E}}$ the ring of stable fractions in $\mathbb{R}(s, \mathrm{e}^{-\theta s})$, in such a way that $\mathcal{P}_{\mathcal{E}}$ is a subring of $\mathcal{S}_{\mathcal{E}}$. Such a pair (p_c, q_c) is called an $\mathcal{S}_{\mathcal{E}}$-coprime factorization of t. Thus, we have the following result.

Theorem 2. *Any transfer function $t(s, \mathrm{e}^{-\theta s}) \in \mathbb{R}(s, \mathrm{e}^{-\theta s})$, which has a denominator with a principal part, admits an $\mathcal{S}_{\mathcal{E}}$-coprime factorization, that is there exists a pair (p_c, q_c) over $\mathcal{S}_{\mathcal{E}}$, such that*

$$t(s, \mathrm{e}^{-\theta s}) = \frac{p_c(s, \mathrm{e}^{-\theta s})}{q_c(s, \mathrm{e}^{-\theta s})},$$

that satisfy the identity $p_c a_c + q_c b_c = 1$, where a_c and b_c are defined over $\mathcal{S}_{\mathcal{E}}$.

Note that the pair (p_c, q_c) can always be taken in the ring $\mathcal{P}_{\mathcal{E}}$. For non formally stable neutral systems, there may be not exist, in general, a proper and stable coprime factorization, as indicates the following example.

Example 1. Let

$$t(s, \mathrm{e}^{-s}) = \frac{1}{s(1 - \mathrm{e}^{-s})}. \tag{28}$$

This transfer function writes $t(s, \mathrm{e}^{-s}) = \frac{n(s, \mathrm{e}^{-s})}{d(s, \mathrm{e}^{-s})}$, where

$$n(s, \mathrm{e}^{-s}) = 1, \quad d(s, \mathrm{e}^{-s}) = s(1 - \mathrm{e}^{-s})$$

are two coprime pseudopolynomials. There exist two pseudopolynomials $x(s, \mathrm{e}^{-s})$ and $y(s, \mathrm{e}^{-s})$ such that $xd + yn = 1$. For instance, take

$$\begin{aligned} x(s, \mathrm{e}^{-s}) &= \tfrac{1 - \mathrm{e}^{-s}}{s}, \\ y(s, \mathrm{e}^{-s}) &= 2\mathrm{e}^{-s} - \mathrm{e}^{-2s}. \end{aligned} \tag{29}$$

Following the proof of Theorem 1, an $\mathcal{S}_{\mathcal{E}}$-coprime factorization $(\tilde{n}, \tilde{d})$ of $t(s, \mathrm{e}^{-s})$ is

$$\tilde{n} = \frac{1}{s+1}, \quad \tilde{d} = \frac{s(1 - \mathrm{e}^{-s})}{s+1}. \tag{30}$$

The pair $(\tilde{n}, \tilde{d})$ satisfies $\tilde{x}\tilde{d} + \tilde{y}\tilde{n} = 1$, where

$$\begin{aligned} \tilde{x} &= \tfrac{\alpha(s, \mathrm{e}^{-s})}{s+1}, \\ \tilde{y} &= \tfrac{(s+1)^2 - \alpha(s, \mathrm{e}^{-s}) s (1 - \mathrm{e}^{-s})}{s+1}, \end{aligned}$$

for any $\alpha(s, \mathrm{e}^{-s}) \in \mathcal{S}_{\mathcal{E}}$. While $\tilde{x}$ can be chosen to be proper, $\tilde{y}$ can never be proper.

For non formally stable neutral systems, there does not exist, in general, a coprime proper and stable factorization. We will see that this property is always valid for any strictly proper neutral system that is not formally stable. It remains now to give necessary and sufficient conditions for the stabilization of such systems. This is done in the next section.

4 Stabilizing compensator

All previous definitions and concepts are easily generalized to the multivariable case. In this section, we describe the synthesis of a stabilizing compensator in the general multivariable case.
Consider the closed-loop system described on Figure 1. Let be given a plant $P \in \mathbb{R}^{p\times m}(s, \mathrm{e}^{-\theta s})$, and a compensator $C \in \mathbb{R}^{m\times p}(s, \mathrm{e}^{-\theta s})$. The external inputs of this loop are u_1 and u_2, e_1 and e_2 are the inputs of C and P respectively, while y_1 and y_2 are the outputs of C and P, respectively.

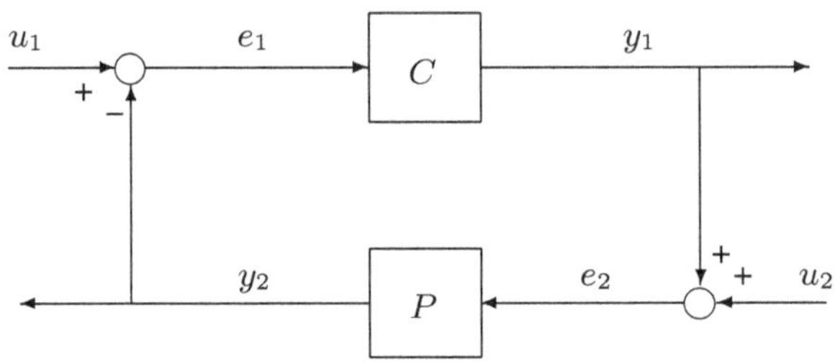

Fig. 1. Closed-loop system.

The relations between u_1, u_2, e_1, e_2 are expressed as

$$\begin{pmatrix} e_1 \\ e_2 \end{pmatrix} = H(P,C) \begin{pmatrix} u_1 \\ u_2 \end{pmatrix}, \tag{31}$$

where

$$H(P,C) = \begin{pmatrix} (I+PC)^{-1} & -(I+PC)^{-1}P \\ C(I+PC)^{-1} & I - C(I+PC)^{-1}P \end{pmatrix}.$$

We will assume in what follows that the condition

$$\det(I+PC) = \det(I+CP) \neq 0 \tag{32}$$

is fulfilled, for some $s \in \mathbb{C}$. This condition is still necessary and sufficient for (31) to have a unique solution.

We say that the pair (P,C) is stable if (32) is satisfied, and $H(P,C)$ is in $\mathcal{S}_{\mathcal{E}}^{(p+m)\times(p+m)}$. We also say that C stabilizes P. This is a natural extension of the well known internal stability [16], [14]. The proofs of the theorems below are quite standard, so they are omitted.

Theorem 3. *Let $P, C \in M(\mathbb{R}(s, \mathrm{e}^{-\theta s}))$, (N_p, D_p), $(\widetilde{D}_p, \widetilde{N}_p)$ be right (left respectively) $\mathcal{S}_{\mathcal{E}}$-coprime factorizations of P, (N_c, D_c), $(\widetilde{D}_c, \widetilde{N}_c)$ be right (left respectively) $\mathcal{S}_{\mathcal{E}}$-coprime factorizations of C. The following assertions are equivalent.*

(i) The pair (P,C) is stable.
(ii) The matrix $\Lambda(P,C) = \widetilde{N}_c N_p + \widetilde{D}_c D_p$ is bistable.
(iii) The matrix $\widetilde{\Lambda}(P,C) = \widetilde{N}_p N_c + \widetilde{D}_p D_c$ is bistable.

Once we have the general characterization of feedback stability for neutral type systems. We can get further insight by the following corollary.

Corollary 1. *Let* $P, C \in M(\mathbb{R}(s, \mathrm{e}^{-\theta s}))$, *and* (N_p, D_p), $(\widetilde{D}_p, \widetilde{N}_p)$ *be right (left, respectively)* $\mathcal{S}_{\mathcal{E}}$*-coprime factorizations of* P. *The following assertions are equivalent.*

(i) *The pair* (P, C) *is stable.*
(ii) C *admits a stable left factorization* $(\widetilde{D}_c, \widetilde{N}_c)$ *such that* $\widetilde{D}_c D_p + \widetilde{N}_c N_p = I$.
(iii) C *admits a stable right factorization* (N_c, D_c) *such that* $\widetilde{D}_p D_c + \widetilde{N}_p N_c = I$.

>From this characterization, it is possible to parameterize all stabilizing compensators, as arises for a broad class of convolution systems of finite or infinite dimension [17], [14]. Denote $S(P)$ the set of all stabilizing compensators that are in $\mathbb{R}(s, \mathrm{e}^{-\theta s})$.

Theorem 4. *Let* $P \in M(\mathbb{R}(s, \mathrm{e}^{-\theta s}))$, *and* (N_p, D_p), $(\widetilde{D}_p, \widetilde{N}_p)$ *be right (left, respectively)* $\mathcal{S}_{\mathcal{E}}$*-coprime factorizations of* P. *There exist stable matrices* X, Y, $\widetilde{X}$, $\widetilde{Y}$ *such that* $XN_p + YD_p = I$ *and* $\widetilde{N}_p\widetilde{X} + \widetilde{D}_p\widetilde{Y} = I$. *Any stabilizing compensator* $C \in S(P)$ *writes*

$$\begin{aligned} C &= (Y - R\widetilde{N}_p)^{-1}(X + R\widetilde{D}_p), \\ &= (\widetilde{X} + D_p Q)(\widetilde{Y} - N_p Q)^{-1}, \end{aligned} \tag{33}$$

where R *and* Q *are any matrices in* $\mathcal{S}_{\mathcal{E}}$ *such that* $\det(Y - R\widetilde{N}_p) \neq 0$ *and* $\det(\widetilde{Y} - N_p Q) \neq 0$.

Example 2. Consider the stabilization problem for the plant in Example 1. An $\mathcal{S}_{\mathcal{E}}$-coprime factorization $(\tilde{n}, \tilde{d})$ of $p(s, \mathrm{e}^{-s})$ was found in (30). Such a factorization satisfies $\tilde{x}\tilde{n} + \tilde{y}\tilde{d} = 1$, with $\tilde{x}$ and $\tilde{y}$ in $\mathcal{S}_{\mathcal{E}}$. All $\mathcal{S}_{\mathcal{E}}$-stabilizing controllers in $\mathbb{R}(s, \mathrm{e}^{-\theta s})$ are parameterized by

$$C(s, \mathrm{e}^{-s}) = \frac{(s+1)^2}{r - \alpha} + s(1 - \mathrm{e}^{-s}).$$

Any stabilizing compensator is not proper, since $\deg(C) \geq 1$, for any stable fractions α and $r \neq \alpha$.

This example is illustrative of what happens for formally unstable neutral time-delay systems that are strictly proper. Indeed, the theorem below generalizes the previous simple example, and is certainly the main result of this paper.

Theorem 5. *Let* $t(s, \mathrm{e}^{-\theta s}) \in \mathbb{R}(s, \mathrm{e}^{-\theta s})$ *be a strictly proper plant that is not formally stable. Then, there does not exist a proper stabilizing compensator in the class of rational fractions in* s *and* $\mathrm{e}^{-\theta s}$.

Proof Let (p, q) be an $\mathcal{S}_{\mathcal{E}}$-coprime factorization of t. There exist stable fractions x and y in $\mathbb{R}(s, \mathrm{e}^{-\theta s})$ such that $px + qy = 1$. For any stable matrix r in $\mathbb{R}(s, \mathrm{e}^{-\theta s})$, we have $px' + qy' = 1$, with $x' = x + qr$ and $y' = y - pr$. In what follows, we show that $\deg(y') < \deg(x')$. Two cases may arise. If $\deg(px') \geq \deg(qy')$, since $\deg(p) < \deg(q)$, we have

$$\begin{aligned} \deg(q) + \deg(y') &\leq \deg(p) + \deg(x') \\ &< \deg(q) + \deg(x'), \end{aligned}$$

so that $\deg(y') < \deg(x')$. Now if $\deg(px') < \deg(qy')$, denote $px' = \varepsilon$, and consequently $qy' = 1 - \varepsilon$. The element ε admits a coprime factorization over $\mathcal{E}$, say $(n_\varepsilon, d_\varepsilon)$. >From the previous degree condition, it follows that $\deg(\varepsilon) < 0$, *i.e.* $\deg(n_\varepsilon) < \deg(d_\varepsilon)$. Hence $1 - \varepsilon = \frac{d_\varepsilon - n_\varepsilon}{d_\varepsilon}$, and consequently,

$$\mathrm{pp}(qy') = \mathrm{pp}(1-\varepsilon) = \mathrm{pp}(d_\varepsilon).$$

By assumption the principal part of q is not stable. This implies that the principal part of d_ε is also unstable, which in turn implies that d_ε has unstable zeros. Hence, the fraction ε is not stable, which is a contradiction, so that we get $\deg(y') < \deg(x')$. From Theorem 4, it follows that any stabilizing compensator writes $x'{y'}^{-1}$, and it is necessarily not proper. □

The previous result is stated in the scalar case, but an extension to the multivariable case holds. For strictly proper and formally unstable neutral time-delay systems, and limiting ourselves to the class of stabilizing compensators that are fractions in $\mathbb{R}(s, \mathrm{e}^{-\theta s})$, it is shown that any such a compensator is not proper. This theorem is then a complement of some previous results, like in [5] or [2] where formally stable neutral systems are considered. However, this theorem does not permit to conclude on the existence of other classes of stabilizing compensators that are for instance non rational fractions involving non entire fractional derivatives. It is also clear that proper and formally unstable neutral systems can always be stabilized by proper fractions in $\mathbb{R}(s, \mathrm{e}^{-\theta s})$. The perspectives of this work include the relationship between the existence of an $\mathcal{S}_\mathcal{E}$-coprime factorization and the BIBO or $\mathcal{H}_\infty$ coprime factorizations, the robustness issue of such a control synthesis, and the practical implementation of the control laws.

References

1. Bellman, R., Cooke, K.L.: Differential-difference equations. Academic Press, New York (1963)
2. Brethé, D.: Stabilization and observation of neutral-type delay systems. In: Proc. IFAC CIS, Belfort, France, pp. 353–358 (1997)
3. Brethé, D., Loiseau, J.J.: Stabilization of time-delay systems. JESA 31, 1025–1047 (1997)
4. Brethé, D., Loiseau, J.J.: Proper stable factorizations for time-delay systems: The multivariable case. In: Proc. IFAC Workshop on Time-Delay Systems, Grenoble, France (1998)
5. Byrnes, C.I., Spong, M.W., Tarn, T.J.: A several complex variables approach to feedback stabilization of neutral delay-differential systems. Math. Syst. Theor. 17, 97–134 (1984)
6. Datko, R.: An example of an unstable neutral differential equation. Int. J. Control 38(1), 263–267 (1983)
7. Emre, E., Knowles, G.J.: Control of linear systems with fixed noncommensurate point delays. IEEE Trans. on Autom. Contr. 29(12), 1083–1090 (1984)
8. Hale, J.K., Verduyn Lunel, S.M.: Introduction to functional differential equations. Springer, New York (1993)
9. Loiseau, J.J., Cardelli, M., Dusser, X.: Neutral-type time-delay systems that are not formally stable are not BIBO stabilizable. IMA J. Math. Contr. Information 19, 217–227 (2002)
10. O'Connor, D., Tarn, T.J.: On stabilization by state feedback for neutral differential difference equations. IEEE Trans. on Autom. Contr. 28(5), 615–618 (1983)
11. Pandolfi, L.: Stabilization of neutral functional differential equations. J. Optim. Theory Appl. 20, 191–204 (1976)
12. Partington, J.R., Bonnet, C.: H_∞ and BIBO stabilization of delay systems of neutral type. Systems & Control Letters 52, 283–288 (2004)
13. Pontryagin, L.S.: On the zeros of some elementary transcendental functions. Amer. Math. Soc. Transl. 2(1), 95–110 (1955)

14. Quadrat, A.: The fractional representation approach to synthesis problems: An algebraic analysis viewpoint. Part II: Internal stabilization. SIAM J. Contr. Optimiz. 42(1), 300–320 (2003)
15. Rabah, R., Sklyar, G.M., Rezounenko, A.V.: Stability analysis of neutral type systems in Hilbert space. J. Differential Equations 214, 391–428 (2005)
16. Vidyasagar, M.: Control System Synthesis. A Factorization Approach. MIT Press, Cambridge (1985)
17. Vidyasagar, M., Schneider, H., Francis, B.A.: Algebraic and topological aspects of feedback stabilization. IEEE Trans. on Autom. Contr. 27(4), 880–894 (1982)

Robust Stabilization and H_∞ Control of Uncertain Distributed Delay Systems

Ulrich Münz[1], Jochen M. Rieber[2] and Frank Allgöwer[1]

[1] Institute for Systems Theory and Automatic Control, University of Stuttgart, Germany, {muenz,allgower}@ist.uni-stuttgart.de

[2] Astrium GmbH, Friedrichshafen, Germany, j.rieber@gmx.de

Summary. Linear matrix inequality (LMI) conditions for the robust stabilization and robust H_∞ control of uncertain linear systems with distributed delays are presented. All system matrices including the delay kernel are uncertain. Yet, the nominal delay kernel is assumed to be a matrix of rational functions, i.e., it can be written in a linear fractional form. The synthesis conditions are derived using a Lyapunov-Krasovskii functional. Techniques from robust control, such as the full-block S-procedure, are used to transform the resulting parametric matrix inequality into an LMI. As an important feature, the controller synthesis algorithm uses explicitly the information about the continuous delay kernel.

1 Introduction

Time-delay systems (TDS) have attracted an increasing interest over the last years, see for example [13, 25, 11, 27, 26] and references therein. Most publications deal with linear systems with *discrete delays*. However, *distributed delays* play an important role in many fields of biology [19, 9, 6] and engineering [28, 20, 21].

The stability of distributed delay systems (DDS) can be analyzed in the frequency domain, see for example [36, 4], in particular if the delay is γ-distributed, e.g., [2, 22]. However, these approaches are not suitable for robust analysis and synthesis problems. For this purpose, Lyapunov-based conditions are more suitable. (Robust) stability and stabilization of DDS have been studied in [29, 5, 14, 18, 17] for constant delay kernels, in [10, 12, 8] for piecewise constant kernels, and in [16] for continuous kernels. Stabilization of DDS for continuous delay kernels has been investigated in [7, 39, 40] using finite dimensional comparison systems. In [41], robust stabilization conditions are presented in terms of parametric matrix inequalities. Finally, H_∞ control of DDS has been studied in [37] for constant delay kernels, in [8] for piecewise constant delay kernels, and in [25] using generalized Popov theory.

The design conditions presented here expand a recently developed robust stability condition [23] towards robust stabilization and robust H_∞ control of uncertain DDS. They require that the nominal delay kernel is a matrix of rational functions, i.e., it can be written in a *linear fractional form*. This assumption is not restrictive as discussed later on. Due to this assumption, it is possible to reformulate the resulting parametric matrix inequality using the full-block S-procedure and a convex hull relaxation into linear matrix inequalities (LMI). The nominal case has been investigated in [24].

J.J. Loiseau et al. (Eds.): Topics in Time Delay Systems, LNCIS 388, pp. 221–231.
springerlink.com

In contrast to most results from the literature, the new synthesis conditions do not require piecewise constant delay kernels. Compared to [7, 39, 40, 25, 16], the controller design applies also for uncertain DDS. Unlike [41], the conditions are formulated as LMIs which considerably simplifies the solution and improves the accuracy, see [23]. At the same time, the nominal delay kernel is used explicitly in the LMI conditions, which is a source of conservatism for the condition in [37] if the delay kernel is known at least approximately. Finally, this work presents the first H_∞ controller design condition for uncertain DDS.

The chapter is structured as follows: We first present the problem statement and some fundamentals on parametric matrix inequalities and uncertain LMIs in Section 2. The main contribution, the new controller synthesis algorithms, are given in Section 3 before the chapter is concluded in Section 4.

Notation: Our notation is standard. $\mathcal{L}_2^\alpha[a,b]$ is the class of all square integrable functions $f\,:\,[a,b]\rightarrow\mathbb{R}^\alpha$. The corresponding $\mathcal{L}_2$-norm is $\|f\|_2=\sqrt{\int_a^b f^T(t)f(t)dt}$. We write $\begin{bmatrix}A & B\\ * & C\end{bmatrix}$ for $\begin{bmatrix}A & B\\ B^T & C\end{bmatrix}$. A block diagonal matrix with entries $M_1,\ldots,M_N$ is denoted diag$(M_1,\ldots,M_N)$. Finally, $\succ,\succeq,\prec$, and $\preceq$ indicate positive and negative (semi-)definiteness, respectively.

2 Problem Statement and Preliminaries

2.1 Problem Statement

We consider the following class of *uncertain distributed delay systems* (DDS)

$$\begin{aligned}\dot{x}(t)&=\tilde{A}(t)x(t)+\int_0^r\tilde{F}(t,\theta)x(t-\theta)d\theta+\tilde{B}_1(t)u(t)+\tilde{B}_2(t)w(t)\\ z(t)&=\tilde{C}(t)x(t)+\tilde{D}_1(t)u(t)+\tilde{D}_2(t)w(t)\\ x(t)&=\phi(t)\,,\quad t\in\Omega=[-r,0],\end{aligned}\tag{1}$$

where $x(t)\in\mathbb{R}^n$, $u(t)\in\mathbb{R}^{q_1}$, $w(t)\in\mathbb{R}^{q_2}$, $z(t)\in\mathbb{R}^p$ are the state, controlled input, disturbance input, and performance output, respectively. The initial condition is ϕ and $r>0$ is the range of the distributed delay. All matrices consist of a known constant part and a time-varying uncertainty, i.e., $\tilde{A}(t)=A+\Delta A(t)$, $\tilde{B}_i(t)=B_i+\Delta B_i(t)$, $\tilde{C}(t)=C+\Delta C(t)$, $\tilde{D}_i(t)=D_i+\Delta D_i(t)$, and $\tilde{F}=F(\theta)+\Delta F(t,\theta)$, where $A\in\mathbb{R}^{n\times n}$, $B_i\in\mathbb{R}^{n\times q_i}$, $C\in\mathbb{R}^{p\times n}$, $D_i\in\mathbb{R}^{p\times q_i}$, and $F:\Omega\rightarrow\mathbb{R}^{n\times n}$ are known matrices and $\Delta A:\mathbb{R}\rightarrow\mathbb{R}^{n\times n}$, $\Delta B_i:\mathbb{R}\rightarrow\mathbb{R}^{n\times q_i}$, $\Delta C:\mathbb{R}\rightarrow\mathbb{R}^{p\times n}$, $\Delta D_i:\mathbb{R}\rightarrow\mathbb{R}^{p\times q_i}$, and $\Delta F:\mathbb{R}\times\Omega\rightarrow\mathbb{R}^{n\times n}$ are time-varying uncertainties. To shorten the notation, we often drop the time dependence of the matrices in (1). The presented synthesis conditions can be easily extended to systems with additional discrete and distributed delays.

The uncertainties are assumed to satisfy the following:

Assumption 1. *The admissible uncertainties are bounded and of the form*

$$\Delta A(t)=U_A\Delta A_0(t)V_A,\quad\forall t\in\mathbb{R}\tag{2}$$
$$\Delta B_i(t)=U_{B_i}\Delta B_{i,0}(t)V_{B_i},\quad\forall t\in\mathbb{R}\tag{3}$$
$$\Delta C(t)=U_C\Delta C_0(t)V_C,\quad\forall t\in\mathbb{R}\tag{4}$$
$$\Delta D_i(t)=U_{D_i}\Delta D_{i,0}(t)V_{D_i},\quad\forall t\in\mathbb{R}\tag{5}$$
$$\Delta F(t,\theta)=U_F\Delta F_0(t,\theta)V_F,\quad\forall t\in\mathbb{R},\theta\in\Omega,\tag{6}$$

where $U_A, U_{B_i}, U_C, U_{D_i}, U_F$ *and* $V_A, V_{B_i}, V_C, V_{D_i}, V_F$ *are known constant matrices of appropriate dimensions. Without loss of generality, we assume moreover that the induced matrix 2-norm of* $\Delta A_0(t), \Delta B_{i,0}(t), \Delta C_0(t), \Delta D_{i,0}(t)$ *and* $\Delta F_0(t,\theta)$ *satisfy* $\|\Delta A_0(t)\| \leq 1$, $\|\Delta B_{i,0}(t)\| \leq 1$, $\|\Delta C_0(t)\| \leq 1$, $\|\Delta D_{i,0}(t)\| \leq 1$ *for all* $t \in \mathbb{R}$ *and* $\|\Delta F_0(t,\theta)\| \leq 1$ *for all* $t \in \mathbb{R}$ *and* $\theta \in \Omega$, *respectively.*

In addition, we assume that the known part of the kernel F is a matrix of rational functions in θ. This is particularly important in order to obtain LMI conditions for the controller synthesis. More formally, we require the following:

Assumption 2. *The matrix function* F *can be written as a* linear fractional representation *(LFR)*

$$F(\theta) = D_F + C_F(I - \theta A_F)^{-1}\theta B_F, \tag{7}$$

with $A_F \in \mathbb{R}^{n_F \times n_F}$ *and* B_F, C_F, *and* D_F *of appropriate dimensions for some positive integer* n_F.

Note that there exist different LFRs of $F(\theta)$. Here, we assume that LFR (7) is minimal in the sense that there are no pole-zero-cancellations. Since F does not have poles on Ω, we know that $\det(I - \theta A_F) \neq 0$ for all $\theta \in \Omega$, i.e., the inverse in (7) exists. Assumption 2 is not restrictive for F. It is a well-known fact that any continuous function on a closed and bounded interval can be approximated by a polynomial of sufficiently high order. Clearly, the set of polynomial functions is a subset of the set of rational functions. Given a non-rational delay kernel F^*, the best rational approximation is given by the Padé approximation, see e.g. [1]. The error of this approximation can be included in the uncertainty ΔF. Note that it is not assumed that F is piecewise constant nor $F(\theta) \geq 0$, $\forall \theta \in \Omega$, nor $\|F(\theta)\| = 1$ as in other publications, e.g. [10, 29, 2].

In order to illustrate Assumption 2, we consider

$$\int_0^r F(\theta)x(t-\theta)d\theta \tag{8}$$

as a finite convolution between the "input" $x(t) \in \mathbb{R}^n$ and the truncated multi-input multi-output (MIMO) "impulse response" $F : \Omega \to \mathbb{R}^{n \times n}$ of some linear system. However, this impulse response is not, as usual, a sum of exponential and sinusoidal functions but a matrix of rational functions, i.e., F takes the form of a MIMO "transfer function" with real variable θ instead of a complex frequency s. This transfer function can be represented by the LFR (7). It is important to note that A_F, B_F, C_F, and D_F have nothing in common with the aforementioned linear system with impulse response F. These matrices can be chosen such that (7) is satisfied.

After describing the considered uncertain distributed delay system, we can not turn to the synthesis problems we are looking at in this contribution. Our goal is to design a static state feedback controller

$$u(t) = Kx(t) \tag{9}$$

that solves the following synthesis problems:

Problem 1 (Stabilization). *Given system* (1) *with* $w \equiv 0$ *that satisfies Assumption 1 and 2, find a state-feedback controller* (9) *that stabilizes the closed loop system for all admissible uncertainties.*

Problem 2 (**H_∞ Control**). *Given system* (1) *that satisfies Assumption 1 and 2, find a stabilizing controller* (9) *that minimizes an upper bound $\gamma > 0$ of the $\mathcal{L}_2$-gain of the closed loop system, i.e.,*

$$\gamma > \sup_{0<\|w\|_2<\infty} \frac{\|z\|_2}{\|w\|_2}, \tag{10}$$

for all admissible uncertainties and all $w \in \mathcal{L}_2^{q_2}[0,\infty)$.

>From [35, 30] we know that (10) is equivalent to system (1) with controller (9) being strictly dissipative with respect to the supply function $s(w,z) = \gamma^2 w^T w - z^T z$, i.e., there exists a storage function V such that

$$\dot{V}(x(t)) + \frac{1}{\gamma} z^T(t)z(t) - \gamma w^T(t)w(t) < 0, \tag{11}$$

for all t and all $w \in \mathcal{L}_2^{q_2}[0,\infty)$. Solutions to these controller synthesis problems are given in Theorem 1 and 2, respectively.

2.2 Preliminaries on Parametric Matrix Inequalities

In the proof of the main results, we use the full-block S-procedure in order to transform parametric matrix inequalities into LMIs. It is used intensively for solving robust analysis and synthesis problems, e.g. [31, 33, 15]. In this subsection, we present some basic results related to this tool.

First, we define a matrix function $G(\delta) : \Delta \to \mathbb{R}^{n_1 \times n_2}$ that is rational in $\delta \in \Delta \subseteq \mathbb{R}$. Hence, G can be written as an LFR

$$G(\delta) = D_G + C_G(I - \delta A_G)^{-1}\delta B_G, \tag{12}$$

with $A_G \in \mathbb{R}^{n_G \times n_G}$ and B_G, C_G, D_G of appropriate dimensions. The LFR (12) is well-posed if $\det(I - \delta A_G) \neq 0$ for all $\delta \in \Delta$, cf. [33, 31]. This is obviously fulfilled if $G(\delta)$ has no poles for $\delta \in \Delta$ and if (12) is a minimal realization without pole-zero-cancellations.

It is possible to simplify some parametric matrix inequalities related to G using the *full-block S-procedure*:

Lemma 1 ([33, 32, 15]). *Suppose $R_p = R_p^T$, $Q_p = Q_p^T$, S_p, $G(\delta)$ according to* (12)*, and a compact set $\Delta \subseteq \mathbb{R}$ are given. Then*

$$\begin{bmatrix} I \\ G(\delta) \end{bmatrix}^T \begin{bmatrix} Q_p & S_p \\ S_p^T & R_p \end{bmatrix} \begin{bmatrix} I \\ G(\delta) \end{bmatrix} \prec 0, \quad \forall \delta \in \Delta, \tag{13}$$

if and only if there exist matrices $Q, R, S \in \mathbb{R}^{n_G \times n_G}$ with $Q = Q^T$, $R = R^T$, satisfying

$$\begin{bmatrix} \delta I \\ I \end{bmatrix}^T \begin{bmatrix} Q & S \\ S^T & R \end{bmatrix} \begin{bmatrix} \delta I \\ I \end{bmatrix} \succeq 0, \quad \forall \delta \in \Delta, \quad \textit{and} \tag{14}$$

$$\begin{bmatrix} I & 0 \\ A_G & B_G \end{bmatrix}^T \begin{bmatrix} Q & S \\ S^T & R \end{bmatrix} \begin{bmatrix} I & 0 \\ A_G & B_G \end{bmatrix} + \begin{bmatrix} 0 & I \\ C_G & D_G \end{bmatrix}^T \begin{bmatrix} Q_p & S_p \\ S_p^T & R_p \end{bmatrix} \begin{bmatrix} 0 & I \\ C_G & D_G \end{bmatrix} \prec 0. \tag{15}$$

Clearly, we still have a parametric matrix inequality (14). However, using the convex hull relaxation from [33, 31], it is possible to transform this into a finite set of non-parametric matrix inequality.

Lemma 2 ([33, 31]). *Suppose that $Q \prec 0$ and Δ is the convex hull of two points $\delta_1,\ \delta_2 \in \mathbb{R}$ with $\delta_1 < \delta_2$, i.e., $\Delta = Co(\{\delta_1,\ \delta_2\}) = [\delta_1,\ \delta_2]$. Then*

$$\begin{bmatrix} \delta I \\ I \end{bmatrix}^T \begin{bmatrix} Q & S \\ S^T & R \end{bmatrix} \begin{bmatrix} \delta I \\ I \end{bmatrix} \succeq 0, \quad \forall \delta \in \Delta \tag{16}$$

if and only if

$$\begin{bmatrix} \delta_i I \\ I \end{bmatrix}^T \begin{bmatrix} Q & S \\ S^T & R \end{bmatrix} \begin{bmatrix} \delta_i I \\ I \end{bmatrix} \succeq 0, \quad i = 1, 2\,. \tag{17}$$

Summarizing, the parametric LMI (13) can be replaced by (15) and (17). The only introduced conservatism is $Q \prec 0$.

2.3 Preliminaries on Uncertain LMIs

In order to deal with the model uncertainties, we use the following result:

Lemma 3 ([34]). *Let U, V, W, and Z be real matrices of appropriate dimensions with Z satisfying $\|Z\| \leq 1$, where $\|\cdot\|$ is the induced matrix 2-norm. Then, we have the following:*

(1) For any real number $\epsilon > 0$, $UZV + (UZV)^T \preceq \epsilon UU^T + \epsilon^{-1}V^TV$.
(2) For any matrix $P \succ 0$ and scalar $\epsilon > 0$ such that $\epsilon I - VPV^T \succ 0$, we have

$$\begin{aligned}(W+UZV)P(W+UZV)^T \preceq\ & WPW^T \\ & + WPV^T(\epsilon I - VPV^T)^{-1}VPW^T + \epsilon UU^T.\end{aligned}$$

(3) For any matrix $P \succ 0$ and scalar $\epsilon > 0$ such that $P - \epsilon UU^T \succ 0$, we have

$$(W+UZV)^T P^{-1}(W+UZV) \preceq W^T(P-\epsilon UU^T)^{-1}W + \epsilon^{-1}V^TV.$$

3 Robust Controller Design

Now, we are ready to present the main results of this contribution, namely the solutions to Problem 1 and 2 in Subsection 3.1 and 3.2, respectively.

3.1 Robust Stabilization of Uncertain Distributed Delay Systems

Problem 1 can be solved using the LMIs in the next theorem.

Theorem 1. *Consider system (1) with $w \equiv 0$ where the uncertainties satisfy Assumption 1 and F satisfies Assumption 2. The static feedback controller (9) solves Problem 1 for all admissible uncertainties if there exist real $\epsilon_1 > 0, \epsilon_2 > 0, \epsilon_3 > 0$ and matrices $\overline{K}, Q_1, \overline{P}_2, \overline{P}_3, R, Q, S$ where Q_1, $\overline{P}_2$, $\overline{P}_3$, R, Q are symmetric and $Q_1 \succ 0$, $\overline{P}_2 \succ 0$, $R \succeq 0$ and $Q \prec 0$ such that*

$$\begin{bmatrix} \Psi_1 & Q_1 V_A^T & \overline{K}^T V_{B_1}^T \\ * & -\epsilon_1 I & 0 \\ * & * & -\epsilon_2 I \end{bmatrix} \prec 0 \tag{18}$$

$$r^2 Q + r(S + S^T) + R \succeq 0 \tag{19}$$

$$\begin{bmatrix} \Psi_2 & B_F Q_1 & SC_F^T + A_F RC_F^T & 0 \\ * & -\overline{P}_2 & Q_1 D_F^T & Q_1 V_F^T \\ * & * & \overline{P}_2 - \overline{P}_3 + \epsilon_3 U_F U_F^T + C_F RC_F^T & 0 \\ * & * & * & -\epsilon_3 I \end{bmatrix} \prec 0\,, \tag{20}$$

is feasible, where $\Psi_1 = AQ_1 + Q_1 A^T + B_1 \overline{K} + \overline{K}^T B_1^T + r\overline{P}_3 + \epsilon_1 U_A U_A^T + \epsilon_2 U_{B_1} U_{B_1}^T$ *and* $\Psi_2 = Q + A_F S^T + SA_F^T + A_F RA_F^T$. *The corresponding controller gain is* $K = \overline{K} Q_1^{-1}$.

Proof. Consider the following Lyapunov-Krasovskii functional candidate, cf. [11], with $P_1 \succ 0$, $P_2 \succ 0$:

$$V(x_t) = x^T P_1 x + \int_0^r \int_{t-\theta}^t x^T(\xi) P_2 x(\xi) d\xi d\theta\,. \tag{21}$$

The derivative of V along solutions of (1) with controller (9) is

$$\begin{aligned} \dot{V}(x_t) &= x^T(t)((\tilde{A} + \tilde{B}_1 K)^T P_1 + P_1(\tilde{A} + \tilde{B}_1 K) + rP_3)x(t) \\ &+ \int_0^r \begin{bmatrix} x(t) \\ x(t-\theta) \end{bmatrix}^T \underbrace{\begin{bmatrix} P_2 - P_3 & P_1 \tilde{F}(\theta) \\ \tilde{F}^T(\theta) P_1 & -P_2 \end{bmatrix}}_{M(\theta)} \begin{bmatrix} x(t) \\ x(t-\theta) \end{bmatrix} d\theta, \end{aligned} \tag{22}$$

where $P_3 = P_3^T \in \mathbb{R}^{n \times n}$. Clearly, $\dot{V} < 0$, i.e., the controller (9) stabilizes the system, if

$$(\tilde{A} + \tilde{B}_1 K)^T P_1 + P_1(\tilde{A} + \tilde{B}_1 K) + rP_3 \prec 0, \tag{23}$$

$$M(\theta) \prec 0\,, \quad \forall \theta \in \Omega\,. \tag{24}$$

The remainder of the proof shows that LMI (18) to (20) imply (23) and (24) for all admissible uncertainties.

First, we show that (18) implies (23). We apply the Schur lemma [3] to (18) and pre- and post-multiply $P_1 = Q_1^{-1}$. With $\overline{K} = KQ_1$ and $\overline{P}_3 = Q_1 P_3 Q_1$, we obtain $\Psi_3 \prec 0$ where $\Psi_3 = (A^T + K^T B_1^T)P_1 + P_1(A + B_1 K) + rP_3 + \epsilon_1 P_1 U_A U_A^T P_1 + \epsilon_1^{-1} V_A^T V_A + \epsilon_2 P_1 U_{B_1} U_{B_1}^T P_1 + \epsilon_2^{-1} K^T V_{B_1}^T V_{B_1} K$. Now, we apply Lemma 3 to (23) and see that

$$(\tilde{A}^T + K^T \tilde{B}_1^T)P_1 + P_1(\tilde{A} + \tilde{B}_1 K) + rP_3 \preceq \Psi_3 \prec 0,$$

for any $\epsilon_1 > 0$ and $\epsilon_2 > 0$, i.e., (18) implies (23).

Next, we show that

$$\begin{bmatrix} -\overline{P}_2 & Q_1 F(\theta)^T & Q_1 V_F^T \\ * & \overline{P}_2 - \overline{P}_3 + \epsilon_3 U_F U_F^T & 0 \\ * & * & -\epsilon_3 I \end{bmatrix} \prec 0. \tag{25}$$

implies (24) for all admissible uncertainties. Therefore, we apply the Schur lemma first on $-\epsilon_3 I$ and then on $-\overline{P}_2 + \epsilon_3^{-1} Q_1 V_F^T V_F Q_1$ and obtain $\Psi_4(\theta) \prec 0$ where $\Psi_4(\theta) = \overline{P}_2 - \overline{P}_3 +$

$\epsilon_3 U_F U_F^T + F(\theta)Q_1(\overline{P}_2 - \epsilon_3^{-1} Q_1 V_F^T V_F Q_1)^{-1} Q_1 F^T(\theta)$. Then, we pre- and post-multiply M with diag(Q_1, Q_1) and apply the Schur lemma. With Lemma 3, we have

$$\overline{P}_2 - \overline{P}_3 + \tilde{F}(\theta) Q_1 \overline{P}_2^{-1} Q_1 \tilde{F}^T(\theta) \preceq \Psi_4(\theta) \prec 0,$$

for any $\epsilon_3 > 0$, where $\overline{P}_2 = Q_1 P_2 Q_1$, i.e., (25) implies (24).

Finally, we transform the parametric matrix inequality (25) into an LMI using the full-block S procedure. Therefore, (25) is formulated as follows

$$\begin{bmatrix} I & 0 & 0 \\ 0 & I & 0 \\ 0 & 0 & I \\ 0 & F^T(\theta) & 0 \end{bmatrix}^T \underbrace{\begin{bmatrix} -\overline{P}_2 & 0 & Q_1 V_F^T & Q_1 \\ 0 & \overline{P}_2 - \overline{P}_3 + \epsilon_3 U_F U_F^T & 0 & 0 \\ V_F Q_1 & 0 & -\epsilon_3 I & 0 \\ Q_1 & 0 & 0 & 0 \end{bmatrix}}_{\tilde{M}} \begin{bmatrix} I & 0 & 0 \\ 0 & I & 0 \\ 0 & 0 & I \\ 0 & F^T(\theta) & 0 \end{bmatrix} \prec 0 .$$

With Assumption 2 and the full-block S-procedure (Lemma 1), we see that this inequality is true for all $\theta \in \Omega$ if and only if there exist $Q = Q^T$, $R = R^T$ and S satisfying $\theta^2 Q + \theta(S + S^T) + R \succeq 0$ for all $\theta \in \Omega$ and

$$\begin{bmatrix} I & 0 & 0 & 0 \\ A_F^T & 0 & C_F^T & 0 \end{bmatrix}^T \begin{bmatrix} Q & S \\ S^T & R \end{bmatrix} \begin{bmatrix} I & 0 & 0 & 0 \\ A_F^T & 0 & C_F^T & 0 \end{bmatrix} + \begin{bmatrix} 0 & I & 0 & 0 \\ 0 & 0 & I & 0 \\ 0 & 0 & 0 & I \\ B_F^T & 0 & D_F^T & 0 \end{bmatrix}^T \tilde{M} \begin{bmatrix} 0 & I & 0 & 0 \\ 0 & 0 & I & 0 \\ 0 & 0 & 0 & I \\ B_F^T & 0 & D_F^T & 0 \end{bmatrix} \prec 0.$$

We can simplify the last LMI and obtain (20). Moreover, we apply the convex hull relaxation (Lemma 2) to $\theta^2 Q + \theta(S + S^T) + R \succeq 0$ with $\delta_1 = 0$ and $\delta_2 = rI$. The result is $R \succeq 0$ and (19). □

3.2 Robust H_∞ Control of Uncertain Distributed Delay Systems

Now, we turn to Problem 2 that can be solved using the LMIs in the following theorem.

Theorem 2. *Consider system* (1) *where the uncertainties satisfy Assumption 1 and F satisfies Assumption 2. The static feedback controller* (9) *solves Problem 2 for all admissible uncertainties if there exist real $\epsilon_1 > 0$, $\epsilon_2 > 0$ and matrices $\overline{K}, Q_1, \overline{P}_2, \overline{P}_3, R, Q, S$ where Q_1, $\overline{P}_2$, $\overline{P}_3$, R, Q are symmetric and $Q_1 \succ 0$, $\overline{P}_2 \succ 0$, $R \succeq 0$, $Q \prec 0$ such that the minimization problem*

$$\min \gamma \tag{26}$$

subject to

$$\left[\begin{array}{cc|ccc|c|ccc} \Psi_5 & B_2 & Q_1 V_A^T & \overline{K}^T V_{B_1}^T & 0 & Q_1 C^T + \overline{K}^T D_1^T & Q_1 V_C^T & \overline{K}^T V_{D_1}^T & 0 \\ B_2^T & -\gamma I & 0 & 0 & V_{B_2}^T & D_2^T & 0 & 0 & V_{D_2}^T \\ \hline * & * & -\epsilon_1 I & 0 & 0 & 0 & 0 & 0 & 0 \\ * & * & 0 & -\epsilon_1 I & 0 & 0 & 0 & 0 & 0 \\ * & * & 0 & 0 & -\epsilon_1 I & 0 & 0 & 0 & 0 \\ \hline * & * & * & * & * & -\gamma I + \epsilon_2 \Sigma_{U_{CD}} & 0 & 0 & 0 \\ \hline * & * & * & * & * & * & -\epsilon_2 I & 0 & 0 \\ * & * & * & * & * & * & 0 & -\epsilon_2 I & 0 \\ * & * & * & * & * & * & 0 & 0 & -\epsilon_2 I \end{array}\right] \prec 0, \tag{27}$$

as well as (19) *and* (20) *is feasible, where* $\Psi_5 = Q_1 A^T + \overline{K}^T B_1^T + AQ_1 + B_1\overline{K} + r\overline{P}_3 + \epsilon_1 \Sigma_{U_{AB}}$ *with* $\Sigma_{U_{AB}} = U_A U_A^T + U_{B_1} U_{B_1}^T + U_{B_2} U_{B_2}^T$ *and* $\Sigma_{U_{CD}} = U_C U_C^T + U_{D_1} U_{D_1}^T + U_{D_2} U_{D_2}^T$. *The corresponding controller gain is* $K = \overline{K} Q_1^{-1}$.

Proof. Consider the Lyapunov-Krasovskii functional candidate (21) as a storage function. Then, this storage function has to satisfy (11), i.e.,

$$\dot{V}(x_t) + \frac{1}{\gamma} z^T(t) z(t) - \gamma w^T(t) w(t) = \int_0^r \begin{bmatrix} x(t) \\ x(t-\theta) \end{bmatrix}^T M(\theta) \begin{bmatrix} x(t) \\ x(t-\theta) \end{bmatrix} d\theta$$
$$+ \begin{bmatrix} x(t) \\ w(t) \end{bmatrix}^T \underbrace{\begin{bmatrix} \Psi_6 & P_1 \tilde{B}_2 + \frac{1}{\gamma}\left(\tilde{C}^T + K^T \tilde{D}_1^T\right)\tilde{D}_2 \\ * & -\gamma I + \frac{1}{\gamma}\tilde{D}_2^T \tilde{D}_2 \end{bmatrix}}_{N} \begin{bmatrix} x(t) \\ w(t) \end{bmatrix} < 0, \quad (28)$$

with M given in (22) and $\Psi_6 = P_1(\tilde{A} + \tilde{B}_1 K) + (\tilde{A}^T + K^T \tilde{B}_1^T) P_1 + rP_3 + \frac{1}{\gamma}(\tilde{C}^T + K^T \tilde{D}_1^T)(\tilde{C} + \tilde{D}_1 K)$. Obviously, (11) is fulfilled if $N \prec 0$ and $M(\theta) \prec 0$ for all $\theta \in \Omega$. $M(\theta) \prec 0$ can be transformed into (19) and (20) as in the proof of Theorem 1. It remains to show that (27) implies $N \prec 0$.

First, we pre- and post-multiply (27) with $\operatorname{diag}(P_1, I, I, I, I, I, I, I, I)$, where $P_1 = Q_1^{-1}$, $\overline{K} = KQ_1$ and $\overline{P}_3 = Q_1 P_3 Q_1$. Then, we apply the Schur lemma on the upper-left block and obtain $\Psi_8 + \Psi_9 \prec 0$ with

$$\Psi_8 = \begin{bmatrix} \Psi_7 & P_1 B_2 \\ * & -\gamma I \end{bmatrix} + \epsilon_1 \begin{bmatrix} P_1 \Sigma_{U_{AB}} P_1 & 0 \\ 0 & 0 \end{bmatrix} + \frac{1}{\epsilon_1} \begin{bmatrix} V_A^T & K^T V_{B_1}^T & 0 \\ 0 & 0 & V_{B_2}^T \end{bmatrix} \begin{bmatrix} V_A & 0 \\ V_{B_1} K & 0 \\ 0 & V_{B_2} \end{bmatrix}$$

$$\Psi_9 = \left[\begin{array}{c|ccc} \overline{C}^T & V_C^T & K^T V_{D_1}^T & 0 \\ D_2^T & 0 & 0 & V_{D_2}^T \end{array}\right] \left[\begin{array}{c|c} (\gamma I - \epsilon_2 \Sigma_{U_{CD}})^{-1} & 0 \\ \hline 0 & \epsilon_2^{-1} I \end{array}\right] \left[\begin{array}{cc} \overline{C} & D_2 \\ \hline V_C & 0 \\ V_{D_1} K & 0 \\ 0 & V_{D_2} \end{array}\right],$$

where $\Psi_7 = (A + B_1 K)^T P_1 + P_1 (A + B_1 K) + rP_3$ and $\overline{C} = C + D_1 K$.

Then, we separate N as $N = N_1 + N_2$ with

$$N_1 = \begin{bmatrix} \Psi_{10} & P_1 \tilde{B}_2 \\ * & -\gamma I \end{bmatrix}, \qquad N_2 = \frac{1}{\gamma} \begin{bmatrix} \tilde{C}^T + K^T \tilde{D}_1^T \\ \tilde{D}_2^T \end{bmatrix} \begin{bmatrix} \tilde{C} + \tilde{D}_1 K & \tilde{D}_2 \end{bmatrix},$$

with $\Psi_{10} = P_1(\tilde{A} + \tilde{B}_1 K) + (\tilde{A}^T + K^T \tilde{B}_1^T) P_1 + rP_3$. Next, we rewrite N_1 as

$$N_1 = \begin{bmatrix} \Psi_7 & P_1 B_2 \\ * & -\gamma I \end{bmatrix} + P_1 \begin{bmatrix} U_A & U_{B_1} & U_{B_2} \\ 0 & 0 & 0 \end{bmatrix} \begin{bmatrix} \Delta A_0 & 0 & 0 \\ 0 & \Delta B_{1,0} & 0 \\ 0 & 0 & \Delta B_{2,0} \end{bmatrix} \begin{bmatrix} V_A & 0 \\ V_{B_1} K & 0 \\ 0 & V_{B_2} \end{bmatrix}$$
$$+ \left(P_1 \begin{bmatrix} U_A & U_{B_1} & U_{B_2} \\ 0 & 0 & 0 \end{bmatrix} \begin{bmatrix} \Delta A_0 & 0 & 0 \\ 0 & \Delta B_{1,0} & 0 \\ 0 & 0 & \Delta B_{2,0} \end{bmatrix} \begin{bmatrix} V_A & 0 \\ V_{B_1} K & 0 \\ 0 & V_{B_2} \end{bmatrix} \right)^T.$$

We apply Lemma 3 and see that $N_1 \preceq \Psi_8$ for any $\epsilon_1 > 0$. Furthermore, note that the first part of N_2 can be written as

$$\begin{bmatrix} \tilde{C}^T + K^T \tilde{D}_1^T \\ \tilde{D}_2^T \end{bmatrix} = \begin{bmatrix} \overline{C}^T \\ D_2^T \end{bmatrix} + \begin{bmatrix} V_C^T & K^T V_{D_1}^T & 0 \\ 0 & 0 & V_{D_2}^T \end{bmatrix} \begin{bmatrix} \Delta C_0^T & 0 & 0 \\ 0 & \Delta D_{1,0} & 0 \\ 0 & 0 & \Delta D_{2,0}^T \end{bmatrix} \begin{bmatrix} U_C^T \\ U_{D_1}^T \\ U_{D_2}^T \end{bmatrix} .$$

Hence, we can apply Lemma 3 also to N_2 and obtain $N_2 \preceq \Psi_9$. Summarizing, we see that $N_1 + N_2 \preceq \Psi_8 + \Psi_9 \prec 0$, i.e., (27) implies $N \prec 0$. □

Note that the applied Lyapunov-Krasovskii functional is very simple. Nonetheless, the robust stability condition in [23] is less conservative than results from the literature. Here, we use similar techniques as in [23] but for the controller design. Hence, we expect also quite accurate results. The conservatism of the synthesis conditions could be further decreased by using more advanced Lyapunov-Krasovskii functional candidates, as those presented for example in [18, 5, 14].

4 Conclusions

We presented two LMI conditions for the robust stabilization and robust H_∞ controller design for uncertain systems with distributed delays. Both conditions are *delay-dependent* because they explicitly depend on the range r of the distributed delay as well as on the shape of the delay kernel, given by the linear fractional representation of the nominal delay kernel F in Assumption 2.

References

1. Baker, G.A.: Essentials of Padé Approximants. Academic Press, New York (1975)
2. Bernard, S., Bélair, J., Mackey, M.C.: Sufficient conditions for stability of linear differential equations with distributed delay. Discrete and Continuous Dynamical Systems – Series B 1, 233–256 (2001)
3. Boyd, S., El Ghaoui, L., Feron, E., Balakrishnan, V.: Linear matrix inequalities in systems and control theory. SIAM, Philadelphia (1994)
4. Breda, D., Maset, S., Vermiglio, R.: Pseudospectral differencing methods for characteristic roots of delay differential equations. SIAM J. Scientific Computing 27(2), 482–495 (2005)
5. Chen, W.H., Zheng, W.X.: Delay-dependent robust stabilization for uncertain neutral systems with distributed delays. Automatica 43(1), 95–104 (2007)
6. Cushing, J.M.: Integrodifferential equations and delay models in population dynamics. Springer, Heidelberg (1977)
7. Fiagbedzi, Y.A., Pearson, A.E.: A multistage reduction technique for feedback stabilizing distributed time-lag systems. Automatica 23(3), 311–326 (1987)
8. Fridman, E., Tsodik, G.: H_∞ control of systems with distributed delay: Discretized Lyapunov functional method. In: Proc. European Control Conf., Kos, Greece, pp. 229–234 (2007)
9. Gopalsamy, K.: Global asymptotic stability in volterra's population systems. J. Math. Bio. 19, 157–168 (1984)
10. Gu, K., Han, Q.L., Luo, A.C.J., Niculescu, S.I.: Discretized Lyapunov functional for systems with distributed delay and piecewise constant coefficients. Int. J. Contr. 74(7), 737–744 (2001)
11. Gu, K., Kharitonov, V.L., Chen, J.: Stability of Time-Delay Systems. Birkhäuser, Boston (2003)

12. Gu, K.: An improved stability criterion for systems with distributed delays. Int. J. Robust and Nonlinear Control 13(9), 819–831 (2003)
13. Hale, J.K., Verduyn Lunel, S.M.: Introduction to Functional Differential Equations. Springer, Heidelberg (1993)
14. Han, Q.L.: A descriptor system approach to robust stability of uncertain neutral systems with discrete and distributed delays. Automatica 40(10), 1791–1796 (2004)
15. Iwasaki, T., Shibata, G.: LPV system analysis via quadratic separator for uncertain implicit systems. IEEE Trans. Autom. Contr. 46(8), 1195–1208 (2001)
16. Kharitonov, V.L.: Lyapunov matrices for a class of time-delay systems. Systems & Control Letters 55(7), 610–617 (2006)
17. Kolmanovskii, V.B., Richard, J.P.: Stability of some linear systems with delays. IEEE Trans. Autom. Contr. 44(5), 984–989 (1999)
18. Li, X.G., Zhu, X.J.: Stability analysis of neutral systems with distributed delays. Automatica 44(8), 2197–2201 (2008)
19. MacDonald, N.: Biological delay systems: linear stability theory. Cambridge University Press, Cambridge (1989)
20. Michiels, W., Van Assche, V., Niculescu, S.I.: Stabilization of time-delay systems with a controlled time-varying delay and applications. IEEE Trans. Autom Contr. 50(2), 493–504 (2005)
21. Michiels, W., Morărescu, I.C., Niculescu, S.I.: Consensus problems for car following systems with distributed delays. In: Proc. European Control Conf., Kos, Greece, pp. 2158–2165 (2007)
22. Morărescu, I.C., Niculescu, S.I., Gu, K.: Stability crossing curves of shifted gamma-distributed delay systems. SIAM J. Applied Dynamical Systems 6(2), 475–493 (2007)
23. Münz, U., Rieber, J.M., Allgöwer, F.: Robust stability of distributed delay systems. In: Proc. 17th IFAC World Congress, Seoul, South Korea, pp. 12354–12358 (2008)
24. Münz, U., Allgöwer, F.: $\mathcal{L}_2$-gain based controller design for linear systems with distributed delays and rational delay kernels. In: Proc. 7th IFAC Workshop on Time-Delay Systems, Nantes, France (2007)
25. Niculescu, S.I.: Delay effects on stability: A robust control approach. Springer, Heidelberg (2001)
26. Niculescu, S.I., Gu, K. (eds.): Advances in time-delay systems. Springer, Heidelberg (2004)
27. Richard, J.P.: Time-delay systems: An overview of some recent advances and open problems. Automatica 39(10), 1667–1694 (2003)
28. Roesch, O., Roth, H., Niculescu, S.I.: Remote control of mechatronic systems over communication networks. In: Proc. Int. Conf. on Mechatronics and Automation, Niagara Falls, Canada, pp. 1648–1653 (2005)
29. Santos, O., Mondié, S., Kharitonov, V.L.: Robust stability conditions for systems with distributed delays. In: Proc. IEEE Conf. Decision and Control, San Diego, USA, pp. 217–222 (2006)
30. Scherer, C.W., Weiland, S.: Linear matrix inequalities in control (2005), `http://www.cs.ele.tue.nl/sweiland/lmi.html`
31. Scherer, C.W.: Robust mixed control and linear parameter-varying control with full block multipliers. In: EL Ghaoui, L., Niculescu, S.I. (eds.) Recent Advances on LMI Methods in Control. SIAM, Philadelphia (2000)
32. Scherer, C.W.: LPV control and full block multipliers. Automatica 37(3), 361–375 (2001)
33. Scherer, C.W.: LMI relaxations in robust control. Europ. J. Contr. 12(1), 3–29 (2006)

34. de Souza, C.E., Li, X.: Delay-dependent robust H-infinity control of uncertain linear state-delayed systems. Automatica 35(7), 1313–1321 (1999)
35. van der Schaft, A.: L_2-gain and passivity techniques in nonlinear control. Springer, Heidelberg (1996)
36. Verriest, E.I.: Linear systems with rational distributed delay: Reduction and stability. In: Proc. European Control Conf., Karlsruhe, Germany (1999)
37. Xie, L., Fridman, E., Shaked, U.: Robust H_∞ control of distributed delay systems with application to combustion control. IEEE Trans. Autom. Contr. 46(12), 1930–1935 (2001)
38. Yue, D., Han, Q.L.: Robust H_∞ filter design of uncertain descriptor systems with discrete and distributed delays. In: Proc. IEEE Conf. Decision and Control, Maui, USA, pp. 610–615 (2003)
39. Zheng, F., Cheng, M., Gao, W.B.: Feedback stabilization of linear systems with distributed delays in state and control variables. IEEE Trans. Autom. Contr. 39(8), 1714–1718 (1994)
40. Zheng, F., Frank, P.M.: Finite dimensional variable structure control design for distributed delay systems. Int. J. Contr. 74(4), 398–408 (2001)
41. Zheng, F., Frank, P.M.: Robust control of uncertain distributed delay systems with application to the stabilization of combustion in rocket motor chambers. Automatica 38(2), 487–497 (2002)

Observer Design for Systems with Non Small and Unknown Time-Varying Delay

Alexandre Seuret[1], Thierry Floquet[1,3], Jean-Pierre Richard[1,3] and Sarah Spurgeon[2,*]

[1] LAGIS UMR CNRS 8146, Ecole Centrale de Lille, BP 48, Cité Scientifique, 59651 Villeneuve-d'Ascq, France,
seuret.alexandre@ec-lille.fr, Thierry.Floquet@ec-lille.fr, jean-pierre.richard@ec-lille.fr

[2] Department of Electronics, University of Kent, UK, S.K.Spurgeon@kent.ac.uk

[3] Équipe Projet ALIEN, Centre de Recherche INRIA Lille-Nord Europe, France.

Summary. This paper deals with the design of observers for linear systems with unknown, time-varying, but bounded delays (on the state and on the input). In this work, the problem is solved for a class of systems by combining the unknown input observer approach with an adequate choice of a Lyapunov-Krasovskii functional for non small delay systems. This result provides workable conditions in terms of rank assumptions and LMI conditions. The dynamic properties of the observer are also analyzed. A 4th-order example is used to demonstrate the feasibility of the proposed solution.

1 Introduction

State observation is an important issue for both linear and nonlinear systems. This work considers the observation problem for the case of linear systems with non small and unknown delay. Several authors proposed observers for delay systems (see, e.g., [15, 16]). Most of the literature, as witnessed in [15], considers that the value of the mainly constant delay can be used in the observer realization. This means that the delay is known or measured. Likewise, what are defined as "observers without internal delay" [3, 4, 7] involve output knowledge both in the present and at delayed instants.

There are currently very few results in which the observer does not assume knowledge of the delay [2, 5, 11, 18, 19]. These interesting approaches consider linear systems and guarantee an H_∞ performance. They are based on stability techniques which are delay independent and lead to the minimization of the state observation error. It is interesting to reduce the probable conservatism of such results by taking into account information on a delay upper-bound and hence derive an asymptotically stable observer.

In [18], the authors design an observer using a computational delay which can be assimilated to estimation of the delay $\hat{h}$. The conditions which guarantee the convergence of the

* Partially supported by EPSRC grant reference EP/E020763/1, entitled "Robust Output Feedback Sliding Mode Control for Time-delay Systems"

J.J. Loiseau et al. (Eds.): Topics in Time Delay Systems, LNCIS 388, pp. 233–242.
springerlink.com

error dynamics developed for this observer do not take into account the value of $\hat{h}$. This means that the estimate of the state is guaranteed whatever the delay estimate, $\hat{h}$. The property is not related to the conservatism of the conditions. The errors between the real and computational delays are controlled by the discontinuous sliding function. The greater the error between the real and computational delays, the greater the gain of the discontinuous function will be. It is thus straightforward to conclude that a worse estimate of the delay can lead to a high gain in the sliding injection.

In this paper, another method is proposed to solve the problem of the observation of linear systems with unknown time-varying delays which are assumed to be "non small" i.e. the delay function lies in an interval excluding 0. The result is based on a combination of results on sliding mode observers (see, e.g., [1, 6, 8, 9, 14]) with an appropriate choice of a Lyapunov-Krasovskii functional. For sake of simplicity, the unknown time delay $\tau(t)$ is assumed to be the same for the state and the input. In order to reduce the conservatism of the developed conditions, the existence of known real numbers d, τ_1 and τ_2 is assumed such that $\forall t \in \mathbf{R}_+$:

$$\begin{array}{c} \tau_1 \leq \tau(t) \leq \tau_2 \\ \dot{\tau}(t) \leq d < 1. \end{array} \tag{1}$$

Here the delay used in the observer is the average of the delay $(\tau_2 + \tau_1)/2$. Then the design of the observer does not require the definition or the computation of a delay estimate and the stability conditions only depend on the parameters of the studied system.

Throughout the article, the notation $P > 0$ for $P \in \mathbf{R}^{n \times n}$ means that P is a symmetric and positive definite matrix. $[A_1|A_2|...|A_n]$ is the concatenated matrix with matrices A_i. I_n represents the $n \times n$ identity matrix.

2 Problem statement

Consider the linear time-invariant system with state and input delay:

$$\left\{ \begin{array}{l} \dot{x}(t) = Ax(t) + A_\tau x(t - \tau(t)) + Bu(t) + B_\tau u(t - \tau(t)) + D\zeta(t) \\ y(t) = Cx(t) \\ x(s) = \phi(s), \quad \forall\, s \in [-\tau_2, 0] \end{array} \right. \tag{2}$$

where $x \in \mathbf{R}^n$, $u \in \mathbf{R}^m$ and $y \in \mathbf{R}^q$ are the state vector, the input vector and the measurement vector, respectively. $\zeta \in \mathbf{R}^r$ is an unknown and bounded perturbation that satisfies:

$$\|\zeta(t)\| \leq \alpha_1(t, y, u), \tag{3}$$

where α_1 is a known scalar function. $\phi \in C^0([-\tau_2, 0], \mathbf{R}^n)$ is the vector of initial conditions. It is assumed that A, A_τ, B, B_τ, C and D are constant known matrices of appropriate dimensions. The following structural assumptions are required for the design of the observer:

A1. $\mathrm{rank}(C[A_\tau|B_\tau|D]) = \mathrm{rank}([A_\tau|B_\tau|D]) \triangleq p$,
A2. $p < q \leq n$,
A3. The invariant zeros of $(A, [A_\tau|B_\tau|D], C)$ lie in $\mathbb{C}^-$.

Under these assumptions and using the same linear change of coordinates as in [6], Chapter 6, the system can be transformed into:

$$\begin{cases} \dot{x}_1(t) = A_{11}x_1(t) + A_{12}x_2(t) + B_1u(t) \\ \dot{x}_2(t) = A_{21}x_1(t) + A_{22}x_2(t) + B_2u(t) + D_1\zeta(t) \\ \qquad +G_1x_1(t-\tau(t)) + G_2x_2(t-\tau(t)) + G_uu(t-\tau(t)) \\ y(t) = Tx_2(t) \end{cases} \tag{4}$$

where $x_1 \in \mathbf{R}^{n-q}$, $x_2 \in \mathbf{R}^{q}$ and where G_1, G_2, G_u, D_1 and A_{21} are defined by:

$$G_1 = \begin{bmatrix} 0 \\ \bar{G}_1 \end{bmatrix}, G_2 = \begin{bmatrix} 0 \\ \bar{G}_2 \end{bmatrix}, G_u = \begin{bmatrix} 0 \\ \bar{G}_u \end{bmatrix},$$
$$D_1 = \begin{bmatrix} 0 \\ \bar{D}_1 \end{bmatrix}, \ A_{21} = \begin{bmatrix} A_{211} \\ A_{212} \end{bmatrix},$$

with $\bar{G}_1 \in \mathbf{R}^{p\times(n-q)}$, $\bar{G}_2 \in \mathbf{R}^{p\times q}$, $\bar{G}_u \in \mathbf{R}^{p\times m}$, $\bar{D}_1 \in \mathbf{R}^{p\times r}$, $A_{211} \in \mathbf{R}^{(q-p)\times(n-q)}$, $A_{212} \in \mathbf{R}^{p\times(n-q)}$ and T an orthogonal matrix involved in the change of coordinates given in [6].

Under these conditions, the system can be decomposed into two subsystems. $A1$ implies that the unmeasurable state x_1 is not affected by the delayed terms and the perturbations. $A3$ ensures that the pair (A_{11}, A_{211}) is at least detectable.

In this article, the following lemma will be used:

Lemma 1. *[12] For any matrices A, $P_0 > 0$ and $P_1 > 0$, the inequality*

$$A^T P_1 A - P_0 < 0,$$

is equivalent to the existence of a matrix Y such that:

$$\begin{bmatrix} -P_0 & A^T Y^T \\ YA & -Y - Y^T + P_1 \end{bmatrix} < 0.$$

3 Observer design

Define the following sliding mode observer:

$$\begin{cases} \dot{\hat{x}}_1(t) = A_{11}\hat{x}_1(t) + A_{12}x_2(t) + B_1u(t) \\ \qquad +(LT^TG_lT - A_{11}L)(x_2(t) - \hat{x}_2(t)) + LT^T\nu(t) \\ \dot{\hat{x}}_2(t) = A_{21}\hat{x}_1(t) + A_{22}x_2(t) + B_2u(t) \\ \qquad +G_1\hat{x}_1(t-h) + G_2x_2(t-h) + G_uu(t-h) \\ \qquad -T^T\nu(t) - (A_{21}L + T^TG_lT)(x_2(t) - \hat{x}_2(t)) \\ \hat{y}(t) = T\hat{x}_2(t) \end{cases} \tag{5}$$

where the linear gain G_l is a Hurwitz matrix and L has the form $\begin{bmatrix} \bar{L} & 0 \end{bmatrix}$ with $\bar{L} \in \mathbf{R}^{(n-q)\times(q-p)}$. The computed delay $h = (\tau_2 + \tau_1)/2$ is an implemented value that is chosen according to the delay definition. It corresponds to the delay average. The discontinuous injection term ν is given by:

$$\nu(t) = \begin{cases} -\rho(t,y,u)\frac{P_2(y(t)-\hat{y}(t))}{\|P_2(y(t)-\hat{y}(t))\|}, \text{if } \ y(t) - \hat{y}(t) \neq 0 \\ 0, \text{otherwise.} \end{cases} \tag{6}$$

where $P_2 > 0$, $P_2 \in \mathbf{R}^{p\times p}$ and where ρ is a nonlinear positive gain yet to be defined. Note that the non delayed terms depending on x_2 are known because $x_2(t) = T^T y(t)$. Define $\mu = (\tau_2 - \tau_1)/2$.

Remark 1. Compared to [18], this observer does not required an artificial delay $\hat{h}$. It only needs to have an average value of the delay. Contrary to the observer proposed in [18], the implemented delay will appear in the conditions which guarantee stability.

Defining the state estimation errors as $e_1 = x_1(t) - \hat{x}_1(t)$ and $e_2 = x_2(t) - \hat{x}_2(t)$, one obtains:

$$\begin{cases} \dot{e}_1(t) = A_{11}e_1(t) - LT^T\left(G_lTe_2(t) + \nu(t)\right) + A_{11}Le_2(t) \\ \dot{e}_2(t) = A_{21}e_1(t) + G_1e_1(t-\tau(t)) + \xi(t) + D_1\zeta(t) \\ \qquad\quad +(T^TG_lT + A_{21}L)e_2(t) + T^T\nu \end{cases} \tag{7}$$

where $\xi : \mathbf{R} \longmapsto \mathbf{R}^p$ is given by:

$$\begin{aligned} \xi(t) = {} & G_1(\hat{x}_1(t-\tau(t)) - \hat{x}_1(t-h)) + G_2(x_2(t-\tau(t)) - x_2(t-h)) \\ & +G_u(u(t-\tau(t)) - u(t-h)). \end{aligned}$$

which can be rewritten as:

$$\xi(t) = \begin{bmatrix} G_1 \; G_2 \; G_u \end{bmatrix} \int_{t-h}^{t-\tau(t)} \begin{bmatrix} \dot{\hat{x}}_1(s) \\ \dot{x}_2(s) \\ \dot{u}(s) \end{bmatrix} ds.$$

The function ξ only depends on the known variables $\hat{x}_1$, x_2 and u and on the unknown delay $\tau(t)$. One can then assume that there exists a known scalar function α_2 such that:

$$\|\xi(t)\| \leq \alpha_2(t, \hat{x}_1, x_2, u). \tag{8}$$

Let us define an expression for ρ in (6) by using results introduced in the case of control law design [10]. Define γ, a real positive number and ρ such that:

$$\rho(t, y, u) = \|D_1\|\alpha_1(t, y, u) + \alpha_2(t, \hat{x}_1, x_2, u) + \gamma. \tag{9}$$

Introduce the change of coordinates

$$\begin{bmatrix} \bar{e}_1 \\ \bar{e}_2 \end{bmatrix} = T_L \begin{bmatrix} e_1 \\ e_2 \end{bmatrix}$$

with

$$T_L = \begin{bmatrix} I_{n-q} & L \\ 0 & T \end{bmatrix}.$$

Using the fact that $LG_1 = LG_2 = LG_u = LD_1 = 0$, one obtains:

$$\begin{cases} \dot{\bar{e}}_1(t) = (A_{11} + LA_{21})\bar{e}_1(t) \\ \dot{\bar{e}}_2(t) = TA_{21}\bar{e}_1(t) + TG_1\bar{e}_1(t-\tau(t)) + G_l\bar{e}_2(t) \\ \qquad\quad -TG_1L\bar{e}_2(t-\tau(t)) + T\xi(t) + TD_1\zeta(t) + \nu \end{cases} \tag{10}$$

Theorem 1. *Under assumptions $A1 - A3$ and (8) and for all Hurwitz matrices G_l, system (10) is asymptotically stable for any delay $\tau(t)$ in $[\tau_1 \;\; \tau_2]$ if there exist symmetric definite positive matrices P_1, R_1 and $R_{1a} \in \mathbf{R}^{(n-q)\times(n-q)}$, $P_2 \in \mathbf{R}^{q\times q}$, symmetric matrices Z_2 and $Z_{2a} \in \mathbf{R}^{q\times q}$ and a matrix $W \in \mathbf{R}^{(n-q)\times(q-p)}$ such that the following Linear Matrix Inequalities (LMI) hold:*

$$\begin{bmatrix} \Psi_0 & \Psi_1 & \Psi_1 & \Psi_2 & 0 \\ * & -2P_1 + hR_1 & 0 & 0 & 0 \\ * & * & -2P_1 + hR_{1a} & 0 & 0 \\ * & * & * & \Psi_3 & -P_2TG_1 \\ * & * & * & * & -P_1 \end{bmatrix} < 0. \tag{11}$$

where

$$\begin{aligned}\Psi_0 &= A_{11}^T P_1 + P_1 A_{11} + A_{211}^T W^T + W A_{211},\\ \Psi_1 &= A_{11}^T P_1 + A_{211}^T W^T,\\ \Psi_2 &= (A_{21} + G_1)^T T^T P_2,\\ \Psi_3 &= G_l^T P_2 + P_2 G_l + hZ_2 + 2\mu Z_{2a} + R_2,\end{aligned}$$

and

$$\begin{bmatrix} -(1-d)R_2 & [W \;\; 0]^T \\ [W \;\; 0] & -P_1 \end{bmatrix} < 0,$$

$$\begin{bmatrix} R_1 & (TG_1)^T P_2 \\ P_2 T G_1 & Z_2 \end{bmatrix} \geq 0, \tag{12}$$

$$\begin{bmatrix} R_{1a} & (TG_1)^T P_2 \\ P_2 T G_1 & Z_{2a} \end{bmatrix} \geq 0.$$

The gain $\bar{L}$ is given by $\bar{L} = P_1^{-1} W$.

Proof. Consider the Lyapunov-Krasovskii functional:

$$\begin{aligned} V(t) &= \bar{e}_1^T(t) P_1 \bar{e}_1(t) + \textstyle\int_{-h}^{0} \int_{t+\theta}^{t} \dot{\bar{e}}_1^T(s) R_1 \dot{\bar{e}}_1(s) ds d\theta \\ &+ \textstyle\int_{-\mu}^{\mu} \int_{t+\theta}^{t} \dot{\bar{e}}_1^T(s) R_{1a} \dot{\bar{e}}_1(s) ds d\theta \\ &+ \bar{e}_2^T(t) P_2 \bar{e}_2(t) + \textstyle\int_{t-\tau(t)}^{t} \bar{e}_2^T(s) R_2 \bar{e}_2(s) ds. \end{aligned} \tag{13}$$

The functional V can be divided into three parts. The first line of (13) is designed to control the errors $e_1(t)$ subject to the constant delay h. The second line presents a functional which takes into account the delay variation around the average delay h. The last part which appears in the last line of (13) controls the error $e_2(t)$.

Using the following transformation

$$\bar{e}_i(t - \tau(t)) = \bar{e}_i(t) - \int_{t-h}^{t} \dot{\bar{e}}_i(s) ds - \int_{t-\tau(t)}^{t-h} \dot{\bar{e}}_i(s) ds,$$

one obtains:

$$\begin{aligned} \dot{V}(t) &= \bar{e}_1^T(t)[(A_{11} + LA_{21})^T P_1 + P_1(A_{11} + LA_{21})]\bar{e}_1(t) \\ &+ 2\bar{e}_2^T(t) P_2 T (A_{21} + G_1) \bar{e}_1(t) \\ &+ \bar{e}_2^T(t)[G_l^T P_2 + P_2 G_l + R_2]\bar{e}_2(t) - 2\bar{e}_2^T(t) P_2 T G_1 L \bar{e}_2(t - \tau(t)) \\ &- (1 - \dot{\tau}(t)) \bar{e}_2^T(t - \tau(t)) R_2 \bar{e}_2(t - \tau(t)) + h \dot{\bar{e}}_1^T(t) R_1 \dot{\bar{e}}_1(t) \\ &- \textstyle\int_{t-h}^{t} \dot{\bar{e}}_1^T(s) R_1 \dot{\bar{e}}_1(s) ds + 2\mu \dot{\bar{e}}_1^T(t) R_{1a} \dot{\bar{e}}_1(t) - \int_{t-\tau_1}^{t-\tau_2} \dot{\bar{e}}_1^T(s) R_{1a} \dot{\bar{e}}_1(s) ds \\ &+ \eta_1(t) + \eta_2(t) + \eta_3(t) - 2\rho(t, y, u) \| P_2 \bar{e}_2(t) \|, \end{aligned}$$

where

$$\begin{aligned} \eta_1(t) &= -2\bar{e}_2^T(t) P_2 T G_1 \textstyle\int_{t-h}^{t} \dot{\bar{e}}_1(s) ds, \\ \eta_2(t) &= -2\bar{e}_2^T(t) P_2 T G_1 \textstyle\int_{t-\tau(t)}^{t-h} \dot{\bar{e}}_1(s) ds, \\ \eta_3(t) &= 2\bar{e}_2^T(t) P_2 \left[T D_1 \zeta(t) + T \xi(t) \right]. \end{aligned}$$

The LMI condition (12) implies that for any vector X:

$$X^T \begin{bmatrix} R_1 & (TG_1)^T P_2 \\ P_2 T G_1 & Z_2 \end{bmatrix} X \geq 0.$$

Developing this relation for $X = \begin{bmatrix} \dot{\bar{e}}_1(s) \\ \bar{e}_2(t) \end{bmatrix}$, the following inequality holds:

$$-2\bar{e}_2(t)P_2G_1\dot{\bar{e}}_1(s) \leq \bar{e}_2(t)^T Z_2\bar{e}_2(t) + \dot{\bar{e}}_1^T(s)R_1\dot{\bar{e}}_1(s).$$

Then, integration with respect to s of the previous inequality leads to an upper bound for $\eta_1(t)$:

$$\begin{array}{l} \eta_1(t) \leq \int_{t-h}^{t} \bar{e}_2^T(t)Z_2\bar{e}_2(t)ds + \int_{t-h}^{t} \dot{\bar{e}}_1^T(s)R_1\dot{\bar{e}}_1(s)ds, \\ \eta_1(t) \leq h\bar{e}_2^T(t)Z_2\bar{e}_2(t) + \int_{t-h}^{t} \dot{\bar{e}}_1^T(s)R_1\dot{\bar{e}}_1(s)ds. \end{array} \tag{14}$$

By using the same techniques, an upper bound for η_2 is found:

$$\eta_2(t) \leq 2\mu\bar{e}_2^T(t)Z_{2a}\bar{e}_2(t) + \int_{t-h-\mu}^{t-h+\mu} \dot{\bar{e}}_1^T(s)R_{1a}\dot{\bar{e}}_1(s)ds. \tag{15}$$

>From (9) and from the orthogonality of the matrix T, the following inequality holds:

$$\eta_3(t) - 2\rho(t,y,u)\|P_2\bar{e}_2(t)\| \leq -2\gamma\|P_2\bar{e}_2(t)\|. \tag{16}$$

Taking into account (14), (15), (16) and the fact that

$$\dot{\bar{e}}_1(t) = (A_{11} + \bar{L}A_{211})\bar{e}_1(t),$$

$\dot{V}$ can be upper bounded as follows:

$$\begin{array}{l} \dot{V}(t) \leq \bar{e}_1^T(t)(P_1\tilde{A}_{11} + \tilde{A}_{11}^T P_1 + h\tilde{A}_{11}^T R_1\tilde{A}_{11} + 2\mu\tilde{A}_{11}^T R_{1a}\tilde{A}_{11})\bar{e}_1(t) \\ \quad -2\gamma\|P_2e_2(t)\| + \bar{e}_2^T(t)(P_2G_l + G_l^T P_2 + R_2 + hZ_2 + 2\mu Z_{2a})\bar{e}_2(t) \\ \quad +2\bar{e}_2^T(t)P_2(A_{21} + G_1)\bar{e}_1(t) \\ \quad -(1-\dot{\tau}(t))\bar{e}_2^T(t-\tau(t))R_2\bar{e}_2(t-\tau(t)) - 2\bar{e}_2^T(t)P_2TG_1L\bar{e}_2(t-\tau(t)) \end{array}$$

where $\tilde{A}_{11} = (A_{11} + \bar{L}A_{211})$.

Then, the last term of this inequality can be upper bounded by noting that:

$$\begin{array}{l} -2\bar{e}_2^T(t)P_2TG_1L\bar{e}_2(t-\tau(t)) \\ \quad \leq \bar{e}_2^T(t)P_2TG_1P_1^{-1}(TG_1)^T P_2\bar{e}_2(t) + \bar{e}_2^T(t-\tau(t))L^T P_1L\bar{e}_2(t-\tau(t)) \\ \quad \leq \bar{e}_2^T(t)P_2TG_1P_1^{-1}(TG_1)^T P_2\bar{e}_2(t) + \bar{e}_2^T(t-\tau(t))(P_1L)^T P_1^{-1}(P_1L)\bar{e}_2(t-\tau(t)), \end{array}$$

which leads to the following upper bound:

$$\dot{V}(t) \leq \begin{bmatrix} \bar{e}_1(t) \\ \bar{e}_2(t) \end{bmatrix}^T \Psi \begin{bmatrix} \bar{e}_1(t) \\ \bar{e}_2(t) \end{bmatrix} - 2\gamma\|P_2\bar{e}_2(t)\| + \bar{e}_2^T(t-\tau(t))\psi_{30}\bar{e}_2(t-\tau(t)), \tag{17}$$

where $\Psi = \begin{bmatrix} \psi_{10} & \Psi_2 \\ * & \psi_{20} \end{bmatrix}$ and :

$$\begin{array}{rl} \psi_{10} = & (A_{11} + \bar{L}A_{211})^T P_1 + P_1(A_{11} + \bar{L}A_{211}) \\ & +h(A_{11} + \bar{L}A_{211})^T R_1(A_{11} + \bar{L}A_{211}) \\ & +2\mu(A_{11} + \bar{L}A_{211})^T R_{1a}(A_{11} + \bar{L}A_{211}), \\ \psi_{20} = & G_l^T P_2 + P_2G_l + R_2 + hZ_2 + 2\mu Z_{2a} \\ & +P_2TG_1P_1^{-1}(TG_1)^T P_2, \\ \psi_{30} = & (1-d)R_2 + (P_1\left[\bar{L}\ \ 0\right])^T P_1^{-1}(P_1\left[\bar{L}\ \ 0\right]). \end{array} \tag{18}$$

This matrix inequality is not an LMI because of the multiplication of matrix variables. Considering ψ_{20} and ψ_{30}, the Schur complement can remove these nonlinearities but for ψ_{10} Lemma 1 is required. As ψ_{10} must be negative definite to have a solution to the problem

(17), the use of Lemma 1 is possible. Applying it twice to ψ_{10}, the nonlinear condition can be expressed as:

$$\begin{bmatrix} \Psi_0 & \bar{A}_{11}^T Y^T & \bar{A}_{11}^T Y_a^T & \Psi_2 & 0 \\ * & \psi_2 & 0 & 0 & 0 \\ * & * & \psi_3 & 0 & 0 \\ * & * & * & \Psi_3 & -P_2 T G_1 \\ * & * & * & * & -P_1 \end{bmatrix} < 0. \tag{19}$$

$$\begin{bmatrix} -(1-d)R_2 & \begin{bmatrix} \bar{L}^T P_1 \\ 0 \end{bmatrix} \\ * & -P_1 \end{bmatrix} < 0. \tag{20}$$

where

$$\begin{aligned} \bar{A}_{11} &= A_{11} + \bar{L}A_{211} \\ \psi_2 &= -Y - Y^T + hR_1 \\ \psi_3 &= -Y_a - Y_a^T + hR_{1a} \end{aligned}$$

Choosing $Y = P_1$, $Y_a = P_1$ and defining $W = P_1\bar{L}$, the LMI conditions from the Theorem appear. Then, if (11) and (12) are satisfied, (19) and (20) are also satisfied. Finally the error dynamics are asymptotically stable and converge to the solution $e(t) = 0$.

Theorem 1 provides conditions for asymptotic stability of the error dynamics. In the following corollary, it is shown that the error dynamic system is in fact driven to the sliding surface $S_0 = \{(\bar{e}_1, \bar{e}_2) : \bar{e}_2 = 0\}$ in finite time and that a sliding motion is maintained thereafter.

Corollary 1. *With the observer gain given in Theorem 1, an ideal sliding motion takes place on S_0 in finite time.*

Proof. Consider the Lyapunov function:

$$V_2(t) = \bar{e}_2^T(t) P_2 \bar{e}_2(t) \tag{21}$$

Differentiating (21) along the trajectories of (10) yields:

$$\begin{aligned} \dot{V}_2(t) = \bar{e}_2^T(t)(G_l^T P_2 + P_2 G_l)\bar{e}_2(t) + 2\bar{e}_2^T(t) P_2 T \left[T^T \nu + A_{21}\bar{e}_1(t) \right. \\ \left. + G_1\bar{e}_1(t-\tau(t)) + G_1 L\bar{e}_2(t-\tau(t)) + D_1\zeta(t) + \xi(t)\right]. \end{aligned}$$

Noting that G_l is Hurwitz and using (6), the following inequality holds:

$$\dot{V}_2(t) \leq 2\|P_2\bar{e}_2(t)\| \left[\|A_{21}\bar{e}_1(t) + G_1\bar{e}_1(t-\tau(t)) + G_1 L\bar{e}_2(t-\tau(t))\| - \gamma\right].$$

>From Theorem 1, the errors $\bar{e}_1$ and $\bar{e}_2$ are asymptotically stable. There thus exist an instant t_0 and a real positive number δ such that $\forall t \geq t_0$, $\|A_{21}\bar{e}_1(t) + G_1\bar{e}_1(t-\tau(t)) + G_1 L\bar{e}_2(t-\tau(t))\| \leq \gamma - \delta$. This leads to:

$$\begin{aligned} \forall t \geq t_0,\ \dot{V}_2(t) &\leq -2\delta\|P_2\bar{e}_2(t)\| \\ &\leq -2\delta\sqrt{\lambda_{min}(P_2)}\sqrt{V_2(t)}. \end{aligned} \tag{22}$$

where $\lambda_{min}(P_2)$ is the lowest eigenvalue of P_2. Integrating the previous inequality shows that a sliding motion takes place on the manifold S_0 in finite time.

In [18], the observer convergence was improved by enforcing exponential convergence (see [13, 17] for definitions). Following the same approach, observer gains can be derived from LMIs in order to ensure the observer error dynamics is similarly exponentially stable in this case.

4 Example

Consider the system with time-varying delay (4) and:

$$A_{11} = \begin{bmatrix} 0 & 0 \\ 0 & -1 \end{bmatrix}, \ A_{12} = \begin{bmatrix} -1 & 0 \\ 0 & 0.1 \end{bmatrix},$$
$$A_{21} = \begin{bmatrix} 2 & 3 \\ 2 & -1 \end{bmatrix}, A_{22} = \begin{bmatrix} -1 & 0 \\ 0 & -1 \end{bmatrix},$$
$$G_1 = \begin{bmatrix} 0 & 0 \\ 0.1 & 0.21 \end{bmatrix}, \ G_2 = \begin{bmatrix} 0 & 0 \\ 0.2 & 1 \end{bmatrix},$$
$$T = \begin{bmatrix} 1 & 0 \\ 0 & 1 \end{bmatrix}, G_u = \begin{bmatrix} 0 \\ 1 \end{bmatrix}, \ D_1 = B_1 = B_2 = \begin{bmatrix} 0 \\ 0 \end{bmatrix},$$

The delay is chosen as $\tau(t) = \tau_0 + \tau_1 sin(\omega_1 t))$, with $\tau_0 = 0.225s$, $\tau_1 = 0.075$ and $\omega_1 = 0.5s^{-1}$. The control law is

$$u(t) = u_0 sin(\omega_2 t)$$

with $u_0 = 2$ and $\omega_2 = 3$ and the Hurwitz matrix G_l is $\begin{bmatrix} -5 & 0 \\ 0 & -3 \end{bmatrix}$. Using Theorem 1, the following observer gain is obtained:

$$\bar{L} = \begin{bmatrix} -0.1144 \\ 0.0280 \end{bmatrix}$$

Since the system (4) is open loop stable, its dynamics are bounded. Thus the function $\alpha_2(t, \hat{x}_1, x_2, u)$ can be chosen as a constant $K = 4$. The simulation results are given in the following figures. Figure 1 shows the observation errors. Figures 2 and 3 show the comparison between the real and observed states.

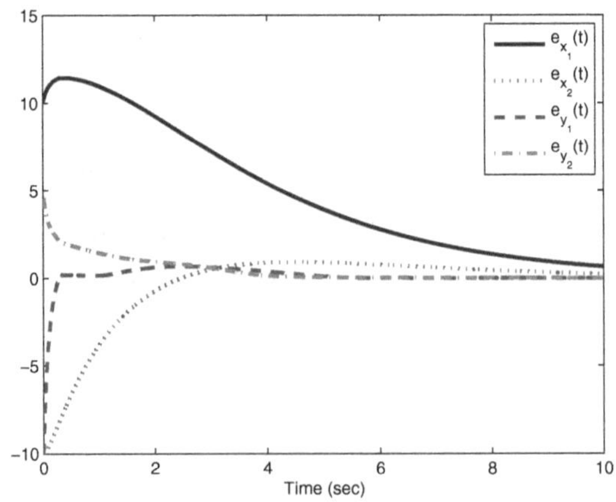

Fig. 1. Observation errors for $\tau_0 = 0.225$ and $\tau_1 = 0.075$.

Figure 1 shows that the system enters a sliding motion at time $t = 2.8s$. The unmeasured variables converge asymptotically to 0.

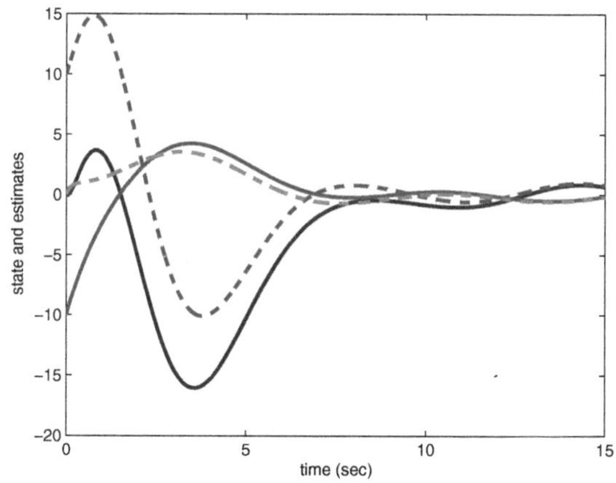

Fig. 2. State x_1 (solid lines) and its estimate $\hat{x}_1$ (dashed lines).

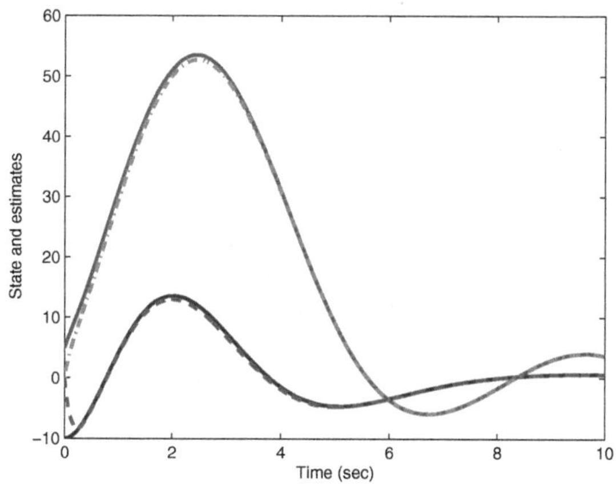

Fig. 3. State x_2 (solid lines) and its estimate $\hat{x}_2$ (dashed lines).

5 Conclusion

This Chapter has considered the problem of designing observers for linear systems with non small and unknown variable delay on both the input and the state. Delay-dependent LMI conditions have been found to guarantee asymptotic stability of the dynamical error system. The conditions only depend on the delay definition and do not incorporate values of an estimated or computed delay. In addition, the dynamic properties of the proposed observer can be characterized.

References

1. Barbot, J.P., Boukhobza, T., Djemaï, M.: Sliding mode observer for triangular input form. In: Proc. 35th IEEE Conference on Decision and Control, Kobe, Japan (1996)
2. Choi, H.H., Chung, M.J.: Robust observer-based H_∞ controller design for linear uncertain time-delay systems. Automatica 33, 1749–1752 (1997)
3. Darouach, M.: Linear functional observers for systems with delays in the state variables. IEEE Trans. on Automatic Control 46, 491–497 (2001)
4. Darouach, M., Pierrot, P., Richard, J.P.: Design of reduced-order observers without internal delays. IEEE Trans. on Automatic Control 44, 1711–1713 (1999)
5. de Souza, C.E., Palhares, R.E., Peres, P.L.D.: Robust H_∞ filtering for uncertain linear systems with multiple time-varying state: An LMI approach. In: Proc. 38th IEEE Conference on Decision and Control, Phoenix, USA, pp. 2023–2028 (1999)
6. Edwards, C., Spurgeon, S.K.: Sliding mode control: Theory and applications. Taylor & Francis, Abington (1998)
7. Fairman, F.W., Kumar, A.: Delayless observers for systems with delay. IEEE Trans. on Automatic Control 31, 258–259 (1986)
8. Floquet, T., Barbot, J.P., Perruquetti, W., Djemaï, M.: On the robust fault detection via a sliding mode disturbance observer. International Journal of Control 77, 622–629 (2004)
9. Floquet, T., Edwards, C., Spurgeon, S.K.: On sliding mode observers for systems with unknown inputs. In: Proc. International Workshop on Variable Structure Systems, Sardinia, Italy (2006)
10. Fridman, E., Gouaisbaut, F., Dambrine, M., Richard, J.P.: Sliding mode control of systems with time-varying delays via a descriptor approach. Int. J. System Sc. 34, 553–559 (2003)
11. Fridman, E., Shaked, U., Xie, L.: Robust H_∞ filtering of linear systems with time-varying delay. IEEE Trans. on Automatic Control 48, 159–165 (2003)
12. Hu, L.S., Huang, J., Cao, H.H.: Robust digital model predicitve control for linear uncertain systems with saturations. IEEE Trans. on Automatic Control 49, 792–796 (2004)
13. Niculescu, S.I., de Souza, C.E., Dugard, L., Dion, J.M.: Robust exponential stability of uncertain systems with time-varying delays. IEEE Trans. on Automatic Control 43, 743–748 (1998)
14. Perruquetti, W., Barbot, J.P. (eds.): Sliding mode control in engineering. Marcel Dekker, New York (2002)
15. Richard, J.P.: Time delay systems: An overview of some recent advances and open problems. Automatica 39, 1667–1694 (2003)
16. Sename, O.: New trends in design of observers for time-delay systems. Kybernetica 37, 427–458 (2001)
17. Seuret, A., Dambrine, M., Richard, J.P.: Robust exponential stabilization for systems with time-varying delays. In: Proc. 5th IFAC Workshop on Time-Delay Systems, Leuven, Belgium (2004)
18. Seuret, A., Floquet, T., Richard, J.P., Spurgeon, S.K.: A sliding mode observer for linear systems with unknown time varying delay. In: Proc. American Control Conference, New York, USA (2007)
19. Wang, Z., Huang, B., Unbehausen, H.: Robust H_∞ observer design for uncertain time-delay systems: I the continuous case. In: Proc. 14th IFAC World Congress, Beijing, China, pp. 231–236 (1999)

Input-Output Representation and Identifiability of Delay Parameters for Nonlinear Systems with Multiple Time-Delays

Milena Anguelova[1,2] and Bernt Wennberg[1,2]

[1] Mathematical Sciences, Chalmers University of Technology, SE-412 96 Gothenburg, Sweden milena@chalmers.se

[2] Mathematical Sciences, Göteborg University, SE-412 96 Gothenburg, Sweden

Summary. We have analyzed the identifiability of time-lag parameters in nonlinear delay systems using an algebraic framework. The identifiability is determined by the form of the system's input-output representation. The values of the time lags can be found directly from the input-output equations, if these can be obtained explicitly. Linear-algebraic criteria are formulated to decide the identifiability of the delay parameters when explicit computation of the input-output relations is not possible.

1 Introduction

Observability and parameter identifiability are important properties of a system where initial state or parameter estimation are concerned. These properties guarantee that the desired quantities can be uniquely determined from the available data.

For nonlinear systems without time delays, these properties are well-characterized, see for instance [7, 13, 14] and the references therein. The characterization of observability and identifiability has now been extended to nonlinear systems with time delays in [16] and [17], using an algebraic approach introduced in [10], and developed in [9]. In these works the time delays themselves are assumed to be known, or multiples of a unit delay. The identifiability of general unknown time-delays has been analyzed only for linear systems [11, 15, 12, 3]. Recently, we used the mathematical setting of [10, 9] and [16] to analyze the identifiability of the time-delay parameter for nonlinear systems with a single unknown constant time delay ([2]).

In this paper we analyze the identifiability of the time-delay parameters for nonlinear control systems with several unknown constant time delays. We show that state elimination produces input-output relations for the system, the form of which decides the identifiability of the delay parameters. The values of the delay parameters can be found directly from the input-output equations, if the latter can be obtained explicitly. We formulate linear-algebraic criteria to check the identifiability of the delay parameters which eliminate the need for an explicit calculation of the input-output relations. The identifiability of the delay parameters can be a necessary but not sufficient condition for the observability of the state variables (and identifiability of the regular parameters in the system). The already established methods for

J.J. Loiseau et al. (Eds.): Topics in Time Delay Systems, LNCIS 388, pp. 243–253.
springerlink.com

testing weak observability and identifiability for nonlinear delay systems alone [16, 17] cannot be used to determine the identifiability of the time-delay parameters and a prior analysis is necessary for the latter.

2 Notation and preliminary definitions

Consider nonlinear time-delay systems of the form:

$$\begin{cases} \dot{x}(t) = f\big(x(t), x(t-\tau_1), \dots, x(t-\tau_\ell), u(t), u(t-\tau_1), \dots, u(t-\tau_\ell)\big) \\ y(t) = h\big(x(t), x(t-\tau_1), \dots, x(t-\tau_\ell)\big) \\ x(t) = \varphi(t), \quad t \in [-\max_i \tau_i, 0] \\ u(t) = u_0(t), \quad t \in [-T, 0] \end{cases} , \tag{1}$$

where $x \in \mathbb{R}^n$ denotes the state variables, $u \in \mathbb{R}^m$ is the input and $y \in \mathbb{R}^p$ is the output. The unknown constant time-delays are denoted by the vector $\boldsymbol{\tau} = (\tau_1, \dots, \tau_\ell)$, $\tau_i \in [0, T), \quad T \in \mathbb{R}$. The entries of f and h are meromorphic in their arguments and it is reasonable to assume that none of the h_i are constants. The unknown continuous function of initial conditions is denoted by $\varphi : [-\max_i(\tau_i), 0] \to \mathbb{R}^n$. The set of initial functions for the variables x is denoted by $C := C([-\max_i(\tau_i), 0], \mathbb{R}^n)$. A meromorphic input function $u(t)$ is called an admissible input if the differential equation above admits a unique solution. The set of all such input functions is denoted by C_U.

One of the objectives of this work is to investigate the property of local identifiability of the delay parametersτ_i. Intuitively, τ_i are identifiable if any two sets can be distinguished by the system's input-output behaviour. A formal definition is as follows:

Definition 1. *The delay parameters $\boldsymbol{\tau}$ are said to be locally identifiable at $\boldsymbol{\tau_0} \in (0, T)^\ell$ if there exists an open set $W \ni \boldsymbol{\tau_0}$, $W \subset [0, T)^\ell$, such that $\forall \boldsymbol{\tau_1} \in W : \boldsymbol{\tau_1} \neq \boldsymbol{\tau_0}$, $\forall \varphi_0, \varphi_1 \in C$, there exist $t \geq 0$ and $u \in C_U$ such that $y(t, \varphi_1, u, \boldsymbol{\tau_1}) \neq y(t, \varphi_0, u, \boldsymbol{\tau_0})$, where $y(t, \varphi, u, \boldsymbol{\tau})$ denotes the parameterized output for the initial function φ, the admissible input u and delays $\boldsymbol{\tau}$.*

Following the notations and algebraic setting of [10, 9, 16] and [17], let $\mathcal{K}$ be the field of meromorphic functions of a finite number of variables from

$$\{x(t - \boldsymbol{i\tau}), u(t - \boldsymbol{i\tau}), \dots, u^{(l)}(t - \boldsymbol{i\tau}), \quad \boldsymbol{i} = (i_1, \dots, i_\ell), i_j, l \in \mathbb{Z}^+\} \quad ,$$

where we have denoted $i_1\tau_1 + \dots + i_\ell\tau_\ell$ by $\boldsymbol{i\tau}$. Let $\mathcal{E}$ be the vector space over $\mathcal{K}$ given by

$$\mathcal{E} = \operatorname{span}_{\mathcal{K}}\{d\xi : \xi \in \mathcal{K}\} \quad . \tag{2}$$

Then $\mathcal{E}$ is the set of linear combinations of a finite number of one-forms from $\{dx(t - \boldsymbol{i\tau}), du(t - \boldsymbol{i\tau}), \dots, du^{(l)}(t - \boldsymbol{i\tau})\}$ with row vector coefficients in $\mathcal{K}$.

Let $\mathcal{K}(\boldsymbol{\delta}]$ denote the set of polynomials in $\delta_1, \dots, \delta_\ell$ with coefficients from $\mathcal{K}$. This set is a noncommutative ring, where addition is defined as usual, while multiplication is defined as follows. Let $a(\boldsymbol{\delta}]$ be a polynomial in $\mathcal{K}(\boldsymbol{\delta}]$, $a(\boldsymbol{\delta}] = \sum_{\boldsymbol{k}} a_{\boldsymbol{k}} \boldsymbol{\delta}^{\boldsymbol{k}}$, where we have denoted $\delta_1^{k_1} \dots \delta_\ell^{k_\ell}$ by $\boldsymbol{\delta}^{\boldsymbol{k}}$, $\boldsymbol{k} = (k_1, \dots, k_\ell)$. Order the different powers $\boldsymbol{k}$ according to the largest k_1, then k_2, etc. According to this order, let the highest degree of $a(\boldsymbol{\delta}]$ be $\boldsymbol{r_a}$, where $\boldsymbol{r_a} = (r_{a,1}, \dots, r_{a,\ell})$ and analogously for another polynomial $b(\boldsymbol{\delta}]$ in $\mathcal{K}(\boldsymbol{\delta}]$. Multiplication of $a(\boldsymbol{\delta}]$ and $b(\boldsymbol{\delta}]$ is then given by

$$a(\delta]b(\delta] = \sum_{k=0}^{r_a+r_b} \sum_{i+j=k}^{i\le r_a, j\le r_b} a_i(t)b_j(t-i\tau)\delta^k \quad . \tag{3}$$

The ring $\mathcal{K}(\delta]$ is Noetherian (see [1] for a proof) and therefore also a (left) Ore domain by Corollary 8.10 in Chapter 0 of [4]. It thus has the invariant basis number (IBN) by Propositions (1.8) and (1.13) in [8] - a free left module over $\mathcal{K}(\delta]$ has a uniquely defined rank and all its bases have the same cardinality. Let $\mathcal{M}$ denote the module $\text{span}_{\mathcal{K}(\delta]}\{d\xi : \xi \in \mathcal{K}\}$. The closure of a submodule $\mathcal{N}$ in $\mathcal{M}$ is the submodule

$$\overline{\mathcal{N}} = \{w \in \mathcal{M} : \exists a(\delta] = a(\delta_1, \ldots, \delta_\ell] \in \mathcal{K}(\delta], \quad a(\delta]w \in \mathcal{N}\} \quad .$$

Equivalently, $\overline{\mathcal{N}}$ is the largest submodule of $\mathcal{M}$ containing $\mathcal{N}$ and having a rank equal to $\text{rank}_{\mathcal{K}(\delta]}\mathcal{N}$ [16].

Differentiation of functions $\phi(x(t-i\tau), u(t-j\tau), \ldots, u^{(l)}(t-j\tau))$ in $\mathcal{K}$ and one-forms $\omega = \sum_i \kappa_x^i dx(t-i\tau) + \sum_{j,r} \nu_j du^{(r)}(t-j\tau)$ in $\mathcal{M}$ is defined in the natural way [16, 17]:

$$\dot{\phi} = \sum_i \frac{\partial \phi}{\partial x(t-i\tau)}\delta^i f + \sum_{j,r} \frac{\partial \phi}{\partial u^{(r)}(t-j\tau)} u^{(r+1)}(t-j\tau) \tag{4}$$

$$\begin{aligned} \dot{\omega} = &\sum_i \dot{\kappa}_x^i dx(t-i\tau) + \sum_{j,r} \dot{\nu}_j du^{(r)}(t-j\tau) + \\ &+ \sum_i \kappa_x^i d\delta^i f + \sum_{j,r} \nu_j du^{(r+1)}(t-j\tau) \quad . \end{aligned} \tag{5}$$

A one-form $\omega \in \mathcal{M}$ is also an element of $\mathcal{E}$ and Poincaré's lemma holds (see Lemma 3 in [9]):

Lemma 1 (Poincaré). *Consider a one-form $\omega \in \mathcal{M}$. Then there exists a function $\xi \in \mathcal{K}$ such that (locally) $\omega = d\xi$ if and only if $d\omega = 0$.*

Define

$$\mathcal{X} = span_{\mathcal{K}(\delta]}\{dx\} \quad , \tag{6}$$

$$\mathcal{Y}_k = span_{\mathcal{K}(\delta]}\{dy, d\dot{y}, \ldots, dy^{(k-1)}\} \quad , \tag{7}$$

$$\mathcal{U} = span_{\mathcal{K}(\delta]}\{du, d\dot{u}, \ldots\} \quad . \tag{8}$$

Then, from [16],

$$(\mathcal{Y}_k + \mathcal{U}) \cap \mathcal{X} = (\mathcal{Y}_n + \mathcal{U}) \cap \mathcal{X}$$

for $k \ge n$ and

$$\text{rank}_{\mathcal{K}(\delta]}(\mathcal{Y}_n + \mathcal{U}) \cap \mathcal{X} \le n \quad .$$

3 State elimination

In this section we consider the problem of obtaining an input-output representation from the state-space form of a time-delay control system. This problem has been treated for polynomial systems in [6]. We show that for a system of the form (1) there always exists, at least locally, a set of input-output delay-differential equations of the form:

$$\begin{aligned} &F(\boldsymbol{\delta}, y, \ldots, y^{(k)}, u, \ldots, u^{(J)}) := \\ &= F(y(t - \boldsymbol{i_0}\boldsymbol{\tau}), \ldots, y^{(k)}(t - \boldsymbol{i_k}\boldsymbol{\tau}), u(t - \boldsymbol{j_0}\boldsymbol{\tau}), \ldots, u^{(J)}(t - \boldsymbol{j_l}\boldsymbol{\tau})) = 0\,, \end{aligned} \tag{9}$$

such that any pair $(y(t), u(t))$ which solves (1), also satisfies (9), for t such that all derivatives involved are continuous. The function F is meromorphic in its arguments.

Theorem 1. *Given a system of the form* (1), *there exists an integer $J \geq 0$ and an open dense subset V of $C \times C_U^{J+1}$, such that in the neighborhood of any point of V, there exists an input-output representation of the system of the form* (9).

Proof: The proof is an adaptation of the proof of Theorem 2.2.1. in [5] for the analogous result for ODE-systems.

Let f be an r-dimensional vector with entries $f_j \in \mathcal{K}$. Let $\frac{\partial f}{\partial x}$ denote the $r \times n$ matrix with entries

$$\left(\frac{\partial f}{\partial x}\right)_{j,i} = \sum_{\boldsymbol{k}} \frac{\partial f_j}{\partial x_i(t - \boldsymbol{k}\boldsymbol{\tau})} \boldsymbol{\delta}^{\boldsymbol{k}} \in \mathcal{K}(\boldsymbol{\delta}]\,, \tag{10}$$

where the sum is over finitely many terms, for f_j is a function of finitely many delays of x_i. Denote by s_1 the least positive integer such that

$$\text{rank}_{\mathcal{K}(\boldsymbol{\delta}]} \frac{\partial(h_1, ..., h_1^{(s_1-1)})}{\partial x} = \text{rank}_{\mathcal{K}(\boldsymbol{\delta}]} \frac{\partial(h_1, ..., h_1^{(s_1)})}{\partial x}\,. \tag{11}$$

Inductively, for $1 < i \leq p$, set $s_i = 0$ if

$$\begin{aligned} &\text{rank}_{\mathcal{K}(\boldsymbol{\delta}]} \frac{\partial(h_1, ..., h_1^{(s_1-1)}, \ldots, h_{i-1}, \ldots, h_{i-1}^{(s_{i-1}-1)})}{\partial x} = \\ &= \text{rank}_{\mathcal{K}(\boldsymbol{\delta}]} \frac{\partial(h_1, ..., h_1^{(s_1-1)}, \ldots, h_{i-1}, \ldots, h_{i-1}^{(s_{i-1}-1)}, h_i)}{\partial x}\,, \end{aligned} \tag{12}$$

and otherwise, let s_i be the least positive integer such that

$$\text{rank}_{\mathcal{K}(\boldsymbol{\delta}]} \frac{\partial(h_1, ..., h_1^{(s_1-1)}, \ldots, h_i^{(s_i-1)})}{\partial x} = \text{rank}_{\mathcal{K}(\boldsymbol{\delta}]} \frac{\partial(h_1, ..., h_1^{(s_1-1)}, \ldots, h_i^{(s_i)})}{\partial x}\,. \tag{13}$$

Let $S = (h_1, ..., h_1^{(s_1-1)}, \ldots, h_p^{(s_p-1)})$, where h_i does not appear if $s_i = 0$. Then

$$\text{rank}_{\mathcal{K}(\boldsymbol{\delta}]} \frac{\partial S}{\partial x} = s_1 + \cdots + s_p \leq n \quad . \tag{14}$$

For simplicity, denote the elements in S by $\tilde{x}_1, \ldots, \tilde{x}_{s_1+\cdots+s_p}$. The definition of s_i (equation (13)) implies that $\frac{\partial h_i^{(s_i)}}{\partial x}$ is in $\overline{\mathcal{M}}$, where

$$\mathcal{M} = \text{span}_{\mathcal{K}(\boldsymbol{\delta}]} \left\{ \frac{\partial(\tilde{x}_1, \ldots, \tilde{x}_{s_1+\cdots+s_i})}{\partial x} \right\} \quad . \tag{15}$$

Thus, there exist nonzero polynomials $a_i(\boldsymbol{\delta}] \in \mathcal{K}(\boldsymbol{\delta}]$, $i = 1, \ldots, p$ such that $a_i(\boldsymbol{\delta}] \frac{\partial h_i^{(s_i)}}{\partial x} \in \mathcal{M}$. Therefore,

$$a_i(\boldsymbol{\delta}] dh_i^{(s_i)} + \sum_{r=1}^{m} \sum_{j=0}^{J} b_{i,j,r}(\boldsymbol{\delta}] du_r^{(j)} - \sum_{j=1}^{s_1+\cdots+s_i} c_{i,j}(\boldsymbol{\delta}] d\tilde{x}_j = 0 \quad , \tag{16}$$

for some $J \geq 0$, where J is the highest derivative of u appearing in the functions in S and $b_{i,j,r}(\boldsymbol{\delta}], c_{i,j}(\boldsymbol{\delta}] \in \mathcal{K}(\boldsymbol{\delta}]$. Since all functions are assumed meromorphic and we have continuous dependence for the output on the input and initial function, the above equality holds on an open dense set of $C \times C_U^{J+1}$. The left hand side of equation (16), being equal to zero, is a closed one-form on $\mathcal{M}$. Applying the Poincaré lemma, Lemma 1, we obtain functions $\xi_i \in \mathcal{K}$ such that

$$d\xi_i = a_i(\boldsymbol{\delta}]dh_i^{(s_i)} + \sum_{r=1}^{m}\sum_{j=0}^{J} b_{i,j,r}(\boldsymbol{\delta}]du_r^{(j)} - \sum_{j=1}^{s_1+\cdots+s_i} c_{i,j}(\boldsymbol{\delta}]d\tilde{x}_j$$

and

$$\xi_i(\boldsymbol{\delta}, h_i^{(s_i)}, \tilde{x}_1, \ldots, \tilde{x}_{s_1+\cdots+s_i}, u, \ldots, u^{(J)}) = 0 \quad . \tag{17}$$

This produces an input-output equation

$$\xi_i(\boldsymbol{\delta}, y_i^{(s_i)}, y_1, \ldots, y_1^{(s_1-1)}, \ldots, y_i^{(s_i-1)}, u, \ldots, u^{(J)}) = 0 \quad , \tag{18}$$

if $s_i \geq 1$ and

$$\xi_i(\boldsymbol{\delta}, y_i, y_1, \ldots, y_1^{(s_1-1)}, \ldots, y_{i-1}^{(s_{i-1}-1)}, u, \ldots, u^{(J)}) = 0 \quad , \tag{19}$$

for $s_i = 0$, where only those y_j for which $s_j \geq 1$ appear in equations (18) and (19).

This is true for each i, $1 \leq i \leq p$, resulting in p input-output equations of the form (9). ■

4 Identifiability of the delay parameters

In this section we formulate criteria for determining the identifiability of the delay parameters τ_i from (1). The starting point is the linear form of the input-output equations (16), whose i-th equation is

$$a_i(\boldsymbol{\delta}]dy_i^{(s_i)} + \sum_{r=1}^{m}\sum_{j=0}^{J} b_{i,j,r}(\boldsymbol{\delta}]du_r^{(j)} = \sum_{j=1}^{s_1+\cdots+s_i} c_{i,j}(\boldsymbol{\delta}]d\tilde{x}_j \quad . \tag{20}$$

With no loss of generality, we assume that the polynomials $a_i(\boldsymbol{\delta}]$, $b_{i,j,r}(\boldsymbol{\delta}]$ and $c_{i,j}(\boldsymbol{\delta}]$ above have an element from $\mathcal{K}$ as greatest common divider. Let

$$a_i(\boldsymbol{\delta}] = \sum_{\boldsymbol{k}} a_{i_{\boldsymbol{k}}}\boldsymbol{\delta}^{\boldsymbol{k}}, \quad b_{i,j,r}(\boldsymbol{\delta}] = \sum_{\boldsymbol{k}} b_{i,j,r_{\boldsymbol{k}}}\boldsymbol{\delta}^{\boldsymbol{k}} \text{ and } c_{i,j}(\boldsymbol{\delta}] = \sum_{\boldsymbol{k}} c_{i,j_{\boldsymbol{k}}}\boldsymbol{\delta}^{\boldsymbol{k}} \quad .$$

Denote all the different monomials $\boldsymbol{\delta}^{\boldsymbol{k}}$ appearing above by $\Delta_{i,1}, \ldots, \Delta_{i,q}$. Each $\Delta_{i,j}$ can be written

$$\Delta_{i,j} = \delta_1^{k_{i,j,1}} \ldots \delta_\ell^{k_{i,j,\ell}} \quad , \tag{21}$$

where $k_{i,j,1}, \ldots, k_{i,j,\ell}$ are nonnegative integers, and $\Delta_{i,j}$ thus represents the (integer) linear combination $T_{i,j} = k_{i,j,1}\tau_1 + \cdots + k_{i,j,\ell}\tau_\ell$ of the different time delays. If any one of the terms in $a_i(\boldsymbol{\delta}]$, $b_{i,j,r}(\boldsymbol{\delta}]$ or $c_{i,j}(\boldsymbol{\delta}]$ is a polynomial in $\boldsymbol{\delta}$ of degree zero, that is, the input-output equations contain un-delayed variables, then we set Δ_{i0} equal to δ_0, where δ_0 denotes the identity operator, and the corresponding $T_{i,0}$ is zero. There are $A_{i,s}$, $B_{i,j,r,s}$ and $C_{i,j,s}$ in $\mathcal{K}$ such that

$$a_i(\boldsymbol{\delta}] = \sum_{s=0}^{q} A_{i,s}\Delta_{i,s}, \quad b_{i,j,r}(\boldsymbol{\delta}] = \sum_{s=0}^{q} B_{i,j,r,s}\Delta_{i,s}, \quad c_{i,j}(\boldsymbol{\delta}] = \sum_{s=0}^{q} C_{i,j,s}\Delta_{i,s}. \tag{22}$$

Let $\bar{\Delta}_i = \delta_1^{\bar{k}_{i,1}} \dots \delta_\ell^{\bar{k}_{i,\ell}}$ be the monomial $\boldsymbol{\delta}^{\boldsymbol{k}}$ in $a_i(\boldsymbol{\delta}]$ with smallest index $\boldsymbol{k} = k_{i,j,1}, \dots, k_{i,j,\ell}$ (ordered after $k_{i,j,1}, \dots, k_{i,j,\ell}$) - it is either equal to $\Delta_{i,0}$ or is among the $\Delta_{i,1}, \dots, \Delta_{i,q}$. Denote its corresponding linear combination of time delays by $\bar{T}_i$. Consider the matrix

$$M = \begin{bmatrix} k_{11,1} - \bar{k}_{1,1} & \cdots & k_{11,\ell} - \bar{k}_{1,\ell} \\ \vdots & \vdots & \vdots \\ k_{1q,1} - \bar{k}_{1,1} & \dots & k_{1q,\ell} - \bar{k}_{1,\ell} \\ \vdots & \vdots & \vdots \\ k_{p1,1} - \bar{k}_{p,1} & \cdots & k_{p1,\ell} - \bar{k}_{p,\ell} \\ \vdots & \vdots & \vdots \\ k_{p1,1} - \bar{k}_{p,1} & \cdots & k_{pq,\ell} - \bar{k}_{p,\ell} \end{bmatrix} , \tag{23}$$

for which we have the equality

$$[T_{11} - \bar{T}_1, \dots, T_{1q} - \bar{T}_1, \dots, T_{p1} - \bar{T}_p, \dots, T_{pq} - \bar{T}_p]^{tr} = M[\tau_1, \dots, \tau_\ell]^{tr} \quad , \tag{24}$$

where $\boldsymbol{v}^{tr}$ denotes the transpose of $\boldsymbol{v}$.

We can now formulate the identifiability criteria for $\tau_1, \dots, \tau_\ell$ in the following proposition:

Proposition 1. *If M is defined as above and $y_i^{(s_i+j)}(t)$ is not identically equal to zero for any $0 \le j < iq - 1$, $i = \{1, \dots, p\}$, then $\tau_1, \dots, \tau_\ell$ are locally identifiable generically, if and only if* $\operatorname{rank}(M) = \ell$.

Proof: Using (22) in (20) and multiplying by $\bar{\Delta}_i^{-1}$ from the left we obtain

$$\begin{aligned} \bar{\Delta}_i^{-1} \textstyle\sum_{s=0}^{q} A_{i,s}\Delta_{i,s} dy_i^{(s_i)} &= \bar{\Delta}_i^{-1} \textstyle\sum_{j=1}^{s_1+\dots+s_i} \sum_{s=0}^{q} C_{i,j,s}\Delta_{i,s} d\tilde{x}_j - \\ &- \bar{\Delta}_i^{-1} \textstyle\sum_{r=1}^{m} \sum_{j=0}^{J} \sum_{s=0}^{q} B_{i,j,r,s}\Delta_{i,s} du_r^{(j)} \quad , \end{aligned} \tag{25}$$

which can be written (with no un-delayed terms $dy_i^{(s_i)}$ on the right-hand side)

$$\begin{aligned} dy_i^{(s_i)} &= \bar{\Delta}_i^{-1} \textstyle\sum_{j=1}^{s_1+\dots+s_i} \sum_{s=0}^{q} C_{i,j,s}\Delta_{i,s} d\tilde{x}_j - \\ &- \bar{\Delta}_i^{-1} \textstyle\sum_{r=1}^{m} \sum_{j=0}^{J} \sum_{s=0}^{q} B_{i,j,r,s}\Delta_{i,s} du_r^{(j)} - \\ &- \bar{\Delta}_i^{-1} \textstyle\sum_{s=1}^{q} A_{i,s}\Delta_{i,s} dy_i^{(s_i)} \quad . \end{aligned} \tag{26}$$

The Poincaré lemma, Lemma 1, then gives locally

$$\begin{aligned} y_i^{(s_i)}(t) = \tilde{f}_i(&\tilde{x}_1(t), \dots, \tilde{x}_{s_1+\dots+s_i}(t), u(t), \dots, u^{(J)}(t), \\ &\tilde{x}_1(t - T_{i1} + \bar{T}_i), \dots, \tilde{x}_{s_1+\dots+s_i}(t - T_{i1} + \bar{T}_i), y_i^{(s_i)}(t - T_{i1} + \bar{T}_i), \\ &u(t - T_{i1} + \bar{T}_i), \dots, u^{(J)}(t - T_{i1} + \bar{T}_i), \\ &\dots, \\ &\tilde{x}_1(t - T_{iq} + \bar{T}_i), \dots, \tilde{x}_{s_1+\dots+s_i}(t - T_{iq} + \bar{T}_i), y_i^{(s_i)}(t - T_{iq} + \bar{T}_i), \\ &u(t - T_{iq} + \bar{T}_i), \dots, u^{(J)}(t - T_{iq} + \bar{T}_i)) \quad . \end{aligned} \tag{27}$$

Let first $\operatorname{rank}(M) = \ell$. Then, $q \ge \ell$ and we thus have at least ℓ different Δ_{iq}:s. Consider those of the equations (27), for which $iq \ge 1$. Evaluated at a fixed time point $t_0 \ge T$, (27) gives an equation for $T_{i1} - \bar{T}_i, \dots, T_{iq} - \bar{T}_i$:

$$
\begin{aligned}
y_i^{(s_i)}(t_0) = \tilde{f}_i(&\tilde{x}_1(t_0),\ldots,\tilde{x}_{s_1+\cdots+s_i}(t_0),u(t_0),\ldots,u^{(J)}(t_0),\\
&\tilde{x}_1(t_0-T_{i1}+\bar{T}_i),\ldots,\tilde{x}_{s_1+\cdots+s_i}(t_0-T_{i1}+\bar{T}_i),y_i^{(s_i)}(t_0-T_{i1}+\bar{T}_i),\\
&u(t_0-T_{i1}+\bar{T}_i),\ldots,u^{(J)}(t_0-T_{i1}+\bar{T}_i),\\
&\ldots,\\
&\tilde{x}_1(t_0-T_{iq}+\bar{T}_i),\ldots,\tilde{x}_{s_1+\cdots+s_i}(t_0-T_{iq}+\bar{T}_i),y_i^{(s_i)}(t_0-T_{iq}+\bar{T}_i),\\
&u(t_0-T_{iq}+\bar{T}_i),\ldots,u^{(J)}(t_0-T_{iq}+\bar{T}_i)) \quad .
\end{aligned}
\tag{28}
$$

If the time-point t_0 is chosen large enough to ensure the existence of all time-derivatives involved (which can be achieved by choosing for example $t_0 \geq (\max_i s_i - 1)T$), then differentiating (28) with respect to time gives new equations for $T_{i1} - \bar{T}_i, \ldots, T_{iq} - \bar{T}_i$ which are independent, since $dy_i^{(j)}, j \geq 0$ are linearly independent over $\mathcal{K}$ due to $iq \geq 1$:

$$
\begin{aligned}
y_i^{(s_i+j)}(t_0) = \tfrac{d^j}{dt^j}\tilde{f}_i(&\tilde{x}_1(t),\ldots,\tilde{x}_{s_1+\cdots+s_i}(t),u(t),\ldots,u^{(J)}(t),\\
&\tilde{x}_1(t-T_{i1}+\bar{T}_i),\ldots,\tilde{x}_{s_1+\cdots+s_i}(t-T_{i1}+\bar{T}_i),y_i^{(s_i)}(t-T_{i1}+\bar{T}_i),\\
&u(t-T_{i1}+\bar{T}_i),\ldots,u^{(J)}(t-T_{i1}+\bar{T}_i),\\
&\ldots,\\
&\tilde{x}_1(t-T_{iq}+\bar{T}_i),\ldots,\tilde{x}_{s_1+\cdots+s_i}(t-T_{iq}+\bar{T}_i),y_i^{(s_i)}(t-T_{iq}+\bar{T}_i),\\
&u(t-T_{iq}+\bar{T}_i),\ldots,u^{(J)}(t-T_{iq}+\bar{T}_i))_{|t_0} \, .
\end{aligned}
\tag{29}
$$

Unless $y_i^{(s_i+j)}(t)$ is identically zero for some $0 \leq j < q-1$, the first q of these equations identify $T_{i1} - \bar{T}_i, \ldots, T_{iq} - \bar{T}_i$ locally (the rest can in some cases be used to analyze global identifiability). Since $\text{rank}(M) = \ell$, equation (24) implies that the time delays τ_i, $i = 1, \ldots, \ell$ are generically defined uniquely by the locally identifiable linear combinations $T_{i1} - \bar{T}_i, \ldots, T_{iq} - \bar{T}_i$ (an obvious exception is the case of commensurate time-delays). Thus, all $\tau_i, i = 1, \ldots, \ell$ are generically locally identifiable, which completes the proof.

Let now $\text{rank}(M) < \ell$, that is, there are infinitely many τ_i:s which give the same linear combinations $\{T_{i1} - \bar{T}_i, \ldots, T_{iq} - \bar{T}_i\}$, $i = 1, \ldots, p$ and the input-output equations (27) cannot be used to identify the τ_i:s. We treat the case $\ell \geq 2$ and refer to [2] for the case of a single delay. If $\text{rank}(M) = 0$, the proof is analogous to the proof of Theorem 2 in [2] and is therefore left out.

We will show that $\tau_1, \ldots, \tau_\ell$ are not locally identifiable generically by showing that (1) can be represented locally as a neutral system with time lags $\tilde{\boldsymbol{\tau}} = (\tilde{\tau}_1, \ldots, \tilde{\tau}_{\tilde{\ell}})$, where the latter are the nonzero elements in $\{T_{i1} - \bar{T}_i, \ldots, T_{iq} - \bar{T}_i\}$, $i = 1, \ldots, p$. To see this, observe that by using equation (27), we obtain the following delay-differential equations for each i with $s_i \geq 1$:

$$
\begin{cases}
\dot{\tilde{x}}_{s_1+\cdots+s_{i-1}}(t) = \tilde{x}_{s_1+\cdots+s_{i-1}+1}(t)\\
\qquad\vdots\\
\dot{\tilde{x}}_{s_1+\cdots+s_i}(t) \;= \tilde{f}_i(\dot{\tilde{x}}_{s_1+\cdots+s_i}(t-\tilde{\tau}_1),\ldots,\dot{\tilde{x}}_{s_1+\cdots+s_i}(t-\tilde{\tau}_{\tilde{\ell}}),\tilde{x}(t),\tilde{x}(t-\tilde{\tau}_1),\\
\qquad\ldots,\tilde{x}(t-\tilde{\tau}_{\tilde{\ell}}),u(t),\ldots,u^{(J)}(t),u(t-\tilde{\tau}_1),\ldots,u(t-\tilde{\tau}_{\tilde{\ell}}),\ldots,\\
\qquad u^{(J)}(t-\tilde{\tau}_1),\ldots,u^{(J)}(t-\tilde{\tau}_{\tilde{\ell}})) \quad .
\end{cases}
\tag{30}
$$

We now combine these delay-differential equations for all $i = 1, \ldots, p$ with $s_i \geq 1$ and set the outputs $\tilde{y}_1, \ldots, \tilde{y}_p$ equal to $y_1, \ldots, y_p$. The outputs y_j for which $s_j \geq 1$ are amongst the $\tilde{x}$, and the rest of the y_j:s (for which $s_j = 0$) are dependent on $\tilde{x}$ as solutions to difference equations according to (27). Using appropriate initial conditions $\tilde{x}(t) = \tilde{\varphi}(t), \quad t \in [t_0 - \max_j \tilde{\tau}_j, t_0]$, we obtain a neutral system. Any input-output pair $(y(t), u(t))$ resulting from the original system (1) also satisfies the above neutral system for $t_0 \geq \max_i \tilde{\tau}_j$ such that all derivatives are continuous. Thus, the input-output behaviour of the system does not distinguish the infinitely many τ_i:s which give the same linear combinations $\tilde{\tau}$ and $\tau_1, \ldots, \tau_\ell$ are not locally identifiable generically. ■

5 Examples

In this section we illustrate the theory by simple examples and show that weak observability (and/or parameter identifiability for the regular model parameters) does not necessarily imply identifiability of the delay parameters, or vice versa. Thus, the already established methods for testing observability and identifiability for nonlinear delay systems [16, 17] cannot be used to determine the identifiability of the delay parameters.

Example 1:

$$\begin{cases} \dot{x}_1(t) &= x_2^2(t-\tau_1) + u(t) \\ \dot{x}_2(t) &= x_1(t-\tau_2)x_2(t) \\ y(t) &= x_1(t) \\ x(t) &= \varphi(t), \quad t \in [-\tau, 0] \end{cases} \tag{31}$$

We have

$$\begin{aligned} \dot{y} &= (\delta_1(x_2))^2 + u \\ \ddot{y} &= 2\delta_1(x_2)\delta_1\delta_2(x_1)\delta_1(x_2) + \dot{u} = 2(\delta_1(x_2))^2\delta_1\delta_2(x_1) + \dot{u} \end{aligned} \tag{32}$$

and

$$\begin{aligned} \begin{bmatrix} dy \\ d\dot{y} - du \\ d\ddot{y} - d\dot{u} \end{bmatrix} &= \frac{\partial(S, h_1^{(s_1)})}{\partial x} \begin{bmatrix} dx_1 \\ dx_2 \end{bmatrix} = \\ &= \begin{bmatrix} 1 & 0 \\ 0 & 2\delta_1(x_2)\delta_1 \\ 2(\delta_1(x_2))^2\delta_1\delta_2 & 4\delta_1\delta_2(x_1)\delta_1(x_2)\delta_1 \end{bmatrix} \begin{bmatrix} dx_1 \\ dx_2 \end{bmatrix} . \end{aligned} \tag{33}$$

Clearly, the matrix $\frac{\partial(S, h_1^{(s_1)})}{\partial x}$ has rank 2 over $\mathcal{K}(\boldsymbol{\delta}]$ and so the system is weakly observable according to the definition in [16], if τ_1 and τ_2 are known. However, τ_1 and τ_2 are not identifiable. The input-output equation in linear form (eq. (20)) is:

$$d\ddot{y} - d\dot{u} + 2\delta_1\delta_2(x_1)du = 2(\delta_1(x_2))^2\delta_1\delta_2 dy + 2\delta_1\delta_2(x_1)d\dot{y}, \tag{34}$$

and we see that there are two monomials, $\bar{\Delta}_1 = \Delta_{10} = \delta_0$, the identity operator and $\Delta_{11} = \delta_1\delta_2$ (with corresponding combinations $\bar{T}_1 = T_{10} = 0$ and $T_{11} = \tau_1 + \tau_2$ of the two time-delays). Thus, $M = \begin{bmatrix} 1 & 1 \end{bmatrix}$ with rank 1, and the time lags are not identifiable.

Following the the first part of the proof of Corollary 1, we can use the change of variables $\tilde{x}_1 = y = x_1, \tilde{x}_2 = \dot{y} = (\delta_1(x_2))^2 + u$ to rewrite the system as

$$\begin{cases} \dot{\tilde{x}}_1(t) &= \tilde{x}_2(t) \\ \dot{\tilde{x}}_2(t) &= 2(\tilde{x}_2(t) - u(t))\tilde{x}_1(t - T_{11}) + \dot{u}(t) \\ y(t) &= \tilde{x}_1(t) \end{cases} . \tag{35}$$

Example 2:

$$\begin{cases} \dot{x}_1(t) &= -x_2(t-\tau_1) \\ \dot{x}_2(t) &= x_1(t-\tau_2) \\ y_1(t) &= x_1(t) \\ y_2(t) &= x_2(t-\tau_2) \\ x(t) &= \varphi(t), \quad t \in [-T, 0] \end{cases} \tag{36}$$

We have

$$\begin{array}{rcl}\dot{y}_1 &=& -\delta_1 x_2 \\ \ddot{y}_1 &=& -\delta_1\delta_2 x_1\end{array} \quad \text{and} \quad \frac{\partial(S, h_1^{(s_1)}, h_2^{(s_2)})}{\partial x} = \begin{bmatrix} 1 & 0 \\ 0 & \delta_1 \\ \delta_1\delta_2 & 0 \\ 0 & \delta_2 \end{bmatrix} \quad . \tag{37}$$

The input-output equations in linear form are

$$\begin{array}{rcl} d\ddot{y}_1 &=& -\delta_1\delta_2 dy_1 \\ \delta_1 dy_2 &=& -\delta_2 d\dot{y}_1 \Leftrightarrow dy_2 = -\delta_1^{-1}\delta_2 d\dot{y}_1 \quad . \end{array} \tag{38}$$

Then, Δ_{10} is the identity δ_0 operator, $\Delta_{11} = \delta_1\delta_2$, $\Delta_{21} = \delta_1$ and $\Delta_{22} = \delta_2$ (with corresponding combinations $T_{10} = 0$, $T_{11} = \tau_1 + \tau_2$, $T_{21} = \tau_1$ and $T_{22} = \tau_2$ of the two time-delays). Thus, $\bar{\Delta}_1 = \delta_0$ and $\bar{\Delta}_2 = \Delta_{21}$ (with $\bar{T}_1 = T_{10} = 0$ and $\bar{T}_2 = T_{21} = \tau_1$) and

$$M = \begin{bmatrix} 1 & 1 \\ -1 & 1 \end{bmatrix} \quad ,$$

which is of rank 2. Thus τ_1 and τ_2 are identifiable.

For this simple example, we can actually calculate the values of the two time lags from the explicit input-output equations $\ddot{y}_1 = -\delta_1\delta_2 y_1$, $y_2 = -\delta_1^{-1}\delta_2\dot{y}_1$. To illustrate, we have carried out a numerical simulation using the parameters $\tau_1 = 1$, $\tau_2 = \sqrt{2}$, $\varphi_1(t) = e^t$ and $\varphi_2(t) = t + 1$. We then plot

$$\mu_1(T_{11} - T_{10})|_{t_0} := \ddot{y}_1(t_0) + y_1(t_0 - T_{11} + T_{10})$$

and

$$\mu_2(T_{22} - T_{21})|_{t_0} := y_2(t_0) + \dot{y}_1(t_0 - T_{22} + T_{21})$$

for $t_0 = 4$. As expected, the functions are zero for $T_{11} - T_{10} = 1 + \sqrt{2}$ and $T_{22} - T_{21} = \sqrt{2} - 1$ and locally these are the only roots, see Fig. 1. Globally, there are also other roots for $\mu_2(T_{22} - T_{21})$ that can be seen in Fig. 1. They can be discarded in this case by plotting

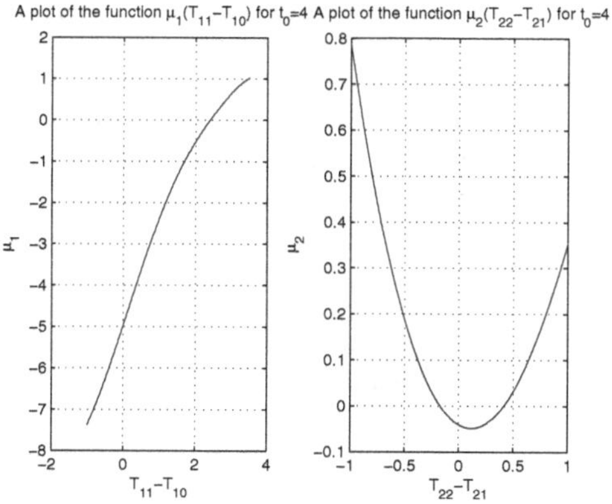

Fig. 1. *The functions* $\mu_1(T_{11} - T_{10})|_4$ *and* $\mu_2(T_{22} - T_{21})|_4$.

$$\dot{\mu}_2(T_{22} - T_{21})|_{t_0} := \frac{d}{dt}(y_2(t_0) + \dot{y}_1(t_0 - T_{22} + T_{21}))|_{t_0}$$

(and other subsequent time-derivatives) since $T_{22} - T_{21}$ must be a root for this, too, see Fig. 2. The simulations were carried out using MATLAB.

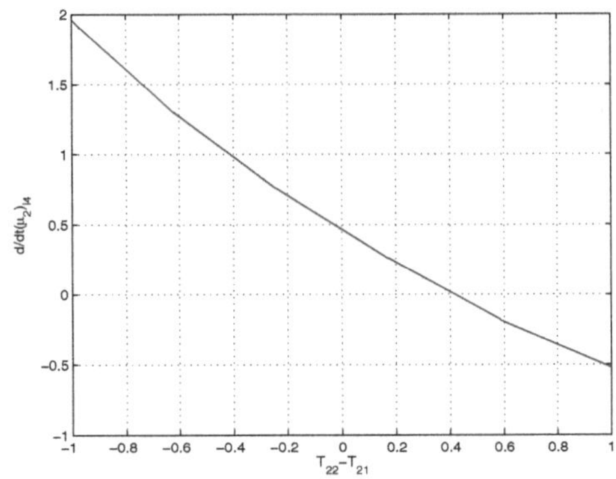

Fig. 2. *The function* $\dot{\mu}_2(T_{22} - T_{21})|_4$.

6 Conclusions

We have analyzed the identifiability of the time-lag parameters in nonlinear systems with multiple constant time delays.

State elimination is shown to yield an external input-output representation of the system, the form of which decides the identifiability of the delay parameters. For simpler models with few variables and parameters, the input-output equations can be used directly to identify the values of the time-lags from measured data.

We have formulated linear-algebraic criteria to check the identifiability of the delay parameters which eliminate the need for an explicit calculation of the input-output relations.

7 Acknowledgements

This work was supported by the National Research School in Genomics and Bioinformatics, the Swedish Research Council, and the Swedish Foundation for Strategic Research via GMMC and CMR.

References

1. Anguelova, M.: Observability and identifiability of nonlinear systems with applications in biology. PhD Thesis. Chalmers University of Technology and Göteborg University (2007)

2. Anguelova, M., Wennberg, B.: State elimination and identifiability of the delay parameter for nonlinear time-delay systems. Automatica 44(5), 1373–1378 (2008)
3. Belkoura, L., Orlov, Y.: Identifiability analysis of linear delay-differential systems. IMA J. Math. Control I. 19, 73–81 (2002)
4. Cohn, P.M.: Free rings and their relations, 2nd edn. Academic Press, London (1985)
5. Conte, G., Moog, C.H., Perdon, A.M.: Nonlinear control systems: An algebraic setting. LNCIS, vol. 242. Springer, London (1999)
6. Forsman, K., Habets, L.: Input-output equations and observability for polynomial delay systems. In: Proc. 33rd IEEE Conference on Decision and Control, Lake Buena Vista, USA, pp. 880–882 (1994)
7. Hermann, R., Krener, A.J.: Nonlinear controllability and observability. IEEE Transactions on Automatic Control 22, 728–740 (1977)
8. Lam, T.Y.: Lectures on modules and rings. Springer, Heidelberg (1999)
9. Márquez-Martínez, L.A., Moog, C.H., Velasco-Villa, M.: The structure of nonlinear time delay systems. Kybernetika 36, 53–62 (2000)
10. Moog, C.H., Castro-Linares, R., Velasco-Villa, M., Márquez-Martínez, L.A.: The disturbance decoupling problem for time-delay nonlinear systems. IEEE Transactions on Automatic Control 45, 305–309 (2000)
11. Nakagiri, S., Yamamoto, M.: Unique identification of coefficient matrices, time delays and initial functions of functional differential equations. Journal of Mathematical Systems, Estimation and Control 5, 1–22 (1995)
12. Orlov, Y., Belkoura, L., Richard, J.P., Dambrine, M.: On identifiability of linear time-delay systems. IEEE Transactions on Automatic Control 47, 1319–1324 (2002)
13. Pohjanpalo, H.: System identifiability based on the power series expansion of the solution. Math. Biosci. 41, 21–33 (1978)
14. Vajda, S., Godfrey, K., Rabitz, H.: Similarity transformation approach to identifiability analysis of nonlinear compartmental models. Math. Biosci. 93, 217–248 (1989)
15. Verduyn Lunell, S.M.: Parameter identifiability of differential delay equations. Int. J. Adapt. Control 15, 655–678 (2001)
16. Xia, X., Márquez, L.A., Zagalak, P., Moog, C.H.: Analysis of nonlinear time-delay systems using modules over noncommutative rings. Automatica 38, 1549–1555 (2002)
17. Zhang, J., Xia, X., Moog, C.H.: Parameter identifiability of nonlinear systems with time-delay. IEEE Transactions on Automatic Control 47, 371–375 (2006)

Time Optimal and Optimal Impulsive Control for Coupled Differential Difference Point Delay Systems with an Application in Forestry

Erik I. Verriest[1] and Pierdomenico Pepe[2]

[1] School of Electrical and Computer Engineering, Georgia Institute of Technology, Atlanta, GA 30332-0250, USA erik.verriest@ee.gatech.edu
[2] Dipartimento di Ingegneria Elettrica, Università degli Studi dell'Aquila, 67040 L'Aquila, Italy pepe@ing.univaq.it

Summary. In this paper we consider the optimal timing and impulsive control for a class of systems described by coupled differential and continuous time difference equations. The necessary conditions for optimality follow the approach for a single time-optimal impulse control, which is derived in full detail, using a streamlined approach with appropriately defined Hamiltonian. An example in regeneration of a deforested area is reported.

1 Introduction

In [6], stability for a class of systems described by coupled differential and continuous time difference equations was studied. In this chapter we continue the study of such coupled systems, and consider the optimal timing and impulsive control for such systems. Optimal impulsive control problems for a system with a single discrete delay were studied in [10, 12], and for distributed delays in [11]. The class of impulsive systems consists of systems where the control consists only of a fixed sequence modulated impulses, the control variables being the impulse times and their magnitudes. Necessary conditions for a stationary solution were derived and shown to extend those of the delay free case. An application of this result in control of epidemics through vaccination is found in [13]. It is therefore natural to extend these results for the class of coupled delay differential difference equations. In the present paper we single out an application in ecology, and more precisely forestry. Indeed, periodic and impulsive control seems to be a 'natural' in this context [14].

It is known that the necessary conditions for the optimal impulsive control are essentially determined by concatenation of the inter-impulse intervals [10]. Hence, as a "warm-up" and illustration of the techniques involved, we fist discuss the optimal stopping problem in Section 2. In Section 3, an application to forestry is given. However, the model involved is typical for a large class of single-species population dynamics and may be applicable to other ecological realms. Possible extensions for general impulsive control are briefly sketched in Section 4.

J.J. Loiseau et al. (Eds.): Topics in Time Delay Systems, LNCIS 388, pp. 255–265.
springerlink.com

2 Optimal Initialization and Stopping

Consider an autonomous system, governed by the equations

$$\dot{\xi} = f(\xi, x_\tau) \tag{1}$$

$$x = g(\xi, x_\tau) \tag{2}$$

The notation x_τ means $x_\tau(t) = x(t-\tau)$, which departs from its usual meaning as the (infinite dimensional) state at time τ, as used in [5]. The reason for this departure is the notational convenience it allows. Of course, without a control of any kind, the entire solution is predetermined, and there would be nothing, except for the stopping time itself to influence the performance. For this reason, we shall assume that the *initial data*, or part of it is also free. This initial data consists of the segment $\{x(\theta) \mid \theta \in (-\tau, 0]\}$, which we assume to be a continuous function, and the initial vector $\xi(0)$. The goal in optimal stopping is to find the $T > 0$ minimizing a performance index, J, equal to [3]

$$\begin{aligned} J = {} & \Psi_i(\xi(0)) + \int_{-\tau}^{0} K_i(x(\theta), \theta)\, \mathrm{d}\theta + \int_0^T L(\xi, x, t, T)\, \mathrm{d}t + \\ & + \Psi_f(\xi(T), T) + \int_{-\tau}^{0} K_f(x(T+\theta), T+\theta)\, \mathrm{d}\theta \end{aligned} \tag{3}$$

with or without state constraints, $M_i(\xi(0)) = 0$ and $M_f(\xi(T), T) = 0$, respectively at the initial and at the final time. Such problems occur for instance in optimal periodic control. As usual, the problem is solved by adjoining the state equations with Lagrange multiplier functions λ, and the final constraints with a Lagrange multiplier vector ζ to form $\overline{J}$

$$\begin{aligned} \overline{J} = {} & \Psi_i(\xi(0)) + \zeta_i' M_i(\xi(0)) + \int_{-\tau}^{0} K_i(x(\theta), \theta)\, \mathrm{d}\theta + \\ & + \int_0^T \Big[L(\xi, x, t, T) + \lambda'(f(\xi, x_\tau) - \dot{\xi})\mu'(g(\xi, x_\tau) - x) \Big]\, \mathrm{d}t + \\ & + \Psi_f(\xi(T), T) + \zeta_f' M_f(\xi(T), T) + \int_{-\tau}^{0} K_f(x(T+\theta), T+\theta)\, \mathrm{d}\theta. \end{aligned} \tag{4}$$

As this does not change the numerical values of the performance index, we get $\overline{J} = J$ numerically. Consider then a perturbation in the free final time and the initial data, all scaled by the parameter ϵ:

$$\begin{aligned} \xi_0 &\to \xi_0 + \epsilon\tilde{\xi}_0 \\ x(\theta) &\to x(\theta) + \epsilon\tilde{x}(\theta), \qquad \theta \in (-\tau, 0] \\ T &\to T + \epsilon\theta. \end{aligned}$$

and their induced perturbations (for $t \in (0, T)$) of the state variables

$$\begin{aligned} \xi &\to \xi + \epsilon\tilde{\xi} \\ x &\to x + \epsilon\tilde{x}. \end{aligned}$$

Define, for simplicity

[3] This is not the most general performance index but it suffices to capture all interesting behavior.

$$\Phi_i(\xi,\zeta) = \Psi_i(\xi) + \zeta' M_i(\xi) \tag{5}$$
$$\Phi_f(\xi,\zeta,T) = \Psi_f(\xi,T) + \zeta' M_f(\xi,T). \tag{6}$$

The induced change in $\overline{J}$ is obtained as

$$\begin{aligned}
\overline{J}_\epsilon = {} & \Phi_i(\xi(0) + \epsilon\tilde{\xi}(0)), \zeta_i) + \int_{-\tau}^{0} K_i(x(\sigma) + \epsilon\tilde{x}(\sigma))\,\mathrm{d}\sigma + \\
& + \int_0^{T+\epsilon\theta} \Big[L(\xi + \epsilon\tilde{\xi}, x + \epsilon\tilde{x}, t, T + \epsilon\theta) + \\
& \qquad + \lambda'[f(\xi + \epsilon\tilde{\xi}, x_\tau + \epsilon\tilde{x}_\tau) - \dot{\xi} - \epsilon\dot{\tilde{\xi}}] + \\
& \qquad + \mu'[g(\xi + \epsilon\tilde{\xi}, x_\tau + \epsilon\tilde{x}_\tau) - x - \epsilon\tilde{x}] \Big] \,\mathrm{d}t + \\
& + \Phi_f(\xi(T + \epsilon\theta) + \epsilon\tilde{\xi}(T + \epsilon\theta), \zeta_f, T + \epsilon\theta) + \\
& + \int_{-\tau}^{0} K_f(x(T + \epsilon\theta + \sigma) + \epsilon\tilde{x}(T + \epsilon\theta + \sigma), T + \epsilon\theta + \sigma)\,\mathrm{d}\sigma.
\end{aligned} \tag{7}$$

Defining the variation

$$\delta J = \frac{\overline{J}_\epsilon - \overline{J}_0}{\epsilon}$$

and using integration by parts, the following expression for δJ is obtained:

$$\begin{aligned}
\delta J = {} & \left.\frac{\partial \Phi_i}{\partial \xi}\right|_0 \tilde{\xi}(0) + \int_{-\tau}^{0} \left.\frac{\partial K_i}{\partial x}\right|_\sigma \tilde{x}(\sigma)\,\mathrm{d}\sigma + \\
& + \theta \Bigg[L(\xi(T), x(T), T) + \left.\frac{\partial \Phi_f}{\partial \xi}\right|_T \dot{\xi}(T) + \left.\frac{\partial \Phi_f}{\partial x}\right|_T \dot{x}(T) + \left.\frac{\partial \Phi_f}{\partial t}\right|_T + \\
& \qquad + \int_0^T \frac{\partial L}{\partial T}\,\mathrm{d}t + \int_{-\tau}^{0} \left(\left.\frac{\partial K_f}{\partial x}\right|_{T+\sigma} \dot{x}(T+\sigma) + \left.\frac{\partial K_f}{\partial t}\right|_{T+\sigma} \right) \mathrm{d}\sigma \Bigg] + \\
& + \int_0^T \left[\frac{\partial L}{\partial \xi} + \lambda' \frac{\partial f}{\partial \xi} + \dot{\lambda}' + \mu' \frac{\partial g}{\partial \xi} \right] \tilde{\xi}\,\mathrm{d}t \; + \; \lambda'(0)\tilde{\xi}(0) + \\
& + \int_0^T \left[\frac{\partial L}{\partial x} - \mu' \right] \tilde{x}\,\mathrm{d}t \; + \; \int_0^T \left[\lambda' \frac{\partial f}{\partial x_\tau} + \mu' \frac{\partial g}{\partial x_\tau} \right] \tilde{x}_\tau\,\mathrm{d}\tau \; + \\
& + \left[\left.\frac{\partial \Phi_f}{\partial \xi}\right|_T - \lambda'(T) \right] \tilde{\xi}(T) \; + \; \int_{-\tau}^{0} \left.\frac{\partial K_f}{\partial x}\right|_{T+\sigma} \tilde{x}(T+\sigma)\,\mathrm{d}\sigma.
\end{aligned} \tag{8}$$

By choosing the Lagrange multiplier function, λ, in the interval $[0, T]$ as the solution to

$$\dot{\lambda} = -\left[\left(\frac{\partial L}{\partial \xi}\right)' + \left(\frac{\partial f}{\partial \xi}\right)' \lambda + \left(\frac{\partial g}{\partial \xi}\right)' \mu \right], \tag{9}$$

the variation of the performance index reduces to

$$
\begin{aligned}
\delta J = \theta[A] &+ \left[\lambda'(0) + \left(\frac{\partial \Phi_i}{\partial \xi}\right)_0\right]\tilde{\xi}_0 + \\
&+ \int_{-\tau}^{0} \left.\frac{\partial K_i}{\partial x}\right|_{\sigma} \tilde{x}(\sigma)\,\mathrm{d}\sigma + \int_0^T \left[\frac{\partial L}{\partial x} - \mu'\right]\tilde{x}\,\mathrm{d}t + \\
&+ \int_0^T \left[\lambda' \frac{\partial f}{\partial x_\tau} + \mu' \frac{\partial g}{\partial x_\tau}\right]\tilde{x}_\tau\,\mathrm{d}t + \left[\left.\frac{\partial \Phi_f}{\partial \xi}\right|_T - \lambda'(T)\right]\tilde{\xi}(T) \\
&+ \int_{-\tau}^{0} \left.\frac{\partial K_f}{\partial x}\right|_{T+\sigma} \tilde{x}(T+\sigma)\,\mathrm{d}\sigma.
\end{aligned} \tag{10}
$$

where we have combined the θ-dependence in

$$
\begin{aligned}
A = L(\xi(T), x(T), T) &+ \left.\frac{\partial \Phi_f}{\partial \xi}\right|_T \dot{\xi}(T) + \left.\frac{\partial \Phi_f}{\partial t}\right|_T + \\
&+ \int_0^T \frac{\partial L}{\partial T}\,\mathrm{d}t + \int_{-\tau}^{0}\left[\left.\frac{\partial K_f}{\partial x}\right|_{T+\sigma} \dot{x}(T+\sigma) + \left.\frac{\partial K_f}{\partial t}\right|_{T+\sigma}\right]\mathrm{d}\sigma.
\end{aligned} \tag{11}
$$

Transform the third integral in (10) as follows

$$
\begin{aligned}
&\int_0^T \left[\lambda' \frac{\partial f}{\partial x_\tau} + \mu' \frac{\partial g}{\partial x_\tau}\right]_t \tilde{x}(t-\tau)\,\mathrm{d}t \\
&= \int_{-\tau}^{T-\tau} \left[\lambda' \frac{\partial f}{\partial x_\tau} + \mu' \frac{\partial g}{\partial x_\tau}\right]_{t+\tau} \tilde{x}(t)\,\mathrm{d}t \\
&= \left(\int_{-\tau}^{0} + \int_0^T - \int_{T-\tau}^{T}\right)\left[\lambda' \frac{\partial f}{\partial x_\tau} + \mu' \frac{\partial g}{\partial x_\tau}\right]_{t+\tau} \tilde{x}(t)\,\mathrm{d}t.
\end{aligned}
$$

Recombining with the other terms in (10) gives

$$
\begin{aligned}
\delta J = \theta[A] &+ \left[\lambda'(0) + \left(\frac{\partial \Phi_i}{\partial \xi}\right)_0\right]\tilde{\xi}_0 + \\
&+ \int_{-\tau}^{0}\left[\left(\lambda' \frac{\partial f}{\partial x_\tau} + \mu' \frac{\partial g}{\partial x_\tau}\right)_{\sigma+\tau} + \left.\frac{\partial K_i}{\partial x}\right|_{\sigma}\right]\tilde{x}(\sigma)\,\mathrm{d}\sigma + \\
&+ \int_0^T \left[\frac{\partial L}{\partial x} - \mu' + \left(\lambda' \frac{\partial f}{\partial x_\tau} + \mu' \frac{\partial g}{\partial x_\tau}\right)_{t+\tau}\right]\tilde{x}(t)\,\mathrm{d}t + \\
&+ \left[\left.\frac{\partial \Phi_f}{\partial \xi}\right|_T - \lambda'(T)\right]\tilde{\xi}(T) + \\
&+ \int_{-\tau}^{0}\left[\left.\frac{\partial K_f}{\partial x}\right|_{T+\sigma} - \left(\lambda' \frac{\partial f}{\partial x_\tau} + \mu' \frac{\partial g}{\partial x_\tau}\right)_{T+\tau+\sigma}\right]\tilde{x}(T+\sigma)\,\mathrm{d}\sigma.
\end{aligned} \tag{12}
$$

Looking at the third integral in (12), and by analogy to the choice of λ, it is crying out loudly to select the multiplier function μ in the interval $[0, T]$ as the solution to the backward *difference* equation

$$
\mu = \left(\frac{\partial L}{\partial x}\right)' + \left(\frac{\partial f^+}{\partial y}\right)' \lambda_\tau^+ + \left(\frac{\partial g^+}{\partial y}\right)' \mu_\tau^+. \tag{13}
$$

We used the notation λ_τ^+ to mean $\lambda_\tau^+(t) = \lambda(t+\tau)$, and similarly for μ_τ^+. Likewise, $\frac{\partial f^+}{\partial y}(t)$ means $\frac{\partial f(x,y)}{\partial y}$ evaluated at $x = \xi(t+\tau)$ and $y = x(t)$.

This reduces the first variation in the performance index further to $\delta J = \theta[A] + B$, where the terms in B are all boundary related terms. Now we note that (13) is a difference equation which requires final data $\mu(T+\sigma)$ for $\sigma \in (-\tau, 0]$, and that (9) is a differential equation requiring a final condition on ξ. The objective is to use all the available freedom to make these terms vanish. Thus let the mixed boundary conditions for $\sigma \in (-\tau, 0]$ be

$$\left[\lambda(0) + \left(\frac{\partial \Phi_i}{\partial \xi}\right)'_0\right] \tilde{\xi}_0 = 0 \tag{14}$$

$$\left[\left.\frac{\partial \Phi_f}{\partial \xi}\right|_T - \lambda'(T)\right] \tilde{\xi}(T) = 0 \tag{15}$$

$$\left[\left(\lambda' \frac{\partial f}{\partial x_\tau} + \mu' \frac{\partial g}{\partial x_\tau}\right)_{\sigma+\tau} + \left.\frac{\partial K_i}{\partial x}\right|_\sigma\right] \tilde{x}(\sigma) \equiv 0, \tag{16}$$

$$\left[\left.\frac{\partial K_f}{\partial x}\right|_{T+\sigma} - \left(\lambda' \frac{\partial f}{\partial x_\tau} + \mu' \frac{\partial g}{\partial x_\tau}\right)_{T+\tau+\sigma}\right] \tilde{x}(T{+}\sigma) \equiv 0. \tag{17}$$

If these conditions can be satisfied, then $\delta J = \theta[A]$, and the necessary condition for stationarity is $A = 0$, which generalizes the *transversality* condition.

Let us focus on the special case where the initial data on x is specified, $\xi(0)$ is a design variable (constrained to lie on the initial manifold) with final condition $\xi(T) = \xi_f$ specified, or constrained to lie on a k_f-dimensional manifold. Let $x(T+\sigma)$ be free for $\sigma \in (-\tau, 0]$. Under these conditions we have $\tilde{x}(\sigma) \equiv 0$, in the interval $(-\tau, 0]$ and (16) is automatically satisfied. If k_i constraints are imposed on the n components of ξ_0, then choose $\lambda(0)$ in terms of the k_i unknowns of ζ_i

$$\lambda'(0) + \left.\frac{\partial \Psi_i}{\partial \xi}\right|_0 + \zeta_i' \left.\frac{\partial M_i}{\partial \xi}\right|_0 = 0.$$

Likewise, if $\xi(T)$ is partially specified or constrained by a k_f-dimensional manifold, express the n components of $\lambda(T)$ in terms of the k_f components of ζ_f.

$$\left.\frac{\partial \Psi_f}{\partial \xi}\right|_T + \zeta_f' \left.\frac{\partial M_f}{\partial \xi}\right|_T - \lambda'(T) = 0.$$

Together, this gives $2n + k_i + k_f$ unknown parameters with $k_i + k_f$ constraints. and $2n$ relations. Now let's look at the difference equation part. Since in this case $\tilde{x}(T+\sigma)$ may be arbitrary, we choose

$$\left.\frac{\partial K_f}{\partial x}\right|_{T+\sigma} - \left(\lambda' \frac{\partial f}{\partial x_\tau} + \mu' \frac{\partial g}{\partial x_\tau}\right)_{T+\tau+\sigma} \equiv 0.$$

If the "future" λ and μ (i.e., in the interval $(T, T+\tau)$ can be chosen to satisfy the above, then for $t = T + \sigma$ in the interval $(T - \tau, T)$, equation (13) is effectively

$$\mu(T+\sigma) = \left.\left(\frac{\partial L}{\partial x}\right)'\right|_{T+\sigma} + \left.\left(\frac{\partial K_f}{\partial x}\right)'\right|_{T+\sigma}. \tag{18}$$

This provides the effective *final data* for the costate μ, which then propagates backwards according to (13) from $t < T - \tau$ to 0.

Summary: Necessary Conditions for Optimality

These choices can be streamlined as follows: Define a *generalized* Hamiltonian, H^o

$$\begin{aligned} &H^o(\xi,\eta,x,y,\lambda,\omega,\mu,\nu) \\ &= L(\xi,x,T)+\lambda' f(\xi,y)+\mu' g(\xi,y)+\omega' f(\eta,x)+\nu' g(\eta,x). \end{aligned} \tag{19}$$

For notational simplicity the explicit dependence on t is suppressed. The state equations are given by:

$$\dot{\xi} = \left(\frac{\partial H^o}{\partial \lambda}\right)' \tag{20}$$

$$x = \left(\frac{\partial H^o}{\partial \mu}\right)' \tag{21}$$

Likewise, the co-state or adjoint equations are:

$$\dot{\lambda} = -\left(\frac{\partial H^o}{\partial \xi}\right)' \tag{22}$$

$$\mu = \left(\frac{\partial H^o}{\partial x}\right)'. \tag{23}$$

The boundary conditions for $\sigma \in (-\tau, 0]$ are given by

$$\left[\lambda(T) - \left(\frac{\partial \Phi_f}{\partial x}\right)'_T\right] = 0 \tag{24}$$

$$\left[\left.\frac{\partial K_f}{\partial x}\right|_{T+\sigma} - \left(\lambda' \frac{\partial f}{\partial x_\tau} + \mu' \frac{\partial g}{\partial x_\tau}\right)_{T+\tau+\sigma}\right] \equiv 0. \tag{25}$$

Finally, consistency requires

$$y(t) = x(t-\tau) \quad ; \quad \omega(t) = \lambda(t+\tau) \tag{26}$$

$$\eta(t) = \xi(t+\tau) \quad ; \quad \nu(t) = \mu(t+\tau). \tag{27}$$

These choices reduce the variation in the performance index simply to $\theta[A]$. The *necessary* conditions for stationarity are therefore

$$\left[\lambda(0) + \left(\frac{\partial \Phi_i}{\partial \xi}\right)'_0\right] = 0 \tag{28}$$

and $[A] = 0$, A given in (10), which is the celebrated *transversality condition.*

Remark 1. This transversality condition reduces to the vanishing of

$$\begin{aligned} &L(\xi(T),x(T),T) + \lambda(T) f(\xi(T),x(T-\tau)) + \frac{\partial \Phi_f}{\partial T} + \\ &+ \int_0^T \frac{\partial L}{\partial T}\, \mathrm{d}t + \int_{T-\tau}^T \frac{\partial K_f}{\partial x}\, \mathrm{d}x(t) + \int_{T-\tau}^T \frac{\partial K_f}{\partial T}\, \mathrm{d}t, \end{aligned} \tag{29}$$

after substituting the boundary conditions. Note that we rewrote the integral involving $\dot{x}$ as a Stieltjes integral, which is more in line with the character of the original state equations.

Remark 2. We have chosen the *running cost* in the form $L(\xi(t), x(t))$. In some problems it may be more appropriate to have $L(\xi(t), x(t-\tau))$ instead. This can easily be converted to the present form as follows. We compute the variation of the total running cost (with free endpoint) in the second case

$$
\begin{aligned}
&\delta \int_0^T L(\xi, x_\tau, t, T)\,\mathrm{d}t \\
&= \int_0^T \left[\frac{\partial L}{\partial \xi}\tilde{\xi} + \left.\frac{\partial L}{\partial y}\right|_t \tilde{x}_\tau\right] \mathrm{d}t + \left[L(\xi(T), x(T-\tau), T) + \int_0^T \frac{\partial L}{\partial T}\,\mathrm{d}t\right] \delta T \\
&= \int_0^T \frac{\partial L}{\partial \xi}\tilde{\xi}\,\mathrm{d}t + \int_{-\tau}^{T-\tau} \left.\frac{\partial L}{\partial y}\right|_{t+\tau} \tilde{x}\,\mathrm{d}t + \left[L(\xi(T), x(T-\tau), T) + \int_0^T \frac{\partial L}{\partial T}\,\mathrm{d}t\right] \delta T \\
&= \int_0^T \frac{\partial L}{\partial \xi}\tilde{\xi}\,\mathrm{d}t + \int_0^T \left.\frac{\partial L}{\partial y}\right|^+ \tilde{x}\,\mathrm{d}t - \int_{T-\tau}^T \left.\frac{\partial L}{\partial y}\right|^+ \tilde{x}\,\mathrm{d}t + \\
&\qquad + \left[L(\xi(T), x(T-\tau), T) + \int_0^T \frac{\partial L}{\partial T}\,\mathrm{d}t\right] \delta T. \qquad (30)
\end{aligned}
$$

We note that if x is specified in $(-\tau, 0)$, then $\tilde{x}$ is zero there, and the integral over this subinterval vanishes. Such a situation may occur if we want to penalize any deviation from non-constancy (the quasi static approach) [9]. In such a case, a suitable running cost may be $\int_0^T \|\dot{\xi}\|^2\,\mathrm{d}t$. Substituting the state equation gives then the corresponding cost rate $L(\xi, x_\tau) = \|f(\xi, x_\tau)\|^2$, which is of the above form.

Remark 3. Approach to periodicity: The condition $\xi(T) = \xi(0)$ is not sufficient for periodicity. However periodicity may be approximated by adding a quadratic cost term to the performance index (i.e., a soft terminal constraint)

$$\int_{-\tau}^0 \|x(T+\sigma) - x(\sigma)\|_Q^2\,\mathrm{d}\sigma.$$

with appropriately chosen weight matrix function $Q(\sigma) > 0$.

3 A Model from Mathematical Ecology

The following is a model for the repopulation of a single species. We may think of a deforested area, where one is interested in reforestation through the planting of saplings. Tilman based a simple one species model on the delay-free logistic or Pearl-Verhulst equation,

$$\dot{y}(t) = ry(t)\left[1 - \frac{y(t)}{K}\right], \qquad (31)$$

with intrinsic growth rate r, and environmental carrying capacity K. Effects of soil depletion and erosion lead to a refinement of the above model to the form

$$\dot{y}(t) = ry(t)\left[1 - \frac{y(t-\tau) + c\dot{y}(t-\tau)}{K}\right]. \qquad (32)$$

The ecological relevance can be found in [7], [3] and [2]. Essentially, it means that crowding subsequently kills trees and new growth especially exhausts resources.

3.1 Coupled Model

Assume that the population of the species in the given ecosystem was extinct before $t = 0$. This means that we have $y(\sigma) \equiv 0$ for $\sigma \in [-\tau, 0]$. At time zero, a mass ξ_0 of the species is introduced (e.g., by planting of saplings). We want to determine the optimal number ξ_0 so that the population increases with minimal ecological impact to its carrying capacity, K, and will be sustained there. Moreover, we want to determine the time this will take. We note that for the population to be sustained at the carrying capacity, we need $\dot{y} = 0$ and $y(T+\sigma) = K$ for $\sigma \in [-\tau, 0]$. Letting $x = y + c\dot{y}$ and $\xi = y$, yields

$$x = \xi(1 + rc) - \frac{rc}{K}\,\xi\, x_\tau = g(\xi, x_\tau)$$

and

$$\dot{\xi} = r\xi\left(1 - \frac{x_\tau}{K}\right) = f(\xi, x_\tau),$$

which are of the coupled form we considered. The initial data for x is specified (zero), while ξ_0 is the only free variable.

For this model [1, Theorem 5.1] determined that if $rc < 1$ and delay $\tau < \tau_0$, the equilibrium is asymptotically stable, and unstable if either $\tau > \tau_0$ or $rc > 1$, where τ_0 is given by

$$\tau_0 = \frac{\sqrt{1-(rc)^2}}{r}\,\operatorname{arccot}\left(-\frac{rc}{\sqrt{1-(rc)^2}}\right). \tag{33}$$

3.2 Choice of Performance Index

If we opt for a sustained equilibrium after some time, T, from a single introduction of the species at time zero, we have $x(\theta) = 0$ for $\theta = (-\tau, 0]$, and $\xi(0) = \xi_0$, the "control" variable. Since the equilibrium value is the carrying capacity K, we use a performance index which penalizes a deviation from a linear growth of the species up to the carrying capacity. This is desirable in order to smoothen the ecological impact of the changing landscape, and allow a smoother adaptation for the insect, bird and herbivore population. In addition, for obvious economic reasons, we want to bound the biomass that should be introduced initially while also not to have to wait too long to reach equilibrium. Instead of choosing the exact final condition $\xi(T) = K$, we use a soft final constraint (for reasons of reachability) by penalizing the deviation $(\xi(T) - K)^2$. Likewise, $x(T + \sigma)$ should remain as close as possible to K for $\sigma \in [-\tau, 0]$. Hence we choose an additional soft constraint with $K_f(x, t) = \beta(x - K)^2$, where β is some weight coefficient. Putting this all together gives the performance index

$$J = \xi_0 + \sigma(\xi(T) - K)^2 + \rho T + \kappa\int_0^T (\xi - tK/T)^2\,\mathrm{d}t + \beta\int_{T-\tau}^T (x(\theta) - K)^2\,\mathrm{d}\theta. \tag{34}$$

Here, σ, ρ, κ and β are positive weight factors. From

$$L(\xi, x, T) = \rho + \kappa\left(\xi(t) - \frac{Kt}{T}\right)^2 \tag{35}$$

we define the generalized Hamiltonian

$$\begin{aligned} H^{o}(\xi,\eta,x,y,\lambda,\omega,\mu,\nu) &= \rho + \kappa\left(\xi - \frac{Kt}{T}\right)^2 + \\ &+ \lambda r\xi\left(1-\frac{y}{K}\right) + \mu\xi\left(1+rc-\frac{rc}{K}y\right) + \\ &+ \omega r\eta\left(1-\frac{x}{K}\right) + \nu\eta\left(1+rc-\frac{rc}{K}x\right). \end{aligned} \tag{36}$$

It follows from the derivations in Section 2 that the necessary conditions for an optimum are:
Co-state equations

$$\begin{aligned} \dot{\lambda}(t) &= -2\kappa\left(\xi(t) - \frac{Kt}{T}\right) - r\lambda(t)\left(1-\frac{x(t-\tau)}{K}\right) \\ &-\mu(t)\left(1+rc-\frac{rc}{K}x(t-\tau)\right), \qquad t\in(0,T) \end{aligned} \tag{37}$$

$$\begin{aligned} \mu(t) &= -\frac{r}{K}\lambda(t+\tau)\xi(t+\tau) - \frac{rc}{K}\mu(t+\tau)\xi(t+\tau), \\ &\qquad\qquad t\in(0,T-\tau) \end{aligned} \tag{38}$$

with the boundary conditions

$$\lambda(T) = 2\sigma(\xi(T)-K) \tag{39}$$

$$\mu(t) = 2\beta(x(t)-K), \qquad t\in(T-\tau,T). \tag{40}$$

The optimality conditions are

$$\lambda(0) = -1, \tag{41}$$

and the transversality condition, for which we get herein terms of the state variables at the final time:

$$\begin{aligned} A(T) &= \rho + \kappa(\xi(T)-K)^2 + 2\sigma r\xi(T)(\xi(T)-K)\left(1-\frac{x(T-\tau)}{K}\right) + \\ &+ \frac{2\kappa K}{T^2}\int_0^T t\left(\xi(t)-\frac{Kt}{T}\right)\mathrm{d}t + 2\beta\int_{T-\tau}^T (x(t)-K)\,\mathrm{d}x(t). \end{aligned} \tag{42}$$

The transversality condition itself being $A(T)=0$.

To solve the problem numerically, assume an initial pair ξ_0 and T, and integrate the state equations by the method of steps (i.e., stagger the solution for ξ and then x in the successive intervals $(k\tau, (k+1)\tau)$). Next solve the costate equations backwards from T, to obtain $\lambda(0)$. The *gradients* of $\overline{J}$ with respect to T and ξ_0 are respectively A and $\lambda(0) + \frac{\partial \Phi_i}{\partial \xi} = \lambda(0)+1$. Finally correct ξ_0 and T by adjustment in the negative gradient direction, choosing an appropriate stepsize. In similar problems we used the Armijo stepsize [13].

3.3 Discussion

In figure 1, we show the optimal initialization and the consequent evolution of ξ for the system with delay $\tau = 1$, parameters $r = c = 0.1$ and carrying capacity $K = 100$. The choices for the cost parameters in (34) were $\kappa = 1$, $\sigma = 10$, $\beta = 1$ and $\rho = 1$. The optimal horizon is $T = 49.8$. The optimal initialization is $\xi(0) = 7.34$. In contrast, figure 2 shows for

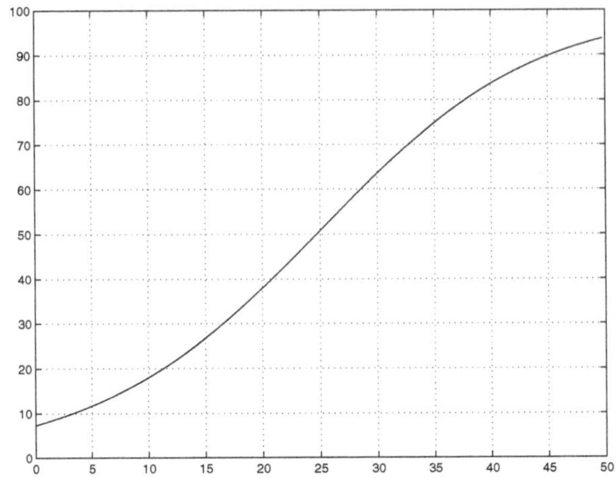

Fig. 1. Optimal population growth for $\tau = 1$.

the same parameters, but with delay $\tau = 10$ the (nonoptimal) initialization with $\xi(0) = 20$ until $T = 100$. (The choice of T does not affect the evolution of the population, but only the PI.) We note a damped oscillation about the carrying capacity, which is expected since the stability bound is $\tau_0 = 16.7$. It turns out that this solution does not differ greatly from the non-neutral case $c = 0$.

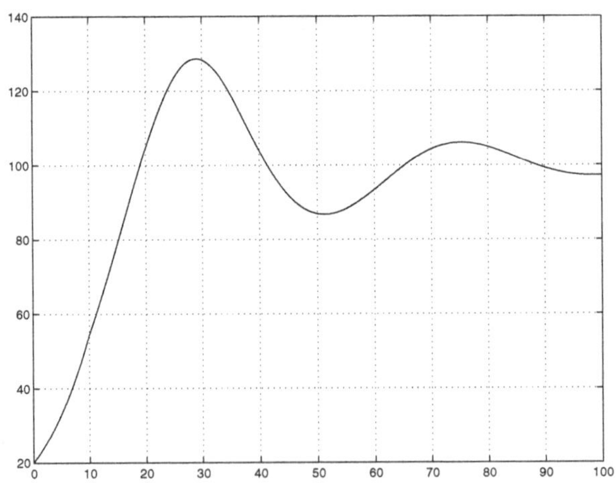

Fig. 2. Non-optimal population growth for $\tau = 10$.

4 Variational Approach to Optimal Impulsive Control

The approach of section 2 is now indicative of how to proceed to the impulsive control. In a way, except for the initial data with each renewal, it simply is a concatenation of several optimal timing problems. But this initial data is still deterministic and depends on the solution obtained in the prior interval. As such it follows with minor changes the derivations given in [10]. It is typical for such problems to have many stationary points. Hence, it is advisable to use several runs with different initial conditions on $\xi(0)$ and T, in order to determine the *local* solutions. Evaluation of J at these local solutions then enables the determination of the global solution.

References

1. Freedman, H.I., Kuang, Y.: Stability switches in linear scalar neutral delay equations. Funkcialaj Ekvacioj 34, 187–209 (1991)
2. Kuang, Y.: Delay differential equations with applications in population dynamics. Academic Press, San Diego (1993)
3. Gopalsamy, K., Zhang, B.G.: On a neutral delay-logistic equation. Dynamics and Stability of Systems 2, 183–195 (1988)
4. Hadeler, K.P., Müller, J.: Optimal harvesting and optimal vaccination. Mathematical Biosciences 206, 249–272 (2007)
5. Hale, J.K., Verduyn Lunel, S.M.: Introduction to functional differential equations. Applied Math. Sciences Series. Springer, New York (1993)
6. Pepe, P., Verriest, E.I.: On the stability of coupled delay differential and continuous time difference equations. IEEE Transactions on Automatic Control 48(8), 1422–1427 (2003)
7. Pielou, E.C.: Mathematical Ecology. Wiley Interscience, New York (1977)
8. Răsvan, V.: Functional differential equations of lossless propoagation and almost linear behavior. In: Proc. 6th IFAC Workshop on Time Delay Systems, l'Aquila, Italy (2006)
9. Verriest, E.I., Yeung, D.: Parity in LQ control: the infinite time limit for terminal control. In: Proc. 2006 American Control Conference, Minneapolis, USA, pp. 1706–1711 (2006)
10. Verriest, E.I., Delmotte, F., Egerstedt, M.: Optimal impulsive control for point delay systems with refractory period. In: Proc. 5th IFAC Workshop on Time Delay Systems, Leuven, Belgium (2004)
11. Verriest, E.I., Delmotte, F.: Optimal control for switched point delay systems with refractory period. In: Proc. 16th IFAC World Congress, Prague, Czeck Republic (2005)
12. Verriest, E.I.: Optimal control for switched distributed delay systems with refractory period. In: Proc. IEEE Conference on Decision and Control, Sevilla, Spain, pp. 1421–1426 (2005)
13. Verriest, E.I., Delmotte, F., Egerstedt, M.: Control of epidemics by vaccination. In: Proc. 2005 American Control Conference, Portland, USA, pp. 985–990 (2005)
14. Xiao, Y., Cheng, D., Qin, H.: Optimal impulsive control in periodic ecosystems. Systems & Control Letters 55, 558–565 (2006)

Part IV

Propagation and Flows

On the Stability of AQM Controllers Supporting TCP Flows

Daniel Melchor-Aguilar

Division of Applied Mathematics, IPICYT, 78216, San Luis Potosí, SLP, México,
dmelchor@ipicyt.edu.mx

Summary. In this chapter we address the local asymptotic stability of some AQM controllers supporting TCP flows. We first consider a PI controller as AQM strategy. The complete set of PI controllers that stabilizes the linearized closed-loop system is obtained. Then, we consider a delay-dependent AQM controller based on control laws for finite spectrum assignment of time-delay systems. Stability conditions for a numerically safe implementation of the controllers are provided.

1 Introduction

Since the dynamic fluid-flow model for describing the behavior of the transmission control protocol (TCP) in computer networks was introduced in [10], several control structures have been proposed as active queue management (AQM) to allow the routers to assist TCP management for congestion avoidance. Thus, proportional (P), proportional-integral (PI) and $\mathcal{H}^\infty$ AQM controllers have been proposed based on the linearization of the model in [3] and [12]. It was shown there that such controllers improve the performance obtained with standard AQM controllers (e.g. based on Random Early Detection (RED)).

Due to their simplicity, the P and PI controllers proposed in [3] has become a reference for the development of new AQM controllers as they are currently implemented in the Network Simulator [11]. However, such controller designs are based only on sufficient conditions for closed-loop stability of the linearization and, therefore, they do not provide the set of all locally stabilizing P and PI gain values. The knowledge of the set of stabilizing controllers results important for the designer on determining some performance objectives as well as on considering system and controller perturbations.

In the recent paper [7], the complete set of P controllers that locally stabilizes the equilibrium point of a simplified version of the model was obtained. Despite this, to the best of the author's knowledge, there are no specific results for the problem of finding the complete set of PI stabilizing controllers, and one of the aims of this chapter is to focus on it.

In this chapter, we first develop a local stability analysis of a simplified version of the model introduced in [10] for a PI control-based AQM strategy. Necessary and sufficient conditions for stability of the closed-loop linearized system are derived. More explicitly, for a given set of network parameters (round-trip time, number of TCP loads and link capacity), we obtain the complete set of PI controllers that locally stabilizes the equilibrium point. As

J.J. Loiseau et al. (Eds.): Topics in Time Delay Systems, LNCIS 388, pp. 269–279.
springerlink.com

a subsidiary result, the complete set of robust stabilizing controllers is also obtained. Then, we suggest a delay-dependent controller for a simplified version the model based on the feedback control laws which assign a finite closed-loop spectrum to delay-systems. Stability conditions for a numerically safe implementation of the controller are provided. It is shown that for any given network parameters there is always a delay-dependent controller for which a safe implementation can be obtained.

The chapter is organized as follows: Section 2, introduces the fluid-flow mathematical model. The main results for PI controllers are presented in section 3. Section 4 is devoted to the design of delay-dependent AQM controllers and their practical implementation. We provide numerical examples where appropriate, and conclude in section 5.

2 Fluid-Flow Mathematical Model

We consider the dynamic fluid-flow model introduced in [10] for describing the behavior of n homogeneous TCP-controlled sources and a single congested router

$$\begin{cases} \dot{w}(t) = \frac{1}{\tau(t)} - \frac{1}{2}\frac{w(t)w(t-\tau(t))}{\tau(t-\tau(t))}p(t-\tau(t)), \\ \dot{q}(t) = n(t)\frac{w(t)}{\tau(t)} - c, \end{cases}$$

where $w(t)$ denotes the average of TCP windows size (packets), $q(t)$ is the average queue length (packets), $\tau(t) = \frac{q(t)}{c} + \tau_p$ is the round-trip time (secs) where τ_p represents the propagation delay, c is the link capacity (packets/secs), $n(t)$ is the number of TCP sessions, and $p(\cdot)$ is the probability of a packet marking which represents the AQM control strategy.

For the study of PI AQMcontrollers we approximate these dynamics by assuming that $n(t) \equiv n$ and $\tau(t) \equiv \tau$ are constants as in [3, 7, 12]. As a result we have the following simplified dynamics:

$$\begin{cases} \dot{w}(t) = \frac{1}{\tau} - \frac{1}{2\tau}w(t)w(t-\tau)p(t-\tau), \\ \dot{q}(t) = \frac{n}{\tau}w(t) - c. \end{cases} \tag{1}$$

For a desired equilibrium queue length q_0, the equilibrium (w_0, q_0, p_0) of (1) is determined by

$$w_0^2 p_0 = 2 \text{ and } w_0 = \frac{\tau c}{n}.$$

In order to get basic results on delay-dependent AQM controller for more general cases, we will consider in section 4 the following dynamic approximation of (1):

$$\begin{aligned} \dot{w}(t) &= \frac{1}{\tau} - \frac{1}{2\tau}w^2(t)p(t-\tau), \\ \dot{q}(t) &= \frac{n}{\tau}w(t) - c. \end{aligned} \tag{2}$$

System (2) approximates the local behavior of (1) about the equilibrium under the assumption $w_0 \gg 1$, see for instance [7] for a mathematical justification.

Our investigation will rely on linearization of the above systems around the equilibrium. Thus, stability will mean local stability near equilibrium, where for simplicity we use stability for the asymptotic stability concept.

3 Proportional-Integral AQM Controller

In order to design a stabilizing PI controller via the linearization of (1) about the equilibrium point, we introduce $\sigma(t) = \int_0^t (q(s) - q_0)\, ds$, and consider the augmented system

$$\begin{cases} \dot{w}(t) = \frac{1}{\tau} - \frac{1}{2\tau} w(t) w(t-\tau) p(t-\tau), \\ \dot{q}(t) = \frac{n}{\tau} w(t) - c, \\ \dot{\sigma}(t) = q(t) - q_0. \end{cases} \tag{3}$$

We now consider a PI controller of the form

$$p(t) = k_p q(t) + \frac{k_p}{I} \sigma(t), \tag{4}$$

where $\frac{k_p}{I} \neq 0$. It can be easily verified that the closed-loop system (3)-(4) has a unique equilibrium point (w_0, q_0, σ_0), where $\sigma_0 = \frac{I}{k_p}(p_0 - k_p q_0)$.

The linearization of the closed-loop system (3)-(4) about the equilibrium (w_0, q_0, σ_0) is

$$\dot{\xi}(t) = A\xi(t) + B\xi(t-\tau), \tag{5}$$

where $\xi(t) = \begin{pmatrix} \tilde{w}(t) \\ \tilde{q}(t) \\ \tilde{\sigma}(t) \end{pmatrix}$, $A = \begin{pmatrix} -\frac{n}{\tau^2 c} & 0 & 0 \\ \frac{n}{\tau} & 0 & 0 \\ 0 & 1 & 0 \end{pmatrix}$, $B = \begin{pmatrix} -\frac{n}{\tau^2 c} & -\frac{\tau c^2}{2n^2} k_p & -\frac{\tau c^2}{2n^2} \frac{k_p}{I} \\ 0 & 0 & 0 \\ 0 & 0 & 0 \end{pmatrix}$, $\tilde{w}(t) = w(t) - w_0$, $\tilde{q}(t) = q(t) - q_0$, and $\tilde{\sigma}(t) = \sigma(t) - \sigma_0$.

Assume for the moment that it is possible to find controller's gains that make (5) stable. Then, it follows that all solutions of (3) starting sufficiently close to (w_0, q_0, σ_0) approach it as t tends to infinity.

It is clear that one cannot investigate the stability of (5) for the delay-free case ($\tau = 0$). This is a particular property of TCP/AQM network systems, where the delay value (round-trip time) cannot be considered zero. Thus, the approach developed in [13], which is based on first determining the set of PI stabilizing controllers for the delay-free system, cannot be directly applied to determine the set of PI stabilizing controllers for (5).

3.1 Stability Analysis

It is well known that (5) is stable if and only if the characteristic function

$$f(s) = s^3 + \frac{n}{\tau^2 c} s^2 + \left[\frac{n}{\tau^2 c} s^2 + \frac{c^2}{2} k_p \left(s + \frac{1}{I} \right) \right] e^{-\tau s}$$

has no zeros with nonnegative real parts [2].

Theorem 1. *System (5) is stable if and only if the controller gains (I, k_p) belong to the stability region $\Phi_{(n,\tau,c)}$, plotted in Fig. 1, whose boundary in the controller gains space (I, k_p) is described by*

$$\partial\Phi_{(n,\tau,c)} = \Bigg\{ (I, k_p) : I = \frac{\omega \cos(\omega\tau) + \frac{n}{\tau^2 c} \sin(\omega\tau)}{\omega \left(\frac{n}{\tau^2 c} (1 + \cos(\omega\tau)) - \omega \sin(\omega\tau) \right)},$$

$$k_p = \frac{2n\omega}{c^2} \left(\omega \cos(\omega\tau) + \frac{n}{\tau^2 c} \sin(\omega\tau) \right), \omega \in (0, \omega^*) \Bigg\},$$

where ω^* is the solution of

$$\tan\left(\frac{\omega\tau}{2}\right) = \frac{n}{\tau^2 c\omega}, \ \omega \in \left(0, \frac{\pi}{\tau}\right). \tag{6}$$

Proof. First observe that since $\frac{k_p}{I} \neq 0$, $s = 0$ is not a zero of $f(s)$. Suppose that $f(s)$ has a pure imaginary zero $s = j\omega \neq 0$. Then, a direct calculation yields

$$\begin{cases} k_p = \frac{2n}{c^2}\omega\left(\omega\cos(\omega\tau) + \frac{n}{\tau^2 c}\sin(\omega\tau)\right), \\ I = \frac{\omega\cos(\omega\tau) + \frac{n}{\tau^2 c}\sin(\omega\tau)}{\omega\left(\frac{n}{\tau^2 c}(1+\cos(\omega\tau)) - \omega\sin(\omega\tau)\right)}. \end{cases} \tag{7}$$

This parameterization defines a countable number of curves in the parameter space (I, k_p) and each one of them is obtained by varying ω in the following intervals: $(0, \omega_0^*)$, $\left(\omega_k^*, \frac{(2k+1)\pi}{\tau}\right)$ and $\left(\frac{(2k+1)\pi}{\tau}, \omega_{k+1}^*\right)$, $k = 0, 1, 2, ...$, where ω_k^* is the solution of

$$\tan\left(\frac{\omega\tau}{2}\right) = \frac{n}{\tau^2 c\omega}, \quad \omega \in \left(\frac{2k\pi}{\tau}, \frac{(2k+1)\pi}{\tau}\right). \tag{8}$$

Since (8) is a transcendental equation we look directly for a numerical solution. This can be found by plotting the two functions $\tan\left(\frac{\omega\tau}{2}\right)$ and $\frac{n}{\tau^2 c\omega}$, see Fig. 2. These curves divide the plane (I, k_p) into a set of connected domains. From the argument principle is easy to show that for all (I, k_p) values inside the open domain $\Phi_{(n,\tau,c)}$, bounded by the curve obtained by varying ω in the interval $(0, \omega_0^*)$ and the coordinate axis $k_p = 0$, the function $f(s)$ has no zeros with strictly positive real part.

Remark 1. When $\tau \to +0$, the stability region $\Phi_{(n,\tau,c)}$ tends to the whole first quadrant of the plane (I, k_p). In other words, for small round-trip time (delay), arbitrarily PI controller's gains locally stabilizes the equilibrium point of (3).

Proof. >From parametrization (7) it is not difficult to see that $I(\omega) \to \frac{\tau^2 c}{2n} + \frac{\tau}{2}$ and $k_p(\omega) \to k_p(0) = 0$ when $\omega \to +0$. On the other hand, it holds that $I(\omega) \to +\infty$ and $k_p(\omega) \to k_p(\omega^*) = \frac{2n}{c^2}(\omega^*)^2$ when $\omega \to -\omega^*$. From the above and the fact that $\omega^* \to +\infty$ when $\tau \to +0$, see Fig. 2, the remark follows.

Remark 2. Given nominal network parameters (n_0, τ_0, c_0) and unknown network parameters (n, τ, c) satisfying

$$n \geq n_0, \tau \leq \tau_0 \text{ and } c \leq c_0, \tag{9}$$

then $\Phi_{(n_0,\tau_0,c_0)} \subseteq \Phi_{(n,\tau,c)}$ holds.

Proof. The remark follows directly from Proposition 2 in [3], which states that stabilizing against the largest expected values of τ and c, and the smallest expected value of n yields a robust stabilizing controller.

3.2 Example

Let us consider nominal network parameters $n_0 = 40$ TCP sessions, $\tau_0 = 0.7$ secs and $c_0 = 300$ packets/secs. As real network parameters we take the following values: $n = 50$ TCP sessions, $\tau = 0.533$ secs and $c = 250$ packets/secs. In Fig. 3 we plot the stability region $\Phi_{(n_0,\tau_0,c_0)}$ in the controller's gains space. In Fig. 4 we plot the response of $q(t)$ for the two pairs of gains $A = (7, 3.5 \times 10^{-4})$ and $B = (15, 4.5 \times 10^{-4})$ inside of $\Phi_{(n_0,\tau_0,c_0)}$, see Fig. 3. The simulations were carried out on the nonlinear model (3). The operation point was chosen as $q_0 = 200$ packets.

It can be seen that the robust stabilization is reached. On the other hand, the responses obtained for the two different pairs of gains show the importance of knowing the complete set of controller parameter values that locally stabilize the closed-loop system.

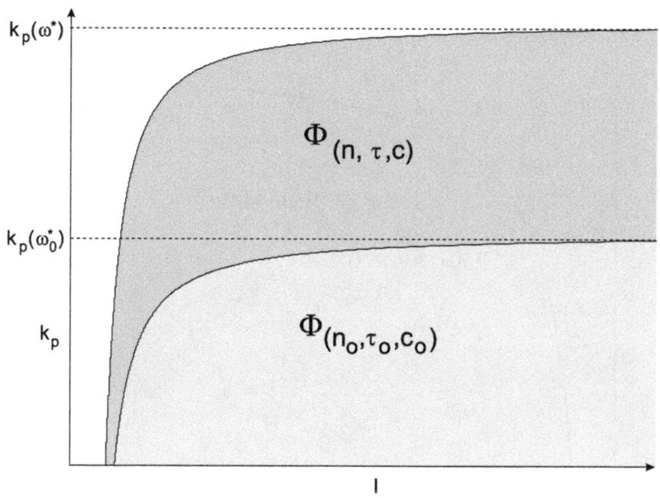

Fig. 1. Stability Regions $\Phi_{(n,\tau,c)}$ and $\Phi_{(n_0,\tau_0,c_0)}$ for (n_0, τ_0, c_0) and (n, τ, c) satisfying (9).

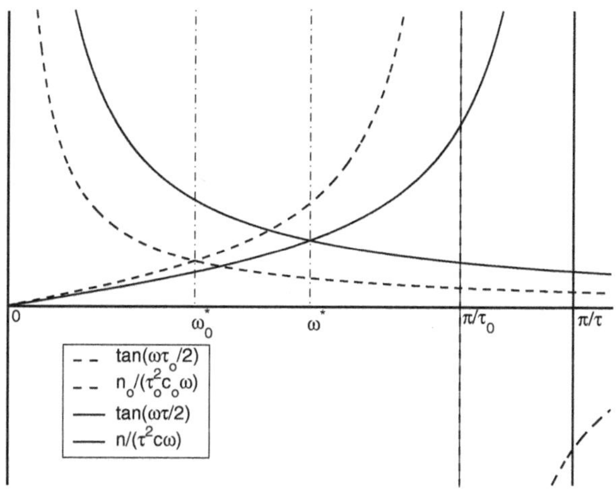

Fig. 2. Numerical solutions ω_0^* and ω^* of (6) for (n_0, τ_0, c_0) and (n, τ, c) satisfying (9).

4 Delay-Dependent AQM Controllers

As it can be seen from the previous section, a PI AQM stabilizing controller results in a closed-loop system governed by a retarded delay differential equation for which controlling its spectrum is not practically feasible. When in addition to stabilization, a desired closed-loop dynamic is required, delay-dependent controllers could result more convenient if certain knowledge of the previous dynamic information and delay are assumed. A delay-dependent AQM controller has been recently proposed in [4]. However, the well-known instability mechanism of the numerical implementation of such kind of controllers was not addressed.

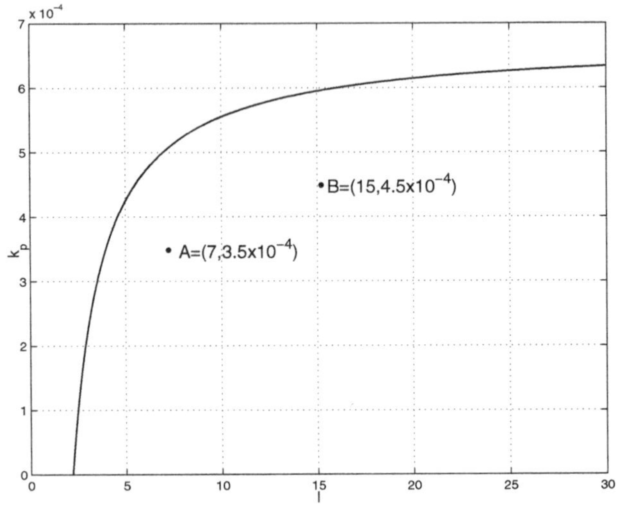

Fig. 3. Stability region $\Phi_{(n_0,\tau_0,c_0)}$ for $n_0 = 40$, $\tau = 0.7$ and $c = 300$

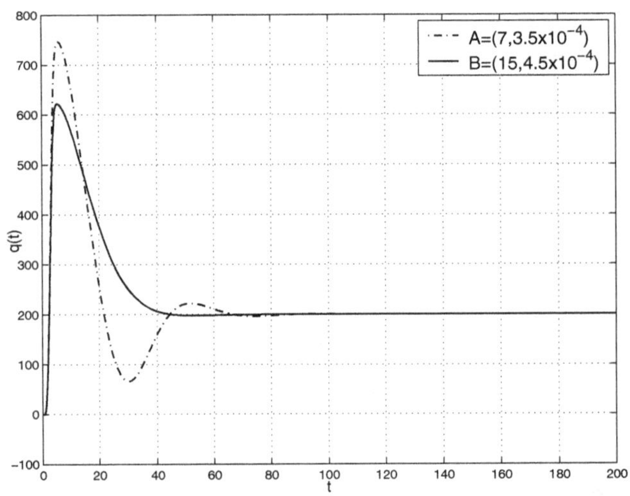

Fig. 4. The response of $q(t)$ for the A and B gains, respectively.

In this section, we suggest the design of a delay-dependent AQM controller based on the feedback control laws proposed by Manitus and Olbrot in [5] for finite spectrum assignment of time-delay systems, and provide stability conditions for a safe numerical implementation of the controller.

4.1 Finite Spectrum Assignment

Let us briefly discuss the feedback control laws for finite spectrum assignment of time-delay systems and their numerical implementation problem.

Consider the linear system with delayed input

$$\dot{x}(t) = Ax(t) + Bu(t-h), \tag{10}$$

where $h > 0$, $x(t) \in \mathbb{R}^n$ and $u(t) \in \mathbb{R}^m$ represent the state and control vectors, and A, B are real constant matrices of appropriate dimensions. The control law

$$u(t) = x(t) + K\int_{-h}^{0} e^{-A(h+\theta)}Bu(t+\theta)d\theta \tag{11}$$

assigns a finite spectrum to the closed-loop system (10)-(11) which coincides with the spectrum of the matrix $A + e^{-Ah}BK$, see [5].

The practical value of such a result is limited by the instability mechanism of the numerical approximation of the integral term in (11), see [1, 8, 14] and the references therein. It has been shown there that if the integral is approximated by a finite sum, then the closed-loop system may become unstable if the controller (11) is not internally stable. So, the internal stability of (11) is an essential condition for its successful implementation.

Motivated from the limitations imposed by the internal stability requirement of (11), the introduction of a low-pass filter (implicitly and/or explicitly) in the control loop has been proposed as remedy to overcome the implementation problems, see for instance [9]. However, such a solution could make the implementation unnecessarily complicated in the case of those parameters of (10) for which the internal dynamics of (11) are stable, see [6].

The internal dynamics of (11) are described by the following integral delay system

$$z(t) = K\int_{-h}^{0} e^{-A(h+\theta)}Bz(t+\theta)d\theta. \tag{12}$$

The characteristic function associated to (12) is

$$f(s) = \det\left(I - K\int_{-h}^{0} e^{-A(h+\theta)}Be^{s\theta}d\theta\right).$$

Here I denote the identity matrix of appropriate dimension. The following result provides simple-to-check stability conditions for (12).

Proposition 1. *System (12) is stable if*

$$\max_{\theta\in[-h,0]} \left\| Ke^{-A(h+\theta)}B \right\| h < 1. \tag{13}$$

Proof. First observe that for any $h > 0$ and any s with $\mathrm{Re}(s) \geq 0$ the following inequality holds:

$$\left| \frac{1 - e^{-hs}}{s} \right| \leq h.$$

Now assume that $f(s)$ has a zero, s_0, with nonnegative real part. Then there exists a complex vector $\nu \neq 0$ such that $\left(I - K\int_{-h}^{0} e^{-A(h+\theta)}Be^{s_0\theta}d\theta\right)\nu = 0$.

It follows that

$$1 \leq \max_{\theta\in[-h,0]} \left\| Ke^{-A(\theta+h)}B \right\| \left| \frac{1 - e^{-hs_0}}{s} \right| \leq \max_{\theta\in[-h,0]} \left\| Ke^{-A(\theta+h)}B \right\| h.$$

The last inequality contradicts the condition of the proposition.

4.2 Controller Design

As mentioned before, in order to get basic results for more general cases we develop the controller design for the linearization of the simplified model (2). The linearization of (2) about the equilibrium (w_0, q_0, p_0) is

$$\dot{\xi}(t) = A\xi(t) + b\tilde{p}(t-\tau), \tag{14}$$

where $\xi(t) = \begin{pmatrix} \tilde{w}(t) \\ \tilde{q}(t) \end{pmatrix}$, $A = \begin{pmatrix} -\frac{2n}{\tau^2 c} & 0 \\ \frac{n}{\tau} & 0 \end{pmatrix}$, $b = \begin{pmatrix} -\frac{\tau c^2}{2n^2} \\ 0 \end{pmatrix}$, $\tilde{w}(t) = w(t) - w_0$, $\tilde{q}(t) = q(t) - q_0$ and $\tilde{p}(t) = p(t) - p_0$. The corresponding control law (11) which assigns a closed-loop finite spectrum to (14) has the following form:

$$\begin{aligned}\tilde{p}(t) = k_1\tilde{w}(t) + k_2\tilde{q}(t) &- \frac{\tau c^2}{2n^2}k_1 \int_{-\tau}^{0} e^{\frac{2n}{\tau^2 c}(\tau+\theta)}\tilde{p}(t+\theta)d\theta \\ &- \frac{\tau^2 c^3}{4n^2}k_2 \int_{-\tau}^{0} \left(1 - e^{\frac{2n}{\tau^2 c}(\tau+\theta)}\right)\tilde{p}(t+\theta)d\theta. \end{aligned} \tag{15}$$

Here we assume that the whole state is accessible and that network parameters are known. The closed-loop ideal spectrum is determined by the zeros of the polynomial

$$m(\lambda) = \lambda^2 + \left[\frac{2n}{\tau^2 c} + \frac{\tau c^2}{2n^2}k_1 e^{\frac{2n}{\tau c}} + \frac{\tau^2 c^3}{4n^2}k_2\left(1 - e^{\frac{2n}{\tau c}}\right)\right]\lambda + \frac{c^2}{2n}k_2.$$

The internal dynamics of (15) are governed by the integral delay system

$$z(t) = -\frac{\tau c^2}{2n^2}k_1 \int_{-\tau}^{0} e^{\frac{2n}{\tau^2 c}(\tau+\theta)} z(t+\theta)d\theta - \frac{\tau^2 c^3}{4n^2}k_2 \int_{-\tau}^{0}\left(1 - e^{\frac{2n}{\tau^2 c}(\tau+\theta)}\right) z(t+\theta)d\theta. \tag{16}$$

Thus, a successful implementation of (15) can be achieved if there exists a pair (k_1, k_2) such that both the polynomial $m(\lambda)$ and system (16) are stable.

The polynomial $m(\lambda)$ is stable if and only if

$$k_2 > 0 \text{ and } \frac{2n}{\tau^2 c} + \frac{\tau c^2}{2n^2}k_1 e^{\frac{2n}{\tau c}} + \frac{\tau^2 c^3}{4n^2}k_2\left(1 - e^{\frac{2n}{\tau c}}\right) > 0. \tag{17}$$

>From (13) we have that (16) is stable if

$$\max_{\theta\in[-\tau,0]} \frac{\tau^2 c^2}{2n^2}\left| e^{\frac{2n}{\tau^2 c}(\tau+\theta)}k_1 + \frac{\tau c}{2}\left(1 - e^{\frac{2n}{\tau^2 c}(\tau+\theta)}\right)k_2\right| < 1. \tag{18}$$

In Fig. 5 we plot the stability regions determined by (17) and (18) in the plane (k_1, k_2). We denote by R_s the intersection of the two regions. The intersection points of the lines determining the boundaries of the stability regions with the coordinate axis are defined by $a = -\frac{8n^3}{\tau^4 c^4}\left(\frac{1}{1-e^{\frac{2n}{\tau c}}}\right)$, $b = -\frac{4n^3}{\tau^3 c^3}e^{-\frac{2n}{\tau c}}$, $d = -\frac{4n^2}{\tau^3 c^3}\left(\frac{1}{1-e^{\frac{2n}{\tau c}}}\right)$, $e = \frac{2n^2}{\tau^2 c^2}e^{-\frac{2n}{\tau c}}$ and $f = \frac{2n^2}{\tau^2 c^2}$. The following relationship

$$\frac{a}{b} = \frac{d}{-e} = \frac{2}{\tau c e^{-\frac{2n}{\tau c}}\left(1 - e^{\frac{2n}{\tau c}}\right)}$$

holds, and taking into account that $w_0 \gg 1$ we have $a < d$ and $b > -e$.

Remark 3. For a given set of network parameters (n, τ, c), there is always a pair (k_1, k_2) of controller's gains for which both system (16) and the ideal closed-loop system (14)-(15) are stable.

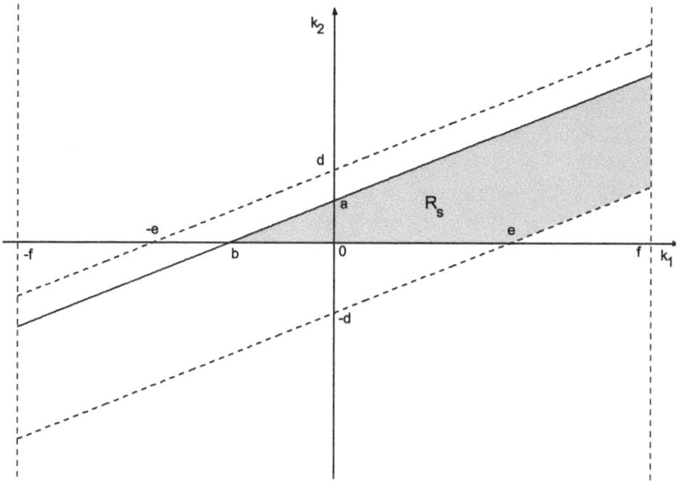

Fig. 5. Stability regions of (16) determined by (18) (- -) and for the ideal closed-loop stability determined by (17) (-). R_s denotes the intersection of the two regions.

4.3 Example

Consider the nominal network parameters considered in section 3. In Fig. 8 we plot the corresponding stability region R_s. In Fig. 9 we present the closed-loop response of $q(t)$ with the approximated control law (15) (by using a trapezoidal rule) for a $(k_1, k_2) = (0.07, 0.6 \times 10^{-3})$ inside of R_s. For these particular values the ideal closed-loop eigenvalues are $\lambda_1 = -1.55$ and $\lambda_2 = -0.43$. The operation point was chosen as $q_0 = 300$ packets with initial condition $q(0) = 400$ packets. The simulations were carried out on the nonlinear system (2).

5 Conclusion

In this chapter we addressed the local stability of two AQM controllers supporting TCP flows. We first considered a PI controller for which we derived necessary and sufficient conditions for closed-loop stability of the linearization. This result provides the complete set of PI AQM controllers that locally stabilizes the equilibrium point in counterpart with the existing works in the literature which give only estimates of this set. Then, we proposed a delay-dependent AQM controller based on feedback control laws for finite spectrum assignment of time-delay systems. Stability conditions for a numerically safe implementation of the controller are given. Numerical examples that illustrate the capabilities of the results for determining performance objectives and for sensitivity analysis with respect to perturbations of the system and controller have been performed.

Acknowledgements

This work was partially funded by the CONACYT grant: *Análisis de Estabilidad y Estabilidad Robusta de Controladores de Flujo en Internet* (2007-2008).

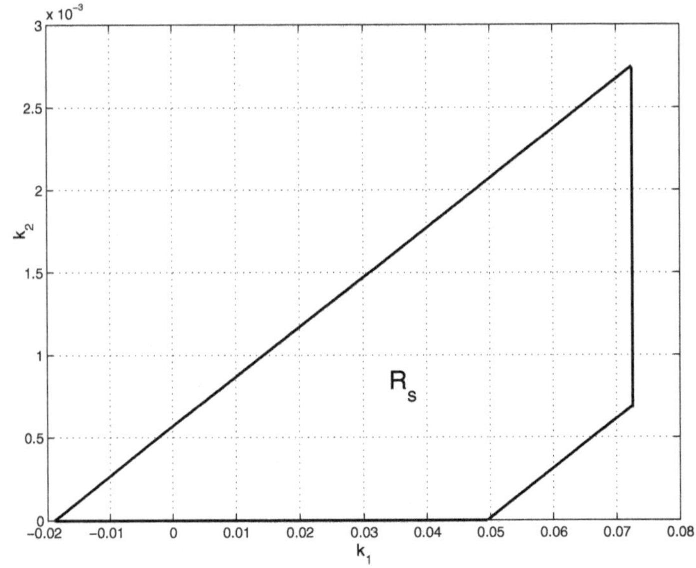

Fig. 6. Stability region R_s for $n = 40$, $\tau = 0.7$ and $c = 300$.

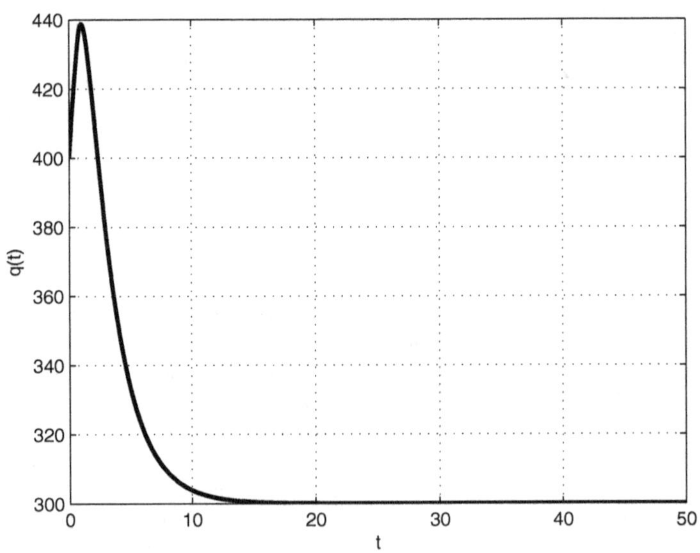

Fig. 7. The numericalsimulation of the closed-loop response of $q(t)$ by approximating the control law (15) with a trapezoidal rule.

References

1. Engelborghs, K., Dambrine, M., Roose, D.: Limitations of a class of stabilization methods for delay systems. IEEE Trans. Automat. Contr. 46(2), 336–339 (2001)
2. Gu, K., Kharitonov, V.L., Chen, J.: Stability of time-delay systems. Birkhäuser, Boston (2003)
3. Hollot, C.V., Misra, V., Towsley, D., Gong, W.B.: Analysis and design of controllers for AQM routers supporting TCP flows. IEEE Trans. Automatic Contr. 47(6), 945–956 (2002)
4. Kim, K.B.: Design of feedback controls supporting TCP based on the state-space approach. IEEE Trans. Automatic Contr. 51(7), 1086–1099 (2006)
5. Manitus, A.Z., Olbrot, A.W.: Finite spectrum assignment problem for systems with delays. IEEE Trans. Automat. Contr. 24(4), 541–553 (1979)
6. Melchor-Aguilar, D., Tristán-Tristán, B.: On the implementation of control laws for finite spectrum assignment: the multiple delays case. In: Proc. of 4th IEEE Conf. on Electrical and Electronics Engineering, México City (2007)
7. Michiels, W., Melchor-Aguilar, D., Niculescu, S.I.: Stability analysis of some classes of TCP/AQM networks. Int. J. Control 79(9), 1136–1144 (2006)
8. Mondié, S., Dambrine, M., Santos, O.: Approximation of control law with distributed delays: A necessary condition for stability. Kybernetica 38(5), 541–551 (2001)
9. Mondié, S., Michiels, W.: Finite spectrum assignment of unstable time-delay systems with a safe implementation. IEEE Trans. Automat. Contr. 48(12), 2207–2212 (2003)
10. Misra, V., Gong, W.B., Towsley, D.: Fluid based analysis of a network of AQM routers supporting TCP flows with an application to RED. In: Proc. of ACM/SIGCOMM (2000)
11. Network Simulator ns-2, Version 2.27, `http://www.isi.edu/nsnam/ns`
12. Quet, P.F., Özbay, H.: On the design of AQM supporting TCP flows using robust control theory. IEEE Trans. Automatic Contr. 49, 1031–1036 (2004)
13. Silva, G.J., Datta, A., Bhattacharyya, S.P.: PID controllers for time-delay systems. Birkhäuser, Boston (2005)
14. Van Assche, V., Dambrine, M., Lafay, J.F.: Some problems arising in the implementation of distributed-delay control laws. In: Proc. of 38th IEEE Conf. Decision Control, Phoenix, AZ (1999)

A Robust State Feedback Control Law for a Continuous Stirred Tank Reactor with Recycle

Pierdomenico Pepe

Dipartimento di Ingegneria Elettrica e dell'Informazione, 67040 Poggio di Roio, L'Aquila, Italy, Fax: +39-0862-434403, e-mail: pepe@ing.univaq.it

Summary. A nonlinear state feedback control law for a stirred tank chemical reactor with recycle is studied in this chapter. The system is described by nonlinear delay differential equations and the projected control law depends on the system variables at present and past times. Such feedback control law drives the output of the system, the reactor temperature, to the desired value, with exponential error decay rate. Moreover, the closed loop system is locally input-to-state stable with respect to a disturbance forcing, as the control law, the jacket temperature. Such disturbance may describe sensor and actuator errors, as well as errors due to parameters uncertainty such as the time-delay.

1 Introduction

The problem of controlling a stirred tank chemical reactor with recycle (see Mahmoud, 2002, Niculescu, 2001) is widely studied in the literature. Conventional control methods based on operating point linearization, which guarantee good performance in a suitable neighborhood of the operating point, are currently mainly used. Nonlinear control laws, which may guarantee a larger range of good performance, as far as the neighborhood of the operating point is concerned, based on the exact I/O linearization methods (Isidori, 1995), do not take into account the recycle time-delay and give poor performance when the coefficient of recirculation is suitably little and such recycle time-delay is suitably large (see fig. 2 in Wu, 1999, where the recirculation coefficient is $\lambda = 0.5$ and the recycle time-delay is $\Delta = 5$). In last years nonlinear control laws which take into account the recycle time-delay have been projected, also because these control laws prove to be more robust with respect to disturbances than delay-free nonlinear control laws, as pointed out in (Wu, 1999, see fig. 4). In (Oguchi & Watanabe, 1999) a nonlinear control law is developed on the basis of recently studied differential geometry tools for time-delay systems (see Oguchi, Watanabe & Nakamizo, 2002, Germani, Manes & Pepe, 2000, 2003). The model presents four state variables and two inputs, no disturbances are considered. A change of variables and an inner feedback are found, by which the system equations are transformed into a linear controllable system, which can be easily controlled by an outer feedback control law. In the paper (Wu, 1999), a Lyapunov-based methodology is presented by which a nonlinear feedback control law is found, driving the output of the system (the reactor temperature) to the desired value. There the model presents three state variables (the reaction conversion, the dimensionless reactor temperature,

J.J. Loiseau et al. (Eds.): Topics in Time Delay Systems, LNCIS 388, pp. 281–291.
springerlink.com

the dimensionless jacket temperature) and one control variable (the dimensionless coolant temperature). In that paper no theoretical results concerning the disturbance are given, but, as said before, simulations showing the better performance with respect to methods based on exact I/O linearization which neglect the recycle time-delay are reported. This chapter focuses on a nonlinear feedback control law which is found by means of the methodologies showed in (Germani, Manes & Pepe, 2000, 2003, Oguchi, Watanabe & Nakamizo, 2002) and on input-to-state stability issues of the closed loop system. The system (Wu, 1999, Mahmoud, 2000) admits non-uniform (type II, see Germani, Manes & Pepe, 2003) relative degree equal to two, that is the relative degree is not full since the state variables are three. Though the output is driven to the desired value, by the feedback control law, with arbitrarily fixed exponential error decay rate, the behavior of the internal dynamics is fundamental to study in order to assure the good performance of the overall closed loop system. A disturbance forcing, as the control input, the jacket temperature dynamics is considered. Such disturbance may describe sensor and actuator errors, as well as errors due to parameters uncertainty such as the time-delay. The input-to-state stability, which implies the stability in the unforced case (disturbance equal to zero), of the overall control system with respect to such disturbance is here studied by means of the Lyapunov-Krasovskii methodologies showed in (Pepe & Jiang, 2006, Pepe, 2007a, Pepe 2007b). Simulations validate the theoretical results.

2 Preliminaries

The symbol $|\cdot|$ stands for the Euclidean norm of a real vector, or the induced Euclidean norm of a matrix. A function $u : R^+ \to R^m$, m positive integer, is said to be *essentially bounded* if $ess\,\sup_{t\geq 0} |u(t)| < \infty$. The essential supremum norm of an essentially bounded function is indicated with the symbol $\|\cdot\|_\infty$. For given times $0 \leq T_1 < T_2$, with $u_{[T_1,T_2)} : [0,+\infty) \to R^m$ the function given by $u_{[T_1,T_2)}(t) = u(t)$ for all $t \in [T_1,T_2)$ and $= 0$ elsewhere is indicated. An input $u : R^+ \to R^m$ is said to be *locally essentially bounded* if, for any $T > 0$, $u_{[0,T)}$ is essentially bounded. For a real $\Delta > 0$, a positive integer n, $C([-\Delta,0], R^n)$ is the Banach space of continuous functions mapping the interval $[-\Delta,0]$ into R^n with the topology of uniform convergence. For a given function $s : [-\Delta,d) \to R^n$, $0 < d \leq +\infty$, the function $s_t : [-\Delta,0] \to R^n, t \in [0,d)$, is defined as $s_t(\tau) = s(t+\tau), \tau \in [-\Delta,0]$. The term ISS stands as usual for input-to-state stable and input-to-state stability. Let us here recall that a function $\gamma : R^+ \mapsto R^+$ is: of class $\mathcal{K}$ if it is zero at zero, continuous and strictly increasing; of class $\mathcal{K}_\infty$ if it is of class $\mathcal{K}$ and it is unbounded; of class $\mathcal{L}$ if it is continuous, decreasing and converges to zero as its argument tends to $+\infty$. A function $\beta : R^+ \times R^+ \mapsto R^+$ is of class $\mathcal{KL}$ if it is of class $\mathcal{K}$ in the first argument and is of class $\mathcal{L}$ in the second argument. Let us report here the definition of input-to-state stability (see Sontag, 1989, see also Pepe & Jiang, 2006).

Definition 1. *A time-delay system, described by*

$$\dot{x}(t) = f(x_t, u(t)), \qquad x_0 = \xi_0 \in C([-\Delta,0]; R^n), \tag{1}$$

is said to be input-to-state stable if there exist a $\mathcal{KL}$ function β and a $\mathcal{K}$ function γ such that, for any initial state ξ_0 and any measurable, locally essentially bounded input u, the solution exists for all $t \geq 0$ and furthermore it satisfies

$$|x(t)| \leq \beta\left(\|\xi_0\|_\infty, t\right) + \gamma\left(\|u_{[0,t)}\|_\infty\right) \tag{2}$$

If the inequality (2) holds for $\|\xi_0\|_\infty < k_1$, $\|u_{[0,\infty)}\|_\infty < k_2$, k_1, k_2 *positive reals, then the time-delay system (1) is said to be locally input-to-state stable.*

3 A Nonlinear Control Law for the CSTR

Let us consider the following model of a continuous stirred tank chemical reactor (Wu, 1999, Mahmoud, 2000):

$$\begin{aligned}
\dot{x}_1(t) &= -\frac{1}{\lambda}x_1(t) + \left(\frac{1}{\lambda} - 1\right)x_1(t-\Delta) + D_a(1 - x_1(t))exp\left(\frac{x_2(t)}{1 + \frac{x_2(t)}{\psi}}\right), \\
\dot{x}_2(t) &= -\frac{1}{\lambda}x_2(t) + \left(\frac{1}{\lambda} - 1\right)x_2(t-\Delta) + \\
&\qquad +BD_a(1 - x_1(t))exp\left(\frac{x_2(t)}{1 + \frac{x_2(t)}{\psi}}\right) - \beta_r(x_2(t) - x_3(t)), \\
\dot{x}_3(t) &= \beta_c(x_2(t) - x_3(t)) + (x_{3f} - x_3(t))u(t) + w(t), \\
y(t) &= x_2(t),
\end{aligned} \tag{3}$$

where x_1, x_2, x_3 are the reaction conversion, the dimensionless reactor temperature and the dimensionless jacket temperature, respectively. The input u is the dimensionless coolant temperature. w is a disturbance. Δ represents the recycle time-delay. The coefficient of recirculation λ varies from 0 to 1. Other parameters are set as follows: $B = 11$, $D_a = 0.135$, $\psi = 20$, $\beta_r = 1.5$, $\beta_c = 1.5$, $x_{3f} = -4$, and the initial conditions are supposed to be constant (all according to (Wu, 1999)). The problem is considered here consists to get the output to track a constant reference y_d. That problem, in the case of no disturbance, has been treated in (Wu, 1999) by means of a Lyapunov based methodology. Here, the methodology developed in (Germani, Manes & Pepe, 2003), which makes use of the standard tools of differential geometry for nonlinear delay-free systems in order to find stabilizing control laws for nonlinear time-delay systems, is exploited. By Definition 2.3 in (Germani, Manes & Pepe, 2003) it follows that system (3) admits type-II relative degree equal to two. By Proposition 2.8 an inner state feedback control law is found by which the input output map is linearized. By an outer state feedback control law, the input output map is described by a linear delay-free asymptotically stable system. The feedback control law which is found by that methodology assumes the following expressions (the term $r(x(t))$ stands for $exp\left(\frac{x_2(t)}{1+\frac{x_2(t)}{\psi}}\right)$):

$$u(t)_{t\in[\Delta,+\infty)} = \frac{1}{\beta_r(x_{3f} - x_3(t))} \cdot \left(-S(x(t), x(t-\Delta), x(t-2\Delta)) + K \begin{bmatrix} x_2(t) - y_d \\ -\frac{1}{\lambda}x_2(t) + (\frac{1}{\lambda} - 1)x_2(t-\Delta) + \\ +BD_a(1 - x_1(t))\cdot \\ \cdot r(x(t)) - \beta_r(x_2(t) - x_3(t)) \end{bmatrix} \right), \tag{4}$$

where the row vector K is easily chosen such that the dynamic matrix

$$A_K = \begin{bmatrix} 0 & 1 \\ 0 & 0 \end{bmatrix} + \begin{bmatrix} 0 \\ 1 \end{bmatrix} K \tag{5}$$

has arbitrarily prescribed eigenvalues in the left half complex plane, and

$$\begin{aligned}
S(x(t),\ & x(t-\Delta), x(t-2\Delta)) = \left(-\frac{1}{\lambda}x_1(t) + \left(\frac{1}{\lambda}-1\right)x_1(t-\Delta) + \right.\\
& \left. +D_a(1-x_1(t))r(x(t))\right)(-BD_a(1-x_1(t))r(x(t))) + \\
& + \left(-\frac{1}{\lambda}x_2(t) + \left(\frac{1}{\lambda}-1\right)x_2(t-\Delta) + \right.\\
& BD_a(1-x_1(t))r(x(t)) - \beta_r(x_2(t)-x_3(t)))\\
& \cdot \left(-\frac{1}{\lambda} + BD_a(1-x_1(t))\left(\frac{1}{1+\frac{x_2(t)}{\psi}}\right)^2 r(x(t)) - \beta_r\right)\\
& +\beta_r\beta_c(x_2(t)-x_3(t)) + \left(\frac{1}{\lambda}-1\right)\cdot\\
& \left(-\frac{1}{\lambda}x_2(t-\Delta) + (\frac{1}{\lambda}-1)x_2(t-2\Delta)\right.\\
& +BD_a(1-x_1(t-\Delta))r(x(t))\\
& -\beta_r(x_2(t-\Delta)-x_3(t-\Delta))) ;
\end{aligned} \tag{6}$$

$$\begin{aligned}
u(t)_{t\in[0,\Delta]} = & \frac{1}{\beta_r(x_{3f}-x_3(t))}\left(M(x(t),x(t-\Delta)) + \right.\\
& \left. +K \begin{bmatrix} x_2(t) - y_d \\ \\ -\frac{1}{\lambda}x_2(t) + \left(\frac{1}{\lambda}-1\right)x_2(t-\Delta) + \\ +BD_a(1-x_1(t))r(x(t)) - \\ -\beta_r(x_2(t)-x_3(t)) \end{bmatrix}\right)
\end{aligned} \tag{7}$$

with

$$\begin{aligned}
M(x(t),\ & x(t-\Delta)) = \left(-\frac{1}{\lambda}x_1(t) + \left(\frac{1}{\lambda}-1\right)x_1(t-\Delta) + \right.\\
& \left. +D_a(1-x_1(t))r(x(t))\right)(-BD_a(1-x_1(t))r(x(t))) + \\
& + \left(-\frac{1}{\lambda}x_2(t) + \left(\frac{1}{\lambda}-1\right)x_2(t-\Delta)\right.\\
& +BD_a(1-x_1(t))r(x(t)) - \beta_r(x_2(t)-x_3(t))) \cdot\\
& \left(-\frac{1}{\lambda} + BD_a(1-x_1(t))\left(\frac{1}{1+\frac{x_2(t)}{\psi}}\right)^2 \cdot\right.\\
& r(x(t)) - \beta_r) + \beta_r\beta_c(x_2(t)-x_3(t))
\end{aligned} \tag{8}$$

In the case of no disturbance, by Proposition 2.8 in (Germani, Manes & Pepe, 2003), such control law yields, provided that $x_3(t) \neq x_{3f}$, $t \geq 0$, the following equation for the variable $e(t) = z(t) - \begin{bmatrix} y_d \\ 0 \end{bmatrix}$, where $z(t) = \begin{bmatrix} y(t) \\ \dot{y}(t) \end{bmatrix}$,

$$\dot{e}(t) = A_K e(t), \qquad t \geq 0 \tag{9}$$

and thus, since A_K is an asymptotically stable matrix, in the case of no disturbance $y(t)$ tracks the desired constant output y_d, with exponential error decay rate arbitrarily fixed by the choice of K.

4 Input-to-State Stability w.r.t. the Disturbance

In the case the disturbance is present, the control law (4) (7) yields the following equation for the variable $e(t)$,

$$\dot{e}(t) = A_K e(t) + \begin{bmatrix} 0 \\ \beta_r \end{bmatrix} w(t), \qquad t \geq 0 \tag{10}$$

The linear system (10) is asymptotically stable in the unforced case and thus input-to-state stable with respect to the disturbance w. Therefore the tracking control problem can still be considered solved, from an input-output point of view, provided the disturbance is suitably bounded. The following inequality holds, for $t \geq t_0 \geq 0$,

$$|e(t)| \leq m_K e^{\lambda_K (t-t_0)} |e(t_0)| - \frac{m_K}{\lambda_K} \beta_r \| w_{[t_0,t)} \|_\infty, \tag{11}$$

where m_K is a positive real and λ_K is a negative real which, as well known, depend on A_K and thus on the choice of K in (5). But, though the output tracks the desired reference signal, nothing can be said yet about the behavior of the overall control system (3), (4), (7). Let, for $t \geq -\Delta$, $\hat{x}(t) = \begin{bmatrix} x_1(t) - x_{1,ref} \\ x_2(t) - x_{2,ref} \\ x_3(t) - x_{3,ref} \end{bmatrix} = x(t) - x_{ref}$, where x_{ref} is the steady-state ($x_{2,ref} = y_d$, $w = 0$) of the closed loop system (3), (4), (7). Let $\xi(t) = \begin{bmatrix} \xi_1(t) \\ \xi_2(t) \\ \xi_3(t) \end{bmatrix} = \begin{bmatrix} \hat{x}_1(t) \\ e(t) \end{bmatrix}$, $t \geq 0$, $\xi_1(\tau) = \hat{x}_1(\tau)$, $\tau \in [-\Delta, 0]$, $\begin{bmatrix} \xi_2(\tau) \\ \xi_3(\tau) \end{bmatrix} = e(0)$, $\tau \in [-\Delta, 0]$. Let, as usual, $\xi_t(\tau) = \xi(t+\tau)$, $\tau \in [-\Delta, 0]$, $t \geq 0$. The overall control system (3), (4), (7) is equivalent to the following cascade system of delay differential equations, forced by the disturbance $w(t)$:

$$\dot{\xi}(t) = \begin{bmatrix} -\frac{1}{\lambda}(\xi_1(t) + x_{1,ref}) + (\frac{1}{\lambda} - 1)\cdot \\ (\xi_1(t-\Delta) + x_{1,ref}) + \\ +D_a(1 - \xi_1(t) - x_{1,ref})\cdot \\ \exp\left(\frac{\xi_2(t) + x_{2,ref}}{1 + \frac{\xi_2(t) + x_{2,ref}}{\psi}} \right) \\ A_K \begin{bmatrix} \xi_2(t) \\ \xi_3(t) \end{bmatrix} \end{bmatrix} + \begin{bmatrix} 0 \\ 0 \\ \beta_r \end{bmatrix} w(t) \tag{12}$$

For, the relation between the variables $\hat{x}$ and ξ, for $t \geq 0$, is the following (note that, in this case, it is not a continuous time difference equation, as usually happens, see Germani, Manes & Pepe, 2000, 2003, see also Pepe, 2003, 2005, Pepe & Verriest, 2003, but just an equality)

$$\hat{x}(t) = \begin{bmatrix} \xi_1(t) \\ \xi_2(t) \\ -x_{3,ref} + \frac{1}{\beta_r}\left(\xi_3(t) + \frac{1}{\lambda}(\xi_2(t) + x_{2,ref}) - \right. \\ -(\frac{1}{\lambda} - 1)(\xi_2(t-\Delta) + x_{2,ref}) - \\ -BD_a(1 - \xi_1(t) - x_{1,ref}) \cdot \\ exp\left(\frac{\xi_2(t)+x_{2,ref}}{1+\frac{\xi_2(t)+x_{2,ref}}{\psi}}\right)\left.\right) + (\xi_2(t) + x_{2,ref}) \end{bmatrix} \tag{13}$$

If $\xi_t = 0$, then $\hat{x}(t) = 0,\ t \geq 0$. Let $r_0 : R \to R$ be the function given by, for $s \in R$,

$$r_0(s) = exp\left(\frac{s + x_{2,ref}}{1 + \frac{s+x_{2,ref}}{\psi}}\right) \tag{14}$$

Let $r_1 : R \to R$ be the function given by, for $s \in R$,

$$r_1(s) = (D_a - D_a x_{1,ref}) r_0(s) - x_{1,ref} \tag{15}$$

The function r_1 is such that $r_1(0) = 0$. For, consider the first equation in (3), set $\dot{x}_1(t) = 0$ and $x(t) = x_{ref}$. Let $a_0 : R \to R$ be the function given by, for $s \in R$,

$$a_0(s) = -\frac{1}{\lambda} - D_a r_0(s) \tag{16}$$

Note that $a_0(s) \leq -\frac{1}{\lambda}$, $\forall\ s \in R$. Let $A_0(s) = \begin{bmatrix} a_0(s) & 0 \\ 0 & A_K \end{bmatrix}$. Let $A_1 = \begin{bmatrix} \frac{1}{\lambda} - 1 & 0 & 0 \\ 0 & 0 & 0 \\ 0 & 0 & 0 \end{bmatrix}$.

Let $\bar{d}(t) = \begin{bmatrix} r_1(\xi_2(t)) \\ w(t) \end{bmatrix}$. Let $F = \begin{bmatrix} 1 & 0 \\ 0 & 0 \\ 0 & \beta_r \end{bmatrix}$. Then, (12) can be rewritten as

$$\dot{\xi}(t) = A_0(\xi_2(t)) \cdot \xi(t) + A_1 \xi(t - \Delta) + F\bar{d}(t) \tag{17}$$

Let us here suppose that the initial tracking error $e(0)$ and the disturbance w are such that $|e(0| \leq \rho_1,\ ess\sup_{t \geq 0} |w(t)| \leq \rho_2$, ρ_1, ρ_2 being positive reals satisfying, for a positive real $\gamma < 1$, $m_K \rho_1 - \frac{\overline{m}_K}{\lambda_K} \rho_2 \leq \gamma |\psi + y_d|$. Then, from (11), (13), it follows that there exist positive reals ρ_0, L, $\hat{L}$ such that

$$\sup_{t \geq 0} |r_1(\xi_2(t))| \leq \rho_0, \quad |r_1(\xi_2(t))| \leq L|\xi_2(t)|, \quad |\hat{x}(t)| \leq \hat{L}\|\xi_t\|_\infty, \quad t \geq 0 \tag{18}$$

Let us replace $\bar{d}$ in (17) with a disturbance $d = \begin{bmatrix} d_1 \\ d_2 \end{bmatrix}$ satisfying $\|d_{1,[0,\infty)}\|_\infty \leq \rho_0$, $\|d_{2,[0,\infty)}\|_\infty \leq \rho_2$ and let us study the input-to-state stability of the resulting system, w.r.t. d, by means of Theorem 3.1 in (Pepe & Jiang, 2006). Let us choose the following Lyapunov-Krasovskii functional $V : C([-\Delta, 0]; R^3) \to R^+$:

$$V(\phi) = \phi^T(0) P \phi(0) + \int_{-\Delta}^{0} \phi^T(\tau) Q(\tau) \phi(\tau) d\tau, \quad \phi \in C([-\Delta, 0]; R^3) \tag{19}$$

where P is a symmetric positive matrix, $Q(\tau) = -\frac{\tau}{\Delta} Q_1 + \frac{\tau+\Delta}{\Delta} Q_2$, Q_1, Q_2 are symmetric positive matrices, $Q_2 > Q_1$. The following equalities/inequalities hold

$$
\begin{aligned}
D^+V(\phi,d) = {} & \phi^T(0)P\left(A_0(\phi_2(0))\cdot\phi(0)+A_1\phi(-\Delta)\right. \\
& \left.+Fd\right)+\left(A_0(\phi_2(0))\cdot\phi(0)+A_1\phi(-\Delta)+Fd\right)^T\cdot \\
& P\phi(0)+\phi^T(0)Q_2\phi(0)-\phi^T(-\Delta)Q_1\phi(-\Delta)- \\
& -\frac{1}{\Delta}\int_{-\Delta}^{0}\phi^T(\tau)(Q_2-Q_1)\phi(\tau)d\tau \le \begin{bmatrix}\phi(0)\\ \phi(-\Delta)\end{bmatrix}^T\cdot \\
& \begin{bmatrix} A_0^T(\phi_2(0))P+PA_0(\phi_2(0))+Q_2 & PA_1 \\ A_1^TP & -Q_1\end{bmatrix}\cdot \\
& \begin{bmatrix}\phi(0)\\ \phi(-\Delta)\end{bmatrix}+2\phi^T(0)PFd-\frac{1}{\Delta}\int_{-\Delta}^{0}\phi^T(\tau)(Q_2-Q_1)\phi(\tau)d\tau
\end{aligned} \tag{20}
$$

Let $P = diag(P_1, P_2)$, with P_1 a positive real, P_2 a symmetric positive matrix in $R^{2\times 2}$. Let $\bar{A}_0 = \begin{bmatrix} \frac{-1}{\lambda} & 0 \\ 0 & A_K \end{bmatrix}$. Now, whenever $\|\phi\|_{M_2} \ge \frac{|d|}{\rho}$, where ρ is a positive real to be chosen, from the inequality

$$|2\phi^T(0)PFd| \le \rho|\phi(0)|^2 + \rho|PF|\|\phi\|_{M_2}^2 \tag{21}$$

and taking into account (16), it follows that, for suitable positive reals ρ, p,

$$D^+V(\phi,d) \le -p\|\phi\|_{M_2}, \quad \forall\phi,\ d:\ \|\phi\|_{M_2} \ge \frac{1}{\rho}|d|, \tag{22}$$

provided that the following LMI is feasible (with $Q_2 > Q_1$)

$$\begin{bmatrix} J_1+Q_2 & J_2 \\ J_2^T & -Q_1 \end{bmatrix} < 0, \tag{23}$$

where $J_1 = \begin{bmatrix} -\frac{2}{\lambda}P_1 & 0 \\ 0 & A_K^TP_2+P_2A_K \end{bmatrix}$, $J_2 = \begin{bmatrix} (\frac{1}{\lambda}-1)P_1 & 0 & 0 \\ 0 & 0 & 0 \\ 0 & 0 & 0 \end{bmatrix}$. By Schur complement (see Corollary B.2, pp. 318, in (Gu, Kharitonov & Chen, 2003)) it follows that the LMI (23) is feasible for any value of the circulation parameter $\lambda \in (0, 1]$. Therefore, by applying results in (Pepe & Jiang, 2006), the conclusion holds that there exist a suitable $\mathcal{KL}$ function α_1 and a $\mathcal{K}$ function β_1 such that the following local ISS inequality holds, for $t \ge t_0 \ge 0$, for system (17), when the particular disturbance $d(t) = \bar{d}(t)$ is chosen,

$$|\xi(t)| \le \alpha_1\left(\|\{\xi_{t_0}\}\|_\infty, t-t_0\right) + \beta_1(\|\bar{d}_{[t_0,t)}\|_\infty) \tag{24}$$

By some tedious computations, from (24), taking into account (11), (13), (18) and that, for a suitable positive real L_0, $\|\xi_0\|_\infty \le L_0\|\hat{x}_0\|_\infty$, the following local ISS inequality holds, for suitable $\mathcal{KL}$ function α and $\mathcal{K}$ function β,

$$|\hat{x}(t)| \le \alpha\left(\|\hat{x}_0\|_\infty, t\right) + \beta(\|w_{[0,t)}\|_\infty) \tag{25}$$

One problem is to be taken into account, that is the fact that the control law (4), (7) is not global, it works if $x_3(t) > x_{3f}$. The inequality (25) holds whenever the control law (4), (7) can be applied, that is whenever $x_3(t) > x_{3f}$. Taking into account this fact, taking into account that $\xi_2(t)$ (appearing in $\bar{d}(t)$) is the first component of $e(t)$ which satisfies (11), the following conclusions hold:

1) if the disturbance is zero, then the reactor temperature converges exponentially to the desired value y_d, and the trivial solution of the system (12) is asymptotically stable, which

means that the steady-state x_{ref} is asymptotically stable for the closed loop system (3), (4), (7), provided that $x_3(t) \neq x_{3f},\ t \geq 0$;

2) if the disturbance w is not zero, then the error between the reactor temperature with its derivative and the desired value with its (zero) derivative ($e(t)$ in (10)) satisfies the inequality (11) and $\hat{x}$ satisfies the local ISS inequality (25), provided that $x_3(t) \neq x_{3f},\ t \geq 0$.

The condition $x_3(t) \neq x_{3f},\ t \geq 0$ is always satisfied in the many performed simulations. Nevertheless, it is assured also theoretically by requiring that $\|\hat{x}_0\|_\infty$ and the disturbance are suitably bounded, taking into account (25).

5 Simulation Results

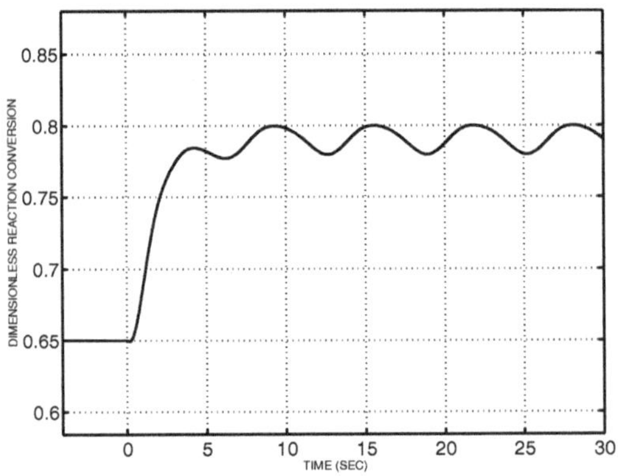

Fig. 1. Reaction Conversion

The theoretical stability results provided in previous sections are validated by simulations. In simulations reported here the row vector K has been chosen such to assign to the error dynamic matrix the eigenvalues -1, -2, y_d has been chosen equal to 4, λ equal to 0.5, Δ equal to $5sec.$ (see Wu, 1999). The initial conditions are chosen as in (Wu, 1999), that is $x(\tau) = \begin{bmatrix} x_1(\tau) \\ x_2(\tau) \\ x_3(\tau) \end{bmatrix} = \begin{bmatrix} 0.65 \\ 3 \\ 0.26 \end{bmatrix}$, $\tau \in [-\Delta, 0]$. The control law (4) is applied from $t = 0$, the initial control law (7) is not used here since it does not give significant improvements. More, in order to apply (4) from t=0, it is supposed that the initial state (constant) function is defined on a double delay interval (see Oguchi, Watanabe & Nakamizo, 2002). The disturbance $w(t)$ has been chosen equal to $0.3sin(2t)$. In figs. $1, 2, 3$ the three state variables are reported, respectively. Recall that the output is the reactor temperature, and it can be seen that the desired value is reached (an oscillation can be seen from fig. 2 around the desired value 4 because of the disturbance). In fig. 4 the signal generated by the control law is reported. The input-to-state stability with respect to the disturbance can be appreciated. The state variables have oscillations around the steady-state (such steady-state is asymptotically reached if the disturbance is not active). The simulations show that the stability of the overall control system is preserved also under the presence of the disturbance.

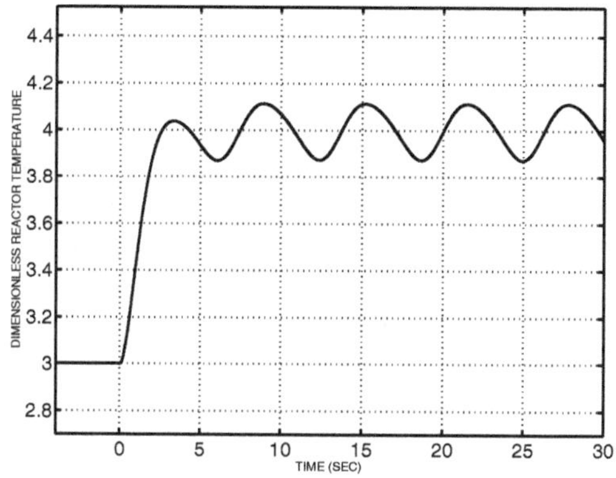

Fig. 2. Reactor Temperature

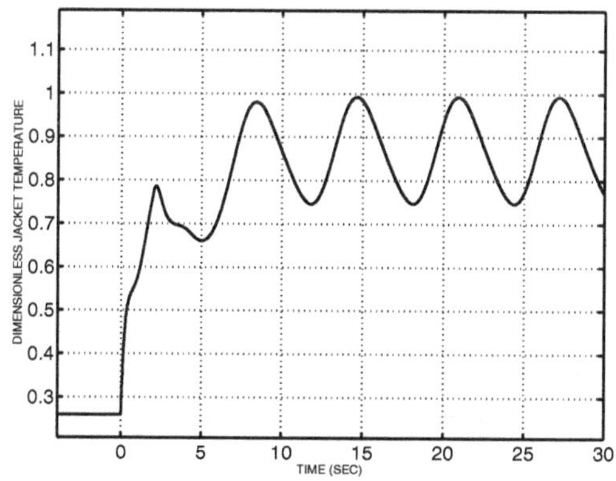

Fig. 3. Jacket Temperature

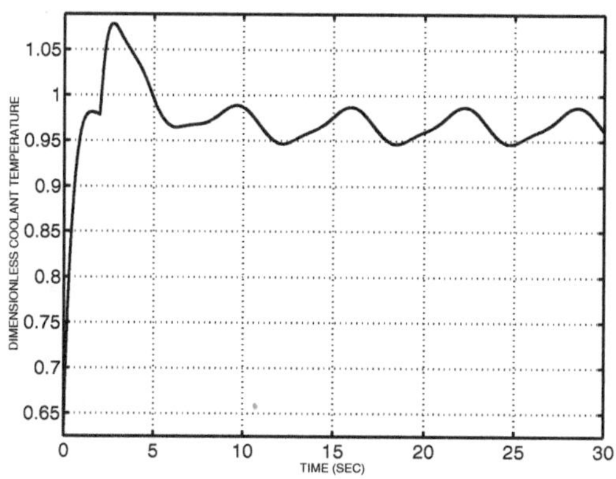

Fig. 4. Control Signal

6 Conclusions

In this chapter a nonlinear feedback control law, developed by means of recent tools for time-delay systems based on differential geometry, for a stirred tank chemical reactor with recycle is studied. Such control law drives the output to the desired value with the error decaying to zero exponentially with prescribed decay rate. Moreover it is shown that the closed loop system is locally input-to-state stable with respect to a disturbance which may describe sensor and actuator errors as well as parameters uncertainty such as the time-delay.

References

1. García-Gabín, W., Normey-Rico, J.E., Camacho, E.F.: Sliding Mode Predictive Control of a Delayed CSTR. In: Proc. 6th IFAC Workshop on Time-Delay Systems, L'Aquila, Italy (2006)
2. Germani, A., Manes, C., Pepe, P.: Local Asymptotic Stability for Nonlinear State Feedback Delay Systems. Kybernetika 36(1), 31–42 (2000)
3. Germani, A., Manes, C., Pepe, P.: Input-Output Linearization with Delay Cancellation for Nonlinear Delay Systems: the Problem of the Internal Stability. Int. Journal of Robust and Nonlinear Control 13(9), 909–937 (2003)
4. Gu, K., Kharitonov, V.L., Chen, J.: Stability of Time Delay Systems. Birkhauser, Boston (2003)
5. Hale, J.K., Verduyn Lunel, S.M.: Introduction to Functional Differential Equations. Springer, New York (1993)
6. Karafyllis, I.: Lyapunov Theorems for Systems Described by Retarded Functional Differential Equations. Nonlinear Analysis: Theory, Methods & Applications 64(3), 590–617 (2006)
7. Luyben, W.L.: Chemical Reactor Design and Control. Wiley-Interscience, Hoboken (2007)
8. Mahmoud, M.S.: Robust Control and Filtering for Time-Delay Systems. Marcel Dekker, New York (2000)
9. Moog, C.H., Castro-Linares, R., Velasco-Villa, M., Marquez-Martinez, L.A.: The Disturbance Decoupling Problem for Time Delay Nonlinear Systems. IEEE Transactions on Automatic Control 45(2), 305–309 (2000)
10. Niculescu, S.I.: Delay Effects on Stability. A Robust Control Approach. LNCIS. Springer, London (2001)
11. Oguchi, T., Watanabe, A.: A finite spectrum assignment for multivariable retarded nonlinear systems. In: 38th SICE Conference, Morioka, Japan (1999)
12. Oguchi, T., Watanabe, A., Nakamizo, T.: Input-Output Linearization of Retarded Nonlinear Systems by Using an Extension of Lie Derivative. Int. J. Control 75(8), 582–590 (2002)
13. Pepe, P.: The Liapunov's Second Method for Continuous Time Difference Equations. International Journal of Robust and Nonlinear Control 13(15), 1389–1405 (2003)
14. Pepe, P., Verriest, E.I.: On the Stability of Coupled Delay Differential and Continuous time Difference Equations. IEEE Transactions on Automatic Control 48(8), 1422–1427 (2003)
15. Pepe, P.: On the Asymptotic Stability of Coupled Delay Differential and Continuous Time Difference Equations. Automatica 41(1), 107–112 (2005)
16. Pepe, P.: a) On Liapunov-Krasovskii Functionals under Carathéodory Conditions. Automatica 43(4), 701–706 (2007)

17. Pepe, P.: b) The Problem of the Absolute Continuity for Liapunov-Krasovskii Functionals. IEEE Trans. on Automatic Control 52(5), 953–957 (2007)
18. Pepe, P., Jiang, Z.P.: A Lyapunov-Krasovskii Methodology for ISS and iISS of Time-Delay Systems. Systems & Control Letters 55(12), 1006–1014 (2006)
19. Sontag, E.D.: Smooth Stabilization Implies Coprime Factorization. IEEE Transactions on Automatic Control 34, 435–443 (1989)
20. Teel, A.R.: Connections between Razumikhin-Type Theorems and the ISS Nonlinear Small Gain Theorem. IEEE Transactions on Automatic Control 43(7), 960–964 (1998)
21. Wu, W.: Lyapunov-Based Design Procedures for a State-Delay Chemical Process. In: Proc. 14th IFAC World Congress, Beijing, China (1999)

Functional Differential Equations Associated to Propagation

Vladimir Răsvan

Dept. of Automatic Control, University of Craiova, A.I.Cuza, 13, Craiova, RO-200585, Romania vrasvan@automation.ucv.ro

Summary. In this paper we aim to point out some problems connected to the approach of associating to propagation systems some functional differential equations. The associated models are slightly different of the standard ones in the sense that they are no longer restricted to lumped (pointwise) delays; this extension is motivated mainly by the physical and engineering significance of the basic (propagation) systems as well as by the significance of the problems. The paper will be focused around two models: the circulating fuel nuclear reactor - where the propagation is no longer lossless - and the overhead crane - where the parameters may be space varying thus inducing distortion in propagation. Basic theory and stability will be considered; in the case of the nuclear reactor a Liapunov functional of special structure is proposed which requires model reformulation.

1 Introduction: State of the art

We have presented several times the strong connection between the so called systems with propagation and some FDE (functional differential equations) of neutral type. By systems with propagation we understand a class of systems with distributed parameters in one dimension, whose models are given by initial/boundary value problems for hyperbolic PDE in two dimensions - time plus one space dimension (the "propagation dimension"); such systems contain long transmission lines for physical signals - electric signals but also water, steam or gas flow/pressure. In various references we had selected and analyzed those systems with propagation to which one may associate FDE of the form

$$\begin{aligned} \dot{x} &= A_0 x(t) + A_1 y(t-\tau) + f_1(t) \\ y(t) &= A_2 x(t) + A_3 y(t-\tau) + f_2(t) \end{aligned} \tag{1}$$

or some nonlinear version of it [19]. The strong motivation of this approach was the one to one correspondence between the two mathematical objects which at its turn offered the possibility of solving various problems for both systems by solving them for the system of FDE (assumed to be simpler). The problems we dealt arose from control: stability and absolute stability of feedback systems, forced oscillations, numerical procedures and control synthesis. From time to time we used to point out this approach and to signal new and open problems - see [14], [16], [18], [19], [20].

J.J. Loiseau et al. (Eds.): Topics in Time Delay Systems, LNCIS 388, pp. 293–302.
springerlink.com

In this paper we aim to point out some other problems connected to this approach. The associated models are slightly different of (1) and its nonlinear counterpart in the sense that they are no longer restricted to lumped (pointwise) delays; this extension is motivated mainly by the physical and engineering significance of the basic (propagation) systems as well as by the significance of the problems.

The paper will be focused around two models: the circulating fuel nuclear reactor - where the propagation is no longer lossless - and the overhead crane - where the parameters may be space varying thus inducing distortion in propagation. Consequently the paper is organized as follows: firstly the models for circulating fuel nuclear reactors are presented - [6], [7], [8], the associated FDE and some stability results; we consider here most interesting the associated Liapunov functional, the "rearrangement" inequality connected to it and the controllability requirement allowing to use this functional. Next the model of the overhead crane is considered with its space varying coefficients as elasticity effect [1]. It is shown that this model can be reduced to a system of FDE provided the coefficients satisfy a condition of distortionless propagation as defined in [2]. The inherent stability is discussed pointing to feedback stabilization. In the final part of the paper there are given some conclusions and are pointed out those problems which are connected with further research on systems with propagation.

2 Models for nuclear reactors with circulating fuel

While the dynamics studies on nuclear reactors with circulating fuel are almost 50 years old and there exist various classes of models, we shall restrict ourselves to the description given in [6], [7]. These descriptions will be completed by a linear system connected in feedback to the reactor model and representing the external (and control) systems; we have adapted this general structure starting from the basic ideas and contributions of V. M. Popov to this field, see [13]

A. The first description is considered to be obtained using a "rough" simplification of the so called most general *age diffusion model* and is given by a standard system of time delay equations of the delayed type; written in deviations with respect to some constant steady state and combined with the feedback block mentioned above, these equations are given below [17]

$$\begin{aligned} \dot{x} &= Ax + b\zeta \\ \dot{\zeta} &= -(c^*x + \alpha\zeta)(1+\zeta) - \sum_1^m \delta_i\xi_i(\zeta - \eta_i) \\ \dot{\eta}_i &= \frac{\sigma_i}{\xi_i}(\zeta(t) - \eta_i(t)) - e^{-\mu\sigma_i}[\eta_i(t) - \eta_i(t-\mu)], \\ i &= 1,\dots,m \end{aligned} \tag{2}$$

where $0 < \xi_i < 1$, $\delta_i > 0$, $\sigma_i > 0$ and A is a Hurwitz matrix. For the sake of completeness we recall here the main results of [17]

1° The system (2) has the invariant set $1+\zeta > 0$, $1+\eta_i > 0$, $i = 1,\dots,m$, accounting for the positivity of the state variables ζ, η_i which represent neutron densities.
2° Under the previous assumptions, if the following Welton type frequency domain inequality holds

$$\alpha + \Re e\, \gamma(\imath\omega) > 0,\ \forall\omega \geq 0 \tag{3}$$

the zero solution of (2) is globally asymptotically stable with respect to the invariant set above; here $\gamma(s) = c^*(sI - A)^{-1}b$ is the transfer function of the external circuits of the reactor.

Worth mentioning that (3) is equivalent to the feasibility of the following LMI (Linear Matrix Inequality)

$$\begin{pmatrix} PA + A^*P & Pb - \frac{1}{2}c \\ (Pb - \frac{1}{2}c)^* & -\alpha \end{pmatrix} \leq 0, \; P > 0 \tag{4}$$

thus ensuring existence of the following Liapunov functional

$$V(x,\zeta,\eta_1,\ldots,\eta_m) = x^*Px + \int_0^{\zeta(0)} \frac{\lambda}{1+\lambda}d\lambda + \sum_1^m \frac{\delta_i\xi_i^2}{\sigma_i}\left[\int_0^{\eta_i(0)} \frac{\lambda}{1+\lambda}d\lambda + \right.$$
$$\left. +e^{-\mu\sigma_i}\int_{-\mu}^0 (\eta_i(\theta) - \ln(1+\eta_i(\theta))d\theta\right] \tag{5}$$

B. The second description is considered an improved approximation of the same *age diffusion model* and is as follows (after some normalization of the dependent as well as of the independent variables):

$$\begin{aligned}
&\dot{x} = Ax + b\zeta \\
&\dot{\zeta} = -(c^*x + \alpha\zeta)(1+\zeta) - \sum_1^m \beta_i\xi_i(\zeta - \bar{y}_i(t)) \\
&\bar{y}_i(t) = \frac{1}{\xi_i}\int_0^h y_i(\eta,t)\varphi(\eta)d\eta \\
&\frac{\partial y_i}{\partial t} + \frac{\partial y_i}{\partial \eta} + \sigma_i y_i = \sigma_i\varphi(\eta)\zeta(t) \\
&y_i(0,t) = y_i(h,t), \; i = 1,\ldots,m
\end{aligned} \tag{6}$$

and the following initial conditions

$$x(0) = x_0, \; y_i(\eta,0) = \gamma_i(\eta), \; 0 \leq \eta \leq h \tag{7}$$

The basic assumptions on system's coefficients are very much alike to those of the previous model: $0 < \xi_i < 1$, $\beta_i > 0$, $\sigma_i > 0$ and A is a Hurwitz matrix; also $\varphi : [0,h] \mapsto [0,1]$ satisfies the scaling condition

$$\int_0^h \varphi^2(\sigma)d\sigma = 1 \tag{8}$$

and is considered as extended to $\mathbb{R}$ by periodicity.

Following the same approach as in the previous applications i.e. integration along the characteristics of the unique characteristics family defined by the differential equation $d\eta/dt = 1$, we may associate the following system of FDE of neutral type

$$
\begin{aligned}
&\dot{x} = Ax + b\zeta \\
&\dot{\zeta} = -(c^*x + \alpha\zeta)(1+\zeta)- \\
&-\sum_1^m \beta_i\xi_i \left[\zeta(t) - \frac{\sigma_i}{\xi_i}\int_{-h}^0 \left(\int_{-\theta}^h \varphi(\eta)\varphi(\eta+\theta)d\eta\right) e^{\sigma_i\theta}\zeta(t+\theta)d\theta\right] + \\
&+\sum_1^m \beta_i \int_0^h e^{-\lambda\sigma_i}\varphi(\lambda)v_i(t-\lambda)d\lambda \\
&v_i(t) = e^{-h\sigma_i}v_i(t-h) + \sigma_i\int_{-h}^0 e^{\lambda\sigma_i}\varphi(\lambda)\zeta(t+\lambda)d\lambda
\end{aligned}
\tag{9}
$$

valid for $t > h$ with the initial conditions defined by

$$
\begin{aligned}
&\dot{x} = Ax + b\zeta,\ x(0) = x_0 \\
&\dot{\zeta} = -(c^*x + \alpha\zeta)(1+\zeta)- \\
&-\sum_1^m \beta_i\xi_i \left[\zeta(t) - \frac{\sigma_i}{\xi_i}\int_0^t \left(\int_0^h \varphi(\eta)\varphi(\eta+\theta-t)d\eta\right) e^{-\sigma_i(t-\theta)}\zeta(t+\theta)d\theta\right] + \\
&+\sum_1^m \beta_i e^{-\sigma_i t}\int_0^h \varphi(t+\lambda)\gamma_i(\lambda)d\lambda \\
&\zeta(0) = \zeta_0, \\
&v_i(t) = e^{-\sigma_i t}\gamma_i(h-t) + \sigma_i\int_0^t e^{-\sigma_i(t-\theta)}\varphi(\theta-t)\zeta(\theta)d\theta
\end{aligned}
\tag{10}
$$

valid for $0 < t < h$.

Between the solutions of (6)-(7) and (9)-(10) there is a one to one correspondence allowing to obtain results for one of them - in fact the system of FDE (9)-(10) - which are projected back on the other. These results are in fact as follows

a) existence and uniqueness of the solutions defined by initial (and boundary) conditions;
b) existence of some invariant sets accounting for positiveness of some physical variables;
c) stability of equilibria.

It is useful to state here the following

Proposition 1. *Under the basic assumptions above the following properties hold: i) if* $1 + \zeta(0) > 0$, $\bar{q}_i + v_i(t) > 0$, $0 \le t \le h$, $i = 1,\dots,m$, *then* $1 + \zeta(t) > 0$, $\bar{q}_i + v_i(t) > 0$, $t \ge h$, $i = 1,\dots,m$ *where*

$$
\bar{q}_i = \frac{e^{-h\sigma_i}}{1-e^{-h\sigma_i}}\int_0^h e^{\theta\sigma_i}\varphi(\theta)d\theta
$$

ii) if $1 + \zeta(0) > 0$, $c_{i0}(\eta) + \gamma_i(\eta) > 0$, $0 \le t \le h$, $i = 1,\dots,m$, *then* $1 + \zeta(t) > 0$, $c_{i0}(\eta) + y_i(\eta,t) > 0$, $0 \le t \le h$, $i = 1,\dots,m$, *where*

$$
c_{i0}(\eta) = \left[\frac{e^{-h\sigma_i}}{1-e^{-h\sigma_i}}\int_0^h e^{\theta\sigma_i}\varphi(\theta)d\theta + \int_0^\eta e^{\theta\sigma_i}\varphi(\theta)d\theta\right]\sigma_i e^{-\eta\sigma_i}
$$

C. For the stability studies concerning the zero solution of (9)-(10) we have at our disposal the Liapunov functional approach; the linearized version of the system has been studied using a Liapunov functional of the quadratic type, see [12]; here we shall consider the nonlinear case, starting from the basic references [6], [7] and giving a criticism of the results published there.

First of all, we use in the difference equation of (9) the formula of variation of parameters - adapted for difference equations and written on intervals $[ph, (p+1)h]$ and substitute the result in the equation for ζ

$$\dot{\zeta} = -(c^*x + \alpha\zeta)(1+\zeta) -$$

$$-\sum_1^m \beta_i\xi_i\left[\zeta(t) - \frac{\sigma_i}{\xi_i}\int_0^t\left(\int_0^h \varphi(\eta)\varphi(\eta+\theta-t)d\eta\right)e^{-\sigma_i(t-\theta)}\zeta(t+\theta)d\theta\right] + \tag{11}$$

$$+\sum_1^m \beta_i e^{-\sigma_i t}\int_0^h \varphi(t+\lambda)\gamma_i(\lambda)d\lambda$$

to which we have to add, as previously, the differential equation for the external circuits of the reactor, which defines the state vector x

$$\dot{x} = Ax + b\zeta \tag{12}$$

Observe that (11) and (12) are exactly the first two equations of (10), which define the initial conditions for $x(t)$ and $\zeta(t)$ on $(0, h)$; this time the equations are valid for all $t > 0$ and account for the main dynamics of the system; the influence of the initial conditions of $v_i(t)$ - the functions $\gamma_i(\eta)$, $0 \leq \eta \leq h$, $i = 1, \ldots, m$ - is incorporated in the corresponding term of (11). This model is very much alike to a model given in [6], [7] which had been obtained in a much less rigorous way. Namely, there were considered the last three equations of (6), defining m linear blocks with space distributed parameters

$$\begin{aligned} &\frac{\partial y_i}{\partial t} + \frac{\partial y_i}{\partial \eta} + \sigma_i y_i = \sigma_i\varphi(\eta)\zeta(t) \\ &y_i(0,t) = y_i(h,t),\ i = 1, \ldots, m \\ &\bar{y}_i(t) = \frac{1}{\xi_i}\int_0^h y_i(\eta,t)\varphi(\eta)d\eta \end{aligned} \tag{13}$$

for $i = 1, \ldots, m$. For the block of this type there was computed the transfer function $H_i(s) = \tilde{\bar{y}}_i(s)/\tilde{\zeta}(s)$ by applying (formally) the standard (one sided) Laplace transform under zero initial conditions. The result,after some manipulation, is

$$H_i(s) = \frac{\sigma_i}{\xi_i}\int_0^\infty e^{-\lambda(s+\sigma_i)}\int_0^h \varphi(\eta)\varphi(\eta-\lambda)d\eta d\lambda \tag{14}$$

One can see that $H_i(s)$ is the Laplace transform of the causal (i.e. identically zero for $t < 0$) impulse response

$$h_i(t) = \frac{\sigma_i}{\xi_i}e^{-\sigma_i t}\int_0^h \varphi(\eta)\varphi(\eta-t)d\eta \tag{15}$$

Therefore

$$\bar{y}_i(t) = \int_0^t h_i(t-\theta)\zeta(\theta)d\theta \tag{16}$$

and substitution in the equation for ζ in (6) will give exactly (11) *but with the last integral being zero since the Laplace transform was applied under zero initial conditions for the "state variables" $y_i(\eta, t)$ of the block (13)* i.e. for $\gamma_i(\eta) \equiv 0, 0 \leq \eta \leq h$.

Strange enough, this obvious fact is neglected in the cited references [6], [7] and the effect of the initial conditions is introduced by replacing (11) by

$$\dot{\zeta} = -(c^*x + \alpha\zeta)(1+\zeta) - \sum_1^m \beta_i\xi_i \left[\zeta(t) - \frac{\sigma_i}{\xi_i}\int_{-\infty}^t h_i(t-\theta)\zeta(t+\theta)d\theta\right] \tag{17}$$

motivated by the "return to the space of the original functions" which will associate to $H_i(s) = \tilde{\bar{y}}_i(s)/\tilde{\zeta}(s)$ the convolution

$$\bar{y}_i(t) = \int_{-\infty}^t h_i(t-\theta)\zeta(\theta)d\theta \tag{18}$$

But the original functions resulting from the inversion of the one sided Laplace transform are identically zero for $t < 0$; therefore the "causal" convolution (16) is valid and not (18)! Nevertheless this replacement is convenient from the point of view of the Liapunov functional, as it will appear next. But we have to insist on the fact that we make basically a change of model because the introduction of (18) instead of (16) is not mathematically justified.

It has to be also mentioned that there are several other ways of associating FDE to the PDE e.g. the so-called *diffusive representation* - [11]; we preferred the direct way in the time domain in order to show how FDE of neutral type are introduced since we consider this to be the best motivation for the neutral equations.

3 A Liapunov functional for a non-rigorous model

In this section we shall consider the system defined by (12) and (17); its state space will be considered $\mathbb{R} \times \mathbf{X}$ where $\mathbf{X}$ is some function space on$(-\infty, 0)$ endowed with a topology obeying the conditions e.g. from the paper of Hale and Kato [9]. On this space we define the state functional which might be a candidate Liapunov functional

$$\begin{aligned} V(z, \psi(\cdot)) = z^*Pz + \int_0^{\psi(0)} \frac{\lambda}{1+\lambda}d\lambda + \\ + \sum_1^m \beta_i\xi_i \left[\int_{-\infty}^0 \frac{\psi(\lambda)}{1+\psi(\lambda)}\left(\psi(\lambda) - \int_{-\infty}^\lambda h_i(\lambda-\theta)\psi(\theta)d\theta\right)d\lambda\right] \end{aligned} \tag{19}$$

where $P = P^*$ will be determined further and $\psi(\lambda) > -1, \forall\lambda < 0$. In order to check positiveness of V as state function we evaluate the last term of the above expression using a simple technical fact [21], a continuous time version of a rearrangement inequality

Proposition 2. *If f is non-decreasing, $f(0) = 0$ and $x(t) \equiv 0$ for $t < 0$ then*

$$\int_0^T f(x(t))(x(t) - x(t-\tau))dt \geq 0$$

for all $\tau > 0$

This result is applied firstly to prove directly that $0 < h_i(t) \leq 1$; here we use also the scaling condition (8). Then it follows, after the application of the Fubini theorem that

$$\int_{-\infty}^{0} \frac{\psi(\lambda)}{1+\psi(\lambda)} \left(\psi(\lambda) - \int_{-\infty}^{\lambda} h_i(\lambda-\theta)\psi(\theta)d\theta \right) \geq$$

$$\geq \int_{0}^{\infty} h_i(\tau) \left(\int_{-\infty}^{0} \frac{\psi(\lambda)}{1+\psi(\lambda)} (\psi(\lambda) - \psi(\lambda-\tau))d\lambda \right) d\tau$$

The same proposition applied again, under a slightly modified form, will give nonnegativeness of the RHS in the above inequality. Consequently the sum in (19) is composed of nonnegative terms and is nonnegative. The second term in (19) is but well known from the standard Liapunov functions associated to nuclear reactor dynamics and is strictly positive. If we assume (3) to hold or, equivalently, choose $P > 0$ from the feasible LMI (4) then V defined by (19) is at least nonnegative definite.

We have now to write down $V^*(t) = V(x(t), \zeta_t(\cdot))$ along the solutions of (12),(17) and differentiate it to find

$$\begin{aligned} \frac{d}{dt}V^*(t) &= x^*(t)P(Ax(t) + b\zeta(t)) + (Ax(t) + b\zeta(t))^* Px(t) - \\ &- \zeta(t)(c^*x(t) + \alpha\zeta(t)) \leq 0 \end{aligned} \tag{20}$$

and stability in the sense of Liapunov follows for the system defined by (12),(17). A Barbašin-Krasovskii-LaSalle like proof will give asymptotic stability. We proved in fact

Theorem 1. *Under the basic assumptions on system (9)-(10) and assuming (3) (or (4)) to hold, the zero solution of (12),(17) is globally asymptotically stable with respect to the invariant set* $1 + \zeta > 0$.

This theorem is again a Welton type criterion but for a model that, as said, is not entirely correct. The result could be nevertheless acceptable provided the following choice of $\zeta(t)$ on $(-\infty, 0)$ would be possible: given the initial conditions $\gamma_i(\eta)$, $0 \leq \eta \leq h$, choose $\zeta_i(\cdot)$ from the condition

$$\sigma_i \int_{-\infty}^{0} \left(\int_{0}^{h} \varphi(\eta)\varphi(\eta+\lambda-t)d\eta \right) e^{\sigma_i\theta}\zeta_i(\theta)d\theta = \int_{0}^{h} \varphi(t+\lambda)\gamma_i(\lambda)d\lambda$$

which reduces to a linear integral convolution equation of the first kind; its solution should allow a convenient (from the stability point of view) dependence of $\zeta_i(\cdot)$ and $\gamma_i(\cdot)$. *This is still an open problem* and the reader is sent to standard references on linear integral equations for possible approaches.

4 The functional differential equations associated to an overhead crane

Following [1] we give below the control model for an overhead crane incorporating the elasticity effect

$$\begin{aligned}&\frac{\partial^2 y}{\partial t^2}-\frac{\partial}{\partial\lambda}\left(a(\lambda)\frac{\partial y}{\partial\lambda}\right)=0\\&\frac{\partial y}{\partial\lambda}(0,t)=0\ ,\ y(L,t)=x_p(t)\\&\ddot{x}_p=K\left(a(L)\frac{\partial y}{\partial\lambda}(L,t)\right)+u(t)\end{aligned}\tag{21}$$

for which a suitable choice of the control $u(t)$ is required [10]. In order to apply previous and known results on this subject, we perform some change of variables first

$$\frac{\partial y}{\partial t}(\lambda,t):=v(\lambda,t)\ ,\ a(\lambda)\frac{\partial y}{\partial\lambda}(\lambda,t):=w(\lambda,t)\tag{22}$$

With these variables system (21) becomes

$$\begin{aligned}&\frac{\partial v}{\partial t}=\frac{\partial w}{\partial\lambda}\ ,\ \frac{\partial w}{\partial t}=a(\lambda)\frac{\partial v}{\partial\lambda}\\&w(0,t)=0\ ,\ v(L,t)=\dot{x}_p(t)\\&\ddot{x}_p=Kw(L,t)+u(t)\end{aligned}\tag{23}$$

We have in (23) again a boundary value problem for hyperbolic PDE in the plane, with boundaries controlled by a system of ODE; unlike other cases, we have here a space varying parameter $a(\lambda)$ and this will bring additional difficulties since the forward and backward waves are not separated in (23) unlike e.g. in [3] or [14]. One may ask whether the model with pointwise delays is applicable here. In order to answer this question we turn to [2] where pointwise delay case is identified to the so called *distortionless propagation* i.e. to solutions of the form

$$v(\lambda,t)\equiv f(\lambda)\varphi(t-\tau(\lambda))\ ,\ w(\lambda,t)\equiv h(\lambda)\varphi(t-\tau(\lambda))\tag{24}$$

Substituting in the PDE of (23) we find

$$\begin{aligned}&f(\lambda)\varphi'(t-\tau(\lambda))\equiv h'(\lambda)\varphi(t-\tau(\lambda))+h(\lambda)\tau'(\lambda)\varphi'(t-\tau(\lambda))\\&h(\lambda)\varphi'(t-\tau(\lambda))\equiv a(\lambda)(f'(\lambda)\varphi(t-\tau(\lambda))+f(\lambda)\tau'(\lambda)\varphi'(t-\tau(\lambda)))\end{aligned}\tag{25}$$

and since φ and φ' are "independent" we find in a rather straightforward way that $a(\lambda)\equiv a_0$ i.e. $a(\lambda)$ has to be constant. In this way we define the backward and forward waves - the Riemann invariants - as below

$$\begin{aligned}&v(\lambda,t)=v_1(\lambda,t)+v_2(\lambda,t)\\&w(\lambda,t)=\sqrt{a_0}(-v_1(\lambda,t)+v_2(\lambda,t))\end{aligned}\tag{26}$$

to obtain

$$\begin{aligned}&\frac{\partial v_1}{\partial t}+\sqrt{a_0}\frac{\partial v_1}{\partial\lambda}=0\ ,\ \frac{\partial v_2}{\partial t}-\sqrt{a_0}\frac{\partial v_2}{\partial\lambda}=0\\&v_1(0,t)-v_2(0,t)=0\ ,\ v_1(L,t)+v_2(L,t)=\dot{x}_p(t)\\&\ddot{x}_p=K\sqrt{a_0}(v_2(L,t)-v_1(L,t))+u(t)\end{aligned}\tag{27}$$

This system has the standard form of all basic references cited previously. We may associate to it the following system of FDE

$$\begin{aligned} \dot{x}_p &= v_p \\ \dot{v}_p &= K\sqrt{a_0}(v_p(t) - 2\eta(t - 2L\sqrt{a_0})) + u(t) \\ \eta(t) &= v_p(t) - \eta(t - 2L\sqrt{a_0}) \end{aligned} \tag{28}$$

It is easy to check its inherent stability i.e. for $u(t) \equiv 0$; it is a linear system of FDE coupled to a difference equation, both linear. The characteristic equation will be

$$s\left[\left(1 - e^{-2sL\sqrt{a_0}}\right)(s - K\sqrt{a_0}) + 2K\sqrt{a_0}\right] = 0 \tag{29}$$

and there are at least two difficulties: there exists a zero root and the difference operator has a single characteristic number equal to 1 - again a critical case, from the robustness point of view. These aspects have to be taken into account in the stabilization solution.Worth mentioning that there exist some earlier results in overhead crane control e.g. [4], [5]. The importance of the application requires some comparison of the approaches.

5 Conclusions and perspectives

Throughout the paper we have discussed two propagation models that display special features thus leading to non-standard FDE. In each case we mentioned difficulties as well as unsolved/open problems. It appears at the same time that additional research is required for model analysis and improvement; it appears also necessary to focus on various functional equations occurring during integration along the characteristics in the PDE. With respect to this it is worth mentioning that there exist still several papers containing various ways of associating functional equations to boundary value problems for hyperbolic PDE which are waiting to be rediscovered; what we want to say is that a sufficient accumulation of basic results exists in order to answer various challenges.

References

1. d'Andrea Novel, B., Boustany, F., Conrad, F., Rao, B.P.: Feedback stabilization of a hybrid PDE-ODE system: Application to an overhead crane. Math. of Control Signals and Syst. 7, 1–22 (1994)
2. Burke, V., Duffin, R.J., Hazony, D.: Distortionless wave propagation in inhomogeneous media and transmission lines. Quart. Appl. Math. 34, 183–194 (1976)
3. Cooke, K.L.: A linear mixed problem with derivative boundary conditions. Seminar on Differential Equations and Dynamical Systems (III) Lecture Series Univ. of Maryland, College Park 51, 11–17 (1970)
4. Fliess, M., Lévine, J., Rouchon, P.: A simplified approach of crane control via a generalized state-space model. In: Proc. 30th IEEE Contr. Dec. Conf. Brighton (1991)
5. Fliess, M., Lévine, J., Rouchon, P.: A generalized state variable representation for a simplified crane description. Int. J. Control 58, 277–283 (1993)
6. Goriačenko, V.D.: Methods of stability theory in the dynamics of nuclear reactors, Atomizdat, Moscow (1971) (in Russian)
7. Goriačenko, V.D.: Methods for nuclear reactor stability studies, Atomizdat, Moscow (1977) (in Russian)
8. Goriačenko, V.D., Zolotarev, S.L., Kolchin, V.A.: Qualitative methods in nuclear reactor dynamics, Energoatomizdat, Moscow (1988) (in Russian)

9. Hale, J.K., Kato, J.: Phase space for retarded equations with infinite delay. Funkcial Ekvac 21, 11–41 (1978)
10. Mondié, S.: Private communication (2006)
11. Montseny, G.: Représentation diffusive. Hermès Science Publications, Paris (2005)
12. Niculescu, S.I., Răsvan, V.l.: Stability of some models of circulating fuel nuclear reactors - a Lyapunov approach. In: Agarwal, R.P., Perera, K. (eds.) Diffential & Difference Equations and Applications. Hindawi Publ. Company, New York (2006)
13. Popov, V.M.: Hyperstability of Control Systems (in Romanian). Editura Academiei, Bucharest (1966); Russian improved version by Nauka Publ. House, Moscow (1970) (Russian); French improved version by Dunod, Paris (1973) (French); English improved version by Springer, Heidelberg (1973)
14. Răsvan, V.l.: A method of distributed parameter control systems and electrical networks analysis. Rev. Roumaine Sci. Techn. Série. Électrotechn. et. Énerg. 20, 561–566 (1975)
15. Răsvan, V.l.: Absolute stability of time lag control systems. Editura Academiei, Bucharest (1975); Russian revised edition by Nauka, Moscow (1983) (Russian)
16. Răsvan, V.I.: Systems with propagation and functional-differential equations. In: Kisielewicz, M. (ed.) Functional-differential equations and related topics. Błażejewko Poland (1981)
17. Răsvan, V.l.: A stability result for a system of differential equations with delay describing a controlled circulating fuel nuclear reactor. Rev. Roumaine Math. Pures. Appl. 39, 365–373 (1994)
18. Răsvan, V.I.: Dynamical systems with lossless propagation and neutral functional differential equations. In: Beghi, A., Finesso, I., Picci, G. (eds.) Mathem Theory Networks Systems Proc. MTNS 1998. Il, Poligrafo Padova (1998)
19. Răsvan, V.I.: Functional Differential Equations of Lossless Propagation and Almost Linear Behavior. In: Manes, C., Pepe, P. (eds.) Proc. IFAC Workshop on Time Delay Systems TDS 2006, l'Aquila (2006)
20. Răsvan, V.I.: Three Lectures on Neutral Functional Differential Equations. Presented at the Department of Automatic Control CINVESTAV Ciudad de Mexico (2006)
21. Willems, J.C., Gruber, M.: Comments on A combined Frequency-Time Domain Stability Criterion for Autonomous Continuous Systems. IEEE Trans. Automat. Control. 12, 217–218 (1967)

Design, Modelling and Control of the Experimental Heat Transfer Set-Up

Tomáš Vyhlídal, Pavel Zítek, Karel Paulů

Centre for Applied Cybernetics, Dept. of Instrumentation and Control Eng., Faculty of Mechanical Engineering, Czech Technical University in Prague, Technická 4, 166 07 Praha 6, Czech Republic
tomas.vyhlidal@fs.cvut.cz, pavel.zitek@fs.cvut.cz

Summary. The paper deals with an experimental heat transfer set-up which has been designed for testing time delay system approaches in modelling and control design. First, the set-up is described and some implementation issues are discussed. Secondly, a time delay model of the set-up is presented, which consists of ten functional differential equations. The state variables of the model are identical with the measured outputs of the system. This convenient feature is used in the control design by state-variable feedback. The results of the set-up control are presented including the feedback design using the continuous pole placement method.

1 Introduction

The progress in the analysis and design of time-delay systems achieved over the last decades and summarized, e.g. in the monographs [1], [2] or [3] is impressive. However, due to the infinite-order nature of time delay systems their identification as models of real plants brings some specific problems. While lumped delays are usually supposed in theoretical investigations, sometimes even considered as commensurate values, the real delay phenomena are of rather complex and distributed nature. For instance the delays arising due to the heat carrying fluid motion are distributed in time, in fact. The final delayed temperature response is then determined not only by the fluid transport, but also by a heat transfer between the flow and its neighbourhood. Various real delay phenomena are of more or less distributed character and their identification with the available structures of time-delay systems becomes a sophisticated issue, see e.g. [4].

That was the motivation for developing a laboratory set-up providing various physical implementations of real delay phenomena in order to prove the applicability of theoretical results. The heat transfer process has been chosen particularly due to the possibility of achieving sufficiently long delays in the obtained responses. The heat transfer relationships are nonlinear and this is also an advantageous feature of the thermal set-up from the point of testing the practical applicability of the theoretical results.

The structure of this contribution is as follows. In Section 2, the set-up and its main subsystems are described in detail. Besides, some implementation issues are mentioned. In Section 3, a linear time delay model (also called anisochronic model [5]) is proposed in order

J.J. Loiseau et al. (Eds.): Topics in Time Delay Systems, LNCIS 388, pp. 303–313.
springerlink.com

to match the dynamics of the set-up for a given operating point and the identification of the system parameters to measured real data is performed. Consequently, the state variable feedback is designed in Section 4 and both simulation and measurement of its implementations are shown. Finally, conclusions are made in Section 5.

2 Description and design of the set-up

The experimental heat transfer set-up, which is shown in Fig. 1, has been designed as a multiple input and multiple output (MIMO) system where the transportation phenomenon plays a significant role in the dynamics.

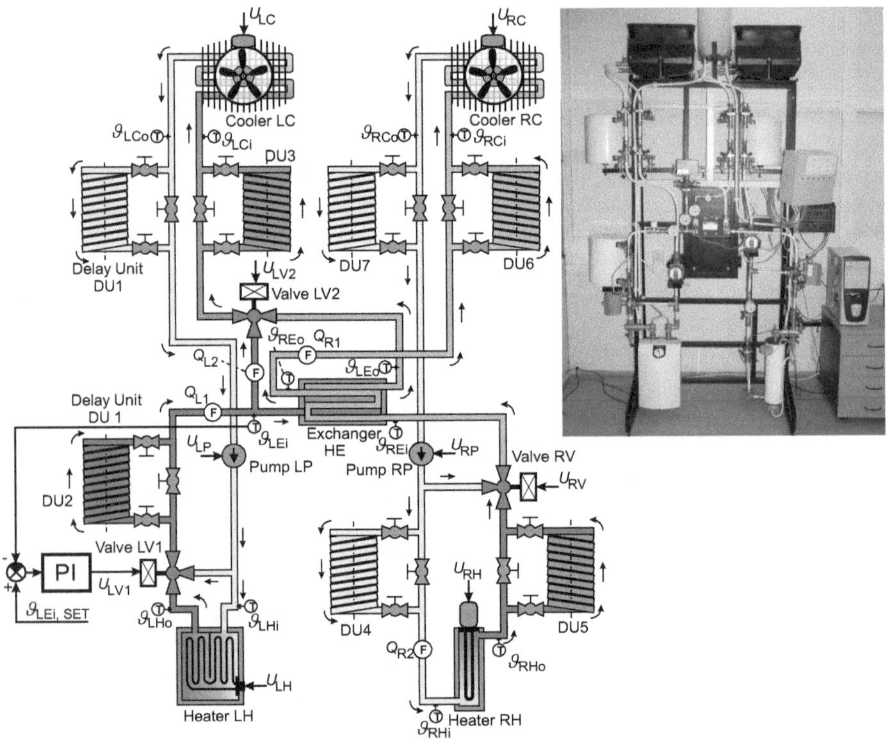

Fig. 1. Scheme and photo of the experimental heat transfer set-up

2.1 Components of the set-up

The set-up consists of two closed independent heating circuits, in which the water is the heat carrying fluid. The main components of the set-up are following (the first letter in brackets indicates the position of each component: L-left circuit, R-right circuit):

- *Accumulation heater* (LH) with volume 5 litres and maximum heating performance 2000 W
- *Flow-through heater* (RH) with volume 1.5 litre and maximum heating performance also 2000 W
- *Two pumps* (LP, RP)
- *Two coolers* (LC, RC) - water/air exchangers from a car heating
- *Three mixing control valves* (LV1, LV2, RV)
- *Multi-plate heat exchanger* (HE) with performance up to 3 kW

As can be seen in Fig. 1, the main components of the set-up are connected by piping lines (with inner diameter 12 mm). Obviously the lengths of the pipes and flow velocities determine the transport delays in the heat transfer throughout the system. Therefore, in order to provide the possibility to enhance the delay effect, seven delay units (DU) have been placed in the set-up. The delay units consist of scrolled pipes with lengths varying from 3 to 30 meters. Since the delay units are interchangeable and they can also be bypassed, the set-up provides broad structural variability for studying the effect of transportation phenomenon in real plants.

2.2 Monitoring and control system

The measurement system of the set-up consists of:

- twelve thermometers placed at the inputs and outputs of the subsystems (their distribution is shown in Fig. 1, symbols ϑ are used for temperature signals)
- four flow meters measuring the flow rate in the branches of the set-up pipe-lines (their positions are shown in Fig. 1, symbols Q are used for the flows rate measurements)

By the control system, the following quantities can be continuously controlled (in brackets, the control signals of the actuators are given, as shown in Fig. 1):

- heat performances of heaters LH and RH (u_{LH}, u_{RH} - the control signals continuously adjust the heat performances of the heating elements in the heaters)
- cooling performances of coolers LC and RC (u_{LC}, u_{RC} - the control signals adjust the air flow rate in the ventilators of the coolers).
- mixing ration in mixing valves LV1, LV2, RV (u_{LV1}, u_{LV2}, u_{RV} - the control signals determine the valve seat positions via servos)

All the above mentioned control signals u and temperature signals ϑ are generated or processed either via PC equipped with data acquisition cards (processed in software Matlab/Simulink with Real Time Toolbox) or via an industrial Programmable Controller. Also, flow rates provided by the pumps (LP, RP) can be adjusted manually.

2.3 Description of the set-up function

In this section, the function of the set-up is described. Let us recall that the symbols and abbreviations, which have been defined above, refer to the set-up scheme shown in Fig. 1.

The ***Left Circuit*** and its function is described as follows. The circulation of the water in the circuit is forced by the pump LP. The flow through the pump is measured by the flow-meter Q_{L1}. The heat source of the circuit is the accumulation heater LH which heats up the water. The inlet and the outlet water temperatures of the heater are measured as ϑ_{LHi} and ϑ_{LHo}, respectively. Even though the performance of the heater can be controlled by the signal u_{LH}, due to relatively large capacity of the heater, control actions made this way are fairly slow. Therefore, the temperature ϑ_{LEi} of the water that enters the heat exchanger HE is controlled by mixing the hot water from the heater LH and the cooled water from the cooler LC in the mixing valve LV1. The valve is controlled by a PI controller with $\vartheta_{LEi,SET}$ as the set point of ϑ_{LEi}. The temperature of the water measured in the outlet of the exchanger in the left circuit is measured as ϑ_{LEo}. The second mixing valve LV2, controlled by the signal u_{LV2}, divides the water flow into two branches. Bypassing the exchanger by LV2 allows us to control the amount of the heat which is transferred between the circuits. The flow through the bypass branch is measured by a flow-meter as Q_{L2}. Consequently, the temperature of the water leaving the mixing valve and forced towards the cooler LC can be controlled by the signal u_{LV2}. The temperatures in the inlet and outlet of the cooler LC are measured as ϑ_{LCi} and ϑ_{LCo}, respectively. The cooling performance of the cooler LC is controlled by signal u_{LC}. The water which leaves the cooler is forced by the pump towards the heater LH and the valve LV1.

Analogously, the ***Right Circuit*** is described as follows. The circulation of the water in the circuit is forced by the pump RP. The flow through the pump is measured by the flow-meter Q_{R1}. The heat source of the circuit is the flow-through heater RH. Its inlet and outlet water temperatures are measured as ϑ_{RHi} and ϑ_{RHo}, respectively. Unlike the dynamics of the heater LH, the dynamics of the heater RH are relatively fast. Therefore, the heater can serve well as an actuator controlling the water temperature entering the exchanger HE, which is measured as ϑ_{REi}. Alternatively, as an actuator by which this temperature may be controlled, the mixing valve RV can be used, which is controlled by the signal u_{RV}. This mixing valve determines the flow passing through the heater measured as Q_{R2}. The temperature of the water leaving the exchanger in the right circuit is measured as ϑ_{REo}. The water is then forced towards the cooler RC. Its inlet and outlet water temperatures are measured as ϑ_{RCi} and ϑ_{RCo}, respectively. The cooling performance of the cooler RC is controlled by the signal u_{RC}. The water which leaves the cooler is transported towards the heater RH and the valve RV.

3 Anisochronic model

Obviously, the transportation phenomenon in pipe-lines as well as distributed parameters in the components play important roles in dynamics of the set-up. Therefore, describing the set-up dynamics using a model with time delays is a natural option. Due to using time-shifted data, this sort of models is referred to as anisochronic [5]. The linear model of the set-up that is given below describes the dynamics of the set-up in the vicinity of a selected operating point, which can be defined by fixed values of all actuating signals u_i. Let us remark that even though the linear model is considered in a deviation form, for the sake of brevity, the symbols Δ are omitted, i.e. $\vartheta \approx \Delta\vartheta$, $u \approx \Delta u$.

In order to avoid critical nonlinearities in the set-up and variability in the delays, control valves LV2 and RV are considered with one branch fully closed and the other fully opened (LV2 - the branch that bypasses the exchanger is closed, RV - the branch that bypasses the heater RH is closed). The valve LV1, which is used for controlling the temperature $\vartheta_{LEi}(t)$ via a PI slave control loop (the parameters of the controller have been adjusted experimentally, [6]) see Fig. 1, is not considered in the model either. Instead, the following model is used

$$T_{LL}\frac{d\vartheta_{LEi}(t)}{dt} = -\vartheta_{LEi}(t-\tau_{LL1}) + \vartheta_{LEi\_SET}(t-\tau_{LL2}) \qquad (1)$$

which describes the dynamics of the whole PI slave control loop.

It is well known that systems with distributed parameters can be described well by models that combine first order dynamics with time delays in the model variables [6], [7]. Such a model in the form

$$T\frac{d\vartheta_o(t)}{dt} = -\vartheta_o(t) + K_1\vartheta_i(t-\tau_1) + K_2u_j(t-\tau_2) \qquad (2)$$

is used for describing the left heater (LH) and both coolers (LC, RC), (ϑ_i, ϑ_o are the input and output temperatures, respectively, and u_j is the control input of the particular unit).

The first-order plus dead time model

$$T_D\frac{d\vartheta_o(t)}{dt} = -\vartheta_o(t) + K_D\vartheta_i(t-\tau_D) \qquad (3)$$

is also used for modelling relations between the input and the output temperatures of the pipe-lines. In fact, the transportation phenomenon in pipe-lines could be, alternatively, described by lumped or distributed delays only. However, model (3) is more appropriate, because it allows description of both the transportation phenomenon and the accumulation effect in the pipe walls and temperature sensors. Moreover, as will be shown later, describing pipe-line dynamics by (3) allows us to consider the outputs ϑ_o as the state variables of the model of the whole set-up.

The dynamics of the heat exchanger (HE) are described by the following equations

$$\begin{aligned} T_{RE}\tfrac{d\vartheta_{REo}(t)}{dt} &= -\vartheta_{REo}(t) + \vartheta_{REi}(t) + \kappa_{RE}\Delta\tilde{\vartheta}(t) \\ T_{LE}\tfrac{d\vartheta_{LEo}(t)}{dt} &= -\vartheta_{LEo}(t) + \vartheta_{LEi}(t) - \kappa_{LE}\Delta\tilde{\vartheta}(t) \end{aligned} \qquad (4)$$

where $\Delta\tilde{\vartheta}(t)$ is a mean temperature difference which can be described either by arithmetic or more accurate logarithmic formula. Due to high efficiency of the multi-plate heat exchanger, the heat loss in the heat exchange can be neglected and the arithmetic formula

$$\Delta\tilde{\vartheta}(t) = \frac{\vartheta_{LEi}(t) - \vartheta_{REi}(t)}{2} + \frac{\vartheta_{LEo}(t) - \vartheta_{REo}(t)}{2} \qquad (5)$$

which is finally used in (4), is fully sufficient for expressing $\Delta\tilde{\vartheta}(t)$.

The anisochronic model of the whole set-up is given by assembling the models of components according to causality relations of the heat transfer within the set-up. In the Laplace operating form, if zero initial conditions are considered, the model is given by

$$\begin{aligned} s\mathbf{x}(s) &= \mathbf{A}(s)\mathbf{x}(s) + \mathbf{B}(s)\mathbf{u}(s) \\ \mathbf{y}(s) &= \mathbf{x}(s) \end{aligned} \qquad (6)$$

where $\mathbf{x}(s)$, $\mathbf{u}(s)$, $\mathbf{A}(s)$, $\mathbf{B}(s)$ are given as follows

$$\mathbf{x}(s) = \begin{bmatrix} \vartheta_{RHi}(s) \\ \vartheta_{RHo}(s) \\ \vartheta_{REi}(s) \\ \vartheta_{REo}(s) \\ \vartheta_{RCi}(s) \\ \vartheta_{RCo}(s) \\ \vartheta_{LEi}(s) \\ \vartheta_{LEo}(s) \\ \vartheta_{LCi}(s) \\ \vartheta_{LCo}(s) \end{bmatrix}, \mathbf{u}(s) = \begin{bmatrix} u_{RH}(s) \\ u_{RC}(s) \\ \vartheta_{LEi,SET}(s) \\ u_{LC}(s) \end{bmatrix} \tag{7}$$

$$\mathbf{A}(s) = \begin{bmatrix} a_{1,1}(s) & \dots & a_{1,10}(s) \\ \vdots & \vdots & \vdots \\ a_{10,1}(s) & \dots & a_{10,10}(s) \end{bmatrix}, \mathbf{B}(s) = \begin{bmatrix} b_{1,1}(s) & \dots & b_{1,4}(s) \\ \vdots & \vdots & \vdots \\ b_{10,1}(s) & \dots & b_{10,4}(s) \end{bmatrix}$$

and where $a_{1,1}(s) = -\frac{1}{T_{RD1}}$, $a_{1,6}(s) = \frac{K_{RD1}\exp(-s\tau_{RD1})}{T_{RD1}}$, $a_{2,1}(s) = \frac{K_{RH1}\exp(-s\tau_{RH1})}{T_{RH}}$, $a_{2,2}(s) = -\frac{1}{T_{RH}}$, $a_{3,2}(s) = \frac{K_{RD2}\exp(-s\tau_{RD2})}{T_{RD2}}$, $a_{3,3}(s) = -\frac{1}{T_{RD2}}$, $a_{4,3}(s) = \frac{1-0.5\kappa_{RE}}{T_{RE}}$, $a_{4,4}(s) = -\frac{1+0.5\kappa_{RE}}{T_{RE}}$, $a_{4,7}(s) = \frac{\kappa_{RE}}{2T_{RE}}$, $a_{4,8}(s) = \frac{\kappa_{RE}}{2T_{RE}}$, $a_{5,4}(s) = \frac{K_{RD3}\exp(-s\tau_{RD3})}{T_{RD3}}$, $a_{5,5}(s) = -\frac{1}{T_{RD3}}$, $a_{6,5}(s) = \frac{K_{RC1}\exp(-s\tau_{RC1})}{T_{RC}}$, $a_{6,6}(s) = -\frac{1}{T_{RC}}$, $a_{7,7}(s) = -\frac{\exp(-s\tau_{LL1})}{T_{LL}}$, $a_{8,3}(s) = \frac{\kappa_{LE}}{2T_{LE}}$, $a_{8,4}(s) = \frac{\kappa_{LE}}{2T_{LE}}$, $a_{8,7}(s) = \frac{1-0.5\kappa_{LE}}{T_{LE}}$, $a_{8,8}(s) = -\frac{1+0.5\kappa_{LE}}{T_{LE}}$, $a_{9,8}(s) = \frac{K_{LD}\exp(-s\tau_{LD})}{T_{LD}}$, $a_{9,9}(s) = -\frac{1}{T_{LD}}$, $a_{10,9}(s) = \frac{K_{LC1}\exp(-s\tau_{LC1})}{T_{LC}}$, $a_{10,10}(s) = -\frac{1}{T_{LC}}$, $b_{2,1}(s) = \frac{K_{RH2}\exp(-s\tau_{RH2})}{T_{RH}}$, $b_{6,2}(s) = \frac{K_{RC2}\exp(-s\tau_{RC2})}{T_{RC}}$, $b_{7,3}(s) = \frac{\exp(-s\tau_{LL2})}{T_{LL}}$, $b_{10,4}(s) = \frac{K_{LC2}\exp(-s\tau_{LC2})}{T_{LC}}$, while the other terms in the matrices $\mathbf{A}(s)$ and $\mathbf{B}(s)$ are equal to zero.

The positive feature of model (6)-(7) is the identifying the state variables with the set-up outputs (measured temperatures). Such a favorable unifying feature is only possible thanks to using the anisochronic structure of the model, i.e. considering time delays in model relationships.

In order to identify parameters of model (6)-(7), series of experiments have been performed. The parameters have been identified from transient responses of the set-up combined with relay-feedback experiments [6] as follows: $T_{RD1} = 5$, $K_{RD1} = 0.973$, $\tau_{RD1} = 15$; $T_{RH} = 25$, $K_{RH1} = 0.96$, $K_{RH2} = 0.536$, $\tau_{RH1} = 23$, $\tau_{RH2} = 11.5$; $T_{RD2} = 5$, $K_{RD2} = 0.983$, $\tau_{RD2} = 5$; $T_{RE} = 3$, $\kappa_{RE} = 0.85$; $T_{RD3} = 5$, $K_{RD3} = 0.975$, $\tau_{RD3} = 3$, $T_{RC} = 17$, $K_{RC1} = 0.9$, $K_{RC2} = 0.043$, $\tau_{RC1} = 5$, $\tau_{RC2} = 6$; $T_{LL} = 63$, $\tau_{LL1} = 29$, $\tau_{LL2} = 7$; $T_{LE} = 3$, $\kappa_{LE} = 0.85$; $T_{LD} = 5$, $K_{LD} = 0.97$, $\tau_{LD} = 5$; $T_{LC} = 15$, $K_{LC1} = 0.92$, $K_{LC2} = 0.042$, $\tau_{LC1} = 5$, $\tau_{LC2} = 6$, where time constants and delays are given in seconds. Let us remark that no extra delay units (DU) were used in the set-up arrangement. This is the reason why delays identified as the direct consequence of the transportation phenomenon are relatively small. As examples of agreement between model and measured responses, responses of the heater (RH), the cooler (LC) and the whole set-up are shown in Fig. 2 and Fig. 3, respectively. In these figures, the thin lines affected by noise are the measured responses while the thick smooth lines are the model responses.

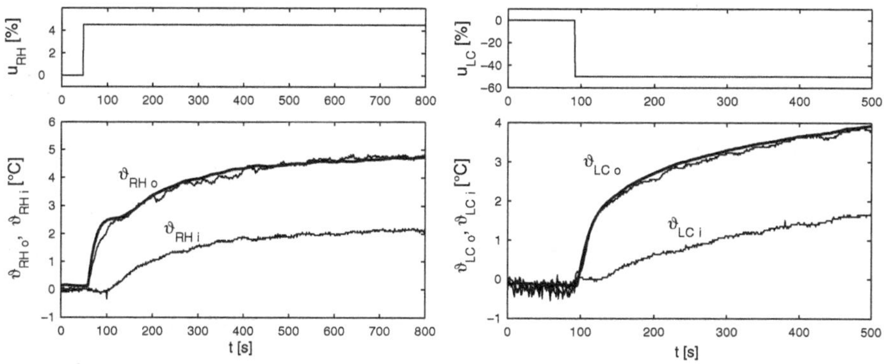

Fig. 2. Left - Transient response of the heater RH. Right - Transient response of the cooler LC, (thick - model responses, thin - measured responses), both figures are in deviation form

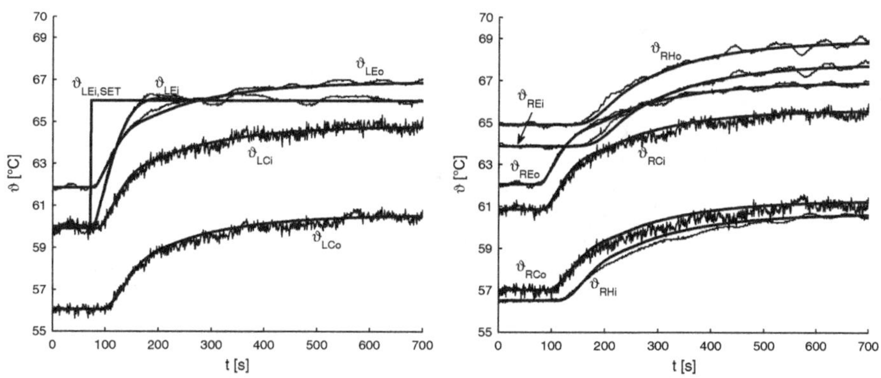

Fig. 3. Transient responses of the whole set-up, left - temperatures measured at the left circuit, right - temperatures measured at the right circuit, (thick - model responses, thin - measured responses)

4 Control design example, state-variable feedback

One of the control schemes that has been tested on the set-up is a cascade scheme with the aim of controlling the temperature $\vartheta_{LCo}(t)$. The master controller of the scheme is given by the state-variable feedback controller

$$\vartheta_{LEi,SET}(t) = -\mathbf{K}\,\bar{\mathbf{x}}(t) \tag{8}$$

which has already been used to control time-delay systems in [5], [8], [9], where $\bar{\mathbf{x}}(t) = [\mathbf{x}(t)^{\,T}\, I(t)]^T$ is the extended state variable vector in which $\mathbf{x}(t)$ is given in (7) and

$$I(t) = \int_0^t (\vartheta_{LCo,SET}(\eta) - \vartheta_{LCo}(\eta))d\eta \tag{9}$$

is an additional state variable introduced for the control purposes. The variable $\vartheta_{LEi,SET}(t)$ in (9) is the set-point of the slave PI control loop, as shown in Fig. 1. Let us repeat that the parameter design of PI controller has been done in an experimental way and it is not an objective of this section.

Since the spectral controllability condition is satisfied

$$Rank([\bar{\mathbf{B}}_L(s),\ \bar{\mathbf{A}}(s)\bar{\mathbf{B}}_L(s),\ \ldots,\ \bar{\mathbf{A}}(s)^{10}\bar{\mathbf{B}}_L(s)]) = 11 \tag{10}$$

where $\bar{\mathbf{A}}(s)$ is the dynamics matrix resulting from the extension of the state vector from $\mathbf{x}(s)$ to $\bar{\mathbf{x}}(s)$ and $\bar{\mathbf{B}}_L(s)$ is the input matrix column of the extended model corresponding to $\vartheta_{LEi,SET}(s)$, feedback (8) can be used to control the system (6)-(7), [10].

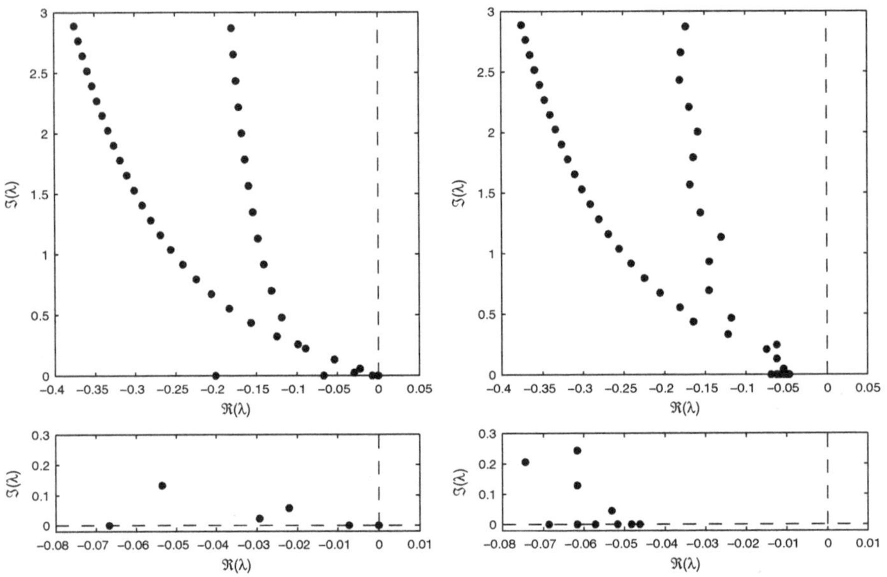

Fig. 4. Left - the spectrum of poles of the extended model with the dynamics matrix $\bar{\mathbf{A}}(s)$, right - the spectrum after applying the feedback setting from the iteration 400 of the continuous pole placement procedure shown in Fig. 5

For assigning parameters of the feedback gain **K**, the method of continuous pole placement (CPP) has been used [9], see also [11]. The method consists of an iteration based continuous shifting of the rightmost poles of the closed loop system to the left via applying small changes in the feedback coefficients. At each iteration, changes in the feedback gain are given by

$$\Delta\mathbf{K} = \mathbf{S}_q^+ \Delta\Lambda \tag{11}$$

where $\Delta\Lambda = [\Delta\lambda_1, \Delta\lambda_2, \ldots\Delta\lambda_q]$ are the desired small displacements of q rightmost poles ($q \leq n$) and $\mathbf{S}_q = [\sigma_{i,j}] \in R^{q\times n}$, with $\sigma_{i,j} = \partial\lambda_i/\partial k_j$, is the sensitivity matrix and + denotes the pseudo-inverse, for more details see [9].

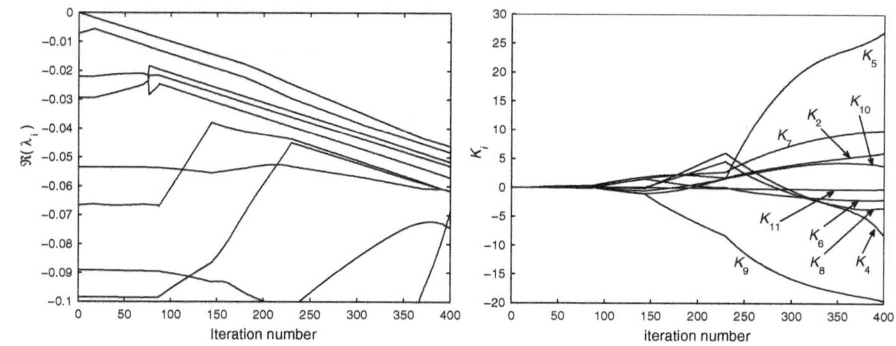

Fig. 5. Real parts of the poles of the set-up model (left) and the values of feedback gain (right) in the continuous pole placement procedure

Table 1. Coefficients of the feedback gain (12), according to the iteration number (i.n.) of the continuous pole placement procedure shown in Fig. 5

i.n.	K_2	K_4	K_5	K_6	K_7	K_8	K_9	K_{10}	K_{11}
140	1.7463	-0.14211	-0.97953	1.3766	1.4114	0.04361	-1.0475	-0.54048	-0.01872
270	3.1265	1.8892	13.1326	-0.99865	5.7015	0.92198	-13.431	2.8531	-0.19551
400	6.0359	-8.1650	26.932	-1.9985	9.8736	-3.4711	-19.615	3.7782	-0.25188

In order to reduce the complexity of numerical computations, only nine out of eleven available feedback loops have been used, i.e.

$$\mathbf{K} = [0\ \ K_2\ \ 0\ \ K_4\ \ K_5\ \ K_6\ \ K_7\ \ K_8\ \ K_9\ \ K_{10}\ \ K_{11}] \tag{12}$$

In Fig. 4 left, the rightmost spectrum of the matrix $\bar{\mathbf{A}}(s)$ is shown computed using quasi-polynomial mapping based rootfinder QPmR [12], [13]. Due to introducing (9) into the open loop model, one system pole is placed at the complex plane origin, i.e., at the stability boundary. In Fig. 5, the results of CPP iterations are shown. Let us remark that only right parts of the poles have been shifted, as proposed in [9].

In Fig. 4 right, the spectrum of the feedback system is shown achieved in the last iteration of CPP. As can be seen, the rightmost poles were shifted relatively far to the left compared to the original spectrum, which results in a fast set-point response as shown in Fig. 6 left. Next to the response for iteration 400, responses with gain coefficients from iterations 270 and 140 are also shown in Fig. 6 left (the gain coefficients are given in Table 1). As can be seen, settings from iterations 270 and 400 result in considerably faster responses than the response for the setting from iteration 140. On the other hand, high values of coefficients for the faster setting result in less robust responses if implemented in the set-up. It is due to high magnification of the measurement and process noise in the feedback loop. In Fig. 6 right, a comparison of set-point responses of the model and the set-up are shown and compared. As can be seen, except the noise, the agreement between the responses is very good.

Let us mention that the control algorithm has been successfully implemented not only via PC with data acquisition cards using Matlab/Simulink but also via the industrial programmable controller, [14].

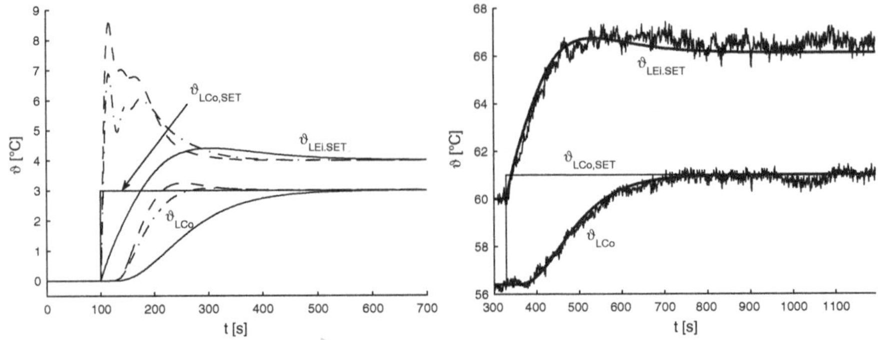

Fig. 6. Left - Set-point responses of the state feedback loop (in deviation form) resulting from the continuous pole placement procedure with feedback gain (12) from iterations: 140 - solid, 270 - dash-dotted, 400 - dashed, with the coefficients given in Table 1. Right - Comparison of the measured and the model set-point responses of the state feedback control for the gain from the iteration 140 of the continuous pole placement procedure, (thin - measured response, thick - model response)

5 Conclusions

As it can be recognized from the measured responses, the delays play important roles in the dynamics of the designed heat transfer set-up. These delays are predominantly induced by the transportation phenomenon in the pipe-lines. However, the delay-effects also play important roles in the dynamics of the set-up heaters and coolers. It is due to the fact that these subsystems are, by their nature, systems with distributed parameters. It has been shown in this paper, and analogously in many examples in the literature, that the dynamics of such subsystems can be effectively described by first order models with time delays. Using this type of simple model units (for modelling the dynamics of heaters and coolers, and also pipelines) the anisochronic model of the set-up has been proposed. A favorable feature of this model is that the state variables are identified with the system outputs (measured temperatures). Consequently, the state-variable feedback can easily be implemented, as no observer is needed for estimating the state variables. On the other hand, the synthesis of the controller is complicated by the fact that the system spectrum is infinite. However, as has been shown, the continuous pole placement method allows considerable shifting of the system's rightmost poles to the left of the stability boundary. Let us point out that the state variable-feedback control has also been successfully tested on the laboratory set-up, obtaining a very good agreement between the model and set-up responses.

Even though the experimental heat transfer set-up has primarily been designed for testing time-delay modelling and control system approaches, it is also being used for testing algorithms developed from different fields of control and system theory. For example, a stochastic based fault detection approach has been tested on the set-up in [15].

Acknowledgement

The presented research has been supported by the Ministry of Education of the Czech Republic under the Project 1M0567.

References

1. Hale, J.K., Verduyn Lunel, S.M.: Introduction to functional differential equations. Springer, New York (1993)
2. Gu, K., Kharitonov, V.L., Chen, J.: Stability of time-delay systems. Birkhauser, Boston (2003)
3. Michiels, W., Niculescu, S.I.: Stability and stabilization of time-delay systems. An eigenvalue based approach. Advances in Design and Control, vol. 12. SIAM Publications, Philadelphia (2007)
4. Niculescu, S.I.: Delay effects on stability. A robust control approach. Springer, London (2001)
5. Zítek, P.: Frequency Domain Synthesis of Hereditary Control Systems via Anisochronic State Space. Int. Journal of Control 66(4), 539–556 (1997)
6. Wang, Q.G., Lee, T.H., Lin, C.: Relay feedback. Springer, London (2003)
7. Zítek, P., Vyhlídal, T.: Low order time delay approximation of conventional linear model. In: 4th Mathmod conference, Vienna (2003)
8. Zítek, P., Vyhlídal, T.: State Feedback Control of Time Delays System: Conformal mapping aided design. In: IFAC Symposium on Linear Time Delay Systems, Ancona (2000)
9. Michiels, W., Engelborghs, K., Vansevenant, P., Roose, D.: Continuous pole placement method for delay equations. Automatica 38(6), 747–761 (2002)
10. Zítek, P.: Time delay control system design using functional state models. CTU Reports, no.1, CTU Prague (1998)
11. Michiels, W., Vyhlídal, T.: An eigenvalue based approach for the stabilization of linear time-delay systems of neutral type. Automatica 41, 991–998 (2005)
12. Vyhlídal, T., Zítek, P.: Mapping the Spectrum of a Retarded Time Delay System Utilizing Root Distribution Features. In: Proc. 6th IFAC Workshop on Time-Delay Systems (2006)
13. Vyhlídal, T., Zítek, P.: Mapping based rootfinder for analysing time delay system spectrum. In: Proc. 4th IFAC Workshop on Time Delay Systems, Rocquencourt (2003)
14. Paulů, K.: Pole placement based control design of the experimental heat transfer set-up. MSc Thesis, Faculty of Mechanical Engineering, Czech Technical University in Prague (2007) (in Czech)
15. Hofreiter, M., Garajayewa, G.: Real-time Fault Diagnosis for Non-linear Stochastic Systems. Advanced Technologies, Research-Development-Applications 1, 393–412 (2006)

Optimal Identification of Delay-Diffusive Operators and Application to the Acoustic Impedance of Absorbent Materials

Céline Casenave[1] and Gérard Montseny[1]

[1] LAAS-CNRS, University of Toulouse, 7 avenue du Colonel Roche 31077 Toulouse cedex 4, France. casenave@laas.fr
[2] LAAS-CNRS, University of Toulouse, 7 avenue du Colonel Roche 31077 Toulouse cedex 4, France. montseny@laas.fr

Summary. We present an original method devoted to the optimal identification of a wide class of complex linear operators involving some delay components, based on suitable infinite dimensional state formulations of diffusive type. Thanks to the intrinsic properties of these state formulations, cheap and precise numerical approximations are straightforwardly obtained, leading to approximate quadratic problems of reasonable dimension. We then propose this method for identification of the acoustic impedance of absorbent materials designed for noise reduction of aircraft motors.

1 Introduction

Identification of dynamic input-output systems is a central problem each time the knowledge of the process under consideration is too imprecise to get sufficiently accurate models from physical analysis (note anyway that such an analysis is sometimes impossible to perform because too complex). However, identification problems can also present some serious difficulties. Namely, when the input-output operator under consideration involves distributed underlying phenomena, significant non rational components, which are associated to infinite-dimensional realizations in the time domain, are quasi systematically generated.

When the operator to be identified is linear, a convenient and rather general approach consists in working in the frequency domain. So, any causal operator can be well-defined by its symbol $Q(t, i\omega)$, that is the Fourier transform of the so-called impulse response $h(t, s)$ of the operator. In the case of convolution operators, the symbol no more depends on t and the problem of identifying $Q(i\omega)$ can be classically solved from physical measurements by means of Fourier techniques. Note however that purely frequency identification presents some well-known shortcomings. In particular, the so-identified symbol is in general ill adapted to the construction of efficient time-realizations. This is partly due to excessive numerical cost of quadrature approximations resulting from the intrinsic convolutive nature of the associated operator, sometimes with long memory [2] or even delay-like behaviors [10]. Another heavy shortcoming is that frequency methods are incompatible with real-time identification (and so with pursuit when the symbol has the ability to evolve slowly). In opposite, time domain

J.J. Loiseau et al. (Eds.): Topics in Time Delay Systems, LNCIS 388, pp. 315–325.
springerlink.com

identification techniques do not present such drawbacks. However, their scope is in general not so wide. See for example [8] for an interesting optimal method based on ARMA lattices.

Nevertheless, the notion of symbol can remain suitable for temporal identification, as it will be shown in this study. The proposed identification method is based on the so-called *diffusive representation* theory [9] and was previously implemented, in a simpler version, on various problems [2], [4]. According to the diffusive representation approach, state formulations of diffusive type are available for a wide class of integral causal operators (including both rational and non rational ones) and the identification problem is carried over the so-called *γ-symbol*, defined as a suitable function (of an auxiliary real variable) easily deduced from the classical symbol. This γ-symbol entirely characterizes the associated operator, up to a suitable infinite-dimensional state equation to be chosen *a priori*. In any case, cheap and precise finite-dimensional approximations of this state representation can be straightforwardly built, running into finite-dimensional problems easy to solve by means of standard techniques. Among the numerous advantages of this approach, we can mention in particular:

- a stable differential input-output time-formulation is available as soon as the γ-symbol is well-defined;
- recursive identification algorithms are easily built and implemented, allowing real-time identification or even pursuit (in that case, the γ-symbol is depending on t);
- similarly to purely frequency methods, no qualitative difference is made between rational and non rational components which can then be identified by the same process.

The paper is organized as follows. In section 2, we first introduce a few fundamental notions of the diffusive representation theory. In section 3, we describe the identification method under consideration, based on the previously introduced notions. We then consider in section 4 a typical problem whose difficulty lies in the fact that both diffusive and delay components are involved but are not separable, and to which the diffusive representation approach is well-adapted: the identification of the acoustic impedance of an absorbent material devoted to aircraft motors noise reduction. Finally, a significant numerical example is given in section 5.

2 Diffusive formulation of causal integral operators

In this section, we present a simplified version of a general methodology introduced and developed in [9] in a general framework.

We consider a causal operator defined, on any continuous function $u : \mathbb{R}^+ \to \mathbb{R}$, by:

$$u \mapsto \int_0^t h(t-s)\, u(s)\, ds. \tag{1}$$

We denote H the Laplace transform of h and $H(\partial_t)$ the convolution operator defined by (1).

Let $u^t(s) = \mathbf{1}_{]-\infty,t]}(s)\, u(s)$ the restriction of u to its past and $u_t(s) = u^t(t-s)$ the history of u. From causality of $H(\partial_t)$, we deduce:

$$\left(H(\partial_t)(u-u^t)\right)(t) = 0 \text{ for all } t;$$

then, we have for any continuous function u:

$$(H(\partial_t)\, u)(t) = \left[\mathcal{L}^{-1}\left(H\, \mathcal{L}u\right)\right](t) = \left[\mathcal{L}^{-1}\left(H\, \mathcal{L}u^t\right)\right](t). \tag{2}$$

We define:

$$\Psi_u(t,p) := \mathrm{e}^{p\,t}\left(\mathcal{L}u^t\right)(p) = \left(\mathcal{L}u_t\right)(-p); \tag{3}$$

by computing $\partial_t(\mathcal{L}u_t)$, Laplace inversion and use of (2), it can be shown:

Lemma 1. *1. The function Ψ_u is solution of the differential equation:*

$$\partial_t \Psi(t,p) = p\,\Psi(t,p) + u, \quad t > 0, \quad \Psi(0,p) = 0.$$

2. For any $b \geqslant 0$,

$$\left(H(\partial_t)\, u\right)(t) = \frac{1}{2i\pi} \int_{b-i\infty}^{b+i\infty} H(p)\, \Psi_u(t,p)\, dp. \tag{4}$$

We denote Ω the holomorphic domain of H (after analytic continuation). Let γ a simple arc in $\mathbb{C}^-$, closed at infinity and separating $\mathbb{C}$ in two open domains denoted Ω_γ^+ and Ω_γ^- such that $\Omega_\gamma^- \subset \mathbb{C}^-$. By use of standard techniques (Cauchy theorem, Jordan lemma [6]), it can be shown:

Lemma 2. *For $\gamma \subset \Omega$ such that H is holomorphic in Ω_γ^+, if $H(p) \to 0$ when $p \to \infty$ in Ω_γ^+, then:*

$$\left(H(\partial_t)\, u\right)(t) = \frac{1}{2i\pi} \int_{\tilde{\gamma}} H(p)\, \Psi_u(t,p)\, dp, \tag{5}$$

where $\tilde{\gamma}$ is any closed[3] simple arc in Ω_γ^+ such that $\gamma \subset \Omega_{\tilde{\gamma}}^-$.

We now suppose that γ and $\tilde{\gamma}$ are defined by functions of the Sobolev space[4] $W^{1,\infty}_{\text{loc}}(\mathbb{R};\mathbb{C})$, also denoted γ, $\tilde{\gamma}$. We use the convenient notation $\langle \mu, \psi \rangle = \int \mu\, \psi\, d\xi$.

Under hypothesis of lemma 2, we have [9]:

Theorem 1. *If the possible singularities of H on γ are simple poles or branching points such that $|H \circ \gamma|$ is locally integrable in their neighborhood, then:*
1. with $\tilde{\mu} = \frac{\tilde{\gamma}'}{2i\pi}\, H \circ \tilde{\gamma}$ and $\tilde{\psi}(t,.) = \Psi_u(t,.) \circ \tilde{\gamma}$:

$$\left(H(\partial_t)\, u\right)(t) = \left\langle \tilde{\mu}, \tilde{\psi}(t,.) \right\rangle; \tag{6}$$

2. with[5] $\tilde{\gamma}_n \to \gamma$ in $W^{1,\infty}_{\text{loc}}$ and $\mu = \frac{\tilde{\gamma}'}{2i\pi} \lim H \circ \tilde{\gamma}_n$ in the sense of measures:

$$\left(H(\partial_t)\, u\right)(t) = \left\langle \mu, \psi(t,.) \right\rangle, \tag{7}$$

where $\psi(t,\xi)$ is solution of the following Cauchy problem on $(t,\xi) \in \mathbb{R}^{*+} \times \mathbb{R}$:

$$\partial_t \psi(t,\xi) = \gamma(\xi)\, \psi(t,\xi) + u(t),\ \psi(0,\xi) = 0. \tag{8}$$

Definition 1. *The measure μ defined in theorem 1 is called the γ-symbol of operator $H(\partial_t)$.*

Such a formulation can be extended to operators of the form $\partial_t \circ H(\partial_t)$ where $H(\partial_t)$ verifies the hypothesis of theorem 1. We indeed have (formally):

$$\begin{cases} \partial_t \psi(t,\xi) = \gamma(\xi)\, \psi(t,\xi) + u(t), \quad \psi(0,\xi) = 0 \\ \partial_t \left(H(\partial_t)\, u\right)(t) = \left\langle \mu, \gamma\, \psi(t,.) + u(t) \right\rangle. \end{cases} \tag{9}$$

---

[3] Possibly at infinity

[4] $W^{1,\infty}_{\text{loc}}(\mathbb{R};\mathbb{C})$ is the topological space of measurable functions $f : \mathbb{R} \to \mathbb{C}$ such that $f, f' \in L^\infty_{\text{loc}}$ (that is f and f' are locally bounded in the almost everywhere sense).

[5] This convergence mode means that $\tilde{\gamma}_{n|K} - \gamma_{|K} \to 0$ and $\tilde{\gamma}'_{n|K} - \gamma'_{|K} \to 0$ uniformly on any bounded set K.

The state equation (8) is infinite-dimensional. To get numerical approximations, we consider a discretization $(\xi_k)_{k=1:n}$ of the variable ξ and approximations μ_n of the γ-symbol μ defined by atomic measures such as:

$$\mu_n = \sum_{k=1}^{n} \mu_n^k \, \delta_{\xi_k}. \tag{10}$$

Let us denote $\mathcal{M}_n$ the space of atomic measures on the mesh $\{\xi_k\}_{k=1:n}$. If $\cup_n \mathcal{M}_n$ is dense in the space of measures, we have [9]:

$$\langle \mu_n, \psi \rangle \underset{n\to+\infty}{\longrightarrow} \langle \mu, \psi \rangle \,, \ \forall \psi \in \mathcal{C}^0(\mathbb{R}).$$

So we have the finite-dimensional approximate state formulation of $\partial_t \circ H(\partial_t)$:

$$\begin{cases} \partial_t \psi(t, \xi_k) = \gamma(\xi_k)\, \psi(t, \xi_k) + u(t), \ k = 1:n \\ \psi(0, \xi_k) = 0, \\ \partial_t \left(H(\partial_t)\, u \right)(t) \simeq \sum_{k=1}^{n} \gamma(\xi_k)\, \mu_n^k\, \psi(t, \xi_k) + \left(\sum_{k=1}^{n} \mu_n^k\right) u(t). \end{cases}$$

Note that when operator $H(\partial_t)$ is not pseudo-differential [11], namely when some delay-type behaviors are present, the function γ must verify [10]:

$$\lim_{\xi\to\pm\infty} \operatorname{Im}(\gamma(\xi)) = \pm\infty$$
$$\text{and } \exists a < 0, \ \operatorname{Re}(\gamma(\xi)) \geqslant a.$$

3 Optimal identification of the γ-symbol

In this section, we focus on the problem of identification of a convolution operator:

$$u \longmapsto y = H(\partial_t)u, \tag{11}$$

where the symbol $H(p)$ is supposed to verify the hypothesis of theorem 1. For simplicity, we present formal developments only; more details will be given in a further paper.

We denote by μ the γ-symbol of $H(\partial_t)$, as defined in section 2. The problem then consists of identifying μ from the data u and y^*, this last term denoting the measurement of y, in general sullied by some additive noise.

Given a suitable γ, we consider the diffusive realization (8,7) of $H(\partial_t)$ and we denote ψ_u the solution of (8). By defining A_u the following linear operator:

$$A_u : \mu \mapsto \langle \mu, \psi_u \rangle \,,$$

we obviously have:

$$y = A_u\, \mu.$$

Given suitable Hilbert spaces $\mathcal{E}, \mathcal{F}$ (not specified here), we consider the problem:

$$\min_{\mu\in\mathcal{E}} \| A_u\, \mu - y^* \|_{\mathcal{F}}^2 \,,$$

the solution of which is classically given by:

$$\mu^* = A_u^\dagger \, y^* \tag{12}$$

where $A_u^\dagger$ denotes the pseudo-inverse of A_u [1]. So, in the sense of the hilbertian norm of $\mathcal{F}$, the estimation μ^* of μ is optimal.

>From the numerical point of view, after suitable time discretization and approximation of μ such as described in section 2, we have:

$$(A_u \, \mu)_i = \sum_{k=1}^{n} \mu_n^k \, \psi(t_i, \xi_k);$$

then the operator A_u can be expressed as a matrix $(\psi_u(t_i, \xi_k))_{i=1:m,k=1:n}$ and its pseudo-inverse is classically expressed[6] (with $m \gg n$):

$$A_u^\dagger = (A_u^* \, A_u + \epsilon I)^{-1} A_u^*.$$

Remark 1. Recursive formulations of (12) can be established under the form (see [4]):

$$\mu_t^* = \mu_{t-\Delta t}^* + K_{t-\Delta t}(y^* - A_u \, \mu_{t-\Delta t}^*)_{|[0,t]};$$

such formulations allow real-time pursuit of μ in case of slowly varying operators $Q(t, \partial_t)$.

4 Application to the impedance of an absorbent material

First, let us present the physical problem under consideration.

In order to reduce the noise generated by aircraft motors, the use of absorbent coverings is an efficient solution. For particular cases where stringent thermal or mechanical constraints are present, such as in hot zones of motors, porous metallic materials have been proposed and studied [5], [7]. Due to the complex structure of such materials, the generated acoustic impedances are non rational complex functions of the frequency with unavoidable delay behaviors due to the presence of propagating waves in such media. In addition, their analytic determination is difficult and possibly poorly reliant. So, experimental measurements followed by identification processes remain necessary when accurate models are expected to carry out numerical simulations. In this way, it becomes possible, for a given material, to numerically compute the aeroacoustic field in a motor cavity equipped with absorbent walls, in aim of evaluating the amount of noise reduction.

Fortunately, the impedance of an homogeneous 2D coating does not depend on the surface parameterization variables and therefore, at least for sufficiently high frequencies, it can be identified from physical measurements relating to a small sample of the material (see figure 1). In the sequel, we show how the method introduced in section 4 can be implemented on such a problem.

For simplicity of the analysis, we suppose that the problem described in figure 1 can be reduced to a 1D problem, which is legitimate if the source and the sensor are sufficiently far from the absorbent sample. This will be sufficient to introduce the principle of the method. For a more realistic analysis, the 3D problem can be similarly treated up to technical adaptations.

Under these conditions, the absorbent material sample can be summarized by an impedance operator $Q(\partial_t)$ at point $x = 1$. The 1D problem of acoustic waves propagation can then be written as[7]:

[6] As usual, ϵ is a small positive parameter devoted to numerical conditioning.
[7] For simplicity the celerity has been taken equal to 1.

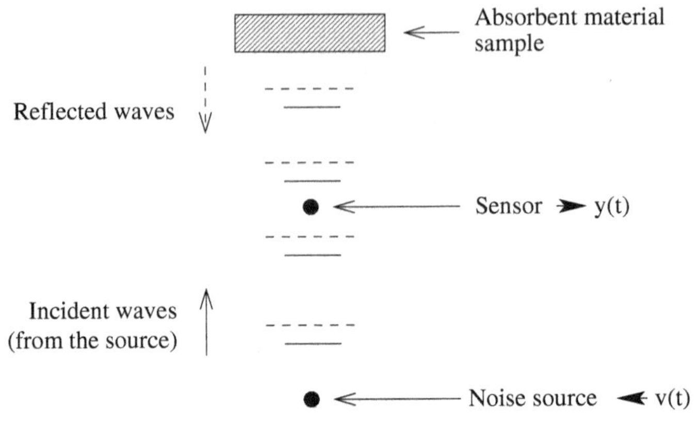

Fig. 1. Physical measurement for the identification of the impedance of an homogeneous absorbing material.

$$\partial_t^2 P = \partial_x^2 P + v \otimes \delta_0 \quad \text{on } x \in]-\infty, 1], \tag{13}$$

where P is the acoustic pressure and $v \otimes \delta_0$ denotes the source at point $x = 0$, with $v(t)$ a known time-function. We have:

$$\partial_x P_{|x=1} = Q(\partial_t)\, \partial_t P_{|x=1}. \tag{14}$$

Finally the acoustic pressure is measured at a point x_m:

$$y = P(t, x_m). \tag{15}$$

Then, the identification problem is to build an approximation of operator $Q(\partial_t)$ from the data y and v (note that in practice, the measurement y can be perturbed by an additive noise, not explicitly specified here).

Let us compute the expression of the operator $P(.,0) \mapsto y$, denoted by $K(\partial_t)$. For that, we must compute the reflection coefficient at $x = 1$ for any frequency ω. Let φ a harmonic solution of (13) with $v = 0$. It can be expressed as $\varphi = \varphi^i + \varphi^r$ where φ^i denotes an incident wave (propagating in the direction of increasing x) and φ^r the associated reflected wave (propagating in the sense of decreasing x), respectively given by:

$$\begin{aligned} \varphi^i(t,\omega,x) &= e^{i\omega(t-x)}, \\ \varphi^r(t,\omega,x) &= k(\omega)\, e^{i\omega(t+x)}, \end{aligned}$$

where $k(\omega)$ is the reflection coefficient at frequency ω, depending on $Q(i\omega)$. At point $x = 1$, the coefficient $k(\omega)$ verifies the impedance relation (14) which leads after simple computations to:

$$k(\omega) = \frac{1 + Q(i\omega)}{1 - Q(i\omega)}\, e^{-2i\omega}.$$

So we have:

$$\varphi(t,\omega,x) = (e^{-i\omega x} + \frac{1 + Q(i\omega)}{1 - Q(i\omega)}\, e^{i\omega(x-2)})\, e^{i\omega t},$$

and then (from (15)):

$$K(i\omega) = e^{-i\omega x_m} + \frac{1+Q(i\omega)}{1-Q(i\omega)}\, e^{i\omega(x_m-2)}.$$

On the other hand, the solution of (13) is classically given by:

$$P(t,x) = \tfrac{1}{2}\,(\partial_t^{-1} v)(t-x);$$

by harmonic synthesis, the input-output relation $v \mapsto y$ can therefore be made explicit in the time domain (τ_a denotes the translation operator $f(t) \mapsto f(t-a)$):

$$y = \tfrac{1}{2}\tau_{x_m} \circ \partial_t^{-1} v + \tfrac{1}{2} M(\partial_t) \circ \tau_{2-x_m} \circ \partial_t^{-1} v \tag{16}$$

with:

$$M(i\omega) := \frac{1+Q(i\omega)}{1-Q(i\omega)}. \tag{17}$$

By defining the new data and notations:

$$\begin{aligned} \tilde{y} &:= y - \tfrac{1}{2}(\tau_{x_m} + \tau_{2-x_m}) \circ \partial_t^{-1} v, \\ \tilde{v} &:= \tfrac{1}{2}\tau_{2-x_m}\, v, \\ \widetilde{M}(\partial_t) &:= M(\partial_t) - 1, \end{aligned}$$

we obtain the suitable relation:

$$\tilde{y} = \partial_t^{-1} \circ \widetilde{M}(\partial_t)\tilde{v} \tag{18}$$

on which we can directly apply the method presented in section 3 to identify the associated γ-symbol.

Let μ denote indifferently the so-identified γ-symbol or the exact γ-symbol of $\partial_t^{-1} \circ \widetilde{M}(\partial_t)$. By considering the following relation deduced from (17):

$$Q(i\omega) = \frac{\widetilde{M}(i\omega)}{\widetilde{M}(i\omega) + 2}$$

and from elementary computations, we then deduce the following input-output state representation of operator $w \mapsto Q(\partial_t)\, w$:

$$\begin{cases} \partial_t \psi = \gamma\, \psi - \dfrac{1}{2}\dfrac{\langle \gamma\mu, \psi\rangle}{\langle \mu, 1\rangle + 1} + \dfrac{1}{\langle \mu, 1\rangle + 1}\, w, \ \psi_0 = 0 \\[2ex] Q(\partial_t) w = \dfrac{1}{2}\dfrac{\langle \gamma\mu, \psi\rangle}{\langle \mu, 1\rangle + 1} + \dfrac{\langle \mu, 1\rangle}{\langle \mu, 1\rangle + 1}\, w, \end{cases} \tag{19}$$

which is exact (infinite-dimensional) if μ is the exact γ-symbol, and approximate (finite-dimensional[8]) if μ is the identified γ-symbol, of the form (10).

Remark 2. A more precise analysis can be performed by replacing (13) by the 3D problem: $\partial_t^2 P = \Delta P + v \otimes \delta_0$. By use of Green functions, similar (but more technical) computations can be achieved and (18) is obtained again while the expression of $\tilde{y}$ and $\tilde{v}$ involves more complex (but explicitly known) operators derived from the Green functions.

[8] In that case, $\langle f, g\rangle = \sum_k f_k\, g_k$.

5 Numerical example and conclusion

We consider the porous material studied in [7], [5], the impedance of which has been analytically computed in [3]. This impedance is expressed:

$$Q(i\omega) = \varphi \sqrt{\frac{\chi_{\text{eff}}(i\omega)}{\rho_{\text{eff}}(i\omega)}} \tanh\left(i\omega \sqrt{\chi_{\text{eff}}(i\omega)\, \rho_{\text{eff}}(i\omega)}\right),$$

where $\varphi = 3.33$ and the functions $\rho_{\text{eff}}(i\omega)$ and $\chi_{\text{eff}}(i\omega)$ are respectively the so-called effective density of Pride *et al.* and the effective compressibility of Lafarge. These functions are expressed [5]:

$$\rho_{\text{eff}}(i\omega) = e\rho\,(1 + a\,\tfrac{(1+b\,i\omega)^{\frac{1}{2}}}{i\omega})$$
$$\chi_{\text{eff}}(i\omega) = e\chi\,(1 - c\,\tfrac{i\omega}{i\omega + a'(1+b'\,i\omega)^{\frac{1}{2}}}),$$

with

$$\rho = \rho_0\,\alpha_\infty,\ \chi = \tfrac{1}{P_0},\ a = \tfrac{8\mu}{\rho_0 \Lambda^2},\ a' = \tfrac{8\mu}{\rho_0 \Lambda'^2},$$
$$b = \tfrac{1}{2a},\ b' = \tfrac{1}{2a'},\ 0 < c = \tfrac{\gamma-1}{\gamma} < 1,$$

where $e, \rho_0, P_0, \mu, \gamma, \alpha_\infty, \Lambda, \Lambda'$ are physical parameters with values $\Lambda = \Lambda' = 0.1\,10^{-3}$m, $\rho_0 = 1.2\,\text{kg.m}^{-3}$, $P_0 = 10^5$ Pa, $\mu = 1.8\,10^{-5}$ kg.m^{-1}.s^{-1}, $\gamma = 1.4$, $\alpha_\infty = 1.3$, $e = 5\,10^{-2}$ m.

>From standard analysis, it can be shown that the analytic continuation of Q presents singularities which can be asymptotically localized near a vertical straight line (see figure 2): operator $Q(\partial_t)$ presents some underlying behaviors of delay type, inherited from propagative modes inside the porous material and from which result complex magnitude and phase variations of Q, making the identification problem possibly sensitive or even ill-posed.

In aim of highlighting the efficiency of the previously introduced method, we consider the problem of numerical identification of such an operator $Q(\partial_t)$ under its state formulation (19) in the ideal situation described in section 4.

The measurement data $y^* = y + \varepsilon$ have been generated on $t \in [0, T]$ from accurate numerical simulation of (16), with v a gaussian white noise and ε an output gaussian white noise with signal-to-noise ratio equal to 26 dB.

The numerical parameters are $\Delta t = 10^{-5}$ s, $T = 0.05$ s, $\xi_{\min} = \xi_1 = 10^3$ rad/s, $\xi_{\max} = \xi_n = 3\,10^5$ rad/s, $\xi_{k+1} = r\,\xi_k$, $k = 1 : n-1$, $n = 100$, $\gamma(\xi) = |\xi|\,\cos\alpha + i\,\xi\,\sin\alpha$, $\alpha = 100^\circ$.

The comparison between the exact[9] and identified frequency responses relating to the impedance $Q(i\omega)$ is available in figures 3 and 4. As expected, the identification is accurate in the frequency band $[10^3$ rad/s, $3\,10^5$ rad/s$]$ covered by the set $\{\gamma(\xi_k)\}_{k=1:n}$. We can note that identification remains correct on the whole audio-frequency band, namely at low frequencies thanks to the asymptotically rational nature of $Q(i\omega)$ which behaves like $i\omega$ when $\omega \to 0$.

In figure 5, we can see the graph of the identified γ-symbol associated to operator $\partial_t^{-1} \circ \widetilde{M}(\partial_t)$ and from which the operator $Q(\partial_t)$ is realized via the input-output formulation (19).

Beyond these quantitative results, we can add that this identification method is robust with respect to the level of the perturbation noise ε, the time of measurements T and the dimension n of the identified model. This last essential property is a consequence of the

[9] Up to the numerical simulation errors, which are negligible in the frequency band under consideration.

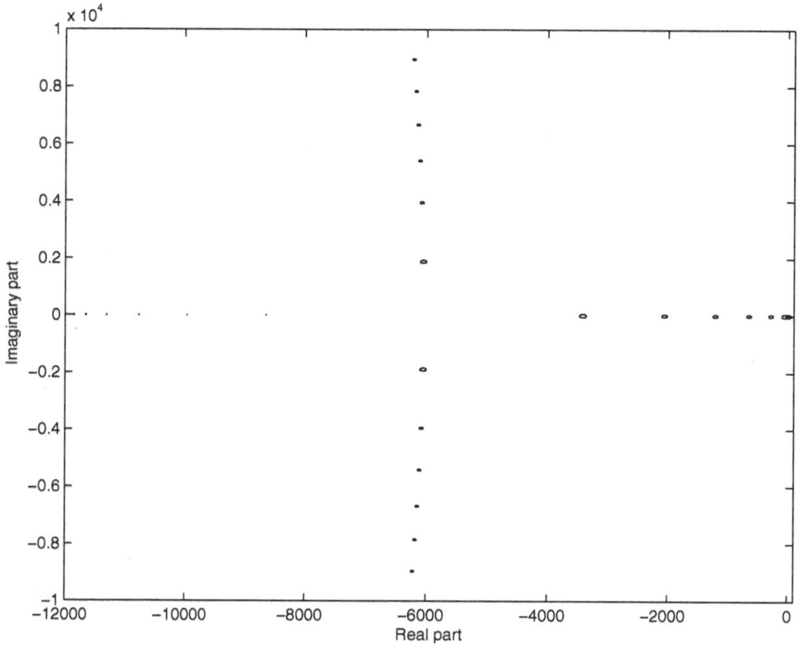

Fig. 2. Singularities of the symbol $Q(p)$.

fact that the poles $\gamma(\xi_k)$ of the identified transfer function are imposed *a priori* by the ξ-discretization of a suitable diffusive input-output model (namely (9)) mathematically well-posed and able to realize a wide class of operators in which belongs the non rational operator under consideration. It follows, thanks to the properties of diffusive representation, that the identified γ-symbol μ converges (in a sense not precised here), when $n \to +\infty$, $\alpha \to \frac{\pi}{2}^+$, $\xi_{\min} \to 0^+$, $\xi_{\max} \to +\infty$ and $T \to +\infty$, to the exact one, and so for the identified impedance $Q(i\omega)$ in the sense of the Fréchet space $L^2_{\text{loc}}(\mathbb{R}_\omega)$.

References

1. Ben-Israel, A., Greville, T.N.E.: Generalized inverses: theory and applications. Springer, New York (2003)
2. Bidan, P., Lebey, T., Montseny, G., Neacsu, C., Saint-Michel, J.: Transient voltage distribution in motor windings fed by inverter: experimental study and modeling. IEEE Trans. on Power Electronics 16(1), 92–100 (2001)
3. Casenave, C., Montseny, E.: Dissipative state formulations and numerical simulation of a porous medium for boundary absorbing control of aeroacoustic waves. In: Proc. 17th IFAC World Congress, Seoul, Korea, July 6-11 (2008)
4. Garcia, G., Bernussou, J.: Identification of the dynamics of a lead acid battery. In: ESAIM: Proceedings, p. 5 (1998)
5. Gasser, S.: Etude des propriétés acoustiques et mécaniques d'un matériau métallique poreux à base de sphères creuses de nickel. PhD Thesis, Grenoble, France (2003)

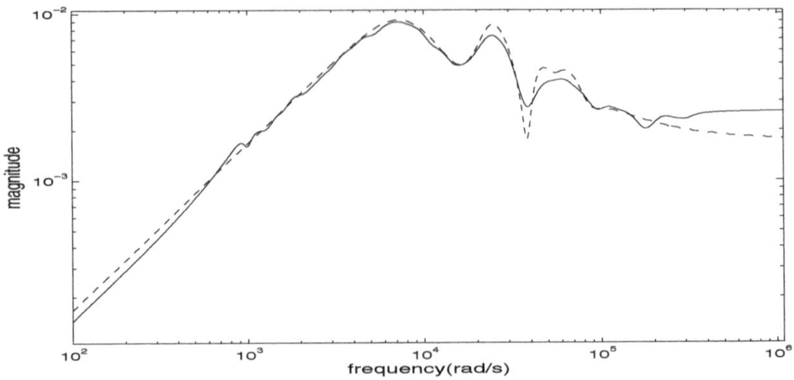

Fig. 3. Identified (—) and exact (- - -) frequency responses of operator $Q(\partial_t)$.

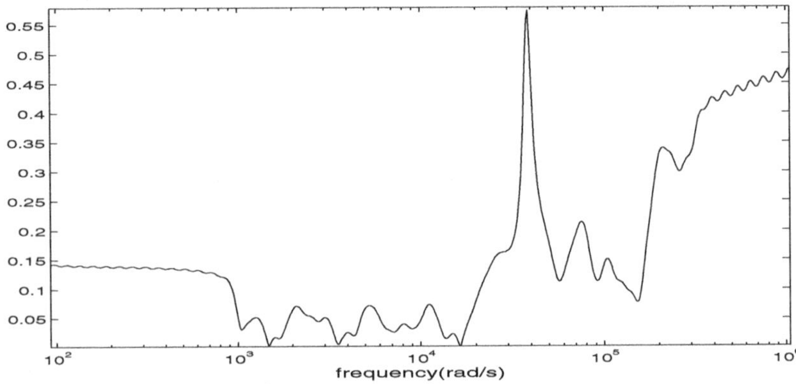

Fig. 4. Relative identification error $|\frac{Q-Q^*}{Q}|$.

6. Lavrentiev, M., Chabat, B.: Méthodes de la théorie des fonctions d'une variable complexe, MIR edn., Moscow, Russia (1977)
7. Mazet, P.A., Ventribout, Y.: Control of Aero-acoustic Propagations with Wall Impedance Boundary Conditions: Application to a Porous Material Model. In: WAVES 2005, Providence, USA (2005)
8. Monin, A., Salut, G.: ARMA Lattice identification: a new hereditary algorithm. IEEE Transactions on Signal Processing 44(2), 360–370 (1996)
9. Montseny, G.: Représentation diffusive. Hermes-Science, Paris, France (2003)
10. Montseny, G.: Diffusive representation for operators involving delays. In: Loiseau, J.J., Chiasson, J. (eds.) Applications of time-delay systems, pp. 217–232. Springer, Heidelberg (2007)
11. Taylor, M.E.: Partial differential equations - II. Applied Mathematical Sciences, p. 116. Springer, Heidelberg (1997)

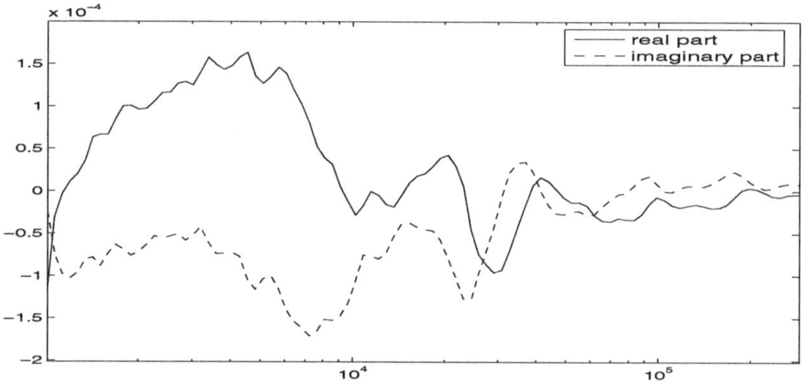

Fig. 5. Identified γ-symbol $\mu(\xi)$, $\xi > 0$.

Part V

Delays in Interconnected Systems and Networks

Controlling Across Complex Networks: Emerging Links Between Networks and Control

Aaron Clauset*[1], Herbert G. Tanner**[2], Chaouki T. Abdallah***[3], and Raymond H. Byrne.†[4]

[1] Santa Fe Institute, 1399 Hyde Park Rd. Santa Fe, NM 87501 USA. aaronc@santafe.edu
[2] Mechanical Engineering Department, MSC01 1150, 1 University of New Mexico, Albuquerque, NM 87131-0001, USA. tanner@unm.edu
[3] Electrical & Computer Engineering Department, MSC01 1100, 1 University of New Mexico, Albuquerque, NM 87131-0001, USA. chaouki@ece.unm.edu
[4] Data Analysis And Exploitation, Department 5535, Sandia National Laboratories, MS 1243, PO Box 5800, Albuquerque, NM 87185-1243I, USA. rhbyrne@sandia.gov

Summary. The presence of communication networks in the feedback path of a control system has led to new problems for control designers. Meanwhile, physicists, computer scientists, and mathematicians have been studying the formation and properties of physical networks under the heading of complex networks. Control engineers use a network model to facilitate controller design, while complex network theorists investigate networks to model their dynamics and growth. Despite the use of distinct analysis tools, network properties such as connectivity, efficiency, and robustness are common to both control and complex networks research. A question that naturally arises is whether ideas used by the complex network community can suggest new control design directions. In this chapter we review the tools from the network theoristÕs arsenal to make them available to control engineers, and describe how ideas developed for complex network research can be exploited within a control systems framework.

1 Introduction

Researchers in control systems have recently focused their attention on the effects caused by the presence of communication networks in the feedback path [1, 2]. Meanwhile, under the heading of complex networks, physicists, computer scientists, and mathematicians have been studying the formation and properties of physical networks as described in [3]. Control engineers use a network model to facilitate controller design, while complex network theorists investigate networks to model their dynamics and growth. Despite the use of distinct analysis tools, network properties such as connectivity, efficiency, and robustness are common

* Work supported in part by NSF grant no.PHY-0200909
** Work supported in part by NSF grant no.0447898
*** Work supported in part by NSF grant no.CNS 0626380
† Work supported by the United States Department of Energy under Contract DE-AC04-94AL8500.

J.J. Loiseau et al. (Eds.): Topics in Time Delay Systems, LNCIS 388, pp. 329–347.
springerlink.com

to both control and complex networks research. A question that naturally arises is whether ideas used by the complex network community can suggest new control design directions. In this chapter we review the tools from the network theorist's arsenal to make them available to control engineers, and describe how ideas developed for complex network research can be exploited within a control systems framework. Consider for example the system depicted

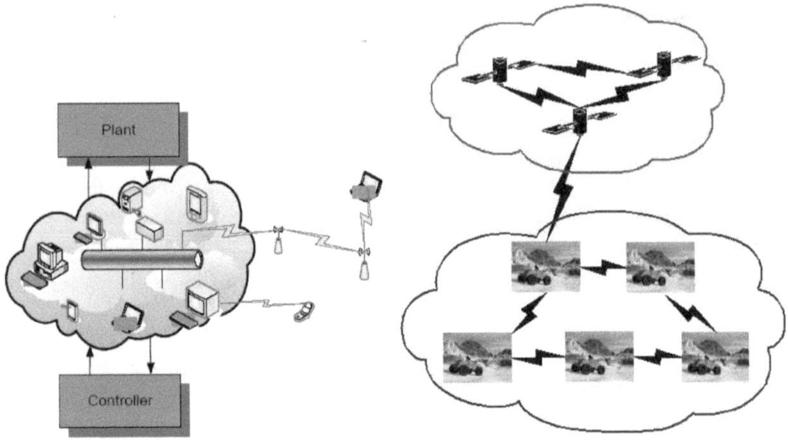

(a) ***Controlling across a network.*** (b) ***A networked control system.***

Fig. 1. Networks in Control.

in Figure 1(a), where a plant is being controlled across a network shared by various systems, computers, and communication devices. Figure 1(b) shows heterogeneous autonomous systems, which communicate locally in an effort to coordinate toward a common objective. From a controls perspective, the communication links of figures 1(a) and 1(b) are a means of information exchange, and as a first approximation all communication is assumed to be instantaneous. In networked control systems research, the impact of the network's connectivity on the closed-loop system performance is emphasized, (see [4, 5, 6, 7, 8]). As we describe in this chapter, connectivity may not be the only critical structural property from the controls perspective. Network theorists, meanwhile, have created tools to characterize the growth of distributed networks [9]. In particular, the relations between navigability, congestion, clustering, and robustness to failure, and the topology of random networks, are generally well understood [10]. A particularly evocative aspect of the work in the complex networks area focuses on understanding how for some systems, such as social networks and the World Wide Web, not only do short paths exist between every pair of nodes, but such paths can be found under certain conditions using only local information [11]. In multi-agent control systems, signals need to flow quickly and efficiently, but agents cannot store enough information to properly model the system's global state. The ability to obtain global stability results by means of a set of distributed controllers that require local information, is therefore important for bounding complexity.

There is actually an elegant symmetry between the perspective of control engineers engaged in studying multi-agent control, and networked control systems describing network

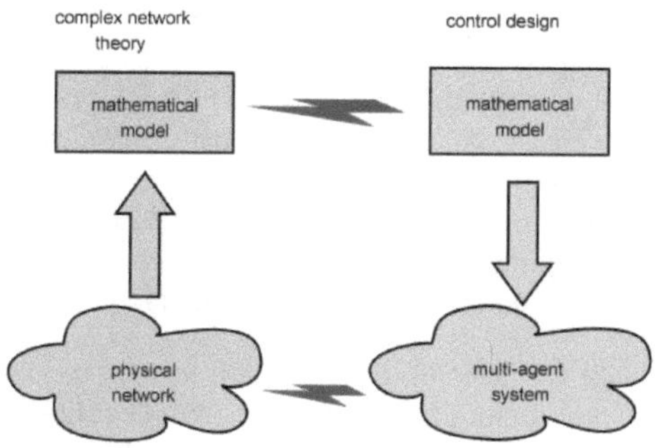

Fig. 2. Complex network and control share domains but follow different directions in analysis.

dynamics (Figure 2). While their starting points and objectives differ, both try to determine what a given network structure may guarantee about a dynamic process running on the network. In control systems, researchers bound the topological effects on capacity, latencies, and lost packets, and try to prove stability, robustness to disturbances, and to guarantee faster response, by means of appropriate network topologies; network theory, on the other hand, focuses on how randomness and topological regularities impact connectivity, navigability, modularity (also called community structure), stability, efficiency and resilience. The tools and models used in the latter can provide a powerful paradigm for understanding networked control performance, and may suggest novel solutions to network control design problems.

This chapter presents an overview of problems at the intersection of control theory and complex networks research. Within a brief review of recent results in cooperative control, we present three arguments that support the hypothesis that issues of interest for the network theorist can also impact control engineering design:

1. Increased network connectivity does not necessarily yield robustly connected networks with respect to node failures;
2. The structure of sensor networks and their algebraic graph properties determine the performance of distributed estimation;
3. Properly interleaving communication and control can protect against the effect of delayed information.

After a brief introduction of the relevant aspects of complex network theory and its methodological differences and similarities to control engineering, we discuss the potential benefits of technology transfer between the two fields. Numerical examples are presented, illustrating the effect of connectivity and navigability on the performance of a simple system controlled across a large network. The chapter concludes by identifying several topics for future joint work between control engineers and network theorists.

2 Network Theoretic Issues

From a consideration of network topology, that is, from an examination of the network as a communication service, three fundamental issues that affect the flow of information over the network are noted:

1. Connectedness, which expresses the existence of a path between the information transmitter and the information receiver.
2. Navigability, which is quantified by the difficulty of finding such a connecting path. Typically, this difficulty depends on whether the path is predetermined, or whether it is discovered in an ad-hoc fashion.
3. Efficiency, as represented by the cumulative latency of each utilized path. This latency, usually a function of the number of hops and the individual link latencies, must be sufficient to guarantee desired end-to-end communication latencies.

All three points are related to the issue of reliability, that is, the robustness of the given network property with respect to node or link failures, and the reliability of the network protocols with respect to various kinds of corruption. Naturally, the relevance of these properties for control depends on the particular network control application.

2.1 Connectedness

Network theorists consider connectedness to be identical to the mathematical notion of percolation. This notion is illustrated as a wildfire, initiated at a source vertex, which spreads across an edge connected to a burning vertex with a fixed probability p. By analyzing the number of vertices reached by the process, it is possible to determine whether there exists a path connecting a given pair of nodes [12]. Percolation is typically analyzed by network theorists in two contexts. The constructive approach answers the question of how many random edges must be successively added to a collection of disconnected vertices before the vast majority of vertices, termed the *giant component*, are connected by some path. In the destructive approach, edges (or nodes) are successively removed until the giant component vanishes and most pairs of vertices are no longer connected by any path. Surprisingly, the appearance and disappearance of the giant component can be quite sudden, and is often a genuine phase transition [10]. For most control system applications, the destructive version of the process is of primary interest since it directly relates to the basic guarantee of end-to-end communication and feedback across the network. However, as more sophisticated ad-hoc applications emerge, the constructive approach is expected to become more relevant.

One common feature of many real world networks is a power-law degree distribution, in which the probability of a randomly chosen vertex have k neighbors scales as $P(k) \propto k^{-\alpha}$, where α is the scaling exponent [3]. The ubiquity of the power-law degree distribution has led network theorists to focus on graph models that exhibit this feature, but whose topological structure is otherwise random. Obviously, a network with many redundant paths between all pairs of vertices becomes more robust to node and edge failures of all kinds. However, theorists are interested in the degree of robustness as characterized by the fraction of vertices p_c that must be removed before the giant component vanishes. Colloquially, this disappearance is said to "shatter" the network.

For a random graph with a power-law degree distribution where a uniformly random fraction of vertices fail (are removed), the after-failure degree distribution $P'(k)$ is given by

$$P'(k) = \sum_{k_O=k}^{\infty} P(k_O) \binom{k_O}{k} (1-p)^k \, p^{k_O-k},$$

where k_O is the degree of a vertex before failure, k is its degree after, and p is the probability of failure. When the scaling exponent α for $P(k_O)$ is larger than 3, [13] shows that the critical threshold for maintaining the giant component is $p_c \approx 0.99$. That is, these random structures are asymptotically robust to random failures. For finite-size networks, the value of p_c is bounded away from 1, and its exact value is related to the size of the graph n. A recent extension of this work shows, however, that the value of p_c can be significantly smaller for a specific subclass of these graphs [14], that is, not all random graphs with a power-law degree distribution are equally robust to random failures.

Unfortunately, randomly removing vertices is not the only kind of failure that networks suffer. For instance, when nodes in these same random graphs are preferentially removed according to some rule (for example removing the 10% of vertices with the highest degree), the network quickly shatters [15], [16]. Furthermore, network theorists in the peer-to-peer research community have considered more subtle forms of failure, in which some fraction of nodes disobey the network communication protocols, possibly in a malicious way. These Byzantine faults have been extensively studied, and continue to drive much of the research in developing secure, and distributed communication protocols [17], [18].

2.2 Navigability

Given that a network is connected, there exist several paths that typically connect a transmitter with a receiver. In network theory, a network's navigability is determined both by how easily such a path can be found, and how many hops such a path ultimately requires. This problem has been extensively studied, and solutions can be grouped into two categories:

1. central authorities, in which the communication path between two vertices is determined by an external source, later mirrored by the network's routers, and
2. decentralized techniques, in which routing decisions are made independently by network routers, possibly in an ad-hoc fashion.

For a static network, namely one whose number of nodes and topology are fixed, a central authority may be easier to construct. In the face of uncertainty, however, such as when routers are added or removed (because of failure) from the network, a decentralized approach may be more favorable. The current standards for routing on Internet-like networks, namely, the Internet Protocol (IP) routing, the open shortest path first (OSPF) [19] protocol, and the border gateway protocol (BGP) [20], are a mixture of both centralized and decentralized techniques. In each, an initial consensus phase allows local connectivity information to propagate throughout the network in a decentralized fashion, until each router has constructed its own view of the network, by which to route packets in the future. After this phase is complete, these views remain fixed until the local topology changes enough to initiate a new consensus phase. While the standard protocols work well for networks that change only occasionally, dynamic networks pose a more challenging problem, as one must balance the overhead of consensus against the efficiency of the network as a communication medium. An alternative approach is to use decentralized or ad-hoc routing strategies, where routing decisions are made on the fly, based on the relative position of the current router, the packet's destination, and perhaps additionally on current local connectivity. If such an approach is to guarantee that a network is navigable, short paths between source and destination must be easily found in a decentralized way. A network is said *efficiently navigable* if the average length of a path T grows sublinearly with the number of vertices in the network, and preferably as a polylogarithm like $O(\log^2 n)$.

Consider a simple generalization of the popular small worlds network model of [21]. In this network, a lattice of vertices have bi-directional local connections to their nearest neighbors (as when computers are connected using optical links), and a single, uni-directional non-local connection to some other node (as in the case of a unidirectional wireless connection). A schematic of such a network is given in Figure 3. Distances between nodes u and v

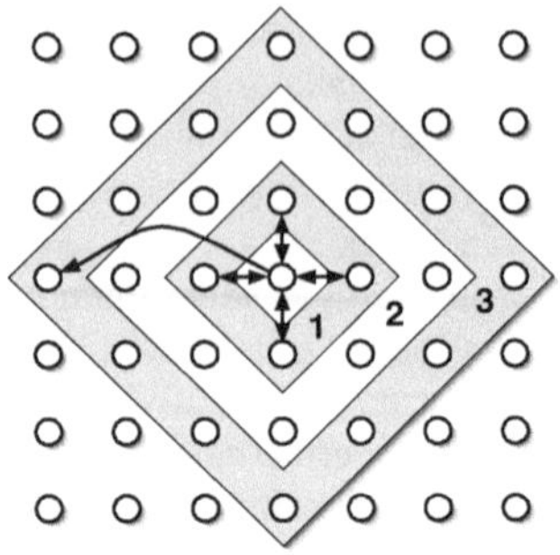

Fig. 3. A small-world graph, after Kleinberg. The nodes inside the shaded diamonds have the same Manhattan distance to the node in the center. This simple lattice network generalizes the small-world model of [21]. Nodes in area 1 are bi-directionally connected to the center node, which is also uni-directionally connected to one node in area 3.

are evaluated using the Manhattan metric, $d(u,v) = \sum_D |u-v|$, where D is the dimensionality of the lattice. To choose the terminator for the non-local link, we first choose its distance ℓ to the linking node from the distribution $P(\ell) \sim \ell^{-D}$, and then, out of all such vertices at this distance, we choose one uniformly at random. In the $D = 2$ network shown in Figure 3, the diamonds illustrate the set of nodes at a fixed distance from the linking node. If each vertex (router) forwards packets to their neighbor with the smallest remaining distance to the packet's destination, then this decentralized routing protocol will, for this particular topology, guarantee packet delivery in an average of $O(\log^2 n)$ steps [22]. To verify this statement, let us assume that a packet travels in phases, and that a phase ends when the remaining distance is halved; thus, there are at most $\log_2 n$ phases in an n nodes network. If the distribution of lengths for the non-local links is a power law with exponent D, the packet visits a router with a non-local neighbor that is roughly half as distant from the destination after $O(\log n)$ trials; thus, the expected routing time is $O(\log^2 n)$.

The algorithm presented in [11] constructs the Kleinberg-routable network through a dynamic, decentralized rewiring process. The algorithm assumes that local connections are fixed, and constructs the correct distribution of non-local link lengths in the following manner. Given a source- destination pair (x, y), route the packet according to the current topology, but choose a threshold t uniformly from the interval $[1, d(x,y)]$. If the routing time of the packet exceeds this threshold $T > t$ at a node z that is not the destination, x rewires its non-local link to terminate at z. That is, x "bookmarks" the location where the packet crossed its threshold. In [11], it is empirically shown that this rewiring algorithm produces the power-law link-length distribution $P(\ell) \sim \ell^{-D}$ that guarantees fast ad-hoc routing over the entire network $T = O(\log^2 n)$, after a modest number of rewiring actions $R \sim n^{1.77}$.

With the availability of global positioning (GPS) systems that provide simple distance measurements, these results could reasonably be adapted as a routing protocol for packets on a wireless array of devices. In such cases, local links are either physical connections or low-power broadcast transmissions, and non-local links are occasional high-power broadcast transmissions or unidirectional long-range transmissions.

The development of dynamic, decentralized routing algorithms that guarantee efficient navigability under a variety of assumptions is an active topic of research in the network theory community. Another interesting ad-hoc routing algorithm is presented in [23], where packets are routed under reasonable assumptions about the connectivity of vertices with similar properties (homophily), and the assumption that higher degree vertices are likely to be closer to the target. In the model used in [23], it is assumed that each node has a set of attributes, and that nodes link to others that are similar to themselves. Thus, a homophily-sensitive algorithm adjusts the routing based on the assumption that a node close to the destination node in their attribute space, is in fact geographically closer to the destination.

2.3 Efficiency

For a network theorist, the concept of efficiency is intimately related to that of *scalability*, which is defined as the cost of some network property as a function of the number of vertices in the network n. Generally, for a property to have a small cost, it should scale sublinearly, and ideally as a polylogarithm $O(\log^k n)$. For example, the decentralized routing algorithm [22] described in the previous section, guarantees that the average number of intermediate nodes through which the message passes is $O(\log^2 n)$. On the other hand, if we prefer that the property be true for as large a portion of the network as possible, we say that it must scale as a constant fraction of the vertices $O(1)$, and ideally to be $1 - o(1)$. For connectedness, the question of efficiency boils down to determining what fraction of the network remains connected, after a fraction of the nodes is removed. In this context, [15] shows that random networks with a power-law degree distribution, are increasingly more efficient at guaranteeing connectedness under random failure as the network grows.

Although these definitions of efficiency are highly applicable for the random graph models that network theorists typically study, they are conceptually different from the idea of efficiency in control systems. Reconciling these differences, and producing network theoretic results for the efficiency questions that control systems require, may be a fertile area of collaboration between researchers in these two fields.

3 The Control Designer's Point of View

Networked control system (NCS) applications such as teleoperation and robot formation control, require measurement and control signals to travel across communication networks. Even when the distance traveled is short (as in the case of a modern car or a smart house), a general purpose communication network introduces new issues into the feedback loop, such as time-varying delays, and the potential loss of information. While some communication applications may suffer from the same limitations, a feedback control system is especially vulnerable, not only to the unavailability of sensory information and control signals, but also to their timing. In particular, in a NCS, the issues of connectedness, navigability, and efficiency of travel manifest themselves as described in the following sections.

3.1 Connectivity, dropped packets, and lost links

From the perspective of control design for networked control systems, connectedness (or connectivity) expresses the ability of two systems to communicate information and actuation signals through the network connecting them. Connectivity is therefore related to the existence of a network path from any node u to any other node v. In recent studies that link the dynamics of the networked systems to the connectivity properties of the network, certain graph algebraic properties of the latter seem to be pervasive.

In [4] the dynamics of the networked system is formally related to the algebraic graph theoretic properties of the interconnection network, and to the graph Laplacian in particular. Connectivity, as expressed by the second smallest Laplacian eigenvalue (known as the *algebraic multiplicity* of the graph), proves to be a crucial network property, allowing the gradual dissemination of local information throughout the network and facilitating stability. In [24] it is shown that if connectivity is permanently lost, stability cannot be guaranteed. If connectivity is regained across a sequence of compact intervals $[t_i, t_{i+1})$, [4] shows that stability in the form of consensus is still guaranteed.

A network may not have a constant topology when communication links are established and lost in an ad-hoc manner [8],[4],[7]. Physical ad-hoc networks are typically modeled by nearest-neighbor type graphs, where nodes are distributed uniformly at random over a certain area, and are assumed connected if they are within a certain distance r_0 from each other. Thus nodes u, and v, are connected if $|u - v| < r_0$, where $|u - v|$ denotes the Euclidean distance between them. The question of whether such an ad-hoc network is connected or not, does not have a definitive answer, especially when the number of nodes grows very large. Results in this area are asymptotic and probabilistic in nature, and are usually related to the minimum degree of the nodes in the network, as exemplified in [25], or to the minimum communication range r_0 [26], [27]. In [25] it is shown that if each node is connected to less than $0.074 \log n$ other nodes, the network is disconnected with probability one, as the total number of nodes n increases. If, on the other hand, each node has more than $5.1774 \log n$ neighbors, the network is asymptotically connected with probability one when n tends to infinity. In [26] it is shown that if the network is to be connected with probability p, the transmission range r_0 must satisfy $r_0 \geq \sqrt{\frac{-\ln(1-p^{1/n})}{\pi\rho}}$, where ρ is the node density.

In networks where information flows in a unidirectional manner, directed graphs are used to capture the network topology. In directed graphs the definition of (strong) connectivity is more strict, requiring the information flow to respect the orientation of the graph edges. However, the existence of a (directed) spanning tree over the union of the graphs that describe the evolution of the network over time [6], may be sufficient for consensus in the network, provided that the graph switching frequency is bounded, on average. This condition is definitely weaker than strong connectivity, though still stronger than weak connectivity (expressed again by the second smallest eigenvalue) – for which edge orientation is irrelevant. The gap between these conditions seems to be the missing piece in a uniform characterization of stability in terms of network topology. Of course, another approach for ensuring stability is to restrict the dynamics, as described in [5], [28].

In most applications connectivity is binary, or unweighted, that is, two nodes are either connected at a particular time instant, or disconnected. In the former case the second smallest eigenvalue is positive, in the latter it is zero. For capturing the quality of a communication link, or the cost of broadcasting information from one node to another, weighted graph models can be used. The edge weights quantify the energy required for a message to be sent over an edge (u, v), usually expressed as $|u - v|^e$, where $e \geq 2$ is a constant, and $|u - v|$ is the Euclidean distance between the physical locations of nodes u, and v. Weighted graphs are not as well understood as their undirected counterparts, but connectivity analysis using the second smallest eigenvalue of the (weighted) Laplacian can be extended to this situation as well.

The effect of network topology and connectivity on the performance of cooperative localization algorithms is pointed out in [29], in which a genetic algorithm is used for choosing network topologies that result in smaller traces of the joint extended Kalman filter (EKF) covariance matrix. The analysis in [29] suggests that increased connectivity can be beneficial for localization accuracy. Intuitively, "the more sensor links between robots, the better."

However, while a genetic algorithm favors complete sensor graphs, other approaches may suggest "cheaper" solutions. In the special case where a landmark's location is accurately known, the expression of the steady-state value of the EKF covariance matrix in [30], contains the eigenvalues of a minor of the sensor graph Laplacian, weighted by the variances of the relative distance measurements. In this case, the (constant) steady-state value of the covariance matrix is [29],[30]

$$P_\infty = Q_o^{1/2} U \mathrm{diag}\left(\frac{1}{2} + (\frac{1}{4} + \frac{1}{\lambda_i})^{\frac{1}{2}}\right) U^T Q_o^{1/2}, \tag{1}$$

where Q_o is a diagonal matrix with entries that depend on the characteristics of the agents' sensors and their speed, U, is the matrix of eigenvectors, and λ_i is the ith eigenvalue of the matrix

$$C = Q_o^{1/2} H_o^T R_o^{-1} H_o Q_o^{1/2}, \tag{2}$$

in which R_o is a diagonal matrix of the noise covariance, and H_o relates to the incidence matrix of the sensor graph. In the case where we have a (single) accurately known landmark, H_o contains a block of zeros that eliminates the graph node that corresponds to the landmark, and $H_o^T R_o^{-1} H_o$ turns out to be a minor of a weighted Laplacian. In view of (1) and (2), and using eigenvalue interlacing theorems, it can be shown that the trace of P_∞ is related to the nonzero (weighted) Laplacian eigenvalues. Thus, forming a complete sensing graph is one way to increase localization accuracy, and more efficient ways may exist to decrease the trace of the covariance matrix.

Network connectivity appears to be a catalyst since without it, messages cannot reach their destination, consensus cannot be achieved, and estimation errors may grow unbounded. Some new insight, however, seems to suggest that perhaps dense connectivity is not all a control designer should strive for. In [31] it is shown that high algebraic connectivity does not necessarily imply a high network's robustness. In other words, while the network connectivity is certainly improved as the second smallest eigenvalue increases, and as the diameter of the network (characteristic path length) decreases, the network remains vulnerable to targeted attacks. In particular, there may be few nodes or links that guarantee the network connectivity, and the removal of as few as one or two such nodes may break the network into disconnected components.

Algebraic connectivity, as determined by the size of the second smallest Laplacian eigenvalue $\lambda_2(G)$, is of great interest because of Fiedler's inequality [32]

$$\lambda_2(G) \leq \nu(G) \leq \eta(G), \tag{3}$$

which states that the algebraic connectivity of a graph G is less than or equal to the node-connectivity which is less than or equal to the edge-connectivity [32]. While increasing the algebraic connectivity increases the lower bound on node-connectivity, it has been shown that, for circular and mesh lattice graphs, an increase in algebraic connectivity often corresponds to a decrease in node-connectivity and edge-connectivity [31].

In order to describe our results, let us consider first the *small-world network* introduced in [21]. This network is based on a one-dimensional lattice on a ring with n nodes where each node is connected to its k nearest neighbors. In [21] it is shown that random rewiring of nodes with a small probability p greatly reduces the characteristic path length resulting in a small-world network. Figure 4 shows the effects of random rewiring for a network with 20 nodes, and $k = 4$. In [33] it is shown that this random rewiring also results in a large increase in algebraic connectivity for ring lattices, and based on this it concludes that the network becomes more robust to node and link failures.

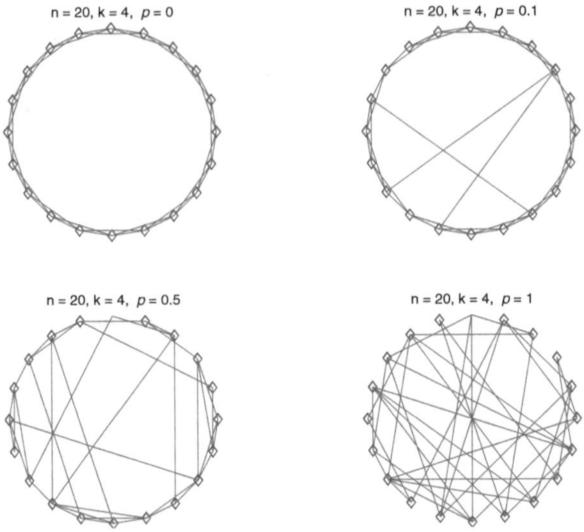

Fig. 4. Random ring lattice graph $G = C(n, k)$ with $n = 20$, $k = 4$, and different edge probabilities. For $p = 0$ a node is only connected to its two closest neighbors along the perimeter. As the probability increases, a larger number of these links are rewired and connect the node to other remote nodes.

Unfortunately, and as described above, large increases in algebraic connectivity for certain types of networks often correspond to a decrease in node-connectivity and edge-connectivity. As an example, the results for a circular random graph with 100 nodes are shown in Figure 6. For this case, we start with a ring lattice with $n = 100$ vertices, and $k = 4$ edges per vertex, then rewire each edge at random with a probability p. As p increases from 0 to 0.9, algebraic connectivity increases sharply, and the mean path length of the network decreases. However, the node-connectivity and edge-connectivity of the network decrease as the probability p increases. Similar results can be obtained for a regular mesh lattice like the one shown in Figure 5 where there are 100 nodes, and each node has a communication radius $R = 1$. The results for this mesh lattice are summarized in Figure 7.

In a system where nodes are redundant or dispensable, improving algebraic connectivity does indeed improve the overall robustness of the network by reducing the characteristic path length. However, in systems where each node is critical, node-connectivity and edge-connectivity are the most important parameters for assessing robustness. In addition, tradeoffs arise when studying robustness: computing the algebraic connectivity is much simpler than computing node-connectivity or edge-connectivity for large networks.

3.2 Navigability: path lengths and hops

During the control design process, and under the assumption of a connected communication network, the designer assumes that there exists a path between a transmitter and a receiver. The problem of determining this path is usually ignored, or assumed solved by those entities that direct the flow of information through the network: the routers.

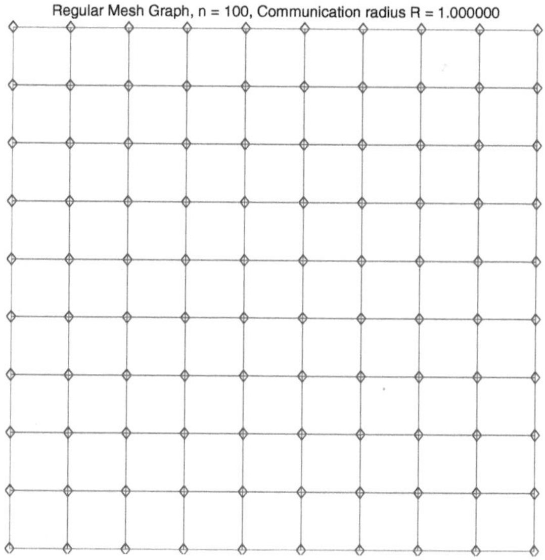

Fig. 5. Regular mesh lattice graph, $N = 100$, Communication radius $R = 1$. In the mesh lattice at the upper left corner, where $p = 0$, nodes are connected horizontally and vertically to their nearest neighbors. As p becomes larger, the edges connecting nearest neighbors are increasingly rewired, and link nodes in remote locations.

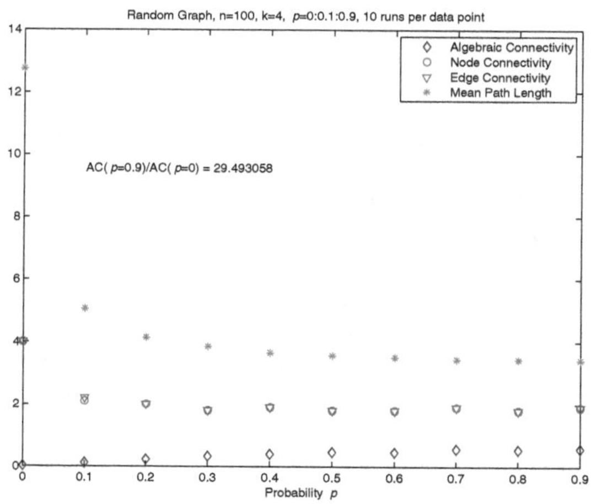

Fig. 6. Results for a ring lattice random graph, $N = 100$, $k = 4$. Although algebraic connectivity increases, node and edge connectivity decreases monotonically with mean path length.

Standard routing protocols make use of assumptions that may not be generally favorable to control system design. For instance, Ethernet is a broadcast protocol, and thus only a limited number of participants can communicate over a given portion of a network. open shortest

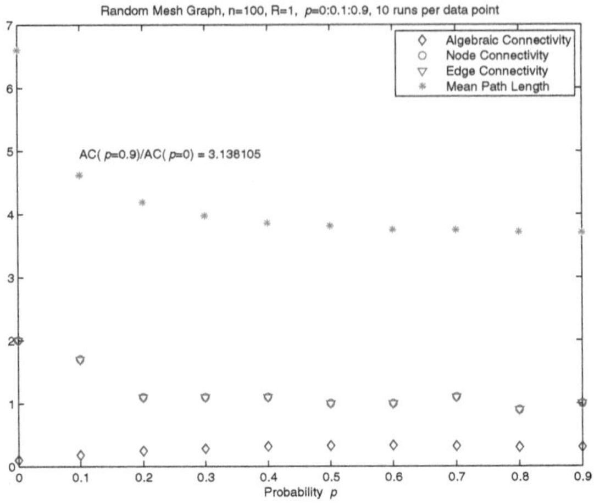

Fig. 7. Results for a mesh lattice graph, $N = 100$, $R = 1$. Although algebraic connectivity increases, node and edge connectivity decreases monotonically with mean path length.

path first (OSPF), as well as BGP, are susceptible to the propagation of corrupt or maliciously faulty information [34].

To provide the most basic packet delivery service, such as on the Internet at the IP-level, protocols like Ethernet, OSPF and BGP, combined with the commodity network hardware do well enough when most nodes are connected, the network is navigable, paths are relatively short, and service is fairly reliable. Such is the case on the Internet, where the average hop-count at the IP-level is at most a few dozens, despite there being potentially billions of routable IP addresses. Notably however, deviations from ideal conditions result in several serious interruptions in global Internet service. For applications such as sensor networks or ad-hoc networks among mobile devices (such as cell phones), all of these issues are active areas of research in both the control systems and network theory communities.

Even if determining the path from source to destination is not a major issue, the lengths of such paths matter, especially when information is processed as it propagates through the nodes of the network. One such example is the case of leader-follower control architectures. When the leading vehicle in a platoon suddenly decelerates, the larger a following vehicle's distance is from the leader, the faster must the follower decelerates. In this context, there comes a point where control signals saturate, stability is lost, and collisions become unavoidable.

String stability research [35] addresses this issue by treating the propagation of information as a disturbance, and ensuring that disturbance signals attenuate as they propagate through the string of interconnected systems. Mesh stability [36] is essentially the generalization of the same concept for strings that run across multiple (physical) dimensions. As each node propagates the disturbance signal to its network neighbors, this signal is attenuated, and systems perturbed by these decaying propagated signals are robust and stable.

When the propagated information is not regulated in terms of its effect on the receiving nodes, it was shown in [37] that in leader-follower control architectures, the distance from the source of the signals (the leader), has an adverse effect on the ability of the follower to track its desired position in the formation. In these applications, the network is described as a directed tree, the root of which is the formation leader. Stability bounds are less conservative for trees of small depth [37]. Therefore, routing the information signals through shorter paths improves stability.

Thus, from a stability viewpoint, the control designer has to either regulate the system dynamics so that it can cope with information traveling over long paths, or make sure that short paths (up to a certain length) can be found. Regarding the latter, ad-hoc routing algorithms that improve the navigability of the network are needed.

3.3 Efficiency: capacity, link quality, and delays

Since for the control engineer, a communication channel is merely a medium for obtaining or sending information (measurement signals, or control commands), what seems to be important is: (i) how much information can be carried, and (ii) how fast can it be transferred.

The first question is related to the channel's capacity. Results linking information theory to control have recently been discovered [38, 39, 40, 41, 42]. While information theory models the communication channel as an information transmitting medium that corrupts portions of the signal, the main issue for control-based applications are the delays (as well as corruption) suffered by the signals as they are carried across the channel. In the case of noiseless channels, a necessary condition for asymptotic observability and stabilizability for linear, time-invariant, discrete-time systems, is that the rate of communication R (which must be less than the capacity C of the channel) is bounded below as $R > \sum_{\lambda_u(A)} \max\{0, \log|\lambda_u(A)|\}$, where $\lambda_u(A)$ are the unstable eigenvalues of the system matrix A. In some cases, this condition is also sufficient. Similar results hold in the case of noisy channels, as described in [43]. In [44] the fundamental limitation of performance for networked feedback systems is addressed. The feedback loop comprises a discrete-time, linear, time-invariant plant, a channel, as well as an encoder and a decoder. The disturbance rejection ability is found to be bounded below by $\sum_{\lambda_u(A)} \max\{0, log|\lambda_u(A)|\} - C$. This particular result shows that the excess capacity $C - \sum_{\lambda_u(A)} \max\{0, log|\lambda_u(A)|\}$ is all that is available for disturbance rejection. These, and other similar results, not only link the feedback control problem to the communication setting, but also provide for fundamental results in feedback control, in the spirit of Shannon's information theory results [44]. The sidedar "Control and Information" expands on links between control and information theory.

The speed at which information can travel from source to destination is usually measured in terms of a communication delay, the time elapsed between transmission, and reception. It is generally recognized that actuation and measurement delays degrade the performance of control systems. A communication delay can cause actuation delays, measurement delays, or both; therefore, it is natural to expect that it must have an adverse effect on the stability of the interconnected system.

Initial investigation seemed to support this claim. In [7], stability analysis in the frequency domain suggests the existence of an upper limit in the (uniform) communication delays that a continuous, nearest neighbor interconnected system can tolerate, before becoming unstable. However, more recent analysis of state space, discrete-time models of interconnected systems, leads to different conclusions.

Several recent independent investigations verify that in some (not so special) cases, arbitrary (but bounded) communication delays may be tolerated at the expense of convergence speed. In [28], the approach of [5] is extended, and concludes that if the agent dynamics are appropriately restricted, stability can be maintained. A different approach in [45] focuses on the communication protocol, and shows that velocity synchronization in a connected group of autonomous mobile agents, may still be achieved when the agent controllers use delayed information, regardless of the size of this delay, if control and communication are used appropriately. In [46] the composition properties of graphs are used to show that under certain assumptions on the communication topology, delays have no effect on the stability of the system. In fact, in a somehow counterintuitive situation, it turns out that longer delays (if used judiciously) can improve the stability of some systems [47].

In all of the cases above, convergence is not as fast as in the case of instantaneous communication. The fact that instability can still be avoided is however a relief. Efficiency in the transmission of information between networked control systems may be desirable, but the critical network property seems to be connectivity, at least when the interconnected systems are stable in the absence of communication delays.

4 What Can Control Learn from Network Theory?

There are two primary benefits for control system designers that may result from their interaction with network theorists. First, network theorists study bulk properties of the network in a way that is fundamentally different from the low-level considerations of packet loss, link failure, latency, and responsiveness. Their analysis provides insights into the robustness of connectedness, and efficiency of routing on networks with certain global topological features, such as networks with power-law degree distributions or small-world networks. Their interest in the scalability of network properties and behavior is particularly relevant for control systems. One of their most robust methods for achieving these scaling properties is through a judicious use of randomness, an approach that has proved powerful in designing algorithms in the face of uncertainty. As control systems become increasingly networked, uncertainty becomes a more significant factor in our designs.

The other area in which control systems may benefit from network theory is in the domain of protocol design for managing the flow of information on the network. Recent work on efficient ad-hoc routing demonstrates that dynamic, decentralized algorithms can provide provably or demonstrably good packet delivery times, often by relying on randomness. Additionally, within the community of peer-to-peer networks, researchers have focused on developing communication protocols that are robust to both failure and corruption. Even in a static network control system such as a modern airplane, software failure can create spurious behavior that must be handled robustly. In a more dynamic network system such as a wireless sensor network, where devices can be installed or removed, such considerations are even more critical. Both for managing uncertainty, and for improving performance, randomness may prove to be a key component in the design of protocols that regulate the flow of information in network control systems. Another particularly interesting notion in computer networks is that of discoverability (not to be confused with reachability). While a node could be reachable on some graph i.e. there exists a path to the node in the graph (whether efficient or not), the same node need not necessarily be discoverable. Briefly, discoverability is reachability with minimal knowledge about the destination's state, whether that be location or some other attribute that the source node uses to identify the destination [48]. As discussed earlier, and while it has been shown by Kleinberg's that complex networks that follow basic but extremely powerful laws (small world) can be efficiently navigable [49], a basic underlying assumption in the model is that some state about the destination is known to the sender *a priori*. For example, the coordinates of the destination on the lattice are known when navigating the latter. In modern networks, such information is not necessarily known and the moment discoverability is to be provided, significant state and complexity is introduced into the network. For example, the recent de-aggregation of the BGP address space has virtually stripped out the location information (a form of state) out of the BGP prefix rendering the latter a flat identifier. Discoverability then in the context of BGP requires that every BGP router in the Default Free Zone (or DFZ) maintains local information about every prefix to deal with the network dynamics. A fundamental tradeoff seems to always exist in the network design process between efficiency, scalability, and dynamism where only two of the three mentioned properties could play well together in any one design.

Controlling across the network can benefit from modeling and accounting for the discoverability issue, or else risk oversimplifying the assumptions that define modern networks and usage models. For example, while the network level end-to-end communication delay could be bounded on the network, instabilities could emerge during the discovery phase. It is known that the impact of the discovery phase, which is generally carried by the first packet, on IP congestion control is significant. More clearly a large first packet end-to-end roundtrip time will hinder the current IP protocol's operation as the latter will tend to treat the delay as congestion.

5 Case Studies

In this section, we briefly delve deeper into several illustrative examples that seem particularly applicable to the development of randomized networked control systems, particularly with respect to connectedness, bandwidth, and navigability.

5.1 Dynamic Ad-hoc Routing

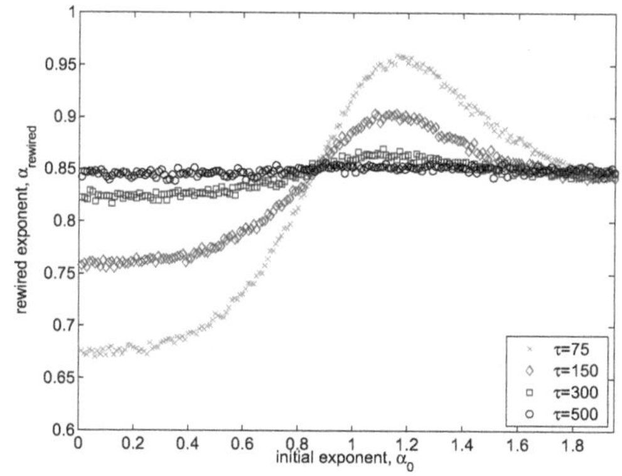

Fig. 8. Convergence of the dynamic, decentralized rewiring algorithm of [11] to the optimum distribution for a variety of initial topologies, and rewiring times (τ). The deviation from the theoretical optimum ($\alpha = 1$ here) is due to finite-size effects.

The dynamic, decentralized routing protocol of [11], introduced above, is intriguing for several reasons. First and foremost, the assumption of an underlying lattice topology is amenable to many existing networked control system frameworks, such as a wireless mesh. Furthermore, the protocol itself is simple, and may be adapted to work within the existing internet protocol framework: given a source node x, a destination node y, and a message-specific threshold t, the protocol works as follows: if the topological length of the route exceeds the message's time-to-live (TTL) t, x rewires its non-local link so as to point to the node at which the threshold is crossed. By choosing the TTL t from the uniform random distribution over the interval $[1, d(x, y)]$, the protocol produces exactly a non-local link topology that minimizes the average transmission time for every node's messages [22]. This protocol has the added benefit that the resulting topology is independent of the initial network topology (see Figure 8), and is self-optimized to account for finite-size effects. From a control systems perspective, these characteristics are attractive in that they reduce the brittleness of the system, while guaranteeing good overall behavior.

5.2 Static Ad-hoc Routing

In networks that lack the distance metric of Kleinberg's small-world network [22], the decentralized routing protocol of [23] yields efficient routing by relying on common topological features of real-world networks. That is, the algorithm capitalizes on the property of *homophily* (the idea that nodes with similar attributes are likely to be connected), and on a right-skewed degree distribution. Each node (router) in the network approximates the full shortest-path routing table $E[\ell_{s\to t}] = \sum x \cdot P(\ell_{s\to t} = x)$, where $\ell_{s\to t}$ is the length of the path between source s and destination t, by simply estimating which of their neighbors is likely to be closer to each possible destination. In practice, nodes accomplish this by calculating the likelihood that two nodes u, and v, are connected: $p_{uv} = 1 - (1 - q_{uv})^k$ where $(1 - q_{uv})^k$ is the probability that none of u's k edges terminate at v. Although this protocol assumes that alternate paths are independent, in practice such correlations in real-world networks are sufficiently small. The clear advantage of this approach is that it utilizes common features of most real-world networks to provide efficient distributed routing in the face of topological uncertainty.

5.3 Gossip Protocols

Next, we consider a communication network in which edges are unreliable, that is, where pairwise connectedness is guaranteed only probabilistically. Although all packet-based networks exhibit this property to some degree, wireless mesh networks exhibit it most strongly. In spite of this intermittent local connectivity, let us attempt to maximize the set of nodes that receive a broadcast message, while simultaneously minimizing the total bandwidth per message, as expressed by the number of messages used. Formally, we wish to find the set of nodes A, to which we initially send the message, such that under some communication protocol ψ, the set of receiving nodes $\psi(A)$ is as large as possible. Gossip protocols, where a node receiving a message attempts to pass it along to its neighbors at most once, thus enforcing strong bandwidth constraints, can provide a provably good solution to this problem, which is otherwise NP-hard [50]. The greedy algorithm presented in [50] iteratively maximizes the marginal gain in $\psi(A)$, and yields a target set that is provably $1 - 1/e$ times size of the optimum. These protocols can also be married with the Kleinberg small-world topology to solve problems such as resource location on a network [51].

6 Conclusions

The control and complex network communities have been looking into similar problems in networked systems from different perspectives. The same concepts and properties appear to be important, both for the control engineer, and for the network theorist. Having established the conceptual link between networked control systems, cooperative control, and complex networks through graph theoretic analysis, the control community may now be in position to capitalize on, and exploit the arsenal available in complex network research and computer science. This chapter offers such a suggestion by highlighting the recently revealed power of randomized algorithms in routing, network design, resource allocation, and game theory.

In the computer science community, much attention has recently turned to mechanism design; this domain seeks to allow selfish individuals to interact in a networked environment in such a way that no outcome is particularly disadvantageous to any of the nodes. Such approaches yield results for routing, network design, and resource allocation, and seem directly applicable to open networked control systems in which the control engineer must ensure that corrupt or misbehaving nodes do not negatively affect the functionality of the system.

References

1. Walsh, G., Ye, H., Bushnell, L.: Stability analysis of networked control systems. IEEE Transactions on Control Systems Technology 10(3), 438–445 (2002)
2. Zhang, W., Branicky, M.S., Phillips, S.M.: Stability of networked control systems. IEEE Control Systems Magazine 21(1), 84–99 (2001)
3. Newman, M.E.J.: The structure and function of complex networks. SIAM Review 45, 167–256 (2003)
4. Jadbabaie, A., Lin, J., Morse, A.S.: Coordination of groups of mobile autonomous agents using nearest neighbor rules. IEEE Transactions on Automatic Control 48(6), 988–1001 (2002)
5. Moreau, L.: Stability of multiagent systems with time-dependent communication links. IEEE Transactions on Automatic Control 50(2), 169–182 (2005)
6. Ren, W., Beard, R.W.: Consensus seeking in multiagent systems under dynamically changing interaction topologies. IEEE Transactions on Automatic Control 50(5), 655–661 (2005)
7. Olfati-Saber, R., Murray, R.M.: Consensus problems in networks of agents with switching topology and time-delays. IEEE Transactions on Automatic Control 49(9), 1520–1533 (2004)
8. Tanner, H.G., Jadbabaie, A., Pappas, G.J.: Stable flocking of mobile agents, part ii: Dynamic topology. In: Proc. of the 44th IEEE Conference on Decision and Control, pp. 2016–2021 (2003)
9. Faloutsos, M., Faloutsos, P., Faloutsos, C.: On power-law relationships of the internet topology. Computer Communications Review 29, 251–262 (1999)
10. Bollobás, B.: Random graphs. Academic Press, London (1995)
11. Clauset, A., Moore, C.: How do networks become navigable? e-print arXiv:cond-mat/0309415 (2003)
12. Stanley, H.E.: Introduction to Phase Transition and Critical Phenomena. Oxford University Press, Oxford (1983)
13. Cohen, R., Erez, K., ben Avraham, D., Havlin, S.: Resilience of the internet to random breakdowns. Physical Review Letters 85, 4626 (2000)
14. Link, H., LaViolette, R.A., Saia, J., Lane, T.: Parameters affecting the resilience of scale-free networks to random failures. e-print arXiv:cs.LG/0511012 (2005)
15. Gallos, L.K., Cohen, R., Argyrakis, P., Bunde, A., Havlin, S.: Stability and topology of scale-free networks under attack. Physical Review Letters 94, 188701 (2005)

16. Guillaume, J.L., Latapy, M., Magnien, C.: Comparison of failures and attacks on random and scale-free networks. In: Higashino, T. (ed.) OPODIS 2004. LNCS, vol. 3544, pp. 186–196. Springer, Heidelberg (2004)
17. Pease, M., Shostak, R., Lamport, L.: Reaching agreement in the presence of faults. Journal of the ACM 27(22), 228–234 (1980)
18. Ben-Or, M., Canetti, R., Goldreich, O.: Asynchronous secure computation. In: Proc. 25th ACM Symposium on Theory of Computing, pp. 52–61 (1993)
19. Internet Engineering Task Force. Rfc 1247 (open shortest path first), `http://www.ietf.org/rfc/rfc1247.txt`
20. Internet Engineering Task Force. Rfc 1771 (border gateway protocol), `http://www.ietf.org/rfc/rfc1771.txt`
21. Watts, D.J., Strogatz, S.H.: Collective dynamics of 'small-world' networks. Nature 393, 440 (1998)
22. Kleinberg, J.: The small-world phenomenon: An algorithmic perspective. Technical Report 99-1776, Computer Science Department, Cornell University (1999)
23. Şimşek, Ö., Jensen, D.: Decentralized search in networks using homophily and degree disparity. In: Proc. 19th International Joint Conference on Artificial Intelligence, pp. 304–310 (2005)
24. Tanner, H.G., Jadbabaie, A., Pappas, G.J.: Stable flocking of mobile agents, part i: Fixed topology. In: Proc. of the 44th IEEE Conference on Decision and Control, pp. 2010–2015 (2003)
25. Xue, F., Kumar, P.R.: The number of neighbors needed for connectivity of wireless networks. Wireless Networks 10(2), 169–181 (2004)
26. Bettstetter, C.: On the minimum node degree and connectivity of a wireless multihop network. In: Proc.3rd ACM Int. Symp. on Mobile Ad Hoc Net. and Comp. (MobiHoc 2002), pp. 80–91 (2002)
27. Santi, P., Blough, D.M.: The critical transmitting range for connectivity in sparse wireless ad hoc networks. IEEE Transactions on Mobile Computing 2(1), 25–39 (2003)
28. Angeli, D., Bliman, P.A.: Extension of a result by Moreau on stability of leaderless multi-agent systems. In: Proc. of the 44th IEEE Conference on Decision and Control, ECC-CDC 2005, pp. 759–764 (2005)
29. Hidaka, Y.S., Mourikis, A.I., Roumeliotis, S.I.: Optimal formations for cooperative localization of mobile robots. In: Proc. of the International Conference on Robotics and Automation, pp. 4137–4142 (2005)
30. Mourikis, A., Roumeliotis, S.: Performance bounds for cooperative simultaneous localization and mapping. In: Thrun, S., Sukhatme, G., Schaal, S., Brock, O. (eds.) Robotics: Science and Systems I, pp. 73–80. MIT Press, Cambridge (2005)
31. Byrne, R., Feddema, J., Abdallah, C.: Algebraic connectivity and graph robustness. Technical report Sandia National Laboratories (2005)
32. Fiedler, M.: Algebraic connectivity of graphs. Czechoslovak Mathematical Journal 23(98), 298–305 (1973)
33. Olfati-Saber, R.: Ultra-fast consensus in small-world networks. In: Proc. of the American Control Conference, pp. 2371–2378 (2005)
34. Nordstrom, O., Dovrolis, C.: Beware of bgp attacks. SIGCOMM Comput. Commun. Rev. 34(2) (2004)
35. Swaroop, D.: A note about the stability of a string of lti systems. ASME Journal of Dynamic Systems, Measurement and Control 124, 472–475 (2002)
36. Pant, A., Seiler, P., Hedrick, K.: Mesh stability of look-ahead interconnected systems. IEEE Transactions on Automatic Control 47, 403–407 (2002)

37. Tanner, H.G., Pappas, G.J., Kumar, V.: Leader-to-formation stability. IEEE Transactions on Robotics and Automation 20(3), 433–455 (2004)
38. Wong, W., Brockett, R.: Systems with finite communication bandwidth constraints i: State estimation problems. IEEE Transactions on Automatic Control 42(9), 1294–1299 (1997)
39. Wong, W., Brockett, R.: Systems with finite communication bandwidth constraints ii: Stabilization with limited information feedback. IEEE Transactions on Automatic Control 44(5), 1049–1053 (1999)
40. Nair, G., Evans, R.: Stabilization with data-rate-limited feedback: Tightest attainable bounds. Systems and Control Letters 41, 49–76 (2000)
41. Ballieul, J.: Feedback designs in information-based control. In: Pasik-Duncan, B. (ed.) Stochastic Theory and Control. Proceedings of Workshop held at the University of Kansas. LNCIS, pp. 35–57. Springer, New York (2002)
42. Brockett, R., Liberzon, D.: Quantized feedback stabilization of linear systems. IEEE Transactions on Automatic Control 45(7), 1279–1289 (2000)
43. Tatikonda, S., Elia, N.: Communication requirements for networked systems. In: Chiasson, J., Trabouriech, S., Abdallah, C.T. (eds.) Advances in Communication Control Networks. LNCIS, pp. 303–326. Springer, New York (2005)
44. Martins, N.C., Dahleh, M.A.: Fundamental limitations of performance in the presence of finite capacity feedback. In: Proc. of the American Control Conference, pp. 79–86 (2005)
45. Tanner, H.G., Christodoulakis, D.: Discrete time flocking with time delays. In: Proc. of the 44th IEEE Conference on Decision and Control, pp. 4945–4950 (2005)
46. Morse, A.S.: Lecture Notes on Logically Switched Dynamical Systems. Springer, Heidelberg (2006)
47. Abdallah, C.T., Dorato, P., Benitez-Read, J., Byrne, R.: Delayed positive feedback can stabilize oscillatory systems. In: Proc. of the American Control Conference, pp. 3106–3107 (1993)
48. Khoury, J., Abdallah, C.T.: Identifier-based discovery mechanism design in large-scale networks. In: Future-Net 2009: Proceedings of the International Workshop on the Network of the Future - IEEE ICC 2009, Dresden, Germany (2009)
49. Kleinberg, J.: Complex networks and decentralized search algorithms. In: Proceedings of the International Congress of Mathematicians, ICM (2006)
50. Kleinberg, J., Kempe, D., Tardos, É.: Influential nodes in a diffusion model for social networks. In: Caires, L., Italiano, G.F., Monteiro, L., Palamidessi, C., Yung, M. (eds.) ICALP 2005. LNCS, vol. 3580, pp. 1127–1138. Springer, Heidelberg (2005)
51. Kempe, D., Kleinberg, J., Demers, A.: Spatial gossip and resource location protocols. In: Proc. ACM Symposium on Theory of Computing, pp. 163–172 (2001)

Inventory Dynamics Models of Supply Chains with Delays; System-Level Connection & Stability

Ismail Ilker Delice and Rifat Sipahi

Department of Mechanical and Industrial Engineering, Northeastern University, Boston, MA 02115. Phone: (617) 373 6011. Email:rifat@coe.neu.edu.

Summary. This paper surveys some deterministic continuous-time inventory dynamics of supply chains which inevitably function with delays due to the presence of manufacturing lead times, decision making, transportation and information flows. Availability of such models opens up new research directions where operations research and the field of time-delay systems can be merged primarily for characterizing inventory oscillations, developing new forecasting techniques and regulating inventories under uncertainties in consumers' buying trends, stochastic nature of delays, market behavior and possible price competitions across enterprizes. This paper also offers one of the first connections between the disciplines via *stability maps*. We show how stability maps can prove to be useful in maintaining the stable nature of inventory oscillations. In a case study we demonstrate that the supply chain manager, by inspecting the stability maps, can be aware of which delay combinations in the supply chain may cause (un)desirable inventory behaviors.

1 Introduction

Supply chains (networks) [5, 8, 9, 24, 31] are an interconnection of various dynamics contributed by customers, suppliers, manufacturing units, enterprizes, assembly lines, parallel running processes and sources. Supplies in supply chains (SC) flow towards the direction of increased demand (from inventories to customers), while the information about the demand flows in the opposite direction (from customer forecasting to company headquarters). Among many objectives in managing SC, one of the most critical ones is to regulate the inventory levels while successfully responding to customer demand. This may sound like a simple task, however, in presence of uncertainties, non-deterministic events, unpredictable seasonal changes (which may affect the sales of a seasonal product) and delays, supply chain management is known to be a challenge, [36].

In this paper, the main focus is on the *delays* in the SC. Delays arise from various different physical reasonings and constraints, such as decision-making, transportation, information flow and manufacturing lead times. Due to the presence of delays, what is currently occurring in the supply chain is the *after-effects* of what has happened earlier. Consequently, any decision based only on what is currently observed in the SC is likely to be unsuccessful as those observations represent the past. The consequences are very well known in operations

J.J. Loiseau et al. (Eds.): Topics in Time Delay Systems, LNCIS 388, pp. 349–358.
springerlink.com

research, business and control theory, [17, 31]. Delays lead to *oscillations, limit cycles, overshoot, excessive/depleted* inventories and *synchronization* problems across parallel-running processes, [1, 2]; and these effects may cost companies billions of dollars, [1, 37].

Delays are not always detrimental either. Decision-making delays may have positive effects to SC management, since waiting enough time gives a clear picture to managers regarding sale trends and the market behavior [31]. This wait time also contains the time for collecting the necessary data to conclude a decision and perception of human behavior towards deciding a new command, [33]. With this fundamental observation, it is clear that it is impossible to conclude intuitively as to how delays may affect inventory behaviors. This observation is the main motivation in many work appearing in both physics, operations research and the field of time delay systems, [11, 17, 20, 21, 22, 27, 32, 35].

In this paper, we survey deterministic continuous-time supply chain models (Section 2), which are in the form of differential equations and have been broadly studied in the literature [20, 21, 27, 31, 35]. Furthermore, various mathematical models of delays depending on the physics of supply chains are covered (Section 3). The interconnection between these models and system-level approach particularly in the field of *time delay systems* (TDS) are demonstrated with an illustrative example (Section 4). In Section 5, we conclude the paper. It is foreseen that this survey will further motivate research in the area of supply chain management "with delays".

2 Mathematical Modeling of Supply Chains without Delays

In this section, we present some ideas developed in the literature for mathematical modeling of supply chains, [11, 20, 32]. The main components that play role in the development of these models are inventories, communication medium, decision-making and human-in-the-loop dynamics, production and transportation medium. It is important to note that the coverage of the models here only pertains to deterministic, continuous time linear time invariant supply chains with time-invariant delays, however, more sophisticated models can be built upon these simpler ones we present.

2.1 Helbing's model, [11, 27]

This model considers a supply chain of n suppliers i delivering products to other suppliers μ or to costumers. The rate at which supplier i delivers products to and consumes product from supplier μ is given by $d_{\mu i}X_i(t)$ and $c_{i\mu}X_i(t)$, respectively, where $X_i(t) > 0$ denotes the production rate. The coefficients $c_{i\mu}$ define an input matrix $\mathbf{C}$ and $d_{i\mu}$ an output matrix $\mathbf{D}$ with $0 \leq d_{i\mu}, c_{i\mu} \leq 1$.

Inventories.

The inventory level $N_i(t)$ at supplier i changes at the rate

$$\frac{d}{dt}N_i = \sum_{\mu=1}^{n_1} (d_{i\mu} - c_{i\mu})\, X_\mu(t) + Y_i(t), \quad i = 1, \ldots, n_2, \tag{1}$$

where the external demand is denoted by $Y_i(t)$. In order to keep the inventory at some desired level $\overline{N}_i$, any changes in the demand $Y_i(t)$ require an *adaptation* of the production rates $X_i(t)$.

Decision-making.

The adaptation is represented by the time constant T_μ, which defines the measure of speed of actual production rate $X_\mu(t)$ converging to a desired one $W_\mu(t)$

$$\frac{dX_\mu(t)}{dt} = \frac{1}{T_\mu}\left(W_\mu(\{N_i(t)\}, \{dN_i(t)/dt\}) - X_\mu(t)\right), \tag{2}$$

which concludes the foundation of Helbing's model (1)-(2). Interested readers are also directed to the work in [27] where a four-delay analysis was conducted on Helbing's model.

2.2 Sterman's model, [32]

The model we present here is Sterman's stock acquisition system. Sterman[1] utilizes two *sequential* inventories representing the supply line SL and the inventory, S, respectively. The rate at which supplies are being delivered from SL to S is the acquisition rate A, and the rate at which supplies leave S is called the loss rate L.

Inventories.

The equations defining the dynamics of the inventory and supply line are

$$S(t) = \int_{t_0}^{t}(A(\kappa) - L(\kappa))d\kappa + S(t_0), \tag{3}$$

$$SL(t) = \int_{t_0}^{t}(O(\kappa) - A(\kappa))d\kappa + SL(t_0), \tag{4}$$

where $O(t)$ is the order rate that is placed by the supply chain managers to the supplier of the products.

Decision-making.

The decision-making utilizes the information pertaining to SL, S and L. Furthermore, desired supply line SL^*, desired inventory S^* and expected loss rate $\hat{L}$, which can be constant or time-varying, are used for comparison with SL, S and L, respectively. This comparison is necessary to re-formulate the order rate $O(t)$ to correct the actual SL, S and L towards maintaining them at SL^*, S^* and $\hat{L}$. The order rate strategy is formulated as,

$$O(t) = \max(0, IO(t)), \tag{5}$$

$$IO(t) = \hat{L} + AS(t) + ASL(t), \tag{6}$$

$$AS(t) = \alpha_S(S^*(t) - S(t)), \tag{7}$$

$$ASL(t) = \alpha_{SL}(SL^*(t) - SL(t)), \tag{8}$$

$$\hat{L} = L^*, \tag{9}$$

where $IO(t)$ is the indicated order rate, 'maximum' between zero and $IO(t)$ assures that $O(t)$ is nonnegative, α_S is the stock adjustment parameter of the inventory, α_{SL} is the fractional adjustment rate of the supply line, AS is the actual stock and ASL is the actual supply line. Furthermore, other than modeling expected loss rate $\hat{L}$ as static, $\hat{L} = L^*$, one can also model it as regressive, adaptive and extrapolative.

[1] In Sterman's model, delays are also considered, but we leave the delay discussions to the following section.

2.3 Riddalls's model, [20]

The model introduced by Riddalls also incorporates time delays. In this model, the inventory dynamics is taken as in (3), while assuming $A(t) = O(t)$, i.e. disregarding (4). In modeling decision-making, Riddalls utilizes (5)-(8), but modifies (9) by proposing a short term forecast/trend detector[2]

$$\hat{L} = \frac{1}{T} \int_{t-T}^{t} L(\kappa) d\kappa, \tag{10}$$

where T is a period of time. The above equation suggests that the expectation is the average of the integration of loss rate over the period T. This averaging is mainly motivated from the fact that decisions based on suddenly changing customer demands are the sources to bullwhip effects and this should be avoided by constructing the decision-making via this averaging (smoothening) formula. Riddalls's model, *without delays*, becomes

$$\frac{dO(t)}{dt} = \alpha_S(L(t) - O(t)). \tag{11}$$

For other work utilizing the extensions of Riddalls's model, more detailed models and applications, see also [4, 9, 10, 13, 15, 21, 31, 35] and the references therein.

3 Mathematical Modeling of Delays in Supply Chains

For accurate understanding of supply chain dynamics, it is crucial to model the delays based on the physics they originate from, see [20, 21, 27, 31, 32, 35]. Various dynamical systems with different delay models are also studied in the field of *time delay systems*, [17, 22, 30]. Therefore, we also aim to point out the link between supply chains and the studies in time delay systems.

3.1 Constant delay model

This model assumes that delay $\tau > 0$ is constant. In operations research, it is also known as *pipeline delay* or *pure delay*, while in time delay systems it is called *discrete delay*. An inflow $i(t)$ through a constant delay model will create an outflow $o(t)$, where $o(t) = i(t - \tau)$. In supply chains, this class of delay model represents the dead-time of humans' decision-making processes. Information flow, data collection and measurement times also cause dead-times and machine set-up lead times, material flow in assembly lines are other examples modeled with constant delays. Discrete delay is also used in Riddalls model where $O(t)$ is replaced by $O(t - \tau)$ in (11). This type of delays is also widely seen in traffic flow behavior, [34]; machine tool chatter, [30]; multi-agent consensus/synchronization problems, [19]; tele-operation, active vibration suppression, [17]. Note however that constant delay only models FIFO (first in, first out) type behavior, but it does not consider any degree of *mixing*, which might be needed in supply chains, biological systems and chemical processes. Incorporating the effects of mixing requires the utilization of distribution functions discussed below.

---

[2] In order to enable easier comparison, we adopt Sterman's notation to express Riddalls work.

3.2 Distributed delay model

In many cases, this type of delay models is used for material delivery delays, mixing of materials, updating of beliefs, adjusting towards a new decision, diffusion in social networks, chemical and biological systems, energy flow delays. Some examples are exponential, gamma (γ) and Erlang distributions, which also arise in biology, [12]; machine tool chatter, [30]; traffic flow, [28] and in chemical process control, [16]. An example supply chain with distributed delays is mass mailing (corresponding to a pulse) where delivery of the mails to various destinations will not be at the same time, thus the number of deliveries will exhibit a distribution with respect to delivery times, [31]. Finally, we note that a distribution function can be in series with a dead-time h, which is nothing but a discrete delay after which deliveries may start to arrive at their destinations. A similar example occurs when modeling the decisions of humans in the study of traffic flow stability [28].

3.3 Other delay models

Other delay models comprise time-varying and state-dependent delays. We refer the readers to the work in [20, 29, 32] for further details.

4 Stability of Inventory Dynamics & Connection with System-Level Approach

In the context of inventory regulation problems and stability issues, we see that operations research and business fields progressed in close connection with control theory. Along these lines, one can find numerous "system thinking" approaches utilized for understanding supply chain dynamics, see [3, 5, 11, 20, 23, 24, 25, 31, 32, 35] and the references therein. At system-level, one considers the supply chain dynamics as a connection of block diagrams representing the suppliers and transportation lines for material deliveries (feed-forward lines), and information flows forming the feedback lines.

4.1 Stability of Equilibrium Dynamics

As it is often implicitly needed in queuing theory, the tendency of supply chain dynamics around an an equilibrium state is of interest, [3, 11, 20, 21, 27, 35]. The linearized dynamics, obtained from the non-linear one, carries rich information as to the fate of the inventory levels and how managerial decisions might be appropriately given in order to maintain the inventories at steady levels. The insight to this can be extracted by analyzing the stability.

The stability analysis requires the study of a class of differential equations that contains delays. The main challenge in the stability analysis is due to the need for analyzing the *infinitely many* dynamic modes that arise in the presence of delays. This in simple terms means that one needs to study a non-linear eigenvalue problem, particularly the locations of infinitely many eigenvalues on the complex plane as a function of the delay parameters in order to assess whether or not the inventory oscillations are stable or unstable. Details on the stability analysis is beyond the scope of this paper. We direct the interested readers to [6, 14, 18, 22, 26, 30].

4.2 Case Study – Supply Chain with Two Delays

In order to demonstrate the connection between operations research and the field of time delay systems, we present below a modified version of the mathematical model in [20]. In this model, the respective authors consider a pure delay h accounting for production lead times. The modification is regarding the transportation of products, which we model with another pure delay τ. This is a case study that considers non-negligible transportation times, for instance in the case of outsourcing.

The objective is then to calculate which lead times and transportation times in the SC causes desirable inventory behavior and what delay values are detrimental. The details of the model and the corresponding differential equation can be found in [29]. We give below the characteristic equation associated with the inventory dynamics,

$$f(s,h,\tau) = s + \alpha_S e^{-(\tau+h)s} + \alpha_{SL}(1 - e^{-hs}) = 0. \tag{12}$$

We remark that the complete stability analysis of the characteristic function above is not trivial due to presence of two delays τ and h. Starting from 1989s, various analytical techniques corresponding to necessary and sufficient conditions of stability in the delay parameter space have been developed, [6, 7, 26, 30].

Utilizing the ideas given in the work of [26], we compute the stability *regions* (gray shaded) of the inventory dynamics on the first quadrant of h versus τ plane, Figure 1, for the choice of $\alpha_S = 0.4$ and $\alpha_{SL} = 0.4$. In Figure 1, the shaded region is the stability region, while the remaining regions indicate instability. In this figure, along the h-axis, i.e. when $\tau = 0$, one recovers the study in [20], while for $\tau \neq 0$ the effects of transportation delays can be seen. It is important to state that the gray shaded region (the stability region) has an intricate geometry; when $h = 10$ and $\tau = 10$, the supply chain dynamics is stable, while *decreasing* transportation delay τ down to $\tau = 5$ will make the inventory oscillations *unstable*.

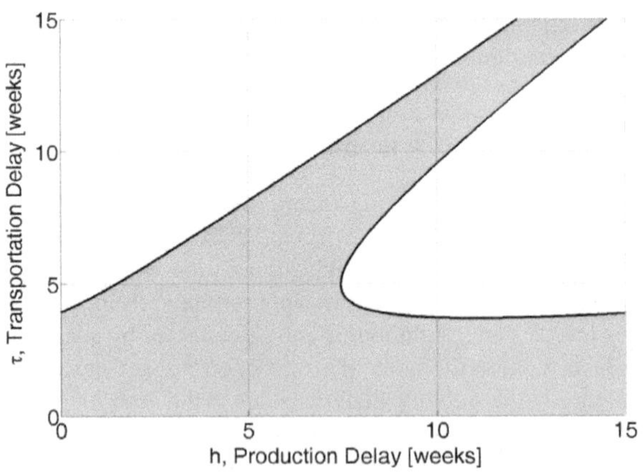

Fig. 1. Stability map of the inventory dynamics on the plane of production and transportation delays. Delays in the shaded regions lead to stable inventory behavior. Supply chain parameters are taken as $\alpha_S = \alpha_{SL} = 0.4$. Only the boundaries separating stable regions from the unstable regions are depicted.

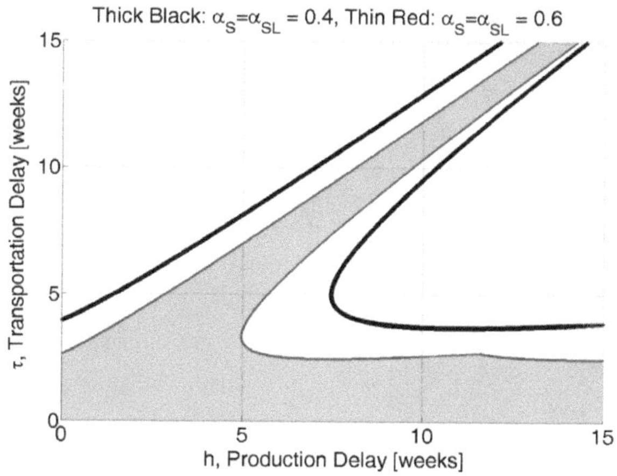

Fig. 2. Comparison of the stability maps for two different parameter settings. The first setting shown with thick black boundaries is found for the parameters $\alpha_S = \alpha_{SL} = 0.4$ and the second setting shown with thin red boundaries corresponds to the case of $\alpha_S = \alpha_{SL} = 0.6$. Only the boundaries separating stable regions from the unstable regions are depicted.

We next perform a comparison study between the case presented in Figure 1 (where $\alpha_S = \alpha_{SL} = 0.4$) and a second case where $\alpha_S = \alpha_{SL} = 0.6$, in order to demonstrate how stability regions are affected. The results are shown in Figure 2 where we observe that stability regions significantly shrink. This observation is consistent with the earlier results found for SC with a single delay [20].

Borrowing from [20], the inventories $i(t)$ are expressed as $\frac{di(t)}{dt} = O(t-h) - d(t)$, where $d(t)$ is the demand of the customers (part of the non-homogeneous terms). Deploying the non-homogenous part of the dynamics from the cited work, we present in Figure 3 and Figure 4 how inventories (where $i(0) = 1000$) behave in response to a step change in *demand* (50 units increase) for the two choices: (a) $\alpha_S = \alpha_{SL} = 0.4$ with $h = 5$ and transportation delay is ignored $\tau = 0$ or considered as $\tau = 5$, and (b) $\alpha_S = \alpha_{SL} = 0.6$ with production delay $h = 5.5$ and three different transportation delays, $\tau = 2$, $\tau = 4$, $\tau = 5$, respectively. The period of time for forecasting is chosen as $T = 10$ and expected delay is $\hat{h} = h$.

5 Conclusion

In this paper, we surveyed three important and widely utilized continuous time deterministic mathematical models of inventory dynamics in supply chains with delays arising from production lead times, material deliveries, information and decision lags and transportation times. These models enable a convenient platform to connect operations research with the time delay system community, and the models serve as the basis to study more complicated supply chain management problems. We presented a case study to demonstrate how one can achieve the connections across the disciplines, and how stability maps can prove to be useful for better managing supply chains.

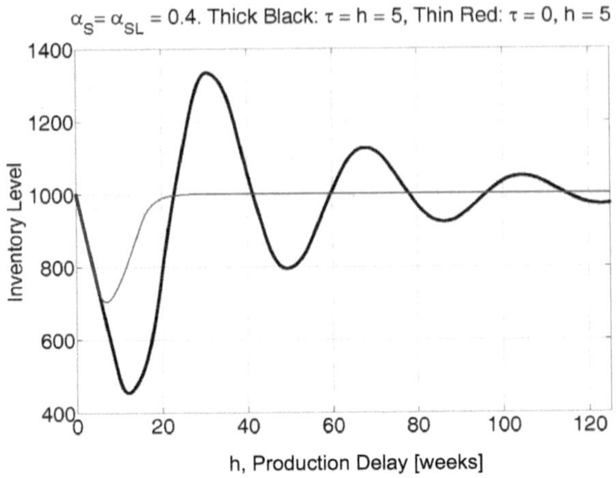

Fig. 3. Ignoring transportation delay τ may significantly affect inventory behavior as shown in the simulation. $\alpha_S = \alpha_{SL} = 0.4$ with $h = 5$ and transportation delay is ignored $\tau = 0$ (thin curve) or considered as $\tau = 5$ (thick curve).

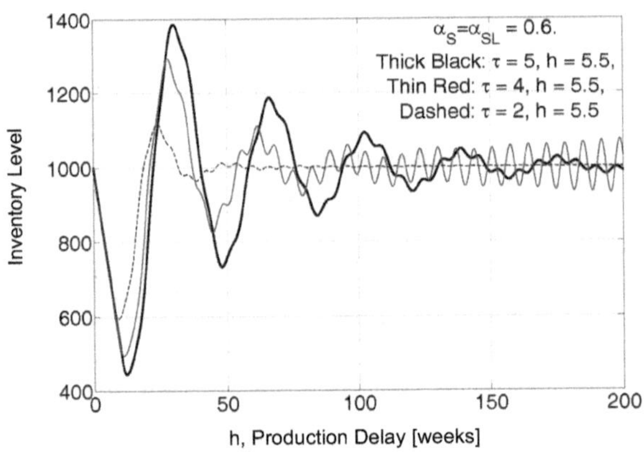

Fig. 4. Different transportation delays τ may result in dramatically different inventory behavior as shown in the simulation. $\alpha_S = \alpha_{SL} = 0.6$ with $h = 5.5$ and transportation delays are $\tau = 2$ (dashed curve), $\tau = 4$ (thin curve), $\tau = 5$ (thick curve).

Acknowledgements

Authors wish to thank Prof. Dirk Helbing at ETH Zurich; Prof. John Sterman at Sloan School of Management of the Massachusetts Institute of Technology, Boston; Prof Silviu-Iulian Niculescu at Laboratoire de Signaux et Systèmes (L2S), Supélec, France; Dr. Stefan Lämmer at Dresden University of Technology, Germany; Prof. Roger D.H. Warburton, Metropolitan

College, Boston University, Boston for fruitful discussions on mathematical modeling of supply networks and of delays in these networks. Prof. Sipahi acknowledges his start-up funds available at Northeastern University.

References

1. Agrawal, M., Kumaresh, T.V., Mercer, G.A.: The false promise of mass customization. In: The McKinsey Quarterly, p. 3 (2001)
2. Ceroni, J.A., Nof, S.Y.: Task parallelism in distributed supply organizations: a case study in the shoe industry. Production Planning & Control 16(5), 500–513 (2005)
3. Dejonckheere, J., Disney, S.M., Lambrecht, M.R., Towill, D.R.: Transfer function analysis of forecasting induced bullwhip in supply chain. International Journal of Production Economics 78, 133–144 (2002)
4. Delice, I.I., Sipahi, R.: Identifying Stability vs. Instability of Inventory Dynamics of Supply Chains in Presence of Delays. In: ASME IMECE Conference, Boston (2008)
5. Forrester, J.W.: Industrial Dynamics. MIT Press, Cambridge (1961)
6. Gu, K., Niculescu, S.I., Chen, J.: On Stability Crossing Curves for General Systems with Two Delays. Journal of Mathematical Analysis and Applications 311, 231–253 (2005)
7. Hale, J.K., Huang, W.: Global Geometry of the Stable Regions for Two Delay Differential Equations. Journal of Mathematical Analysis and Applications 178, 344–362 (1993)
8. Helbing, D., Armbuster, D., Mikhailov, A.S., Lefeber, E.: Information and Material Flows in Complex Networks. Physica A 363, xi–xvi (2006)
9. Helbing, D.: Modelling supply networks and business cycles as unstable transport phenomena. New Journal of Physics 5, 90.1 (2003)
10. Helbing, D., Lämmer, S., Witt, U., Brenner, T.: Network-induced oscillatory behavior in material flow networks and irregular business cycles. Physical Review E 70, 056118 (2004)
11. Helbing, D., Lämmer, S., Seidel, T., Šeba, P., Płatkowski, T.: Physics, Stability and Dynamics of Supply Networks. Physical Review E 70, 066116 (2004)
12. Kuang, Y.: Delay Differential Equations with Applications in Population Dynamics. Academic Press, San Diego (1993)
13. Marion, T.J., Sipahi, R., Delice, I.I.: Characterizing Stability of Inventories in Supply Chains with Delays in Early-Stage Firms. In: IEEE International Engineering Management Conference, Portugal (2008)
14. Michiels, W.: Stability and stabilization of time-delay systems. Ph.D. Thesis at Katholieke Universiteit Leuven, Belgium (2002)
15. Nagatani, T., Helbing, D.: Stability Analysis and Stabilization Strategies for Linear Supply Chains. Physica A 335, 644–660 (2004)
16. Niculescu, S.I.: Delay Effects on Stability: A Robust Control Approach. LNCIS, p. 269. Springer, Heidelberg (2001)
17. Niculescu, S.I., Gu, K. (eds.): Advances in Time-Delay Systems. Springer, Heidelberg (2004)
18. Olgac, N., Sipahi, R.: An exact method for the stability analysis of time-delayed LTI systems. IEEE Trans. Automatic Contrl. 47, 793–797 (2002)
19. Ren, W., Beard, R.W., Atkins, E.M.: A Survey of Consensus Problems in Multi-agent Coordination. In: American Control Conference, Portland, OR (2005)
20. Riddalls, C.E., Bennett, S.: The stability of supply chains. International Journal of Production Research 40(2), 459–475 (2002)

21. Riddalls, C.E., Bennett, S.: Production-inventory system controller design and supply chain dynamics. International Journal of Systems Science 33(3), 181–195 (2002)
22. Richard, J.P.: Time-Delay systems: an overview of some recent advances and open problems. Automatica 39, 1667–1694 (2003)
23. Sarimveis, H., Patrinos, P., Tarantilis, C.D., Kiranoudis, C.T.: Dynamic modeling and control of supply chain systems: A review. Computers and Operations Research 35(11), 3530–3561 (2008)
24. Simchi-Levi, D., Kaminsky, P., Simchi-Levi, E.: Designing & Managing the Supply Chain. McGraw-Hill, New York (2003)
25. Simon, H.A.: On the application of servomechanism theory in the study of production control. Econometrica 20(2), 247–268 (1952)
26. Sipahi, R., Olgac, N.: Complete Stability Robustness of Third-Order LTI Multiple Time-Delay Systems. Automatica 41, 1413–1422 (2005)
27. Sipahi, R., Lammer, S., Helbing, D., Niculescu, S.I.: On Stability Problems of Supply Networks Constrained with Transport Delays. ASME Journal on Dynamic Systems, Measurement and Control 131, 021005 (2009)
28. Sipahi, R., Atay, F., Niculescu, S.I.: Stability of Traffic Flow with Distributed Delays Modeling the Memory Effects of the Drivers. SIAM Applied Mathematics 68(3), 738–759 (2007)
29. Sipahi, R., Delice, I.I.: Delay Modeling in Supply Network Dynamics; Performance, Synchronization, Stability. In: 17th IFAC World Congress, Seoul, Korea (2008)
30. Stepan, G.: Retarded Dynamical Systems: Stability and Characteristic Function. Longman Scientific & Technical, New York; co-publisher John Wiley & Sons Inc., US (1989)
31. Sterman, J.D.: Business Dynamics. Systems Thinking and Modeling for a Complex World. McGraw-Hill, Boston (2000)
32. Sterman, J.D.: Modeling Managerial Behavior: Misperceptions of Feedback in a Dynamics Decision Making Experiment. Management Science 35(3), 321–339 (1989)
33. Sterman, J.D.: Modeling Managerial Behavior: Misperceptions of Feedback in a Dynamic Decision Making Experiment. Management Science 35(3), 321–339 (1989)
34. Treiber, M., Kesting, A., Helbing, D.: Delays, Inaccuracies and Anticipation in Microscopic Traffic Models. Physica A 360(1), 71–88 (2006)
35. Warburton, D.H.R.: An exact analytical solution to the production inventory control problem. International Journal of Production Economics 92(1), 81–96 (2004)
36. Siegele, L.: Chain Reaction: Managing a Supply Chain is Becoming a Bit Like Rocket Science. The Economist 362(8258), 13–15 (2002)
37. CIO Magazine, What Went Wrong at Cisco

Some Remarks on Delay Effects in Motion Synchronization in Shared Virtual Environments

Joono Cheong[1], Silviu-Iulian Niculescu[2], Yonghwan Oh[3], and Irinel Constantin Morărescu[4]

[1] Department of Control and Instrumentation Engineering, Korea University, Jochiwon, South Korea jncheong@korea.ac.kr
[2] L2S (UMR CNRS 8506), CNRS-Supélec,
3, rue Joliot Curie, 91192, Gif-sur-Yvette Silviu.Niculescu@lss.supelec.fr
[3] Bio-Mimetic Robotics Laboratory, Korea Institute of Science and Technology (KIST) oyh@kist.re.kr
[4] INRIA Rhône Alpes, BIPOP project
Inovalée, 655 avenue de l'Europe, 38330, Montbonnot, France
constantin.moraresc u@inrialpes.fr

Summary. This paper addresses the *motion synchronization* problem in shared virtual environments in the presence of *communication delays*. More precisely, we consider the case of multiple users interacting with the same dynamics. Unlike the conventional synchronization, the technological attempt we are interested in pursues a more *robust* and *better synchronization* that gives an almost concurrent evolution of motions between the distributed systems in absolute time-frame (earth's time). Physically, the existence of time delay prevents immediate information exchange, which disables concurrent motions between the distributed systems. Using the delay information available, the proposed controller preserves natural local dynamics and compensate for de-synchronization error caused by mismatched initial conditions. Simulation tests are conducted in order to validate the considered methodology.

1 Introduction

It is well-known that the interconnection of two or more dynamical systems leads to an increasing complexity of the overall system's behavior due to the effects induced by the emerging dynamics (in the presence or not of feedback loops) in strong interactions (sensing, communication) with the environment changes. Decision making in such systems is challenging and is subject to multiple competitive objectives. The development of technology in the last years is accompanied with increasing computing, sensing, communications in decision-making systems and processes. Among these systems, there exists a lot of examples where the *control* (or the decision) is based on the *information* changed and transferred between systems (units) or sub-systems (sub-units). As examples, we can cite: teleoperation, networked control systems (NCS), and shared virtual environment. Without any lack of generality, such systems are simply called *"information-based systems"*. Further details and various references on such topics can be found in [12].

J.J. Loiseau et al. (Eds.): Topics in Time Delay Systems, LNCIS 388, pp. 359–369.
springerlink.com

One of the major problems appearing in such information-based systems is related to the *propagation*, *transport*, and *communication delays* acting "through" and "inside" the interconnections. The *origin* of such *delays* can be: the explicit physical separation between the systems defining the interconnections, or the presence of the human factor in the decision process, or finally the presence of some hierarchy, and synchronization at the lowest levels in the decision process in real-time.

This paper addresses the analysis of *delay effects* in a particular class of *information-based interconnected systems*, namely the shared virtual environment simulation [1]. An extremely brief presentation of the synchronization methodology for these interconnected systems is proposed in Section 2. The construction of the controller and related closed-loop stability analysis are presented in Section 3. A particular attention will be paid to the sensitivity of the scheme with respect to the overall delay parameter (round-trip time). Some simulation results are illustrated in section 4, and finally, some concluding remarks end the paper.

2 Shared Virtual Environment Simulation and Delay

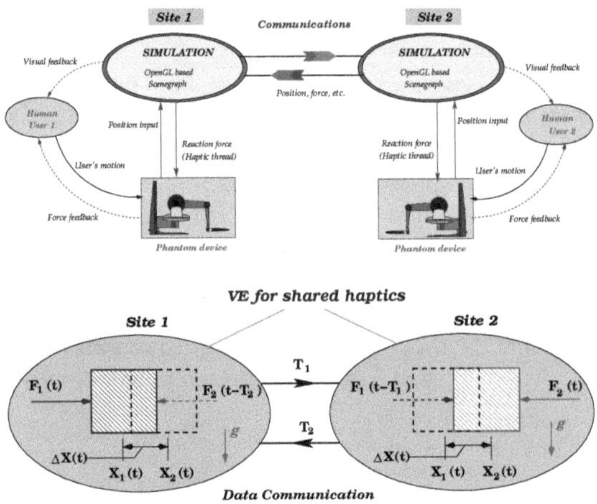

Fig. 1. Schematic of Shared virtual environment with haptic force input. (Up) Implementation of peer-to-peer shared virtual environment; (Down) Time delay and graphic de-synchronization [1].

Shared virtual environment requires synchronized visualization of virtual environment and real-time stable haptic interaction between separate users to carry out collaborative tasks in virtual assembly, CAD modeling, or medical training [15]. The use of communication networks complicates the task since we need to consider the *communication constraints*, and, in particular, the existence of communication time-delays. In the context of shared virtual environment applications, time-delay in the data communication becomes the most difficult part so as to meet synchronized visualization and immediate response from user interaction. Due to the time-delay, a change of a virtual environment in one site cannot be immediately displayed in the remote site, and de-synchronized graphic display between users may lead to unstable interaction between them [6]. Furthermore, in the case when users are interacting through mechanical haptic interface, the instability can cause damage to the device and the users also.

For illustration, consider a shared virtual environment with solid cube as shown in Fig.1. Two remote users at sites 1 and 2 are interacting with the cube at the same time. The *challenging problem* here is the *difficulty of synchronizing the virtual environment at both sites.* Because of the communication delays, the information of local force $F_1(t)$ of site 1 is available after some latency T_1, and similarly the information of local force $F_2(t)$ of site 2 is available after latency some T_2. Thus, the motion of the cube at site 1, or $X_1(t)$, is computed by the sequences of forces $F_1(t)$ and $F_2(t-T_2)$ following physics laws, while the motion of the cube at site 2, or $X_2(t)$, is computed by the sequences of $F_1(t-T_1)$ and $F_2(t)$. With no doubt, the motions of the two sites will be deviated as time goes on, because of the different input histories. Such a deviation results in distinct graphical scenes between users, and collaborative operation in the virtual environment is no longer possible. In the worst case, interaction forces between users can easily be out-of-phases, possibly causing a destructive interaction. Thus, a special care must be paid when dealing with shared interactive system in the presence of time-delay.

For synchronization of shared virtual environment in the *"TransAtlantic Haptic Project"* [7], a long distance haptic experiment was done, and a motion synchronization scheme was further combined to better achieve consistency between users, based on a feedback control using the so-called *Smith principle* [1]. The scheme took into consideration the possibility of delay variation, and the robustness bound of the variation was computed. The analysis result showed that, for large controller gain, the synchronization ability was nice but the tolerance level of delay variation was low and vice versa. However, direct user-to-user interaction became easily unstable under much smaller amount of delay like 150ms. And, though the analysis considered the variability of delay, Cheong *et al.* [1] did not preform real experimentation in environment of variable delay.

Some related works can be found in collaborative visualization and simulations. A method for collaborative visualization was addressed by [9] to present the synchronized view of the virtual environment. They first considered how interaction of dynamic object is perceived by the remote users and, accordingly, a trajectory was extrapolated using the information of motion velocity and polynomial based motion model. This work did not consider the possibility of delay variation contrary to the reality of Internet connections. Mauve *et al.*[10] proposed a local-lag technique to enhance the state synchronization. According to this method, the local simulation is held up for a amount of time (local delay) against local user's action, while the applied force information is to be immediately sent out to remote users. If the local delay is larger than the transmission delay, all sites can virtual have exactly the same force information for the simulation. A drawback is that it is hard to guarantee the sameness of simulations between sites when there is packet loss.

3 Motion synchronization in Delayed Media

3.1 Synchronization controller

The synchronization scheme addressed in this section is directly related to the shared virtual environment, where multiple users are interacting with the same dynamics. However, unlike the conventional synchronization, the technological attempt we are interested in pursues a more robust and better synchronization that gives an almost concurrent evolution of motions between the distributed systems in absolute time-frame (earth's time). Physically, the

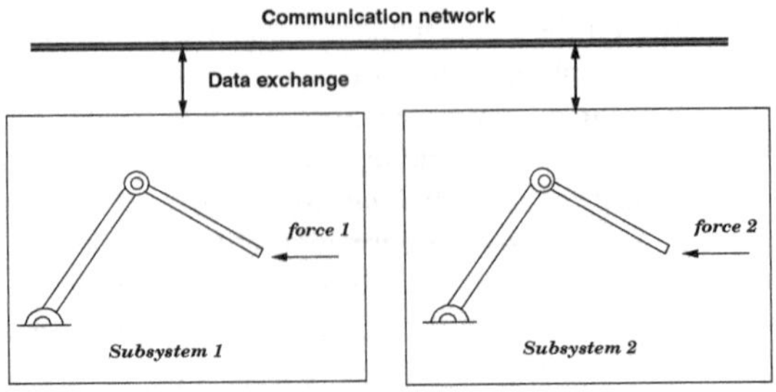

Fig. 2. Configuration of distributed systems via network communication

existence of time-delays prevents immediate information exchange, which disables concurrent motions between the distributed systems. For example, two mechanical systems shown in Fig.2 cannot yield concurrent evolution of motion because of the time-delay between them, while a strict motion synchronization (i.e., the concurrency) is very much necessary for a stable direct user-to-user interaction. Physically, the concurrent evolution of motion does not seem possible, but a sophisticated utilization of Smith principle, disturbance estimation and time-delay analysis in the communication channel, and optimized prediction of input sequences may overcome physical delays and allows a near concurrent evolution of motion between the systems.

First, we define two coupled, but distributed systems to be synchronized (shown in Fig.2), modeled as a simple rigid body with viscous damping as follows:

$$\begin{aligned} m\ddot{x}_1(t) + b\dot{x}_1(t) &= f_1(t) + f_2(t - T_2) \\ m\ddot{x}_2(t) + b\dot{x}_2(t) &= f_1(t - T_1) + f_2(t), \end{aligned} \tag{1}$$

where m and b are mass and damping coefficient of the systems, $x_1(t)$ and $x_2(t)$ denote positions of the systems in sites 1 and 2, respectively, and $f_1(t)$ and $f_2(t)$ are input forces acting on sites 1 and 2, respectively. Constant time-delays, T_1 and T_2, represent unidirectional delays for data communication from site 1 to site 2 and from site 2 to site 1, respectively. At this stage we assume that the delay is constant, but the later we deals with the effect of some variable but rather smooth delay. To overcome possible de-synchronization between the sites, we develop a motion synchronization controller which is of the structure shown in Fig.3. This is robust under data loss or any corruption during the communication since we feed back signals and continuously compensate for de-synchronization. This structure also shows the property that natural dynamics of the given system is not affected by the addition of motion control [2]. The controller, $C_i(s)$, in site i consists of some primary compensator, $K_i(s)$, that generates ultimate control command and the internal model of dynamics that produces estimated states. Two kinds of state estimations, that is, the current state and the state delayed by R time units, are generated through the internal dynamics with exact knowledge of dynamic parameters and time-delay, similar to the standard *Smith predictor* construction [16]. However, the structure is not just a copy of the conventional Smith predictor, but rather we utilize its principle so as to enforce exquisite timing between signals of feed-forward and feedback information. For example, our Smith principle is for the canceling feedback and reference input by utilizing the internal model with the knowledge of plant dynamics and amount of delay, with careful consideration of signal timing.

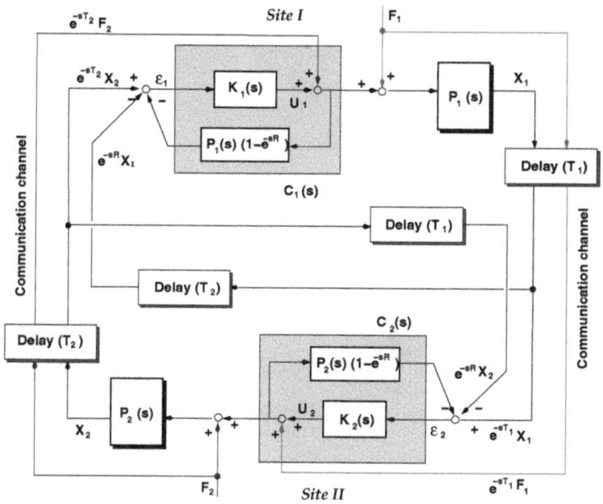

Fig. 3. Synchronization control scheme with two collaborators

To be more specific, two equations are obtained from the structure as:

$$\begin{aligned} u_i(s) &= K_i(s)\left(X_j(s)e^{-sT_j} - X_i(s)e^{-sR}\right) - \\ &\quad K_i(s)P_i(s)\left(1-e^{-sR}\right)\left(u_i(s)+F_j(s)e^{-sT_j}\right) \end{aligned} \tag{2}$$

$$X_i(s) = P(s)\left(u_i(s) + F_j(s)e^{-sT_j} + F_i(s)\right) \tag{3}$$

for $i, j = 1, 2,$ and $i \neq j$, and combining these yields to

$$\begin{aligned} &K(s)X_i(s) + P^{-1}(s)X_i(s) - K(s)X_j(s)e^{-sT_j} = \\ &F_i(s) + K(s)P(s)(1-e^{-sR})F_i(s) + F_j(s)e^{-sT_2}, \end{aligned} \tag{4}$$

by assuming $K_1(s) = K_2(s) = K(s)$ and $P_1(s) = P_2(s) = P(s)$. Be rewriting this in a matrix form after a simple matrix manipulation, we get

$$\begin{bmatrix} X_1 \\ X_2 \end{bmatrix} = \begin{bmatrix} P(s) & P(s)e^{-sT_2} \\ P(s)e^{-sT_1} & P(s) \end{bmatrix} \begin{bmatrix} F_1 \\ F_2 \end{bmatrix}. \tag{5}$$

As shown, the closed-loop system is exactly the same as the natural motion given in (1). This implies that the closed-loop coupled system follows the behavior as natural dynamics of the coupled system shows, while any disturbance effect during communication can be overcome.

This nice and strange property is due to the pole/zero cancelation of the following form of the quasi-polynomial factor:

$$\Phi(s,R) := \alpha(s) + \beta(s)e^{-sR} = \left\{P^{-1}(s) + K(s)\right\}^2 - K^2(s)e^{-sR}. \tag{6}$$

The stability condition of the whole system is, thus, equivalent to finding the *delay margin* R_m, for a given plant and appropriate controller parameters (see, for instance, the analysis suggested by [13]). However, the variation of delay creates uncertainty in the above quasi-polynomial and the stability bound in the worst case must be tremendously reduced.

For an illustration, refer to Fig.4, where numerical values of maximum allowable delay (i.e. delay margin) are computed for the plant $P(s) = 1/(s^2 + 0.01s)$ with different sets of control gains k_v and k_p. The results show that R_m tends to increase as the

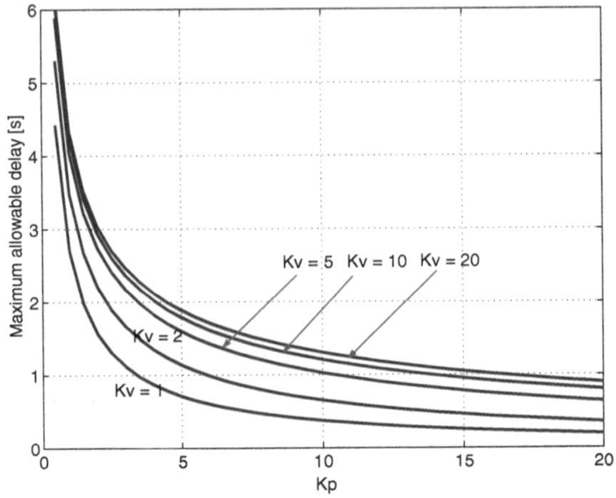

Fig. 4. Numerical values of R_m for $P(s) = 1/(s^2 + 0.01s)$ and $K(s) = k_v s + k_p$.

proportional gain becomes smaller and velocity gain becomes larger. It is well known that a linear time-invariant delay system is stable if and only if all the roots of its characteristic quasi-polynomial have negative real parts. According to the continuity properties of zeros with respect to the delay parameters [3] (see also [13], the number of roots in the right-half plane (RHP) can change only when some zeros appear and cross the imaginary axis. Thus, it is natural to consider the so-called *frequency crossing set* (see also, [11] and [5]) Ω consisting of all real positive ω (obviously, $\omega \in \Omega \Leftrightarrow -\omega \in \Omega$) such that there exist at least one pair (k_v, k_p) for which

$$H(j\omega, k_v, k_p, R) := \left(P^{-1}(j\omega) + K(j\omega)\right)^2 - K^2(j\omega)\mathrm{e}^{-j\omega R} = 0. \tag{7}$$

Using the modulus we arrive to:

$$\left|P^{-1}(j\omega) + K(j\omega)\right|^2 = \left|K^2(j\omega)\right|. \tag{8}$$

In conclusion, Ω consists of the values ω such that $\dfrac{P^{-1}(j\omega)}{K(j\omega)}$ belongs to the circle with radius 1 and centered in $(-1, 0)$. Next,

$$k_p = -\frac{|P^{-1}(j\omega)|^2 + 2k_v Im(P^{-1}(j\omega))}{2Re(P^{-1}(j\omega))}. \tag{9}$$

Remark 1. Since $\left|\dfrac{P^{-1}(j\omega)}{K(j\omega)} + 1\right| \to \infty$, when ω approaches ∞ one obtains that there exists some real $M > 0$ such that $\Omega \subset (0, M]$.

On the other hand, from equation (7) we can derive the following expression:

$$\angle(P^{-1}(j\omega)\bar{K}(j\omega) + |K(j\omega)|^2) = -\frac{\omega R}{2}, \tag{10}$$

where $\angle(z)$ denotes the argument of the complex number z. Next, using (9), one can replace k_p in (10) and thus, we obtain a simple polynomial equation of the form:

$$A(\omega)k_v^2 + B(\omega)k_v + C(\omega) = 0,$$

where

$$A(\omega) = 8\omega^2 \tan\frac{\omega R}{2},$$
$$B(\omega) = 8\omega\left(Im(P^{-1}(j\omega))\tan\frac{\omega R}{2} - Re(P^{-1}(j\omega))\right),$$
$$C(\omega) = 2\left|P^{-1}(j\omega)\right|^2 - 4Re^2(P^{-1}(j\omega))\tan\frac{\omega R}{2} - 4Re(P^{-1}(j\omega))Im(P^{-1}(j\omega))$$

are polynomial functions of ω and $\tan\frac{\omega R}{2}$. The corresponding expression of the k_v-gain is given by:

$$k_v = \frac{-\tan\frac{\omega R}{2} Im(P^{-1}(j\omega)) + Re(P^{-1}(j\omega))\left(1 \pm \sqrt{1+\tan^2\frac{\omega R}{2}}\right)}{2\omega\tan\frac{\omega R}{2}}.$$

When ω sweeps Ω the corresponding pair (k_v, k_p) defined by (9) and (10) moves on some *stability crossing curves*. Every time, the pair of controller parameters (k_v, k_p) crosses such a curve the number of characteristic roots in RHP changes. Finally, we note that for a fixed value of k_v when R increases, the stability bounds in terms of k_p become smaller. The same result is depicted in Fig.4.

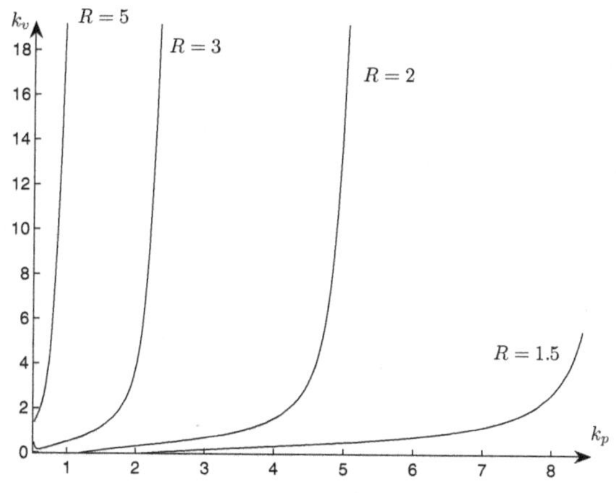

Fig. 5. The geometry of the crossing curves in the parameter space (k_v, k_p)

Now let us find out the effect of unreliability in the data communication. If there is a data loss in the network or disturbance in any form, subsidiary responses from these uncertainties are created and superimposed to the ideal response in (5), which makes unexpected de-synchronization. The proposed controller then compensates for the de-synchronization error, where we can observe the transient behavior that the characteristic modes of (6) are involved in. Assume that $f_j(t)$ is the input force at site j and $f_j(t - T_j)$ is the available information of $f_j(t)$ at site i with communication delay, T_j. If $f_j(t-T_j)$, for some reason, is disturbed to $\tilde{f}_j(t-T_j) = f_j(t-T_j) + \Delta f_j(t-T_j)$, where $\Delta f_j(t-T_j)$ is the amount of disturbance, the closed-loop input-output equation becomes

$$\begin{bmatrix} X_1 \\ X_2 \end{bmatrix} = \begin{bmatrix} P(s) & P(s)e^{-sT_2} \\ P(s)e^{-sT_1} & P(s) \end{bmatrix} \begin{bmatrix} F_1 \\ F_2 \end{bmatrix} + \begin{bmatrix} \frac{K(s)e^{-sT_2}}{\Phi(s,R)} & \frac{(P^{-1}(s)+K(s))}{\Phi(s,R)} \\ \frac{(P^{-1}(s)+K(s))}{\Phi(s,R)} & \frac{K(s)e^{-sT_1}}{\Phi(s,R)} \end{bmatrix} \cdot \begin{bmatrix} e^{-sT_1}\Delta F_1 \\ e^{-sT_2}\Delta F_2 \end{bmatrix} \tag{11}$$

where ΔF_j is the Laplace transform of $\Delta f_j(t)$. In the above, the output response for the transient and finite disturbance dies out whenever $\Delta f_j(t)$ is transient and $\Phi(s, R)$ of (6) is asymptotically stable. So, we can say $\Phi(s, R)$ is the *de facto* characteristic function.

Remark 2. To implement the synchronization controller using the network medium, we need a data packet having the following information fields:

```
PACKET = {
        Subsystem 1: time, state, force;
        Subsystem 2: time, state, force;
        }
```

where data field values of time, state, and force at a certain subsystem refer to the corresponding local values of the site. Packets of this simple form are continuously being sent and received via network.

4 Simulation study

A set of basic simulation is done to examine two elementary abilities of the proposed synchronization scheme: (i) to preserve natural local dynamics and (ii) to compensate for de-synchronization error caused by mismatched initial conditions. The considered system comprises two connected identical subsystems modeled as $P(s) = 1/(s^2 + 0.01s)$ with $T_1 = T_2 = 0.15$s, and the synchronization controller is designed following the structure shown in Fig.3 with $K(s) = 2s + 2$. Note that this is exactly the same example as in [1], but we are proposing in the sequel a further analysis with deeper understanding.

First, we assess the sameness between the uncontrolled natural response and the response with the proposed synchronization scheme to the sinusoidal forces given by

$$f_1(t) = \sin(t),$$
$$f_2(t) = \begin{cases} -\sin(0.4t+1) & 0 < t \le 32\text{s} \\ 0 & t > 32\text{s} \end{cases} .$$

We set up the system so that the initial condition is the same and no information loss occurs during data communication. According to the analysis, the controlled response must be the same as that of the uncontrolled natural response, which is verified in Fig.6.

Second, we simulate the case where both subsystems have different initial conditions such that

$$x_1(0) = 1,\ x_2(0) = 0, \text{ and } \dot{x}_1(0) = 0,\ \dot{x}_2(0) = 0,$$

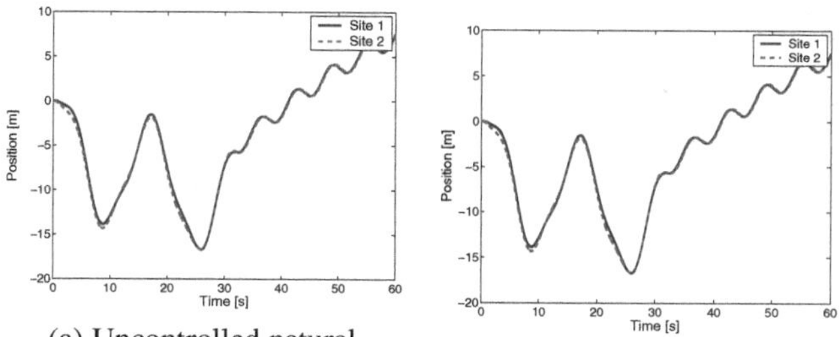

(a) Uncontrolled natural response

(b) Controlled response

Fig. 6. Comparison of natural and controlled responses

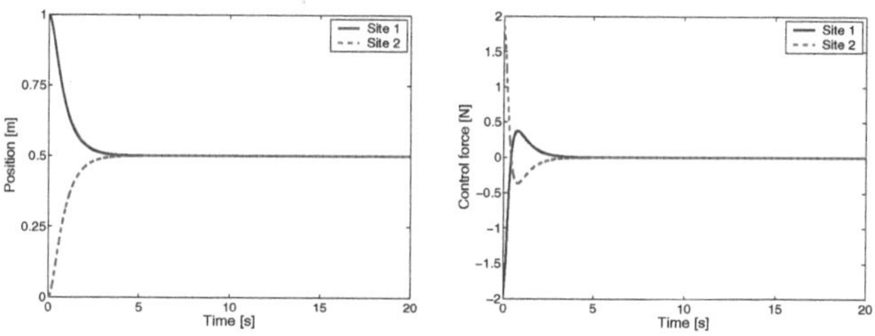

(a) Position

(b) Control input

Fig. 7. Controlled motion for different initial positions

and no external force is applied here. With no doubt, the natural response without control remains to the initial state, but in the controlled response, however, the differences of initial states are overcome and they become synchronized. Fig.7 shows the transient behavior of synchronization and the applied control forces for the case of controlled motion. The speed and shape of the transient behavior are governed by the roots of quasi-polynomial:

$$\Phi(s, R) = (s^2 + 2.01s + 2)^2 - (2s + 2)^2 e^{-0.3s} .$$

By applying a 5-th order Padé approximation [4] for delay $e^{-0.3s}$, we get nine closed-loop poles. Among them, the slowest mode, responsible for the sluggish behavior in Fig.7, is from the pole located at $s = -0.2417$. If k_p gain is increased to 4, the speed of response will be faster because the slowest mode becomes located at $s = -0.4822$. However a further increase brings some (other) pairs of complex poles near the imaginary axis, resulting in oscillatory response. A remedy for the oscillation is that the k_v-gain must be increased simultaneously together with the k_p-gain. In doing so, we make sure that, under the chosen gains, the maximum allow delay is sufficiently larger than the current amount of delay.

5 Concluding remarks

This chapter focused on a motion control scheme for synchronization of distributed subsystems connected via communication network. Remarkably, the scheme enables us to achieve the property of invariant local dynamics of each subsystem under the operation of feedback control. Due to this property, a near concurrent evolution of motion between subsystems was possible, even in the presence of communication time-delays in the network, by some particular appealing way of a combined utilization of the proposed scheme and an input prediction algorithm. The shared virtual environment system can be benefitted from the control scheme. Some illustrative example concluded our presentation.

Acknowledgements

The work of J. CHEONG and S.-I. NICULESCU was partially funded by the French-South Korean bilateral cooperation project PHC Star 16578ZC (2007-2008), entitled *Delay effects in information-based interconnected systems. Analysis, control and applications.*

References

1. Cheong, J., Niculescu, S.I., Annaswamy, A.M., Srinivasan, M.A.: Motion synchronization in virtual environments with shared haptics and large time delays. In: Proc. of Symp. on Haptic Interfaces for Virtual Environment and Teleoperator Systems, pp. 277–282 (2005)
2. Cheong, J., Lee, S., Kim, J.: Motion duplication control for coupled dynamic systems by natural damping. In: Proc. of IEEE Int. Conf. on Robotics and Automation, pp. 387–392 (2006)
3. Datko, R.: A procedure for determination of the exponential stability of certain differential-difference equations. Quart. Appl. Math. 36, 279–292 (1978)
4. Franklin, G.F., Powell, J.D., Emami-Naeini, A.: Feedback Control of Dynamic Systems. Addison-Wesley, Reading (1994)
5. Gu, K., Niculescu, S.I., Chen, J.: On stability crossing curves for general systems with two delays. J. Math. Anal. Appl. 311, 231–253 (2005)
6. Katz, A., Graham, K.: Dead Reckoning for Airplanes in Coordinated Flight. In: Proc. of the Tenth Workshop on Standards for the Interoperability of Defense Simulations, vol. II, pp. 5–13 (1994)
7. Kim, J., Kim, H., Manivannan, M., Srinivasan, M.A., Jordan, J., Mortensen, J., Oliveira, M., Slater, M.: Transatlantic Touch: A study of haptic collaboration over long distance. Presence: Teleoperators and Virtual Environments 13, 328–337 (2004)
8. Lawrence, D.A.: Stability and Transparency in Bilateral Teleoperation. IEEE Trans. on Robotics and Automation 9, 624–637 (1993)
9. Li, L.W.F., Li, F.W.B., Lau, R.W.H.: A Trajectory-Preserving Synchronization Method for Collaborative Visualization. IEEE Transactions on Visualization and Computer Graphics 12, 989–996 (2006)
10. Mauve, M., Vogel, J., Hilt, V., Effelsberg, W.: Local-lag and timewarp: providing consistency for replicated continuous applications. IEEE Trans. on Multimedia 6, 47–57 (2004)

11. Morărescu, C.I.: Qualitative analysis of distributed delay systems: Methodology and algorithms. Ph.D. thesis, University of Bucharest/Université de Technologie de Compiègne (2006)
12. Murray, R.M.: Control in information rich world, Report of the panel on future directions in control, dynamics and systems (2002)
13. Niculescu, S.I.: Delay effects on stability of time delay systems. A robust control approach. LNCIS, p. 269. Springer, Heidelberg (2001)
14. Rodriguez-Angeles, A., Nijmeijer, H.: Mutual synchronization of robots via estimated state feedback: a cooperative approach. IEEE Trans. on Control Systems Technology 12, 542–554 (2004)
15. Singhal, S., Zyda, M.: Networked Virtual Environments: Design and Implementation. Addison-Wesley, Reading (1999)
16. Smith, O.J.M.: Closer Control of Loops with Dead Time. Chem. Eng. Prog. 53, 217–219 (1957)
17. Taoutaou, D., Niculescu, S.I., Gu, K.: Closed-loop stability of a teleoperation control scheme subject to time-varying communication delays. In: Niculescu, S.I., Gu, K. (eds.) Advances in time-delay systems. LNCSE, vol. 38, pp. 327–338. Springer, Berlin (2004)

Control of Teleoperators with Time-Delay: A Lyapunov Approach

Emmanuel Nuño[1,3], Luis Basañez[1], and Romeo Ortega[2]

[1] Institute of Industrial and Control Engineering, Technical University of Catalonia. 08028 Barcelona, Spain. {emmanuel.nuno; luis.basanez}@upc.edu

[2] Laboratoire des Signaux et Systèmes at CNRS-SUPÉLEC. 91192 Gif-sur-Yvette, France. ortega@lss.supelec.fr

[3] Department of Computer Science. University of Guadalajara. Av. Revolución 1500, 44430 Guadalajara, Mexico.

Summary. This chapter aims to take a first step towards the unification of the stability analysis for teleoperators with time-delay. It proposes a *general* Lyapunov–like function that, upon slight modification, generates different control schemes, ranging from constant to variable time delays, with or without the scattering transformation and with or without position tracking. It also presents examples of the design of the corresponding functional for some control schemes based on PD and on the scattering transformation for variable time-delays.

1 Introduction

A teleoperator is commonly referred as the interconnection of five elements: a *human operator* that exerts force on a *local manipulator* connected through a *communication channel* to a *remote manipulator* that interacts with an *environment*. The application of such a system spans multiple fields, the most illustrative being space, underwater, medicine, and, in general, tasks with hazardous environments.

Communications often involve large distances or impose limited data transfer between the local and the remote sites. Such situations can result in substantial delays between the time a command is introduced by the operator and the time the command is executed by the remote manipulator. This time-delay affects the overall stability of the teleoperator. Anderson and Spong [1] presented the basis of the ubiquitous wave variables of Niemeyer and Slotine [2], which are the beginnings of a series of developments on bilateral teleoperators control (*cf.* [3] and [4] for two interesting surveys along this research line).

This work is intended to give a step further on the design of a suitable controller for teleoperators. It tries to unify the stability analysis of some group of controllers, providing a *general* Lyapunov–like function candidate that can be adapted to different control schemes, ranging from constant to variable time delays, with or without the scattering transformation and with or without position tracking.

A key observation here is that for some schemes, like the analyzed in this Chapter, the aforementioned function is *not* a Lyapunov function because its derivative cannot be shown to be nonpositive. However, the analysis can be completed via integration and some standard

J.J. Loiseau et al. (Eds.): Topics in Time Delay Systems, LNCIS 388, pp. 371–381.
springerlink.com

signal chasing. This novel procedure to prove stability, avoiding the nonpositive requirement, was first used in [5] in the context of scattering–based teleoperation and later used in [6] to prove, for the first time, stability of PD–based teleoperation.

Notation. $\mathbb{R} := (-\infty, \infty)$, $\mathbb{R}^+ := (0, \infty)$, $\mathbb{R}_0^+ := [0, \infty)$. $\lambda_m\{\mathbf{A}\}$ and $\lambda_M\{\mathbf{A}\}$ represent the minimum and maximum eigenvalue of the symmetric positive definite matrix $\mathbf{A}$, respectively. $|\cdot|$ stands for the Euclidean norm and $\|\cdot\|_2$ for the $\mathcal{L}_2$ norm. The argument of all functions will be omitted (*e.g.* $\dot{\mathbf{q}} \equiv \dot{\mathbf{q}}(t)$), except for those which appear delayed (*e.g.* $\dot{\mathbf{q}}(t - T(t))$). The argument of signals inside the integrals will be omitted, and it is supposed that is equal to the variable on the differential, unless otherwise noted (*e.g.* $\int_0^t \mathbf{x} d\sigma \equiv \int_0^t \mathbf{x}(\sigma) d\sigma$). The subscript i takes the values l and r for local and remote robot manipulators, respectively.

2 Mathematical Model of the Teleoperator

The local and remote manipulators together with the human and environment interactions, that conform the teleoperator, are modeled as a pair of n–DOF serial links with *revolute* joints. Their corresponding nonlinear dynamics are

$$\begin{aligned} \mathbf{M}_l(\mathbf{q}_l)\ddot{\mathbf{q}}_l + \mathbf{C}_l(\mathbf{q}_l, \dot{\mathbf{q}}_l)\dot{\mathbf{q}}_l + \mathbf{g}_l(\mathbf{q}_l) &= \boldsymbol{\tau}_l^* - \boldsymbol{\tau}_h \\ \mathbf{M}_r(\mathbf{q}_r)\ddot{\mathbf{q}}_r + \mathbf{C}_r(\mathbf{q}_r, \dot{\mathbf{q}}_r)\dot{\mathbf{q}}_r + \mathbf{g}_r(\mathbf{q}_r) &= \boldsymbol{\tau}_e - \boldsymbol{\tau}_r^*, \end{aligned} \tag{1}$$

where: $\ddot{\mathbf{q}}_i, \dot{\mathbf{q}}_i, \mathbf{q}_i \in \mathbb{R}^n$ are the joint acceleration, velocity and position; $\mathbf{M}_i(\mathbf{q}_i) \in \mathbb{R}^{n \times n}$ the inertia matrices; $\mathbf{C}_i(\mathbf{q}_i, \dot{\mathbf{q}}_i) \in \mathbb{R}^{n \times n}$ the Coriolis and centrifugal effects; $\mathbf{g}_i(\mathbf{q}_i) \in \mathbb{R}^n$ the gravitational forces; $\boldsymbol{\tau}_i^* \in \mathbb{R}^n$ the controllers; and $\boldsymbol{\tau}_h \in \mathbb{R}^n$, $\boldsymbol{\tau}_e \in \mathbb{R}^n$ the forces exerted by the human and the environment.

The robot manipulator's dynamical models in (1) have some important well known properties (*cf.* [7] and [8]).

P1. The inertia matrix for manipulators with revolute joints is lower and upper bounded. Thus, $0 < \lambda_m(\mathbf{M}_i(\mathbf{q}_i))\mathbf{I} \leq \mathbf{M}_i(\mathbf{q}_i) \leq \lambda_M(\mathbf{M}_i(\mathbf{q}_i))\mathbf{I} < \infty$.

P2. The Coriolis matrix $\mathbf{C}_i(\mathbf{q}_i, \dot{\mathbf{q}}_i)$ is defined with Christoffel symbols of the first-kind, and it is given by

$$C_i^{jk}(\mathbf{q}_i, \dot{\mathbf{q}}_i) = \sum_{m=1}^{n} \frac{1}{2}\left[\frac{\partial M_i^{jk}}{\partial q_i^m} + \frac{\partial M_i^{jm}}{\partial q_i^k} - \frac{\partial M_i^{km}}{\partial q_i^j}\right].$$

P3. The Coriolis and inertia matrices are related as

$$\mathbf{x}^\top[\dot{\mathbf{M}}_i(\mathbf{q}_i) - 2\mathbf{C}_i(\mathbf{q}_i, \dot{\mathbf{q}}_i)]\mathbf{x} = 0 \ \ \forall \mathbf{x} \in \mathbb{R}^n$$

and, $\dot{\mathbf{M}}_i(\mathbf{q}_i) = \mathbf{C}_i(\mathbf{q}_i, \dot{\mathbf{q}}_i) + \mathbf{C}_i^\top(\mathbf{q}_i, \dot{\mathbf{q}}_i)$.

P4. $\forall\ \mathbf{q}_i, \mathbf{x}, \mathbf{y} \in \mathbb{R}^n\ \exists k_{c_i} \in \mathbb{R}^+$ s.t. $|\mathbf{C}_i(\mathbf{q}_i, \mathbf{x})\mathbf{y}| \leq k_{c_i}|\mathbf{x}||\mathbf{y}|$ [9]. Consequently, $|\mathbf{C}_i(\mathbf{q}_i, \dot{\mathbf{q}})\dot{\mathbf{q}}| \leq k_{c_i}|\dot{\mathbf{q}}|^2$.

Proposition 1 (Boundedness of the Coriolis Matrix' Derivative). *Consider a robot manipulator with only*[4] *revolute joints. Assume that $\dot{\mathbf{q}}_i, \ddot{\mathbf{q}}_i \in \mathcal{L}_\infty$, then, the time derivative of its Coriolis matrix is bounded.*

[4] The restriction of *only* revolute joints can be relaxed for a certain manipulators class with revolute and prismatic joints (*cf.* [10]).

Proof. A sketch of the proof for this remark is as follows: the elements $\frac{d}{dt}C_i^{jk}(\mathbf{q}_i, \dot{\mathbf{q}}_i)$ are given by combinations of $\frac{\partial^2 M_i^{jk}(\mathbf{q}_i)}{\partial q_i^j \partial q_i^k}$ (*cf.* P2) which, for manipulators with revolute joints, are products of velocities, accelerations, $\alpha \sin(\cdot)$ and $\beta \cos(\cdot)$, where $\alpha, \beta \in \mathbb{R}^+$ are physical manipulator constants, i.e. link's mass distribution and length. Hence, if $\dot{\mathbf{q}}_i, \ddot{\mathbf{q}}_i \in \mathcal{L}_\infty$ the term $\frac{d}{dt}\mathbf{C}_i(\mathbf{q}_i, \dot{\mathbf{q}}_i)$ is bounded.

3 Preliminaries

Before going through the stability analysis some assumptions and Lemmas are outlined. The Lemmas are the cornerstones of the stability proofs, since, they provide the conditions for stability of the teleoperator.

3.1 General Assumptions

A1. The human operator and the environment define passive, force to velocity, maps, that is, $\exists\ \kappa_i \in \mathbb{R}_0^+$ s.t. $\forall t \geq 0$

$$\int_0^t \dot{\mathbf{q}}_l^\top \boldsymbol{\tau}_h d\sigma \geq -\kappa_l, \qquad -\int_0^t \dot{\mathbf{q}}_r^\top \boldsymbol{\tau}_e d\sigma \geq -\kappa_r. \tag{2}$$

A2. The gravitational forces are pre-compensated by the controllers $\boldsymbol{\tau}_i^*$, *i.e.*, $\boldsymbol{\tau}_l^* = \boldsymbol{\tau}_l + \mathbf{g}_l(\mathbf{q}_l)$ and $\boldsymbol{\tau}_r^* = \boldsymbol{\tau}_r - \mathbf{g}_r(\mathbf{q}_r)$. Thus, the dynamical models (1) become

$$\begin{aligned}\mathbf{M}_l(\mathbf{q}_l)\ddot{\mathbf{q}}_l + \mathbf{C}_l(\mathbf{q}_l, \dot{\mathbf{q}}_l)\dot{\mathbf{q}}_l &= \boldsymbol{\tau}_l - \boldsymbol{\tau}_h \\ \mathbf{M}_r(\mathbf{q}_r)\ddot{\mathbf{q}}_r + \mathbf{C}_r(\mathbf{q}_r, \dot{\mathbf{q}}_r)\dot{\mathbf{q}}_r &= \boldsymbol{\tau}_e - \boldsymbol{\tau}_r.\end{aligned} \tag{3}$$

A3. It is assumed that the variable time-delay has a known upper bound ${}^*T_i$. *i.e.* $T_i(t) \leq {}^*T_i < \infty$.

A4. The time derivative of the variable time-delay does not grow or decrease faster than time itself is considered. That is, $|\dot{T}_i(t)| < 1$.

3.2 Instrumental Lemmas

Lemma 1 (Teleoperator's Velocity Zero-convergence). *Consider the teleoperator in (3) when $\boldsymbol{\tau}_h = \boldsymbol{\tau}_e = \mathbf{0}$, written as*

$$\begin{aligned}\ddot{\mathbf{q}}_l &= -\mathbf{M}_l^{-1}(\mathbf{q}_l)[\mathbf{C}_l(\mathbf{q}_l, \dot{\mathbf{q}}_l)\dot{\mathbf{q}}_l - \boldsymbol{\tau}_l] \\ \ddot{\mathbf{q}}_r &= -\mathbf{M}_r^{-1}(\mathbf{q}_r)[\mathbf{C}_r(\mathbf{q}_r, \dot{\mathbf{q}}_r)\dot{\mathbf{q}}_r + \boldsymbol{\tau}_r].\end{aligned} \tag{4}$$

Assume that $\dot{\mathbf{q}}_i \in \mathcal{L}_\infty \cap \mathcal{L}_p\ \forall p \in (1, \infty)$ and $\boldsymbol{\tau}_i \in \mathcal{L}_\infty$. Then $|\dot{\mathbf{q}}_i| \to 0$ as $t \to \infty$.

Proof. Properties P1 and P4, and the fact that $\dot{\mathbf{q}}_i, \boldsymbol{\tau}_i \in \mathcal{L}_\infty$ ensures that $\ddot{\mathbf{q}}_i \in \mathcal{L}_\infty$. Thus, under the additional assumption that $\dot{\mathbf{q}}_i \in \mathcal{L}_p\ \forall p \in (1, \infty)$, Barbălat's Lemma [11] allows to prove the asymptotic zero-convergence of the teleoperator's velocities, *i.e.*, $|\dot{\mathbf{q}}_i| \to 0$ as $t \to \infty$. □

Lemma 2 (Teleoperator's Acceleration Zero-convergence). *Additionally to the assumptions in Lemma 1, suppose that the time derivatives of the local and remote controllers are bounded. i.e.,* $\dot{\boldsymbol{\tau}}_i \in \mathcal{L}_\infty$. *Under these conditions* $\ddot{\mathbf{q}}_i$ *in (4) are uniformly continuous, and* $|\ddot{\mathbf{q}}_i| \to 0$ *as* $t \to \infty$.

Proof. For the system (4), Lemma 1 establishes that if $\dot{\mathbf{q}}_i \in \mathcal{L}_\infty \cap \mathcal{L}_p \; \forall p \in (1, \infty)$ and $\boldsymbol{\tau}_i \in \mathcal{L}_\infty$. Then, $|\dot{\mathbf{q}}_i| \to 0$ as $t \to \infty$. If accelerations are proved to be uniformly continuous, using Barbălat's Lemma, it can be proved that $|\ddot{\mathbf{q}}_i| \to 0$. A sufficient condition for a differentiable function to be uniformly continuous is that its derivative be bounded (pp. 123 of [12]). Thus, the proof of this Lemma will be established if $\frac{d}{dt}\ddot{\mathbf{q}}_i \in \mathcal{L}_\infty$.

Differentiating (4) two types of terms are recovered: the first term consists of $\frac{d}{dt}\mathbf{M}_i^{-1}(\mathbf{q}_i)$ times a bounded term, and the second term, the product of $\mathbf{M}_i^{-1}(\mathbf{q}_i)$ times the time derivative of the term in brackets. $\frac{d}{dt}\mathbf{M}_i^{-1}(\mathbf{q}_i)$ with P3, yields $\frac{d}{dt}\mathbf{M}_i^{-1} = -\mathbf{M}_i^{-1}\dot{\mathbf{M}}_i\mathbf{M}_i^{-1} = -\mathbf{M}_i^{-1}\left[\mathbf{C}_i + \mathbf{C}_i^\top\right]\mathbf{M}_i^{-1}$. Properties P1, P4 and the boundedness of velocities, ensure that this term is bounded. Remark 1 and the assumption that $\dot{\boldsymbol{\tau}}_i \in \mathcal{L}_\infty$ allows to show that the time derivative of the term in brackets in (4), is also bounded. Consequently, $\frac{d}{dt}\ddot{\mathbf{q}}_i \in \mathcal{L}_\infty$ as required, and $\ddot{\mathbf{q}}_i$ in (4), are uniformly continuous. Continuity of these signals implies that its integral exists and is given by $\int_0^t \ddot{\mathbf{q}}_i(\sigma)d\sigma = \dot{\mathbf{q}}_i(t) - \dot{\mathbf{q}}_i(0)$. Taking the limit as $t \to \infty$ and using the fact that $|\dot{\mathbf{q}}_i| \to 0$ one gets $\int_0^\infty \ddot{\mathbf{q}}_i(\sigma)d\sigma = -\dot{\mathbf{q}}_i(0)$, which is clearly bounded. This completes the proof. □

The following Lemma is stated without proof, the interested reader can find it in [13].

Lemma 3 (Bound on Delayed Signals). $\forall\, \mathbf{x}, \mathbf{y} \in \mathbb{R}^n$, $\forall\, T(t)$ *s.t.* $0 \le T(t) \le {}^*T < \infty$ *and* $\alpha \in \mathbb{R}^+$, *the following inequality holds*

$$-\int_0^t \mathbf{x}^\top(\sigma) \int_{-T(\sigma)}^0 \mathbf{y}(\sigma + \theta)d\theta d\sigma \le \frac{\alpha}{2}\|\mathbf{x}\|_2^2 + \frac{{}^*T^2}{2\alpha}\|\mathbf{y}\|_2^2. \tag{5}$$

4 Lyapunov–like Function Candidate

This part of the Chapter focuses on presenting the *general* Lyapunov–like function candidate for nonlinear teleoperators with time-delay in the communications and then, in Section 5, use it to analyze the stability of some control schemes.

Let us propose a continuous and differentiable functional $V : \mathbb{R}^n \times \mathbb{R}^n \to \mathbb{R}$,

$$V(\mathbf{q}_i, \dot{\mathbf{q}}_i) = V_1(\mathbf{q}_i, \dot{\mathbf{q}}_i) + V_2(\dot{\mathbf{q}}_i) + V_3(\mathbf{q}_i, \dot{\mathbf{q}}_i). \tag{6}$$

The functional V has to be designed positive definite and radially unbounded. V_1 is the kinetic energy for the local and remote manipulators given by

$$V_1(\mathbf{q}_i, \dot{\mathbf{q}}_i) = \frac{\beta_l}{2}\dot{\mathbf{q}}_l^\top \mathbf{M}_l(\mathbf{q}_l)\dot{\mathbf{q}}_l + \frac{\beta_r}{2}\dot{\mathbf{q}}_r^\top \mathbf{M}_r(\mathbf{q}_r)\dot{\mathbf{q}}_r. \tag{7}$$

For any $\beta_i \in \mathbb{R}^+$, by the property P1, $V_1(\mathbf{q}_i, \dot{\mathbf{q}}_i)$ is positive definite, and is radially unbounded in $\dot{\mathbf{q}}_i$. Differentiating (7), using the skew-symmetric property of $\dot{\mathbf{M}}_i(\mathbf{q}_i) - 2\mathbf{C}_i(\mathbf{q}_i, \dot{\mathbf{q}}_i)$, evaluating it on the system trajectories (3), and doing some manipulations, yields

$$\dot{V}_1(\mathbf{q}_i, \dot{\mathbf{q}}_i) = -\beta_l\dot{\mathbf{q}}_l^\top[\boldsymbol{\tau}_h - \boldsymbol{\tau}_l] - \beta_r\dot{\mathbf{q}}_r^\top[\boldsymbol{\tau}_r - \boldsymbol{\tau}_e]. \tag{8}$$

$V_2(\dot{\mathbf{q}}_i) \in \mathbb{R}^+$ is the human and environment interactions with the local and remote manipulators, respectively, that is

$$V_2(\dot{\mathbf{q}}_i) = \beta_l \int_0^t \dot{\mathbf{q}}_l^\top \boldsymbol{\tau}_h d\sigma + \beta_l \kappa_l - \beta_r \int_0^t \dot{\mathbf{q}}_r^\top \boldsymbol{\tau}_e d\sigma + \beta_r \kappa_r > 0, \tag{9}$$

its time derivative is clearly given by

$$\dot{V}_2(\dot{\mathbf{q}}_i) = \beta_l \dot{\mathbf{q}}_l^\top \boldsymbol{\tau}_h - \beta_r \dot{\mathbf{q}}_r^\top \boldsymbol{\tau}_e, \tag{10}$$

The functional V_3 has to be designed for each control scheme using either a Lyapunov-Krasovskiĭ or a Lyapunov-Razumikhin functional, and it is precisely the selection of this function that allows to prove the stability of the control scheme —the details will be spelled out in the next Section.

5 Control Schemes

The control schemes analyzed in this work belong to the class of controllers that provide position tracking for constant or variable time-delays.

5.1 PD+d Controller

A Proportional Derivative plus damping controller, for variable time-delays, is given by

$$\begin{aligned} \boldsymbol{\tau}_l &= K_d[\gamma_r \dot{\mathbf{q}}_r(t - T_r(t)) - \dot{\mathbf{q}}_l] + K_l[\mathbf{q}_r(t - T_r(t)) - \mathbf{q}_l] - B_l \dot{\mathbf{q}}_l \\ \boldsymbol{\tau}_r &= K_d[\dot{\mathbf{q}}_r - \gamma_l \dot{\mathbf{q}}_l(t - T_l(t))] + K_r[\mathbf{q}_r - \mathbf{q}_l(t - T_l(t))] + B_r \dot{\mathbf{q}}_r, \end{aligned} \tag{11}$$

where $K_d, K_i, B_i \in \mathbb{R}^+$ and $\gamma_i(t) : \mathbb{R}_0^+ \to \mathbb{R}^+$ defined as $\gamma_i^2 = 1 - \dot{T}_i(t)$. Correspondingly, for constant time-delays they simplify to

$$\begin{aligned} \boldsymbol{\tau}_l &= K_d[\dot{\mathbf{q}}_r(t - T_r) - \dot{\mathbf{q}}_l] + K_l[\mathbf{q}_r(t - T_r) - \mathbf{q}_l] - B_l \dot{\mathbf{q}}_l \\ \boldsymbol{\tau}_r &= K_d[\dot{\mathbf{q}}_r - \dot{\mathbf{q}}_l(t - T_l)] + K_r[\mathbf{q}_r - \mathbf{q}_l(t - T_l)] + B_r \dot{\mathbf{q}}_r, \end{aligned} \tag{12}$$

with $K_d, K_i, B_i \in \mathbb{R}^+$.

The Lyapunov-like functional V_3 for this scheme is given by

$$V_3 = \frac{K_l}{2}|\mathbf{q}_l - \mathbf{q}_r|^2 + \frac{K_d K_l}{2K_r} \int_{t-T_l(t)}^t |\dot{\mathbf{q}}_l(\theta)|^2 d\theta + \frac{K_d}{2} \int_{t-T_r(t)}^t |\dot{\mathbf{q}}_r(\theta)|^2 d\theta. \tag{13}$$

Proposition 2 (PD+d Controllers for Variable Time-delays [13]). *Consider the system (3) controlled by (11). Set the control gains as*

$$4B_l B_r > ({}^*T_l^2 + {}^*T_r^2) K_l K_r, \tag{14}$$

with any arbitrary positive K_d, $B_r \geq B_l$ and $K_r \geq K_l$. Then, under assumptions A1–A4,

a) velocities and position error are bounded i.e. $\dot{\mathbf{q}}_i \in \mathcal{L}_\infty \cap \mathcal{L}_2$ $\mathbf{q}_l - \mathbf{q}_r \in \mathcal{L}_\infty$,
b) in the case that the human does not move the local manipulator and the remote manipulator does not touch anything, i.e. $\boldsymbol{\tau}_h = \boldsymbol{\tau}_e = \mathbf{0}$, velocities asymptotically converge to zero and position tracking is achieved:

$$|\dot{\mathbf{q}}_i| \to 0 \quad |\mathbf{q}_l - \mathbf{q}_r(t - T_r(t))| \to 0 \quad t \to \infty.$$

Proof. Set the constants $\beta_l = 1$ and $\beta_r = \frac{K_l}{K_r}$ for V_1 and V_2 in (7) and (9), respectively. Which together with the functional (13) make V in (6) positive definite and radially unbounded. The time derivative of V_3 is given by

$$\begin{aligned}\dot{V}_3 = \quad & K_l[\mathbf{q}_l - \mathbf{q}_r]^\top[\dot{\mathbf{q}}_l - \dot{\mathbf{q}}_r] + \frac{K_d K_l}{2K_r}[|\dot{\mathbf{q}}_l|^2 - \gamma_l^2|\dot{\mathbf{q}}_l(t - T_l(t))|^2] + \\ & + \frac{K_d}{2}[|\dot{\mathbf{q}}_r|^2 - \gamma_r^2|\dot{\mathbf{q}}_r(t - T_r(t))|^2].\end{aligned}$$

Substituting the control laws (11) on (8) plus (10) ($\dot{V}_1$ plus $\dot{V}_2$) and the bounds $2\dot{\mathbf{q}}_l^\top\gamma_r\dot{\mathbf{q}}_r(t - T_r(t)) \leq |\dot{\mathbf{q}}_l|^2 + \gamma_r^2|\dot{\mathbf{q}}_r(t - T_r(t))|^2$ and $2\dot{\mathbf{q}}_r^\top\gamma_l\dot{\mathbf{q}}_l(t - T_l(t)) \leq |\dot{\mathbf{q}}_r|^2 + \gamma_l^2|\dot{\mathbf{q}}_l(t - T_l(t))|^2$, and using the transformation

$$\mathbf{q}_i - \mathbf{q}_i(t - T_i(t)) = \int_{-T_i(t)}^{0} \dot{\mathbf{q}}_i(t + \theta)d\theta, \tag{15}$$

yields

$$\begin{aligned}\dot{V} \leq -& \left[B_l - \frac{K_d}{2}\left(\frac{K_l}{K_r} - 1\right)\right]|\dot{\mathbf{q}}_l|^2 - K_l\dot{\mathbf{q}}_l^\top\int_{-T_r(t)}^{0}\mathbf{q}_r(t+\theta)d\theta - \\ -& \left[B_r - \frac{K_d}{2}\left(\frac{K_l}{K_r} - 1\right)\right]|\dot{\mathbf{q}}_r|^2 - K_l\dot{\mathbf{q}}_r^\top\int_{-T_r(t)}^{0}\mathbf{q}_l(t+\theta)d\theta.\end{aligned} \tag{16}$$

Integrating from 0 to t and invoking Lemma 3 to the integral signals, returns

$$\begin{aligned}V(t) - V(0) \leq -& \left[B_l - \frac{K_d}{2}\left(\frac{K_l}{K_r} - 1\right) - \frac{K_l}{2}\left(\alpha_l + \frac{{}^*T_l^2}{\alpha_r}\right)\right]\|\dot{\mathbf{q}}_l\|_2^2 - \\ -& \left[\frac{K_l B_r}{K_r} - \frac{K_d}{2}\left(\frac{K_l}{K_r} - 1\right) - \frac{K_l}{2}\left(\alpha_r + \frac{{}^*T_r^2}{\alpha_l}\right)\right]\|\dot{\mathbf{q}}_r\|_2^2.\end{aligned} \tag{17}$$

If the terms inside brackets are positive for $\alpha_i > 0$, then $\exists\lambda_i > 0$ s.t.

$$-V(0) \leq -\lambda_l\|\dot{\mathbf{q}}_l\|_2^2 - \lambda_r\|\dot{\mathbf{q}}_r\|_2^2,$$

thus, $\{\dot{\mathbf{q}}_l, \dot{\mathbf{q}}_r\} \in \mathcal{L}_2$. Therefore, solving for positive α_i it is possible to find that, for any arbitrary positive K_d, $K_r \geq K_l$ and $B_r \geq B_l$, there will exist a solution if $4B_lB_r > ({}^*T_l^2 + {}^*T_r^2)K_lK_r$. Hence, setting the control gains as (14) will ensure that $\{\dot{\mathbf{q}}_l, \dot{\mathbf{q}}_r\} \in \mathcal{L}_2$. Positive definiteness of V and the property P1, sustain $\{\dot{\mathbf{q}}_l, \dot{\mathbf{q}}_r, \mathbf{q}_l - \mathbf{q}_r\} \in \mathcal{L}_\infty$. This completes the proof of part a).

The method to establish the proof of part b) is as follows: first it is proved that $\boldsymbol{\tau}_i \in \mathcal{L}_\infty$, and when $\boldsymbol{\tau}_h = \boldsymbol{\tau}_e = \mathbf{0}$ Lemma 1 proves the zero-convergence of velocities. Then, it is proved that $\dot{\boldsymbol{\tau}}_i \in \mathcal{L}_\infty$ allowing to invoke Lemma 2 to establish the zero-convergence of accelerations and ensure position tracking.

Towards this end, it has been shown that $\dot{\mathbf{q}}_i \in \mathcal{L}_\infty \cap \mathcal{L}_2$ and $\mathbf{q}_l - \mathbf{q}_r \in \mathcal{L}_\infty$. Rewriting $\mathbf{q}_l - \mathbf{q}_r(t - T_r(t))$ as $\mathbf{q}_l - \mathbf{q}_r(t - T_r(t)) = \mathbf{q}_l - \mathbf{q}_r + \mathbf{q}_r - \mathbf{q}_r(t - T_r(t))$ and the bound

$$\mathbf{q}_r - \mathbf{q}_r(t - T_r(t)) = \int_0^{T(t)}\dot{\mathbf{q}}_r(t - \theta)d\theta \leq {}^*T_r^{\frac{1}{2}}\|\dot{\mathbf{q}}_r\|_2 \tag{18}$$

(Schwartz), it is concluded that $\mathbf{q}_l - \mathbf{q}_r(t - T_r(t)) \in \mathcal{L}_\infty$. It can also be proved that $\mathbf{q}_r - \mathbf{q}_l(t - T_l(t)) \in \mathcal{L}_\infty$. Assumption A4 allows to state that $0 < \gamma_i^2 = 1 - \dot{T}_i(t) < 2$, hence,

$\gamma_i \in \mathcal{L}_\infty$. This is sufficient to state that $\boldsymbol{\tau}_i \in \mathcal{L}_\infty$. Thus, Lemma 1 guarantees that $|\dot{\mathbf{q}}_i| \to 0$ as $t \to \infty$.

In order to establish $\dot{\boldsymbol{\tau}}_i \in \mathcal{L}_\infty$, it suffices to proof that $\dot{\gamma}_i \in \mathcal{L}_\infty$, due to boundedness of $\dot{\mathbf{q}}_i$ and $\ddot{\mathbf{q}}_i$. $\dot{\gamma}_i = \frac{-\ddot{T}_i(t)}{[1-\dot{T}_i(t)]^{\frac{1}{2}}}$ is clearly bounded by assumption A4 and the piecewise continuity of $T_i(t)$. Lemma 2 supports the zero-convergence of accelerations. Rewriting (4) with the PD controllers as

$$\begin{aligned}\ddot{\mathbf{q}}_l &= -\mathbf{M}_l^{-1}([\mathbf{C}_l + B_l]\dot{\mathbf{q}}_l + K_d[\dot{\mathbf{q}}_l - \gamma_r\dot{\mathbf{q}}_r(t-T_r(t))] + K_l[\mathbf{q}_l - \mathbf{q}_r(t-T_r(t))])\\ \ddot{\mathbf{q}}_r &= -\mathbf{M}_r^{-1}([\mathbf{C}_r + B_r]\dot{\mathbf{q}}_r + K_d[\dot{\mathbf{q}}_r - \gamma_l\dot{\mathbf{q}}_l(t-T_l(t))] + K_r[\mathbf{q}_r - \mathbf{q}_l(t-T_l(t))]),\end{aligned}$$

is obvious that when $|\ddot{\mathbf{q}}_i| \to |\dot{\mathbf{q}}_i| \to 0$ as $t \to \infty$ then $|\mathbf{q}_l - \mathbf{q}_r(t-T_r(t))| \to 0$. This completes the proof. □

Proposition 3 (PD+d for Constant Time-Delays). *Conclusions a) and b) in Proposition 2 hold, if its controllers are replaced by (12) for constant time-delays with* $T_i(t) = constant$.

The interested reader can found its proof in [6]. It follows *verbatim* the proof of Proposition 2, with

$$V_3 = \frac{K_l}{2}|\mathbf{q}_l - \mathbf{q}_r|^2 + \frac{K_d K_l}{2K_r}\int_{t-T_l}^{t} |\dot{\mathbf{q}}_l(\theta)|^2 d\theta + \frac{K_d}{2}\int_{t-T_r}^{t} |\dot{\mathbf{q}}_r(\theta)|^2 d\theta.$$

5.2 P+d Controller

In this scheme the forces applied by the controllers on both, the local and the remote manipulators, are proportional to their position errors plus a damping injection term. These control laws are easier to implement and analyze compared with the PD+d laws.

Proposition 4 (P+d Controller for Variable Time-delays [13]). *Conclusions a) and b) of Proposition 2 hold, if its controllers (11), are replaced by*

$$\begin{aligned}\boldsymbol{\tau}_l &= K_l[\mathbf{q}_r(t-T_r(t)) - \mathbf{q}_l] - B_l\dot{\mathbf{q}}_l\\ \boldsymbol{\tau}_r &= K_r[\mathbf{q}_r - \mathbf{q}_l(t-T_l(t))] + B_r\dot{\mathbf{q}}_r,\end{aligned} \tag{19}$$

for $\{K_i, B_i\} \in \mathbb{R}^+$, *and constant or variable time-delays* $T_i(t)$ *fulfilling condition (14).*

Proof. Take the Lyapunov-Razumikhin functional V_3, as

$$V_3 = \frac{K_l}{2}|\mathbf{q}_l - \mathbf{q}_r|^2, \tag{20}$$

set the constants $\beta_l = 1$ and $\beta_r = \frac{K_l}{K_r}$, which provide a positive definite and radially unbounded function V in (6). Its time derivative along the system trajectories, defined by (3) and (19) with the property P3, is given by

$$\dot{V} = -B_l|\dot{\mathbf{q}}_l|^2 - \frac{K_l B_r}{K_r}|\dot{\mathbf{q}}_r|^2 + K_l\dot{\mathbf{q}}_l^\top[\mathbf{q}_r(t-T_r(t)) - \mathbf{q}_r] + K_l\dot{\mathbf{q}}_r^\top[\mathbf{q}_l(t-T_l(t)) - \mathbf{q}_l]$$

Using expression (15) for the terms inside brackets, after integration from 0 to t, and invoking Lemma 3 on the delayed signals, yields

$$V(t) - V(0) \leq -\left[B_l - \frac{K_l}{2}\left(\alpha_l + \frac{{}^*T_l^2}{\alpha_r}\right)\right]\|\dot{\mathbf{q}}_l\|_2^2 - \left[\frac{K_l B_r}{K_r} - \frac{K_l}{2}\left(\alpha_r + \frac{{}^*T_r^2}{\alpha_l}\right)\right]\|\dot{\mathbf{q}}_r\|_2^2.$$

Solving for positive α_i it is found that if $4B_lB_r > ({}^*T_l^2 + {}^*T_r^2)K_lK_r$ is satisfied then $\dot{\mathbf{q}}_i \in \mathcal{L}_2$. Boundedness of V together with property P1 imply that $\{\dot{\mathbf{q}}_i, \mathbf{q}_l - \mathbf{q}_r\} \in \mathcal{L}_\infty$, which completes the proof of part a).

The proof of part b) follows the same method that in Proposition 2. For this control scheme boundedness of $\boldsymbol{\tau}_i$ and $\dot{\boldsymbol{\tau}}_i$ is implied by $\{\dot{\mathbf{q}}_i, \mathbf{q}_l - \mathbf{q}_r\} \in \mathcal{L}_\infty$ and (18). Lemma 1 proves zero-convergence of velocities and Lemma 2 supports the zero-convergence of accelerations. Therefore, when $|\ddot{\mathbf{q}}_i| \to |\dot{\mathbf{q}}_i| \to 0$ as $t \to \infty$, position tracking is achieved. The case of constant time-delays follows *verbatim* this proof with $T_i(t) = constant$. □

5.3 Scattering-based Controller

The scattering transformation [2] is given by

$$\begin{aligned} \mathbf{u}_l &= \tfrac{1}{\sqrt{2b}}[\boldsymbol{\tau}_{ld} - b\dot{\mathbf{q}}_{ld}] & \mathbf{u}_r &= \tfrac{1}{\sqrt{2b}}[\boldsymbol{\tau}_{rd} - b\dot{\mathbf{q}}_{rd}] \\ \mathbf{v}_l &= \tfrac{1}{\sqrt{2b}}[\boldsymbol{\tau}_{ld} + b\dot{\mathbf{q}}_{ld}] & \mathbf{v}_r &= \tfrac{1}{\sqrt{2b}}[\boldsymbol{\tau}_{rd} + b\dot{\mathbf{q}}_{rd}] \end{aligned} \tag{21}$$

where b is the *virtual* impedance of the transmission line and the subscript id means local or remote desired signals. For constant time-delays (T_l and T_r) in the forward and the backward paths, respectively, the local and the remote manipulators are interconnected with

$$\mathbf{u}_l = \mathbf{u}_r(t - T_l) \quad \mathbf{v}_l = \mathbf{v}_r(t - T_r) \tag{22}$$

Using transformations (21) and (22), it has been shown that the total energy in the communications is stored within signal transmissions as

$$E_{comm} = \frac{1}{2}\int_{t-T_l}^{t} |\mathbf{u}_l|^2 d\sigma + \frac{1}{2}\int_{t-T_r}^{t} |\mathbf{v}_r|^2 d\sigma \geq 0 \tag{23}$$

which means that the communications are passive for any constant time-delay. However, for variable time-delays this statement does not longer apply, and the communications are, in general, not passive. The use of a time varying gain γ_i, in the local and remote interconnection, as

$$\mathbf{u}_r = \gamma_l \mathbf{u}_l(t - T_l(t)) \quad \mathbf{v}_l = \gamma_r \mathbf{v}_r(t - T_r(t)) \tag{24}$$

where $\gamma_i^2 \leq 1 - \dot{T}_i(t)$, provides a positive definite energy storage function for the communications, that is

$$\begin{aligned} E_{comm} = &\ \frac{1}{2}\int_{t-T_l(t)}^{t} |\mathbf{u}_l|^2 d\sigma + \frac{1 - \dot{T}_l(t) - \gamma_l^2}{2 - 2\dot{T}_l(t)} \int_0^{t-T_l(t)} |\mathbf{u}_l|^2 d\sigma + \\ &+ \frac{1}{2}\int_{t-T_r(t)}^{t} |\mathbf{v}_r|^2 d\sigma + \frac{1 - \dot{T}_r(t) - \gamma_r^2}{2 - 2\dot{T}_r(t)} \int_0^{t-T_r(t)} |\mathbf{v}_r|^2 d\sigma. \end{aligned} \tag{25}$$

This means that choosing an appropriate function γ_i, the communications will not generate energy. If $\gamma_i^2 = 1 - \dot{T}_i(t)$, (25) transforms to (23).

The following Proposition presents a scattering based controller that provides position tracking in the presence of variable time-delays. Its proof is established using the functional

$$V_3 = \frac{K_l}{2}|\mathbf{q}_l - \mathbf{q}_r|^2 + \int_0^t (\beta_r \dot{\mathbf{q}}_{rd}^\top \boldsymbol{\tau}_{rd} - \beta_l \dot{\mathbf{q}}_{ld}^\top \boldsymbol{\tau}_{ld}) d\sigma. \tag{26}$$

Proposition 5 (Scattering-based Control Scheme [13]). *Consider the system (3) controlled by*

$$\begin{aligned} \boldsymbol{\tau}_l &= \boldsymbol{\tau}_{ld} + K_l[\mathbf{q}_r(t - T_r(t)) - \mathbf{q}_l] - B_l\dot{\mathbf{q}}_l \\ \boldsymbol{\tau}_r &= \boldsymbol{\tau}_{rd} + K_r[\mathbf{q}_r - \mathbf{q}_l(t - T_l(t))] + B_r\dot{\mathbf{q}}_r, \end{aligned} \tag{27}$$

where $\boldsymbol{\tau}_{ld} = -K_{dl}[\dot{\mathbf{q}}_l - \dot{\mathbf{q}}_{ld}]$ *and* $\boldsymbol{\tau}_{rd} = K_{dr}[\dot{\mathbf{q}}_r - \dot{\mathbf{q}}_{rd}]$*, the desired velocities are codified with the scattering transformation (21) and (24) with* $\gamma_i^2 = 1 - \dot{T}_i(t)$*. For any* $K_{di} \in \mathbb{R}^+$ *set the control gains* $\{K_i, B_i\} \in \mathbb{R}^+$ *as*

$$4B_lB_r > ({}^*T_l^2 + {}^*T_r^2)K_lK_r. \tag{28}$$

Then, under assumptions A1–A4,

a) $\dot{\mathbf{q}}_i \in \mathcal{L}_\infty \cap \mathcal{L}_2$ $\mathbf{q}_l - \mathbf{q}_r \in \mathcal{L}_\infty$,
b) if additionally, $\boldsymbol{\tau}_h = \boldsymbol{\tau}_e = \mathbf{0}$*, then*

$$|\dot{\mathbf{q}}_i| \to 0 \qquad |\mathbf{q}_l - \mathbf{q}_r(t - T_r(t))| \to 0 \qquad t \to \infty.$$

Proof. The scattering transformation and $\gamma_i^2 = 1 - \dot{T}_i(t)$ ensure that the integral term in (26) is positive definite (*cf.* (25)), which together with assumption A1, property P1 and constants $\beta_l = 1$, $\beta_r = \frac{K_l}{K_r}$, prove that the general Lyapunov function candidate V in (6) is positive definite and radially unbounded for variable time-delays. Using property P3, its time derivative along the system trajectories (3) and (27) is given by

$$\begin{aligned} \dot{V} = &- B_l|\dot{\mathbf{q}}_l|^2 - \frac{B_rK_l}{K_r}|\dot{\mathbf{q}}_r|^2 - K_{dl}|\dot{\mathbf{q}}_l - \dot{\mathbf{q}}_{ld}|^2 - \frac{K_{dr}K_l}{K_r}|\dot{\mathbf{q}}_r - \dot{\mathbf{q}}_{rd}|^2 - \\ &- K_l\dot{\mathbf{q}}_l^\top[\mathbf{q}_r - \mathbf{q}_r(t - T_r(t))] - K_l\dot{\mathbf{q}}_r^\top[\mathbf{q}_l - \mathbf{q}_l(t - T_l(t))], \end{aligned}$$

after applying the transformations (15), integrating from 0 to t, and invoking Lemma 3 on the last two terms, yields

$$\begin{aligned} V(t) - V(0) \le &- \left[B_l - \frac{K_l}{2}\left(\alpha_l + \frac{{}^*T_l^2}{\alpha_r}\right)\right]\|\dot{\mathbf{q}}_l\|_2^2 - K_{dl}\|\dot{\mathbf{q}}_l - \dot{\mathbf{q}}_{ld}\|_2^2 - \\ &- \left[\frac{B_rK_l}{K_r} - \frac{K_l}{2}\left(\alpha_r + \frac{{}^*T_r^2}{\alpha_l}\right)\right]\|\dot{\mathbf{q}}_r\|_2^2 - \frac{K_{dr}K_l}{K_r}\|\dot{\mathbf{q}}_r - \dot{\mathbf{q}}_{rd}\|_2^2. \end{aligned}$$

Setting the controller gains with condition (14) guarantees that $\dot{\mathbf{q}}_i \in \mathcal{L}_2 \cap \mathcal{L}_\infty$ and $\mathbf{q}_l - \mathbf{q}_r \in \mathcal{L}_\infty$. This completes the proof of part a). In order to establish part b), two conditions have to be proved: $\boldsymbol{\tau}_i \in \mathcal{L}_\infty$ in (27) and $\dot{\boldsymbol{\tau}}_i \in \mathcal{L}_\infty$. For the first condition it suffices to prove that $\dot{\mathbf{q}}_{id} \in \mathcal{L}_\infty$, due to (18). Taking (21) and (24), the desired velocities become

$$\begin{aligned} \dot{\mathbf{q}}_{ld} &= \frac{1}{b + K_{dl}}\left[K_{dl}\dot{\mathbf{q}}_l + \gamma_r K_{dr}\dot{\mathbf{q}}_r(t - T_r(t)) + \gamma_r[b - K_{dr}]\dot{\mathbf{q}}_{rd}(t - T_r(t))\right] \\ \dot{\mathbf{q}}_{rd} &= \frac{1}{b + K_{dr}}\left[K_{dr}\dot{\mathbf{q}}_l + \gamma_l K_{dl}\dot{\mathbf{q}}_l(t - T_l(t)) + \gamma_l[b - K_{dl}]\dot{\mathbf{q}}_{ld}(t - T_l(t))\right]. \end{aligned}$$

These difference equations are stable if $\dot{T}_i(t) < 1$, thus, assumption A3 ensures that $\dot{\mathbf{q}}_{id} \in \mathcal{L}_\infty$, implying that $\boldsymbol{\tau}_i \in \mathcal{L}_\infty$. This is sufficient to invoke Lemma 1 and prove zero-convergence of velocities and boundedness of accelerations. Finally, $\dot{\boldsymbol{\tau}}_i \in \mathcal{L}_\infty$ if $\dot{\gamma}_i \in \mathcal{L}_\infty$, which was proved in the proof of part b) of the Proposition 1. The use of Lemma 2 completes the proof, when acceleration converges to zero, position error also converges to zero. □

Additional Remarks

Although this work is focused on controllers that provide position tracking despite variable time-delays, there are other schemes that fit in this analysis framework, such as the control schemes, using the scattering transformation without position tracking (*cf.* [1]). In this case, part V_3 of V in (6) could be

$$V_3 = \int_0^t (\beta_r \dot{\mathbf{q}}_{rd}^\top \boldsymbol{\tau}_{rd} - \beta_l \dot{\mathbf{q}}_{ld}^\top \boldsymbol{\tau}_{ld}) d\sigma,$$

For control schemes without the scattering transformation and interconnecting the delayed passive output $\dot{\mathbf{q}}_i$ of the local and remote manipulators (*cf.* [14, 15, 16, 17, 18]), part V_3 could be

$$V_3 = \beta_r \int_{t-T_l(t)}^t |\dot{\mathbf{q}}_l(\theta)|^2 d\theta + \beta_l \int_{t-T_r(t)}^t |\dot{\mathbf{q}}_r(\theta)|^2 d\theta.$$

6 Conclusions

This Chapter presents some control schemes, namely P+d, PD+d, and scattering based controllers, that provide position tracking despite variable time-delays. The zero-convergence of velocities and position error has been established using a *general* Lyapunov–like function V in (6), by replacing one of its components V_3, by, for example, expression (13), (20) or (26). When the sign of $\dot{V}$ is undefined, the stability conclusions can be drawn after its integration applying Barbălat's Lemma. This approach can be seen as a first step towards the unification of the stability analysis for teleoperators with time-delay.

Simulations and experiments that validate the control schemes presented in this article can be found in [13, 6, 19, 20, 17]. Some of the experiments have been performed using the Internet between the Coordinated Science Laboratory (UIUC - CSL), Urbana-Champaign USA, and the Robotics Laboratory at the Institute of Industrial and Control Engineering (UPC - IOC), Barcelona Spain. The experimental testbed on the local site consists of a PHANToM Desktop™ with six sensed DOF and three actuated DOF, and on the remote site, of a Six DOF TX-90 Stäubli robot with a CS8–C™ Stäubli controller.

Acknowledgements

This work has been partially supported by the Spanish CICYT projects: DPI2005-00112, DPI2007-63665 and DPI2008-02448, the FPI program grant BES-2006-13393, and also by the Mexican CONACyT grant-93422.

References

1. Anderson, R.J., Spong, M.W.: Bilateral control of teleoperators with time delay. IEEE Transactions on Automatic Control 34(5), 494–501 (1989)
2. Niemeyer, G., Slotine, J.J.: Stable adaptive teleoperation. IEEE Journal of Oceanic Engineering 16(1), 152–162 (1991)
3. Arcara, P., Melchiorri, C.: Control schemes for teleoperation with time delay: A comparative study. Robotics and Autonomous Systems 38(1), 49–64 (2002)
4. Hokayem, P.F., Spong, M.W.: Bilateral teleoperation: An historical survey. Automatica 42, 2035–2057 (2006)

5. Chopra, N., Spong, M.W., Ortega, R., Barabanov, N.: On tracking performance in bilateral teleoperation. IEEE Transactions on Robotics 22(4), 861–866 (2006)
6. Nuño, E., Ortega, R., Barabanov, N., Basañez, L.: A globally stable PD controller for bilateral teleoperators. IEEE Transactions on Robotics 24(3), 753–758 (2008)
7. Kelly, R., Santibáñez, V., Loria, A.: Control of robot manipulators in joint space. Springer, Heidelberg (2005)
8. Spong, M.W., Hutchinson, S., Vidyasagar, M.: Robot Modeling and Control. Wiley, Chichester (2005)
9. Mulero-Martínez, J.I.: Uniform bounds of the coriolis/centripetal matrix of serial robot manipulators. IEEE Transactions on Robotics 23(5), 1083–1089 (2007)
10. Ghorbel, F., Srinivasan, B., Spong, M.W.: On the uniform boundedness of the inertia matrix of serial robot manipulators. Journal of Robotic Systems 15(1), 17–28 (1998)
11. Teel, A.R.: Asymptotic convergence from $\mathcal{L}_p$ stability. IEEE Transactions on Automatic Control 44(11), 2169–2170 (1999)
12. Slotine, J.J., Li, W.: Applied Nonlinear Control. Prentice-Hall, Englewood Cliffs (1991)
13. Nuño, E., Basañez, L., Ortega, R., Spong, M.W.: On position tracking for nonlinear teleoperators with variable time-delay. The International Journal of Robotics Research, (in Press) (2009)
14. Chopra, N., Spong, M.W.: Passivity-Based Control of Multi-Agent Systems. In: Advances in Robot Control From Everyday Physics to Human-Like Movements, pp. 107–134. Springer, Heidelberg (2007)
15. Chopra, N., Spong, M.W.: Adaptive Synchronization of Bilateral Teleoperators with Time Delay. In: Advances in Telerobotics, pp. 257–270. Springer, Heidelberg (2007)
16. Chopra, N., Spong, M.W.: Delay independent stability for interconnected nonlinear systems with finite $\mathcal{L}_2$ gain. In: Proc. of the IEEE Conference on Decision and Control, pp. 3847–3852 (2007)
17. Nuño, E., Basañez, L., Rodríguez-Seda, E., Spong, M.W.: Bilateral teleoperation experiments: Scattering transformation and passive output synchronization revisited. In: Proc. of the IFAC World Congress, Seoul, Korea (2008)
18. Wang, W., Slotine, J.J.: Contraction analysis of time-delayed communications and group cooperation. IEEE Transactions on Automatic Control 51(4), 712–717 (2006)
19. Nuño, E., Basañez, L., Ortega, R.: Passive bilateral teleoperation framework for assisted robotic tasks. In: Proc. of the IEEE International Conference on Robotics and Automation, pp. 1645–1650 (2007)
20. Nuño, E., Ortega, R., Barabanov, N., Basañez, L.: A new proportional controller for nonlinear bilateral teleoperators. In: Proc. of the IFAC World Congress, Seoul, Korea (2008)

Synchronization and Amplitude Death in Coupled Limit Cycle Oscillators with Time Delays

Fatihcan M. Atay

Max-Planck-Institute for Mathematics in the Sciences
Inselstr. 22, Leipzig 04103, Germany atay@member.ams.org

Summary. A pair of coupled van der Pol oscillators is considered whose interaction involves time delays. Using averaging theory, synchronous periodic solutions are determined, where the phase difference between the oscillators remains fixed in time. Parameter ranges for the coupling strength and delay are calculated such that the oscillators exhibit stable in-phase or anti-phase synchronized oscillations with identical amplitudes. It is shown that for any value of the delay there exists either an in-phase or an anti-phase synchronized solution which is asymptotically stable. In addition, parameter values are found for which the system is multistable, where both type of oscillations can be observed depending on the initial conditions, or it experiences amplitude death, where all oscillations are quenched.

1 Introduction

One of the remarkable dynamical phenomena that has attracted much recent scientific interest is synchronization, where several systems interacting with each other exhibit a common behavior. Many examples in the physical and biological sciences are now known, and the literature is vast in this area (see [1, 2, 3] and the references therein). Synchronization is perhaps more noteworthy when the information flow between the interacting systems are not instantaneous but happens after some time delay, due to finite transmission or information processing speeds. Delays are often a rule rather than an exception in many practical situations, and it is of interest to investigate how synchronization is affected by their presence.

In this chapter we consider the familiar oscillator introduced by Balthasar van der Pol [4],

$$\ddot{y} + \varepsilon(y^2 - 1)\dot{y} + y = 0 \tag{1}$$

to study the synchronization of interacting limit-cycle oscillators. It is known that for each positive value of the parameter ε, system (1) has a periodic solution which attracts all initial conditions, except the unstable equilibrium at zero [6]. The question is how these attracting solutions behave when two such oscillators are diffusively coupled:

$$\begin{aligned} \ddot{y}_1 + \varepsilon(y_1^2 - 1)\dot{y}_1 + y_1 &= \varepsilon\kappa(\dot{y}_2(t-\tau) - \dot{y}_1(t)) \\ \ddot{y}_2 + \varepsilon(y_2^2 - 1)\dot{y}_2 + y_2 &= \varepsilon\kappa(\dot{y}_1(t-\tau) - \dot{y}_2(t)). \end{aligned} \tag{2}$$

J.J. Loiseau et al. (Eds.): Topics in Time Delay Systems, LNCIS 388, pp. 383–389.
springerlink.com

(We mention e.g. [5] for another type of coupling, namely the so-called direct coupling.) Here $\kappa \geq 0$ measures the strength of the coupling and $\tau \geq 0$ is the time it takes for the information to reach from one oscillator to the other one. (For notational simplicity, time dependence of the variables is explicitly indicated only when necessary.) The analysis of (2) is complicated by the presence of delays, which gives rise to an infinite dimensional system. Here we present an analytical treatment using averaging theory, which is essentially a center-manifold reduction to obtain a finite-dimensional system, and leads to rigorous results for small ε. This is in contrast with many works in the literature that assume that the coupling is weak, i.e. κ is small, so that a reduced system is obtained which describes each oscillator only by its phase. Such approaches have the disadvantage of neglecting the effects of coupling on the amplitude of oscillations, and fail to address such phenomenon as amplitude death. By contrast, here we treat the amplitude effects in addition to phase-related behavior like synchronization. We show that the oscillators typically synchronize, and oscillate either with exactly the same phase or with a phase difference of 180°. For certain parameter values both type of solutions can co-exist, and the system is multistable. At certain other parameter ranges no periodic solutions exist and the zero solution is stable, leading to the amplitude death of oscillators. Numerical solutions are presented showing agreement with the analysis.

2 Analysis

Introducing $x_i = (y_i, \dot{y}_i) \in \mathbb{R}^2$, $i = 1, 2$, the system (2) can be equivalently written as

$$\begin{aligned} \dot{x}_1 &= Jx_1 + \varepsilon f(x_1) + \varepsilon\kappa C(x_2(t-\tau) - x_1(t)) \\ \dot{x}_2 &= Jx_2 + \varepsilon f(x_2) + \varepsilon\kappa C(x_1(t-\tau) - x_2(t)) \end{aligned} \tag{3}$$

where

$$J = \begin{pmatrix} 0 & 1 \\ -1 & 0 \end{pmatrix}, \quad C = \begin{pmatrix} 0 & 0 \\ 0 & 1 \end{pmatrix},$$

and

$$f\begin{pmatrix} a \\ b \end{pmatrix} = \begin{pmatrix} 0 \\ (1-a^2)b \end{pmatrix} \quad \text{for } (a,b) \in \mathbb{R}^2. \tag{4}$$

In order to use averaging theory, we let $\Phi(t) = \exp(tJ)$ and introduce the variables

$$u_i = \Phi^{-1}(t)x_i, \quad i = 1, 2. \tag{5}$$

Using the identity $\Phi(t-\tau) = \Phi(t)\Phi(-\tau)$, it is seen that the u_i satisfy the equations

$$\begin{aligned} \dot{u}_1(t) &= \varepsilon\Phi^{-1}(t)f(\Phi(t)u_1) + \varepsilon\kappa\Phi^{-1}C\left(\Phi(t)\Phi(-\tau)u_2(t-\tau) - \Phi(t)u_1(t)\right) \\ \dot{u}_2(t) &= \varepsilon\Phi^{-1}(t)f(\Phi(t)u_2) + \varepsilon\kappa\Phi^{-1}C\left(\Phi(t)\Phi(-\tau)u_1(t-\tau) - \Phi(t)u_2(t)\right) \end{aligned}$$

which are slowly changing for small ε, and thus are suitable for averaging. Each time-dependent term is averaged by integration over time as usual [7]. Thus, the average of the term $\Phi^{-1}f(\Phi u)$ is given by

$$\bar{f}(u) = \lim_{T\to\infty} \frac{1}{T}\int_0^T \Phi^{-1}(t)f(\Phi(t)u)\,dt.$$

Since $\Phi(t)$ is 2π-periodic in t, one can take $T = 2\pi$ in the above limit, and calculate by (4) that

$$\bar{f}(u) = \frac{1}{8}(4 - \|u\|^2) \cdot u. \tag{6}$$

Similarly, by Lemma 1 in [8], the average of the terms $\Phi^{-1}C\Phi$ is equal to $\frac{1}{2}I$. In this manner one obtains the averaged equations, which are the ordinary differential equations given by

$$\begin{aligned} \dot{u}_1(t) &= \varepsilon\bar{f}(u_1) + \tfrac{1}{2}\varepsilon\kappa\,(u_1 + \Phi(-\tau)u_2) \\ \dot{u}_2(t) &= \varepsilon\bar{f}(u_2) + \tfrac{1}{2}\varepsilon\kappa\,(\Phi(-\tau)u_1 + u_2)\,. \end{aligned} \tag{7}$$

For details of averaging in the presence of delays, the reader is referred to [9, 8].

The synchronization properties of the system (7) is better displayed in polar coordinates

$$u_i = (r_i\cos\theta_i, r_i\sin\theta_i). \tag{8}$$

After some algebra, the following equations are obtained for the amplitudes

$$\begin{aligned} \dot{r}_1 &= \varepsilon\tfrac{1}{8}(4 - r_1^2)r_1 + \tfrac{1}{2}\varepsilon\kappa\,(-r_1 + r_2\cos(\theta_2 - \theta_1 + \tau)) \\ \dot{r}_2 &= \varepsilon\tfrac{1}{8}(4 - r_2^2)r_2 + \tfrac{1}{2}\varepsilon\kappa\,(-r_2 + r_1\cos(\theta_1 - \theta_2 + \tau)) \end{aligned}$$

and the phases

$$\begin{aligned} r_1\dot{\theta}_1 &= \tfrac{1}{2}\varepsilon\kappa r_2\sin(\theta_2 - \theta_1 + \tau) \\ r_2\dot{\theta}_2 &= \tfrac{1}{2}\varepsilon\kappa r_1\sin(\theta_1 - \theta_2 + \tau). \end{aligned}$$

Since it is only the difference of the phases that appear on the right-hand sides, the dimension of the system can be reduced by letting $\alpha = \theta_1 - \theta_2$, so that

$$\dot{r}_1 = \varepsilon\tfrac{1}{8}(4 - r_1^2)r_1 + \tfrac{1}{2}\varepsilon\kappa\,(-r_1 + r_2\cos(\tau - \alpha)) \tag{9}$$

$$\dot{r}_2 = \varepsilon\tfrac{1}{8}(4 - r_2^2)r_2 + \tfrac{1}{2}\varepsilon\kappa\,(-r_2 + r_1\cos(\tau + \alpha)) \tag{10}$$

$$\dot{\alpha} = \tfrac{1}{2}\varepsilon\kappa\,[(r_2/r_1)\sin(\tau - \alpha) - (r_1/r_2)\sin(\tau + \alpha)] \tag{11}$$

provided that the r_i do not vanish. For synchronous solutions we seek equilibria of this system of the form

$$(r_1(t), r_2(t), \alpha(t)) = (r^*, r^*, \alpha^*). \tag{12}$$

It follows by averaging theory and the transformations (5) and (8) that such equilibria correspond to almost sinusoidal solutions of the original equations (2) where the oscillators oscillate at the same amplitude and their phase difference is unchanging in time.

Substitution of (12) into (9)–(10) and (11), respectively, gives the conditions

$$0 = \cos(\tau - \alpha^*) - \cos(\tau + \alpha^*) = 2\sin\tau\sin\alpha^*$$

and

$$0 = \sin(\tau - \alpha^*) - \sin(\tau + \alpha^*) = -2\cos\tau\sin\alpha^*$$

which together imply that $\sin\alpha^* = 0$. Hence, there are two types of synchronized solutions, depending on whether $\alpha^* = 0$ (in-phase synchronization) or $\alpha^* = \pi$ (anti-phase synchronization). On the other hand, by (9) the equilibria (12) satisfy

$$\tfrac{1}{8}(4 - (r^*)^2)r^* = \tfrac{1}{2}\kappa r^*\,(1 - \cos(\tau - \alpha^*))$$

which always has the solution $r^* = 0$. In addition, there exists a nontrivial solution

$$r^* = 2\sqrt{1 - \kappa(1 - \cos(\tau - \alpha^*))} \tag{13}$$

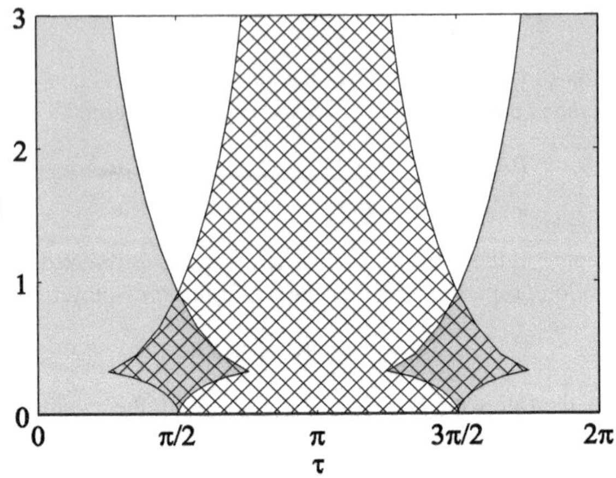

Fig. 1. Parameter values for asymptotically stable synchronized periodic solutions. The shaded and hatched regions correspond to in-phase and anti-phase synchronized solutions, respectively. Their intersection is the region of multistability where both type of stable solutions exist.

provided the radicand is nonnegative. This latter condition requires that

$$\kappa < \frac{1}{1 - \cos(\tau - \alpha^*)} \tag{14}$$

for the existence of synchronized periodic solutions. Note that (13) shows the dependence of oscillation amplitude on the coupling strength and the delay. In particular, if (14) is not satisfied for either $\alpha^* = 0$ or $\alpha^* = \pi$, that is if

$$\kappa > \frac{1}{1 - |\cos\tau|},$$

then there are no periodic solutions of the type sought. This last condition can be shown to imply that the origin is stable and the oscillations of the coupled system are quenched [8].

3 Synchronized behavior

For parameter ranges where (14) is satisfied, the stability of the synchronized solution can be determined from the eigenvalues of the Jacobian of the vector field in (9)–(11) at the corresponding fixed point (12). This makes it feasible to search the parameter space for the type of synchronized behavior that can be observed in the coupled system (2). The results are shown in Figure 1. It is seen that synchronization occurs for a large set of parameter values. In fact, if $0 < \kappa < 1$ then either an in-phase or an anti-phase synchronized stable periodic solution always exists regardless of the value of the delay. If the coupling strength is larger than 1 then it is possible for certain delay values to have a stable zero solution instead of periodic solutions, inside the white regions in Figure 1. This is the region of amplitude death, where oscillatory activity is quenched as the system evolves to a stationary state, and is possible only under the presence of positive delays when the oscillators are identical [10]. Finally, it is seen from the figure that there are parameter ranges for which both in-phase and anti-phase synchronized solutions are stable, and the initial conditions determine the actual observed behavior. In this case, noise or other external influences can move the system from one type of synchronization to the other.

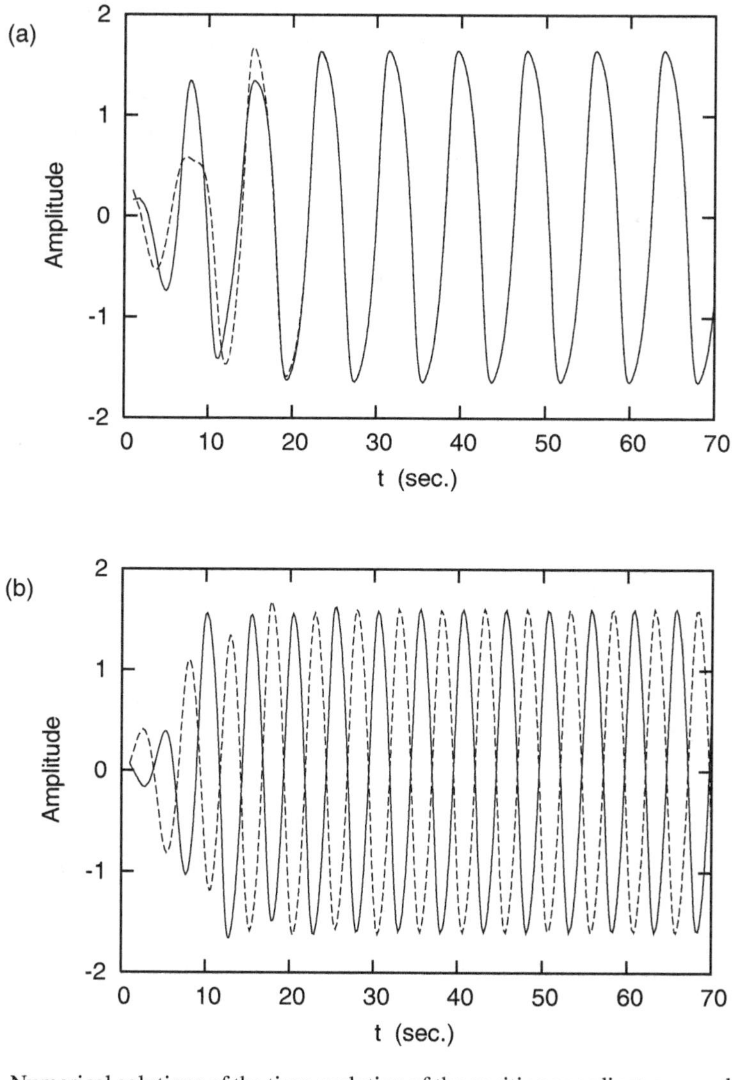

Fig. 2. Numerical solutions of the time evolution of the position coordinates y_1 and y_2, shown with solid and dashed lines, respectively. Here, asymptotically stable in-phase (a) and anti-phase (b) synchronized solutions co-exist for the parameter values $\kappa = 0.5, \tau = 1.5$, and $\varepsilon = 1$.

By averaging theory, the results obtained for the averaged equations (7) are rigorously justified for the original equations (2) for all sufficiently small values of the parameter ε [7, 11]. Nevertheless, these results often remain valid also for larger ε, at least qualitatively, so that they are useful for gaining insight into the type of dynamics that can occur. The parameter regions calculated by averaging are also confirmed through the numerical simulation of the equations (2). For example, Figure 2 shows the two type of synchronized periodic solutions for the coupled system. Note that the value of ε is not taken to be small in these simulations; nevertheless, the synchronized solutions are still found at the predicted parameter values. In fact, the oscillators synchronize faster at larger values of ε. Similarly amplitude death is also confirmed numerically, as shown in Figure 3. However, this region is more sensitive to high values of ε. The reason is that already strong coupling is needed for amplitude death, so the factor $\varepsilon\kappa$ in the equations can become quite large when both parameters are chosen large, moving far away from the domain where perturbation results are valid. Furthermore, amplitude death is typically observed for a rather limited range of parameters when the oscillators are identical and the delays are discrete [12], as is the case here. On the other hand, it has been shown that when the coupling delays are *distributed* rather than concentrated at a fixed value, the region of amplitude death is greatly enlarged [13]. For a study of distributed versus discrete delays, the reader is referred to [14].

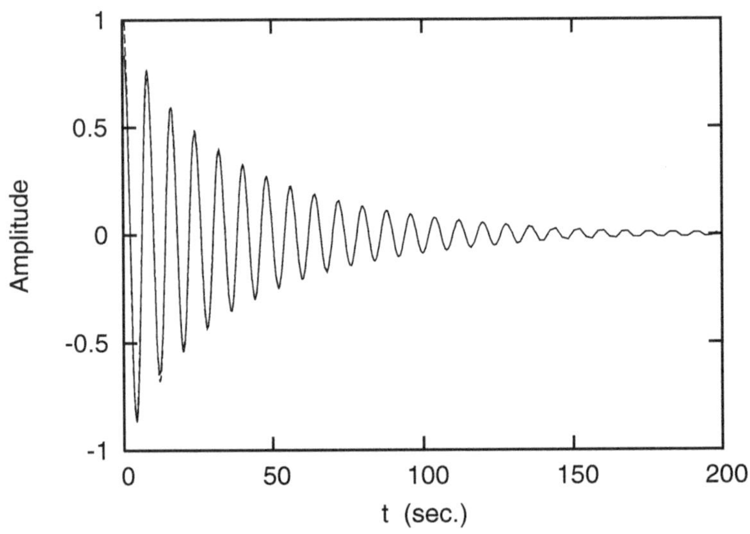

Fig. 3. Amplitude death of the oscillators. The parameter values are $\kappa = 2$, $\tau = 1.5$, $\varepsilon = 0.25$.

4 Conclusion

It is interesting that delay-coupled systems can synchronize, in view of the fact that each one sees only a past state of the others. Using a prototypical oscillator system, we have shown that this is not only possible, but it also happens in a robust manner with respect to the choice of parameters. In fact, the results indicate that, for sufficiently small ε, it is typical for the oscillators to synchronize, either in exact phase or with a 180° phase difference, or else to experience amplitude death. The numerical simulations suggest that a large ε need not invalidate this conclusion; however, care should be taken to extrapolate the results beyond the domain of the perturbation analysis. Furthermore, although the van der Pol oscillator is prototypical, other oscillatory systems with a different character might possibly lead to a wider range of interesting dynamical behavior. Nevertheless, the analysis presented here serves as a further explanation and example of why synchronous behavior is so commonly observed in nature.

References

1. Winfree, A.T.: The Geometry of Biological Time. Springer, New York (1980)
2. Murray, J.D.: Mathematical Biology, 2nd edn. Springer, Heidelberg (1993)
3. Pikovsky, A., Rosenblum, M., Kurths, J.: Synchronization – A Universal Concept in Nonlinear Science. Cambridge University Press, Cambridge (2001)
4. van der Pol, B.: Forced oscillations in a circuit with nonlinear resistance (receptance with reactive triode). London, Edinburgh, and Dublin Phil. Mag. 3, 65–80 (1927)
5. Wirkus, S.R.: The dynamics of two coupled van der Pol oscillators with delay coupling. Nonlinear Dynamics 30, 205–221 (2002)
6. Verhulst, F.: Nonlinear Differential Equations and Dynamical Systems. Springer, Berlin (1990)
7. Guckenheimer, J., Holmes, P.: Nonlinear Oscillations, Dynamical Systems, and Bifurcations of Vector Fields. Springer, New York (1983)
8. Atay, F.M.: Total and partial amplitude death in networks of diffusively coupled oscillators. Physica D 183(1-2), 1–18 (2003)
9. Atay, F.M.: Delayed-feedback control of oscillations in non-linear planar systems. Internat. J. Control 75(5), 297–304 (2002)
10. Ramana Reddy, D.V., Sen, A., Johnston, G.L.: Time delay induced death in coupled limit cycle oscillators. Phys. Rev. Lett. 80(23), 5109–5112 (1998)
11. Hale, J.K.: Theory of Functional Differential Equations. Springer, Heidelberg (1977)
12. Ramana Reddy, D.V., Sen, A., Johnston, G.L.: Time delay effects on coupled limit cycle oscillators at Hopf bifurcation. Physica D 129(1-2), 15–34 (1999)
13. Atay, F.M.: Distributed delays facilitate amplitude death of coupled oscillators. Phys. Rev. Lett. 91(9), 094101 (2003)
14. Atay, F.M.: Delayed feedback control near Hopf bifurcation. Discrete and Continuous Dynamical Systems, Series S 1(2), 197–205 (2008)

Synchronization of Bidirectionally Coupled Nonlinear Systems with Time-Varying Delay

Toshiki Oguchi[1], Takashi Yamamoto[1], and Henk Nijmeijer[2]

[1] Department of Mechanical Engineering, Tokyo Metropolitan University
1-1, Minami-Osawa, Hachioji-shi, Tokyo 192-0397 Japan.

[2] Department of Mechanical Engineering, Eindhoven University of Technology
P.O. Box 513, 5600 MB Eindhoven, The Netherlands.

Summary. This chapter considers the synchronization problem for coupled nonlinear systems with time-varying delay. In previous work, we have derived a sufficient condition for synchronization and boundedness of two identical strictly semi-passive systems coupled using state feedback with time-delay. This condition, however, requires that all state components are mutually coupled and the coupling delay is constant. In this chapter we extend the conditions to identical systems coupled using output feedback with time-varying delay where a bound on the length of the delay and an upper bound of its time-derivative are known. Firstly, we show, using the small-gain theorem, that the trajectories of coupled strictly semi-dissipative systems converge to a bounded region. Then we derive a sufficient condition for synchronization of the systems coupled with time-varying delay by using a delay range dependent stability criterion.

1 Introduction

Synchronization phenomena are of interest of researchers in applied physics, biology, social sciences, engineering and interdisciplinary fields, and the notion of synchronization has have been investigated in order to clarify the mechanism of synchronization [7, 11, 12, 14, 15, 17]. More recently, applications of these phenomena to engineering have also been considered and analyzed via control theory [3, 6, 8, 13]. On the other hand, in practical situations, time-delays caused by the signal transmission affect the behavior of coupled systems. It is therefore important to study the effect of time-delay in existing synchronization schemes. Although the effect of time-delay in the synchronization of coupled systems has been investigated both numerically and theoretically by a number of researchers, these works mostly concentrate on synchronization of systems with the coupling term typically described by $K(x_i(t-\tau) - x_j(t-\tau))$ or $K(Cx_i(t-\tau) - Cx_j(t-\tau))$ [1] and there are few results for the case in which the coupling term is described by $K(x_i(t) - x_j(t-\tau))$. For the latter case, as the coupling term does not vanish if the systems synchronize,, i.e. $x_i(t) = x_j(t)$, therefore even if uncoupled cell systems are bounded, coupled systems are not necessarily bounded. In our previous work [10, 16], we have derived a sufficient condition for boundedness and synchronization of two identical strictly semi-passive systems coupled using state feedback with time-delay. This method, however, requires that all states of each system are coupled with the other system and the coupling delay is constant. In this chapter we extend the conditions to identical systems

J.J. Loiseau et al. (Eds.): Topics in Time Delay Systems, LNCIS 388, pp. 391–401.
springerlink.com

coupled using output feedback with time-varying delay where a bound on the length of the delay and an upper bound of the time-derivative of the time-varying time delay are known.

We conclude this introduction with an outline of the chapter. In Section 2 we define the strict semi-passivity and the strict semi-dissipativity and show the small-gain theorem for coupled nonlinear systems with time-varying delay. In Section 3 we show, using the small-gain theorem, that the trajectories of coupled strictly semi-dissipative systems converge to a bounded region. Then we derive a sufficient condition for synchronization of the systems coupled with time-varying delay in terms of linear matrix inequalities (LMIs). An illustrative example is presented in Section 4 and some concluding remarks end the chapter.

2 Preliminaries

Throughout this chapter, $\|\cdot\|$ denotes the Euclidean norm. For a vector function $v(t) : [0,\infty) \to \mathbb{R}^n$, if $\|v\|_\infty := \sup_{t\geq 0} \|v(t)\| < \infty$, then we denote $v \in \mathcal{L}_\infty^n$. In addition, define a continuous norm by $\|\phi\|_c := \max_{-h_1\leq\theta\leq 0} \|\phi(\theta)\|$ for a vector function $\phi : [-h_1, 0] \to \mathbb{R}^n$. In this section, we review some results derived in our previous work [10, 16].

2.1 Strict semi-passivity and strict semi-dissipativity

Consider the nonlinear system

$$\dot{x}_i(t) = f_i(x_i, u_i), \qquad y_i(t) = C_i x_i \qquad (t \geq 0) \tag{1}$$

with state $x_i \in \mathbb{R}^n$, input $u_i \in \mathbb{R}^p$, output $y_i \in \mathbb{R}^m$, $C_i \in \mathbb{R}^{m\times n}$ and $f_i : \mathbb{R}^n \times \mathbb{R}^p \to \mathbb{R}^n$. Then we introduce strict semi-passivity and strict semi-dissipativity as follows.

Definition 1 (strict semi-passivity). *Set $p = m$. System (1) is said to be strictly semi-passive, if there exist a C^1-class function $V_i : \mathbb{R}^n \to \mathbb{R}$, class-$\mathcal{K}_\infty$ functions $\underline{\alpha}_i(\cdot)$, $\overline{\alpha}_i(\cdot)$ and $\alpha_i(\cdot)$ satisfying*

$$\begin{aligned} \underline{\alpha}_i(\|x_i\|) \leq V_i(x_i) \leq \overline{\alpha}_i(\|x_i\|) \\ \dot{V}_i(x_i) \leq -\alpha_i(\|x_i\|) - H_i(x_i) + y_i^T u_i \end{aligned} \tag{2}$$

for all $x_i \in \mathbb{R}^n, u_i \in \mathbb{R}^m, y_i \in \mathbb{R}^m$, where the function $H_i(x_i)$ satisfies the following condition:

$$\|x_i\| \geq \eta_i \quad \Rightarrow \quad H_i(x_i) \geq 0 \tag{3}$$

for a positive real number η_i.

Definition 2 (strict semi-dissipativity). *System (1) is said to be strictly semi-dissipative with respect to the supply rate $q_i(u, y)$, if there exist a C^1-class function $V_i : \mathbb{R}^n \to \mathbb{R}$, class-$\mathcal{K}_\infty$ functions $\underline{\alpha}_i(\cdot)$, $\overline{\alpha}_i(\cdot)$ and $\alpha_i(\cdot)$ satisfying (2) and*

$$\dot{V}_i(x_i) \leq -\alpha_i(\|x_i\|) - H_i(x_i) + q_i(u_i, y_i) \tag{4}$$

for all $x_i \in \mathbb{R}^n, u_i \in \mathbb{R}^p, y_i \in \mathbb{R}^m$, where the function $H_i(x_i)$ satisfies (3).

Remark 1. The system is strictly semi-passive if the supply rate is given as $q_i(u_i, y_i) = y_i^T u_i$.

Here we decompose u_i into l_i blocks such as $u_i = \mathrm{col}(u_{i1}, \ldots, u_{il_i})$ with $u_{ij} \in \mathbb{R}^{p_j}$ and $\sum_{j=1}^{l_i} p_j = p$. For a strictly semi-dissipative system, the following lemma can be proved in a similar way as the argument of the input-to-state stability (ISS) in [4].

Lemma 1. *Suppose that system (1) is strictly semi-dissipative with respect to* $q_i(u_i, y_i) \leq \sum_{j=1}^{l_i} \beta_{ij}(\|u_{ij}\|) - \epsilon(\|y_{ij}\|)$ *where* $\beta_{ij}, \epsilon \in \mathcal{K}$. *This means that there exist* C^1*-class function* $V_i : \mathbb{R}^{n_i} \to \mathbb{R}$, *class-*$\mathcal{K}_\infty$ *functions* $\underline{\alpha}_i(\cdot)$, $\overline{\alpha}_i(\cdot)$ *and* $\alpha_i(\cdot)$ *such that*

$$\underline{\alpha}_i(\|x_i\|) \leq V_i(x_i) \leq \overline{\alpha}_i(\|x_i\|)$$

$$\dot{V}_i(x_i) \leq -\alpha_i(\|x_i\|) - H_i(x_i) + \sum_{j=1}^{l_i} \beta_{ij}(\|u_{ij}\|)$$

where $H_i(x_i)$ *satisfies the property (3). Then the trajectory* $x_i(t)$ *of the system (1) satisfies the following inequalities:*

$$\begin{cases} \|x_i\|_\infty \leq \max_{1 \leq j \leq l_i} \{\rho(\|x_i^0\|), \gamma_{ij}(\|u_{ij}\|_\infty), \rho_i(\eta_i)\} \\ \limsup_{t\to\infty} \|x_i(t)\| \leq \max_{1 \leq j \leq l_i} \{\gamma_{ij}(\limsup_{t\to\infty} \|u_{ij}(t)\|), \rho_i(\eta_i)\} \end{cases} \tag{5}$$

for any inputs $u_{ij} \in \mathcal{L}_\infty^{p_j}$ *and any initial state* $x_i^0 \in \mathbb{R}^n$, *where*

$$\begin{aligned} &\rho_i(\cdot) = \underline{\alpha}_i^{-1} \circ \overline{\alpha}_i(\cdot) \\ &\gamma_{ij}(\cdot) = \underline{\alpha}_i^{-1} \circ \overline{\alpha}_i \circ \alpha_{ij}^{-1} \circ \kappa\beta_{ij}(\cdot) \qquad (j = 1, \ldots, l_i) \end{aligned} \tag{6}$$

with $\sum_{j=1}^{l_i} \alpha_{ij}(\cdot) = \alpha_i(\cdot)$ *and* $\kappa > 1$, *respectively.*

Next we consider the case in which two systems (1) with $i = 1, 2$ are bidirectionally coupled by the following couplings containing time-varying delay,

$$\begin{aligned} u_{11}(t) &= y_2(t - \tau(t)) = C_2 x_2(t - \tau(t)) \\ u_{21}(t) &= y_1(t - \tau(t)) = C_1 x_1(t - \tau(t)) \end{aligned} \tag{7}$$

where $\tau(t)$ is a time-varying delay and the initial conditions of x_i for $i = 1, 2$ are respectively given by

$$\begin{cases} x_i(\theta) = \phi_i(\theta) \qquad (-h_1 \leq \theta \leq 0) \\ x_i(0) = \phi_i(0) = x_0^i \end{cases} \tag{8}$$

where $\phi_i : [-h_1, 0] \to \mathbb{R}^n$. Define class-$\mathcal{K}$ functions as

$$\begin{aligned} \pi_{12}(r) &= \gamma_{11}(\sigma_{\max}(C_2) \cdot r) \\ \pi_{21}(r) &= \gamma_{21}(\sigma_{\max}(C_1) \cdot r) \end{aligned} \qquad (r \geq 0) \tag{9}$$

where $\gamma_i(\cdot)$ are defined as equation (6) and $\sigma_{\max}(\cdot)$ denotes the maximum singular value of a matrix.

Then from Lemma 1, we obtain the following small-gain theorem.

Theorem 1. *For two systems (1) coupled by (7), if the functions* $\pi_{12}(\cdot)$ *and* $\pi_{21}(\cdot)$ *defined as (9) satisfy*

$$\pi_{12} \circ \pi_{21}(r) < r, \qquad \forall r > 0,$$

and

$$\pi_{12} \circ \pi_{21}(0) = 0,$$

then the trajectories $x_1(t)$ and $x_2(t)$ satisfy

$$\begin{cases} \limsup_{t\to\infty} \|x_1(t)\| \le \max\{\zeta_1, \pi_{12}(\zeta_2)\} \\ \limsup_{t\to\infty} \|x_2(t)\| \le \max\{\zeta_2, \pi_{21}(\zeta_1)\} \end{cases}$$

where $\zeta_i := \max_{2\le j\le l_i}\{\gamma_{ij}(\limsup_{t\to\infty}\|u_{ij}\|), \rho_i(\eta_i)\}$.

Proof. This theorem is proved in a similar way as the case in which the coupling term contains a constant delay [10] by using the definition of the continuous norm defined above.

3 Synchronization

3.1 Problem formulation

We consider two identical systems

$$\Sigma_i : \begin{cases} \dot{x}_i(t) = Ax_i + f(x_i) + Bu_i \\ y_i(t) = Cx_i \qquad\qquad (t \ge 0) \end{cases} \tag{10}$$

with $i = 1, 2$, in which $x_i \in \mathbb{R}^n$, $u_i \in \mathbb{R}^m$, $y_i \in \mathbb{R}^m$, $A \in \mathbb{R}^{n\times n}$, $B \in \mathbb{R}^{n\times m}$, $C \in \mathbb{R}^{m\times n}$, $f : \mathbb{R}^n \times \mathbb{R}^m \to \mathbb{R}^n$ denotes a higher-order term and is Lipschitz continuous, and CB is nonsingular. The initial condition for each system is given by (8).

Now we assume that each system is strictly semi-passive and the systems are coupled with each other by the following controller as shown in Fig. 1.

$$u_i(t) = K_{ij}(y_i(t) - y_j(t - \tau_i(t)))\ (i, j = 1, 2, i \ne j) \tag{11}$$

where $\tau_1(t) = \tau_2(t) := \tau(t)$ is a time-varying delay satisfying

$$0 \le \tau(t) \le h_1, \quad \|\dot{\tau}(t)\| \le d$$

and K_i are gain matrices such as $K_{ij} = K_{ji} := K < 0$.

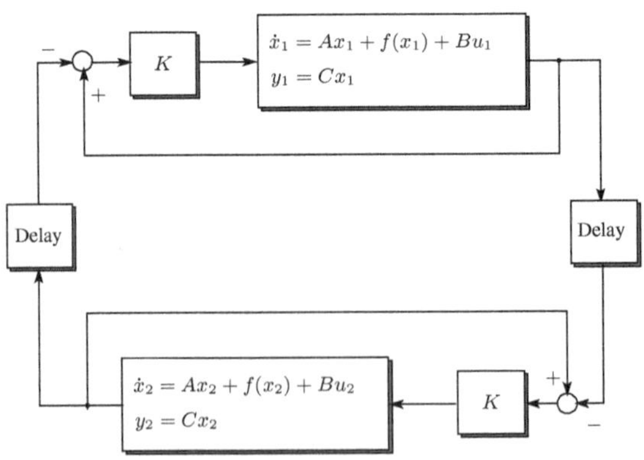

Fig. 1. Coupled systems

Next we formulate synchronization of coupled systems as follows.

Definition 3. If there exists a positive real number r such that the trajectories $x_i(t)$ of the systems (10), (11) with initial conditions ϕ_i such that
$\|\phi_1 - \phi_2\| \leq r$ satisfy

$$\|x_1(t) - x_2(t)\| \to 0 \quad as \quad t \to \infty,$$

then the coupled systems (10) with coupling inputs (11) are asymptotically synchronized.

Therefore our goal in this chapter is to derive synchronization conditions for systems (10) connected with (11).

3.2 Boundedness of the trajectories

Firstly we show under suitable assumptions the boundedness of the coupled systems (10) with coupling inputs (11). Since the matrix CB is nonsingular, the system (10) can be transformed to the following normal form [5].

$$\dot{y}_i(t) = a(y_i, z_i) + CBu_i \tag{12}$$
$$\dot{z}_i(t) = q(y_i, z_i) \tag{13}$$

for $i = 1, 2$, where $z_i \in \mathbb{R}^{n-m}$ and

$$\begin{bmatrix} y_i \\ z_i \end{bmatrix} = \Phi x_i$$

for a nonsingular matrix $\Phi := \begin{bmatrix} C^T & N^T \end{bmatrix}^T$ with $N \in \mathbb{R}^{(n-m)\times n}$ such that $NB = 0$ and functions $a : \mathbb{R}^m \times \mathbb{R}^{n-m} \to \mathbb{R}^m$ and $q : \mathbb{R}^m \times \mathbb{R}^{n-m} \to \mathbb{R}^{n-m}$ are Lipschitz continuous. Here the first equation can be recognized as a system with input $v_i^T = (u_i^T, z_i^T)$ and output y_i. At this point, we assume that

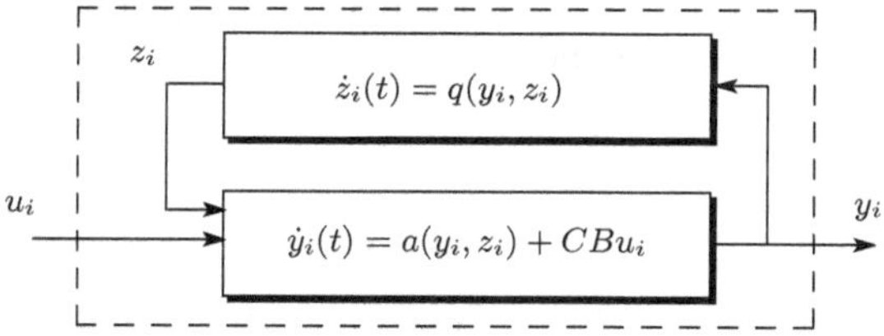

Fig. 2. Decomposition of each system

- The system (12) is strictly semi-dissipative with respect to the supply rate $q(v_i, y_i) \leq \beta_y(\|z_i\|) + y_i^T u_i$, where $\beta_y \in \mathcal{K}$, i.e. there exist a positive definite C^1-class functions V_y, class $\mathcal{K}_\infty$ functions $\underline{\alpha}_y$, $\overline{\alpha}_y$ and ϵ_y satisfying

$$\underline{\alpha}_y(\|y_i\|) \leq V_y(y_i) \leq \overline{\alpha}_y(\|y_i\|)$$
$$\dot{V}_y(y_i) \leq -\epsilon_y(\|y_i\|) - H_y(y_i) + \beta_y(\|z_i\|) + y_i^T u_i \tag{14}$$

for all $y_i \in \mathbb{R}^m, u_i \in \mathbb{R}^m, z_i \in \mathbb{R}^{n-m}$, where the function $H_y(y_i)$ satisfies that $H_y(y_i) \geq 0$ if $\|y_i\| \geq \eta_y$ for some positive real number η_y.

- The system (13) is strictly semi-dissipative with respect to the supply rate $q(y_i, z_i) \leq \beta_z(\|y_i\|) - \epsilon(\|z_i\|)$, where $\beta_z, \epsilon \in \mathcal{K}$, i.e. there exist a positive definite C^1-class functions V_z, class $\mathcal{K}_\infty$ functions $\underline{\alpha}_z$, $\overline{\alpha}_z$ and α_z satisfying

$$\begin{aligned}\underline{\alpha}_z(\|z_i\|) &\leq V_z(z_i) \leq \overline{\alpha}_z(\|z_i\|)\\ \dot{V}_z(z_i) &\leq -\alpha_z(\|z_i\|) - H_z(z_i) + \beta_z(\|y_i\|)\end{aligned}$$

for all $z_i \in \mathbb{R}^{n-m}, y_i \in \mathbb{R}^m$, where the function $H_z(z_i)$ satisfies that $H_z(z_i) \geq 0$ if $\|z_i\| \geq \eta_z$ for some positive real number η_z.

Substituting (11) in (14), we obtain

$$\dot{V}_y(y_i) \leq -\alpha_i(\|y_i\|) - H_y(y_i) + \beta_y(\|z_i\|) + \beta_{ij}(\|y_{j,\tau}\|)$$

where $\alpha_i(r) = \epsilon_y(r) - \frac{1}{2}\lambda_{max}(K)r^2$, $\beta_{ij}(r) = -\frac{1}{2}\lambda_{min}(K)r^2$ and $y_{j,\tau}$ denotes $y_j(t - \tau(t))$. Therefore, from Lemma 1,

$$\begin{cases}\limsup\limits_{t\to\infty} \|y_i(t)\| \leq \max\{\gamma_y(\limsup\limits_{t\to\infty} \|z_i\|), \gamma_{ij}(\limsup\limits_{t\to\infty} \|y_j\|), \rho_y(\eta_y)\}\\ \limsup\limits_{t\to\infty} \|z_i(t)\| \leq \max\{\gamma_z(\limsup\limits_{t\to\infty} \|y_i\|), \rho_z(\eta_z)\}\end{cases}$$

hold, where

$$\begin{aligned}\rho_y(\cdot) &= \underline{\alpha}_y^{-1} \circ \overline{\alpha}_y(\cdot)\ ,\ \gamma_y(\cdot) = \rho_y \circ \alpha_y^{-1} \circ \kappa\beta_y(\cdot)\\ \rho_z(\cdot) &= \underline{\alpha}_z^{-1} \circ \overline{\alpha}_z(\cdot)\ ,\ \gamma_z(\cdot) = \rho_z \circ \alpha_z^{-1} \circ \kappa\beta_z(\cdot)\\ \gamma_{ij}(\cdot) &= \rho_y \circ \alpha_{ij}^{-1} \circ \kappa\beta_{ij}(\cdot)\end{aligned}$$

with $\alpha_y(r) + \alpha_{ij}(r) = \alpha_i(r)$ and $\alpha_{ij}(r) = \epsilon_y(r) - \alpha_y(r) - \frac{1}{2}\lambda_{max}(K)r^2$ such that $\alpha_{ij} \in \mathcal{K}_\infty$. Note that γ_{ij} corresponds to the gain function from y_j to y_i.

As a result, we obtain the following theorem.

Theorem 2. *For all $r > 0$, if*

$$\begin{cases}\gamma_y \circ \gamma_z(r) < r\\ \gamma_{ij}(r) < r\end{cases} \tag{15}$$

hold, then the trajectories of the system (10) converge to a set

$$\Omega := \left\{x_i \in \mathbb{R}^n \,|x_i = \Phi^{-1}\begin{bmatrix} y_i\\ z_i\end{bmatrix}, \|y_i\| \leq s_y \text{ and } \|z_i\| \leq s_z\right\} \tag{16}$$

where $s_y := \max\{\rho_y(\eta_y), \gamma_y \circ \rho_z(\eta_z)\}$ and $s_z := \max\{\rho_z(\eta_z), \gamma_z \circ \rho_y(\eta_y)\}$.

3.3 A Synchronization Condition

From the above discussion, each coupled system converges to the set (16) under the above assumptions. By using this property, we consider a synchronization condition for coupled systems with time-varying delay.

The dynamics of the error $e(t) := x_1(t) - x_2(t)$ is given by

$$\dot{e}(t) = (A + BKC)e(t) + BKCe(t - \tau(t)) + \varphi(e, x_2) \tag{17}$$

where $\varphi(e, x_2) = f(e + x_2) - f(x_2)$. If the error system (17) has $e = 0$ as an asymptotically stable equilibrium, the behaviors of the two systems (10) synchronize. Therefore the synchronization problem can be reduced to the stability problem of the error dynamics (17).

From Lyapunov's indirect method, if the linearized system around the origin:

$$\begin{aligned}\dot{e}(t) &= (A + BKC + D(x_2))\, e(t) + BKCe(t - \tau(t)) \\ &:= A_0(x_2)e(t) + A_1 e(t - \tau(t))\end{aligned} \tag{18}$$

where $D(x_2) := \left(\frac{\partial\varphi(e,x_2)}{\partial e}\right)\Big|_{e=0}$, $A_0(x_2) := A + BKC + D(x_2)$ and $A_1 := BKC$, is asymptotically stable, the origin of the original error dynamics (17) is locally asymptotic stable. In addition, from the discussion in Section 3.2, note that if the conditions (15) hold, x_2 converges to Ω.

Theorem 3. *For all $x_2 \in \Omega$, if there exist $n \times n$ matrices $P_1 = P_1^T > 0$, P_2, P_3, W_1, W_2, $R = R^T \geq 0$, $S = S^T \geq 0$ satisfying the following linear matrix inequality:*

$$\begin{bmatrix} \Psi_1 & \Psi_2 & h_1\Phi_1 & -W_1^T A_1 \\ \Psi_2^T & \Psi_3 & h_1\Phi_2 & -W_2^T A_1 \\ h_1\Phi_1^T & h_1\Phi_2^T & -h_1 R & 0 \\ -A_1^T W_1 & -A_1^T W_2 & 0 & -(1-d)S \end{bmatrix} < 0 \tag{19}$$

where

$$\begin{aligned}\Psi_1 &= (A_0(x_2) + A_1)^T P_2 + P_2^T (A_0(x_2) + A_1) + W_1^T A_1 + A_1^T W_1 + S \\ \Psi_2 &= P_1 - P_2^T + (A_0(x_2) + A_1)^T P_3 + A_1^T W_1 \\ \Psi_3 &= -P_3 - P_3^T + h_1 A_1^T R A_1 \\ \Phi_1 &= W_1^T + P_2^T,\ \Phi_2 = W_2^T + P_3^T,\end{aligned}$$

then $e = 0$ of (17) is locally asymptotically stable.

Proof. This theorem is a special case of the result derived by Fridman *et al.* [2] and can be proved in a similar way as Theorem 1 in [2].

As the LMI (19) is affine with respect to the system matrices $A_0(x_2)$ and A_1, this result can be extended to a stability criterion for polytopic systems.

Since x_2 is bounded, each element of $D(x_2)$ is also bounded. As a result, the approximated error dynamics (18) can be rewritten by the following polytopic system:

$$\dot{e}(t) = \sum_{i=1}^{m} p_i A_0^i e(t) + A_1 e(t - \tau(t))$$

where $A_0^i = A_0 + D_i$ are constant matrices and $p_i(x_2) \in [0, 1]$ are polytopic coordinates satisfying the convex sum property $\sum_{i=1}^m p_i(x_2) = 1$. In other words, the system (18) can be rewritten by a convex combination of m systems $\dot{e} = A_o^i e(t) + A_1 e(t - \tau(t))$ for $i = 1, \ldots, m$. Using these systems, we can obtain the following polytopic linear differential inclusion (PLDI)

$$\dot{e}(t) \in \mathrm{Co}\Big\{A_0^1 e(t) + A_1 e(t - \tau(t)), \cdots, A_0^m e(t) + A_1 e(t - \tau(t))\Big\} \tag{20}$$

where Co denotes the convex hull. Therefore we can obtain the following synchronization criterion.

Theorem 4. *Consider the PLDI (20). If there exist* $n \times n$ *matrices* $P_1^i = P_1^{iT} > 0$*,* P_2*,* P_3*,* W_1^i*,* W_2^i*,* $R = R^T \geq 0$ *and* $S = S^T \geq 0$ *for* $i = 1, 2, \ldots, m$ *such that*

$$\begin{bmatrix} \Psi_1^i & \Psi_2^i & h_1\Phi_1 & -W_1^{iT}A_1 \\ \Psi_2^{iT} & \Psi_3 & h_1\Phi_2 & -W_2^{iT}A_1 \\ h_1\Phi_1^T & h_1\Phi_2^T & -h_1R & 0 \\ -A_1^TW_1^i & -A_1^TW_2^i & 0 & -(1-d)S \end{bmatrix} < 0 \tag{21}$$

are feasible for every $i = 1, \ldots, m$*, where*

$$\begin{aligned} \Psi_1^i &= (A_0^i + A_1)^T P_2 + P_2^T(A_0^i + A_1) + W_1^{iT}A_1 + A_1^TW_1^i + S^i \\ \Psi_2^i &= P_1^i - P_2^T + (A_0^i + A_1)^T P_3 + A_1^T W_1^i \\ \Psi_3 &= -P_3 - P_3^T + h_1 A_1^T R A_1 \\ \Phi_1 &= W_1^{iT} + P_2^T,\ \Phi_2 = W_2^{iT} + P_3^T, \end{aligned}$$

then $e = 0$ *of the error dynamics is locally asymptotically stable and synchronization can be accomplished.*

4 Illustrative Example

Consider two coupled Lorenz systems.

$$\dot{x}_i(t) = \begin{bmatrix} \sigma(x_{i2} - x_{i1}) \\ rx_{i1} - x_{i2} - x_{i1}x_{i3} \\ -bx_{i3} + x_{i1}x_{i2} \end{bmatrix} + Bu_i\ ,\ y_i = Cx_i \tag{22}$$

where $\sigma = 10$, $r = 28$, $b = 8/3$ and $B^T = C = \begin{bmatrix} 0 & 1 & 0; 0 & 0 & 1 \end{bmatrix}$. The inputs are defined by $u_i(t) = K(y_i(t) - y_j(t - \tau(t)))$ for $(i, j) \in \{(1, 2), (2, 1)\}$. For a real number $k > 0$, set coupling gain K as $K = -kI_{2\times 2}$ and time-delays $\tau(t)$ satisfies $0 \leq \tau(t) < 0.05$ and $\dot{\tau}(t) \leq 0.05$.

Now we decompose each system into two subsystems by defining $\tilde{y}_i$ and z_i as follows.

$$\tilde{y}_i = \begin{bmatrix} x_{i2}, & x_{i3} - r \end{bmatrix}^T,\ \ z_i = x_{i1}$$

In addition, we define storage functions as $V_y(\tilde{y}_i) := \frac{1}{2}\tilde{y}_i^T\tilde{y}_i$ and $V_z(z_i) := \frac{1}{2}z_i^2$ for the corresponding subsystems. Then the time derivative of $V_y(\tilde{y}_i)$ along the trajectory of (22) satisfies $\dot{V}_y(\tilde{y}_i) \leq -\alpha_i(\|\tilde{y}_i\|) - H_y(\tilde{y}_i) + \beta_{ij}(\|\tilde{y}_{j,\tau}\|)$ where

$$\begin{aligned} \epsilon_y(r) &= \varepsilon r^2,\ \ H_y(\tilde{y}_i) = (1-\varepsilon)\tilde{y}_{i1}^2 + (b-\varepsilon)\tilde{y}_{i2}^2 + br\tilde{y}_{i2} \\ \alpha_i(r) &= \left(\varepsilon - \frac{\lambda_{max}(K)}{2}\right) r^2 = \left(\varepsilon + \frac{k}{2}\right) r^2 \\ \beta_{ij}(r) &= -\frac{\lambda_{min}(K)}{2} r^2 = \frac{k}{2} r^2 \end{aligned}$$

with $\varepsilon = 0.01$. Here, for $\eta_y = 28.9$, $H_y(\tilde{y}_i)$ satisfies that if $\|\tilde{y}_i\| \geq \eta_y$, then $H_y(\tilde{y}_i) \geq 0$ holds. Similarly, for the function $V_z(z_i)$,

$$\dot{V}_z(z_i) = -\sigma z_i^2 + \sigma\tilde{y}_{i1}z_i \leq -\alpha_z(\|z_i\|) + \beta_z(\|\tilde{y}_i\|)$$

holds, where $\alpha_z(r) = \frac{\sigma}{2}r^2$ and $\beta_z(r) = \frac{\sigma}{2}r^2$. This means that $\eta_z = 0$. In addition, from the definitions of the storage functions $V_y(\tilde{y}_i)$ and $V_z(z_i)$, we can choose ρ_y and ρ_z as $\rho_y(r) = \rho_z(r) = r$ for any $r \geq 0$. Therefore $\gamma_{ij}(r) = \rho_y \circ \alpha_{ij}^{-1} \circ \kappa\beta_{ij}(r)$. As a result, γ_{ij} are given by $\gamma_{ij}(r) = \sqrt{\frac{\kappa\frac{k}{2}}{\frac{k}{2}+\varepsilon}}\, r$ satisfying $\gamma_{ij} < r$ with κ sufficiently close to 1. Furthermore, $\gamma_z(r) = r, \forall r \geq 0$ and since $\beta_y(\|z\|) = 0$, $\gamma_y(r) = 0$ which means (15) holds. Finally, we obtain $s_y = s_z = \eta_y = 28.9$. Therefore, from Theorem 2, the trajectories $x_i(t)$ converge to the set

$$\Omega = \{x_i \in \mathbb{R}^3 \mid (x_{i2}^2 + (x_{i3} - r)^2)^{\frac{1}{2}} \leq 28.9 \text{ and } \|x_{i1}\| \leq 28.9\}.$$

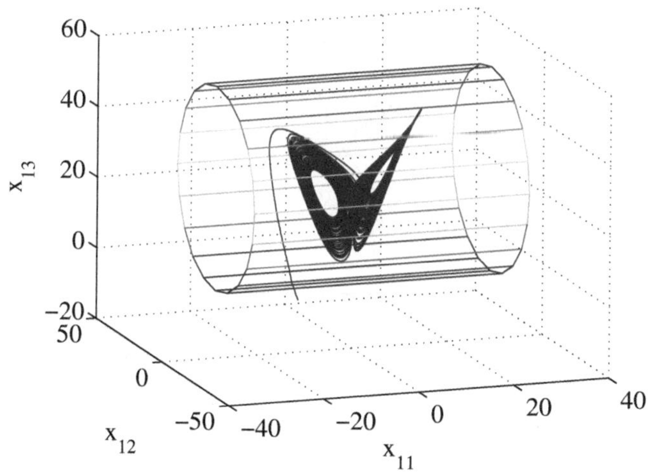

Fig. 3. Phase portrait of system 1

Fig. 3 shows the behavior of $x_1(t)$. In this figure, the cylinder illustrates the estimated boundary of the set Ω. From this figure, we know that the trajectories converge to the set Ω. In this simulation, the time-varying delay is given by $\tau(t) = 0.01 + 0.002 \sin 3t \sin 7t$. Setting $k = 20$, there exist 3×3 matrices $P_1^i = P_1^{iT} > 0, P_2, P_3, W_1^i, W_2^i, R = R^T \geq 0$ and $S = S^T \geq 0$ for $i = 1, 2, \ldots, 8$ satisfying the LMI condition (21) for all vertex systems. Fig. 4 and 5 show the behavior of each component of $x_1(t)$ and $e(t)$, respectively. We know that the error e converges to zero, which means the synchronization of these systems is perfectly accomplished.

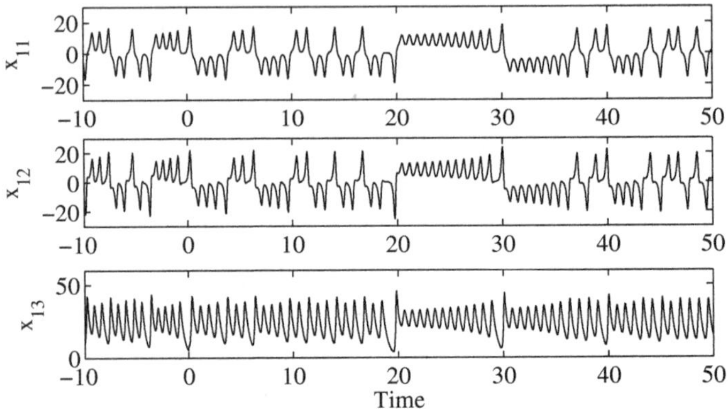

Fig. 4. Behaviors of the state $x_1(t)$

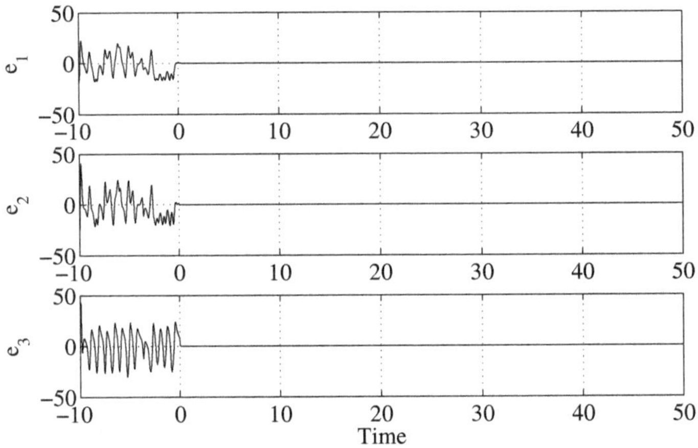

Fig. 5. Behaviors of the error $e(t)$

5 Concluding Remarks

In this chapter, we have considered synchronization of nonlinear systems coupled with time-varying delay. By extending our previous result, we obtained a sufficient condition for synchronization of two systems coupled using output feedback with time-varying delay. If the network structure is symmetric, the analysis introduced in this chapter can be applied to larger number of coupled systems. However, if $\tau_1(t)$ is not equal to $\tau_2(t)$, the symmetric structure collapses and the error dynamics does not have a trivial solution. For such a case, we can consider the coupled system as a system with uncertainties and will be able to apply our previous work [9] in order to estimate an upper bound of the synchronization error.

Acknowledgment

This work was supported in part by the Japan Society for the Promotion of Science (JSPS) Grant-in-Aid for Scientific Research (No. 18560441).

References

1. Amano, M., Luo, Z., Hosoe, S.: Graph dependent sufficient condition of dynamic network system with time-delay. In: Proc. of the 4th IFAC Workshop on Time-Delay Systems, CD-ROM, Rocquencourt, France (2003)
2. Fridman, E., Shaked, U.: Delay-dependent stability and H_∞ control: constant and time-varying delays. Int. J. of Control 76(1), 48–60 (2003)
3. Huijberts, H., Nijmeijer, H., Oguchi, T.: Anticipating synchronization of chaotic Lur'e systems. Chaos 17, 013117 (2007)
4. Isidori, A.: Nonlinear control systems II. Springer, London (1999)
5. Lozano, R., Brogliato, B., Egeland, O., Maschke, B.: Dissipative systems analysis and control. Springer, London (2000)
6. Nijmeijer, H., Rodriguez-Angeles, A.: Synchronization of Mechanical Systems. World Scientific, London (2003)
7. Nijmeijer, H., Mareels, I.M.Y.: An observer looks at synchronization. IEEE Trans. Circ. Systems I 44, 882–890 (1997)
8. Oguchi, T., Nijmeijer, H.: Prediction of chaotic behavior. IEEE Trans. Circ. Systems I: Regular Papers 52(11), 2464–2472 (2005)
9. Oguchi, T., Nijmeijer, H.: Anticipating synchronization of nonlinear systems with uncertainties. In: Proc. of the 6th IFAC Workshop on Time-Delay Systems. CD-ROM, L'Aquila, Italy (2006)
10. Oguchi, T., Nijmeijer, H., Yamamoto, T.: Synchronization in networks of chaotic systems with time-delay coupling. Chaos 18, 039108 (2008)
11. Pecora, L.M., Carroll, T.L.: Synchronization in chaotic systems. Physical Review Letters 64, 821–824 (1990)
12. Pikovsky, A., Rosenblum, M., Kurths, J.: Synchronization: Universal concept in nonlinear science. Cambridge University Press, Cambridge (2001)
13. Pogromsky, A., Santoboni, G., Nijmeijer, H.: Partial synchronization: from symmetry towards stability. Physica D 172, 65–87 (2002)
14. Strogatz, S.H.: From Kuramoto to Crawford: exploring the onset of synchronization in populations of coupled oscillators. Physica D 143, 1–20 (2000)
15. Strogatz, S.H., Stewart, I.: Coupled oscillators and biological synchronization. Scientific American 269, 102–109 (1993)
16. Yamamoto, T., Oguchi, T., Nijmeijer, H.: Synchronization of coupled nonlinear systems with time delay. In: Proc. of European Control Conference 2007, Kos, Greece, pp. 3056–3061 (2007)
17. Wu, C.W.: Synchronization in coupled chaotic circuits and systems. World Scientific, London (2002)

Master-Slave Synchronization for Two Inverted Pendulums with Communication Time-Delay

Héctor Javier Estrada-García[1,2], Luis Alejandro Márquez-Martínez[1], and Claude H. Moog[2]

[1] CICESE Km. 107 Carr. Tijuana - Ensenada C.P. 22860, Ensenada, B.C. Mexico.
hestrada@salamanca.ugto.mx, lmarquez@cicese.mx

[2] IRCCyN 1, rue de la Noë - BP 92 101 - 44321 Nantes, France.
Claude.Moog@irccyn.ec-nantes.fr

Summary. Two inverted pendulums with a constant time-delay in the transmission of their measurements are considered. The control objective for the slave is to track the reference trajectory imposed by the master, with asymptotic stability. A causal solution is designed in this paper, for the synchronization of the two inverted pendulums, which are under-actuated. The time delay in the data transmission is assumed to be constant and the control problem is solved so that the error fulfills a third-order linear delay differential equation.

1 Introduction

Time delays have normally been considered undesirable in control theory, because of their tendency to reduce stability margins of the system. Some control textbooks, (e.g [9] [14]), show the effect of time delays on the stability of a system: they reduce the phase margin. On the other hand, a great amount of literature exists which deal with the analysis and design of controllers to stabilize systems that present instability due to time delays ([6], [7]). Teleoperation allows users to manipulate and interact with remote environments via master and slave robotic mechanisms. It enables remote operation and can also scale human forces and motions to achieve stronger, larger, or smaller interfaces. Applications range from classic nuclear material handling and space exploration, where users are protected from dangerous conditions, to telemanipulation for example minimally invasive surgery, where small tools allow less traumatic access to a patient. These telerobotic systems often face a key challenge: the presence of communication delay between the master and slave sites.

In recent years, a lot of efforts have been made to study the dynamical behaviors and stability of systems with commensurate and incommensurate time-delays (see e.g. [9] and references therein). A control scheme that guarantees, under certain assumptions, stability of a bilaterally controlled teleoperator for any communication delay was first presented in [2]. Since then, the problem of stabilization of bilaterally controlled teleoperators in presence of communication delay has attracted considerable attention in the literature ([1], [10], [13], [16],

J.J. Loiseau et al. (Eds.): Topics in Time Delay Systems, LNCIS 388, pp. 403–413.
springerlink.com

[12]). A comparative study of different control schemes for teleoperation with communication delay can be found in [3]. In this paper, we study the synchronization of two inverted pendulums, which is based on maximal linearization and a causal control law is obtained by solving delay-differential error equations. The obtained results improve and extend earlier works.

2 Preliminaries

2.1 System Under Consideration

In this work, we consider the inverted pendulum 505 from ECP company [17], which is an underactuated system, with 2 d.o.f. and one actuator. The dynamics of such a system includes a nonlinear coupling between the actuated and the unactuated degrees of freedom, non holonomic constraints, and unstable zero dynamics with respect to standard articular positions.

The system consists of a pendulum rod which supports a sliding balance rod; this last one is driven via a belt and pulley, which in turn is driven by a shaft connected to a DC servo motor below the pendulum rod.

This inverted pendulum is not like the conventional rod-on-cart inverted pendulum, but rather, it steers a horizontal balancing rod in the presence of gravity to control the vertical pendulum rod. The center of gravity of the pendulum rod may be altered by adjusting the counter weight at the bottom (brass counter masses), and therefore the system dynamics can be altered. The position of the sliding rod is measured by means of an encoder positioned at the back of the motor, and the angular position of the pendulum rod angle is measured by an other encoder connected to the pivoting base as in Figure 1.

2.2 Equations of Motion

Neglecting friction effects, the plant may be modeled using the Euler - Lagrange equations. The kinetic energy of a rigid solid is the sum of two terms when it is referred to the center of mass of the system. One term is the translation kinetic energy and the second term is the rotation kinetic energy. Setting $\theta = q_1(t)$ and $x = q_2(t)$ in Figure 1, the Lagrangian equations for the ECP 505 system are defined by equations,

$$\begin{aligned} -C_2\ddot{q}_2(t) + 2m_2q_2(t)\dot{q}_1(t)\dot{q}_2(t) + J_0\ddot{q}_1(t) \\ -C_4 g \sin q_1(t) + m_2 g q_2(t)\cos q_1(t) &= 0 \end{aligned} \tag{1}$$

and

$$m_2\ddot{q}_2(t) - C_2\ddot{q}_1(t) - m_2q_2(t)\dot{q}_1(t)^2 + m_2 g \sin q_1(t) = \tau \tag{2}$$

where $C_2 = l_0 m_2,\ C_3 = m_1 l_c,\ C_4 = C_2 + C_3$, and can be represented as the Lagrangian system:

$$M(q)\ddot{q} + C(q,\dot{q})\dot{q} + g(q) = \tau. \tag{3}$$

From (1) and (2), it is easy to prove that

$$\begin{aligned} \ddot{q}_1(t) &= J\left(q_2(t)\right)v + R\left(q_1(t), q_2(t)\right) \\ \ddot{q}_2(t) &= v, \end{aligned} \tag{4}$$

where v is a new control input, $J\left(q_2(t)\right)$ is equal to

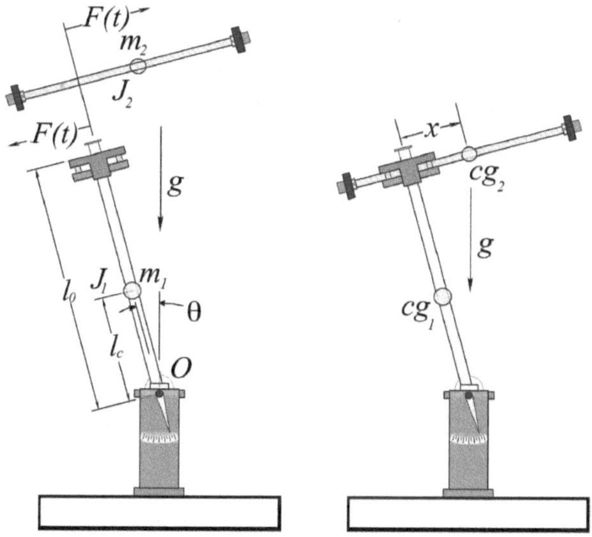

Fig. 1. ECP 505 diagram

$$J\left(q_2(t)\right) = \frac{l_0 m_2}{m_2 q_2(t)^2 + C_1}, \tag{5}$$

and $R\left(q_1(t), q_2(t)\right)$ is equal to

$$R\left(q_1(t), q_2(t)\right) = \frac{-2m_2 q_2(t)\dot{q}_1(t)\dot{q}_2(t)}{m_2 q_2(t)^2 + C_1} + \frac{(C_4) g \sin q_1(t)}{m_2 q_2(t)^2 + C_1} - \frac{m_2 g q_2(t) \cos q_1(t)}{m_2 q_2(t)^2 + C_1}, \tag{6}$$

System (4) is a partial feedback linearized (as defined in [15]) representation of the ECP505 pendulum.

2.3 Maximal Linearization

To achieve the synchronization, first we need to obtain a control law for the ECP505; we use the method proposed in [8]. The objective is to determine an output that gives a small dimension of the zero dynamics, and if this dimension is different from zero, to assure that the zero dynamics is stable. The control objective will be to maximally linearize the system to facilitate the construction of a trajectory that transfers the state of the system from one point to another in finite time.

Mechanical systems are said to be simple if the kinetic energy is quadratic in the velocities and the potential energy depends only on the configuration variables. Denote the Lagrangian by $\mathcal{L} = K - V$, and assume that the system is actuated such that

$$\frac{d}{dt}\frac{\partial \mathcal{L}}{\partial \dot{q}_k} - \frac{\partial \mathcal{L}}{\partial q_k} = \begin{cases} 0, & \text{for i=0} \\ u_k & \text{for i=1,...,N-1} \end{cases} \tag{7}$$

with u_k taking values in $\mathbb{R}$.

In this case, conditions will be identified under which the set of outputs $y = kp + \sigma$, where $k \in \mathbb{R}$ is a constant, yields an exponentially minimum phase system. More precisely, conditions are given such that the zero dynamics is well defined in a neighborhood of the given equilibrium point, has dimension one, and is exponentially stable for all $k > 0$.

We can see that the Lagrangian equations, (1) and (2), fulfills the conditions of equation (7). For more details see [8].

Then, applying the above method, we obtain the output equation $y = kp + \sigma$ for the ECP505:

$$y = K\left(q_1(t) - \frac{C_2 \arctan\left(\frac{m_2 q_2(t)}{\sqrt{m_2 C_1}}\right)}{\sqrt{m_2 C1}}\right) - C_2\dot{q}_2(t) + \left(m_2 q_2^2(t) + C_1\right)\dot{q}_1(t) \tag{8}$$

the first derivative for the output y

$$\dot{y} = \left(\dot{q}_1(t) - \frac{C_2\dot{q}_2(t)}{C1 + m_2 q_2(t)^2}\right)K + +C_4 g sin q_1(t) - m_2 g q_2(t)\cos q_1(t) \tag{9}$$

For the second derivative of y we have:

$$\ddot{y} = K\left(R\left(q_1(t), q_2(t)\right) - \frac{\partial J(q_2(t))}{\partial q_2(t)}\dot{q}_2^2(t)\right) + \frac{\partial\dot{\sigma}}{\partial q_1(t)}\dot{q}_1(t) + \frac{\partial\dot{\sigma}}{\partial q_2(t)}\dot{q}_2(t) \tag{10}$$

Now, the third derivative of y is of the form:

$$y^{(3)} = \left(\Psi(q,\dot{q})J(q_2(t)) + \xi(q,\dot{q})\right)v + \Psi(q,\dot{q})\left(R(q_1,q_2)\right) + \phi(q,\dot{q}) \tag{11}$$

where $\Psi(q,\dot{q}), \xi(q,\dot{q}), \phi(q,\dot{q})$ are suitable functions. The relative degree of the output y is $\rho = 3$, because $y^{(3)}(t)$ depends explicitly on the input term. The formal definition is now recalled.

Definition 1. *[11] A time-delay system is said to have a relative degree ρ if there exists a nonnegative integer ρ such that*

$$\rho = min\left\{k \mid \frac{\partial y^{(k)}(t)}{\partial u(t-\tau)} \not\equiv 0\right\}, \textit{for some } \tau, k \in \mathbb{N}.$$

If, for all $(k,\tau) \in \mathbb{N}^2$, $\partial y^{(k)}(t)/\partial u(t-\tau) \equiv 0$ we set $\rho = \infty$.

2.4 Delay-differential equations

Let us consider the third-order linear delay differential equations of the form

$$\gamma^{(3)}(t) = a\ddot{\gamma}(t-\tau) + b\dot{\gamma}(t-\tau) + c\gamma(t-\tau) \tag{12}$$

where $\tau > 0$ is a constant time delay, and a, b, c are constants. Necessary and sufficient conditions for the asymptotic stability of the zero solution are given in Theorem 3.7 of [5]. With these conditions, it is possible to guarantee the asymptotic stability of equation (12) for any constant delay τ. Note, however, that if τ is increased, then the transitory response time will increase. A description and some results on the robust stability of this kind of equations can be found in [12]. In the rest of the paper we consider $\tau = 0.3$, a_m, b_m and c_m are the parameters of the master pendulum and a_s, b_s and c_s are the parameters for the slave. Using this result we can present our main contributions.

2.5 Definition of synchronization

Next we present the definition of synchronization that is considered. This definition is based on the concepts presented in [4]. Consider k dynamical systems

$$S_i = \{T, U_i, X_i, Y_i, \phi_i, h_i\}, i = 1, \ldots, k, \tag{13}$$

where T is the set of time instances; U_i, X_i, Y_i are the sets of inputs, states and outputs, respectively; $\phi_i : T \times X_i \times U_i \rightarrow X_i$ are the transition maps, and $h_i : T \times X_i \times U_i \rightarrow Y_i$ are the outputs maps.

Suppose l functionals $\kappa_j : \mathcal{F}_1 \times \mathcal{F}_2 \times \cdots \times \mathcal{F}_k \times T \rightarrow \mathbb{R}, j = 1, \ldots, l$ are given. $\mathcal{F}_i$ are the sets of all functions from T into Y_i, that is $\mathcal{F}_i = \{y : T \rightarrow Y_i\}$.

We take as time set T either $T = \mathbb{R}$ or $T = \{t : 0 \leq t < \infty\}$. For any $\zeta \in T$ the shift operator $\sigma_\zeta : \mathcal{F}_i \rightarrow \mathcal{F}_i$ is defined as $(\sigma_\zeta y)(t) = y(t + \sigma_\zeta)$ for all $y \in \mathcal{F}_i$ and all $t \in T$. Let $x^{(1)}(t), \ldots, x^{(k)}(t)$ be solutions of the systems $S_1, \ldots, S_k, x^{(i)}(\cdot) \in X_i$ with initial states $x^{(1)}(0), \ldots, x^{(k)}(0)$ respectively, well defined for all $t \in T$.

Assume for simplicity that all S_i, $i = 1, \ldots, k$ are smooth finite dimensional systems, described by differential equations with a finite-dimensional input. The system models with interconnections are given by

$$\begin{cases} \dfrac{dx_i}{dt} = F_i(x_i, t) + \tilde{F}_i(x_0, x_1, \ldots, x_k, u, t), \; i = 1, \ldots, k, \\ \dfrac{dx_0}{dt} = F_0(x_0, x_1, \ldots, x_k, u, t), \end{cases} \tag{14}$$

where $x_i \in X_i$ and $u = u(t) \in \mathbb{R}^m$ is the input. The vector field F_0 describes the dynamics of the interconnection system, and $\tilde{F}_i$ are vector fields of the interconnections.

Definition 2. *The problem of controlled approximate synchronization with respect to the functionals κ_j, $j = 1, \ldots, l$, is to find a control u as a feedback function of the states $x_0, x_1, \ldots, x_k$, and time, such that there exist an $\epsilon > 0$ and $\zeta_1, \ldots, \zeta_k \in T$ which fulfil*

$$\left|\kappa_j(\sigma_{\zeta_1} y_1(\cdot), \ldots, \sigma_{\zeta_k} y_k(\cdot), t\right| \leq \epsilon, j = 1, \ldots, l. \tag{15}$$

Controlled asymptotic synchronization is defined as

$$\lim_{t \to \infty} \left(\kappa_j(\sigma_{\zeta_1} y_1(\cdot), \ldots, \sigma_{\zeta_k} y_k(\cdot), t\right) = 0, \; j = 1, \ldots, l. \tag{16}$$

3 Main results

3.1 Problem statement

Given two inverted pendulum systems with delay in the communication channel, and a C^∞ reference trajectory $y_{ref}(t)$, $t \geq 0$, find, if possible, a control law that synchronizes both pendulums in a master-slave configuration, such that the difference $e(t)$ between the output $y_2(t)$ and the output $y_1(t)$ is asymptotically stable, and the error between $y_1(t)$ and $y_{ref}(t)$ is asymptotically stable, where $y_1(t)$ is the output of the master pendulum and $y_2(t)$ is the output of the slave pendulum.

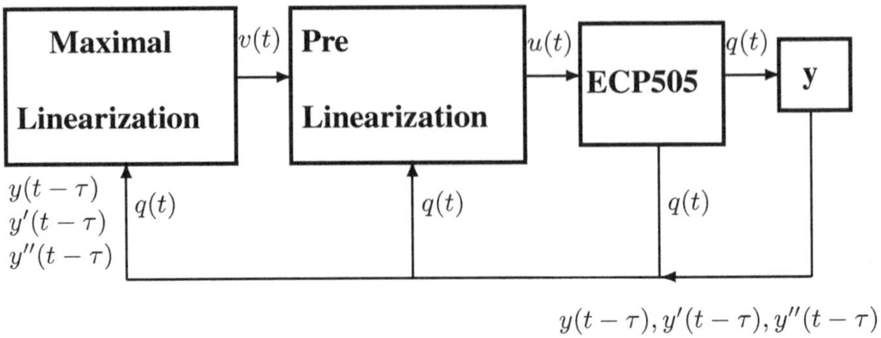

Fig. 2. Trajectory tracking scheme

3.2 Causal Solution

For the synchronization of the two pendulums, we need to derive a local control for the master system. First define the error signal:

$$e(t) = y_1(t) - y_{ref}(t) \tag{17}$$

where $y_1(t)$ is the output of the system, and $y_{ref}(t)$ is the reference trajectory. Setting $\gamma(t) = e(t)$ in (12) we obtain

$$e^{(3)}(t) + a_m\ddot{e}(t-\tau) + b_m\dot{e}(t-\tau) + c_m e(t-\tau) = 0 \tag{18}$$

Applying equation (17) in equation (18):

$$\begin{aligned} y_1^{(3)}(t) - y_{ref}^{(3)}(t) + a_m\big(\ddot{y}_1(t-\tau) - \ddot{y}_{ref}(t-\tau)\big) + \\ + b_m\big(\dot{y}_1(t-\tau) - \dot{y}_{ref}(t-\tau)\big) + c_m\big(y_1(t-\tau) - y_{ref}(t-\tau)\big) = 0. \end{aligned} \tag{19}$$

With the conditions given in [5], we can obtain a set of coefficients for the zero solution of (19) to be asymptotically stable.

Now, set $y_1^{(3)}(t) = w$, and from (11) obtain

$$y_1^{(3)}(t) = \omega_m = \big(\Psi(q,\dot{q})J(q_2(t)) + \xi(q,\dot{q})\big)v_m + \Psi(q,\dot{q})\big(R(q_1,q_2)\big) + \phi(q,\dot{q}) \tag{20}$$

where

$$\begin{aligned} \omega_m = & -a_m\big(\ddot{y}_1(t-\tau) - \ddot{y}_{ref}(t-\tau)\big) - b_m\big(\dot{y}_1(t-\tau) - \dot{y}_{ref}(t-\tau)\big) - \\ & -c_m\big(y_1(t-\tau) - y_{ref}(t-\tau)\big) + y_{ref}^{(3)}(t) \end{aligned} \tag{21}$$

Solving for v_m

$$v_m = \frac{\omega_m - \Psi(q,\dot{q})\big(R(q_1,q_2)\big) - \phi(q,\dot{q})}{\Psi(q,\dot{q})J(q_2(t)) + \xi(q,\dot{q})} \tag{22}$$

we obtain the new control input that solves the trajectory tracking problem for the master system. In Figure 3 we present the simulation results for this control law.

For the synchronization, sketched in Figure 5, we use the same method displayed above. In this case, the error signal is given by

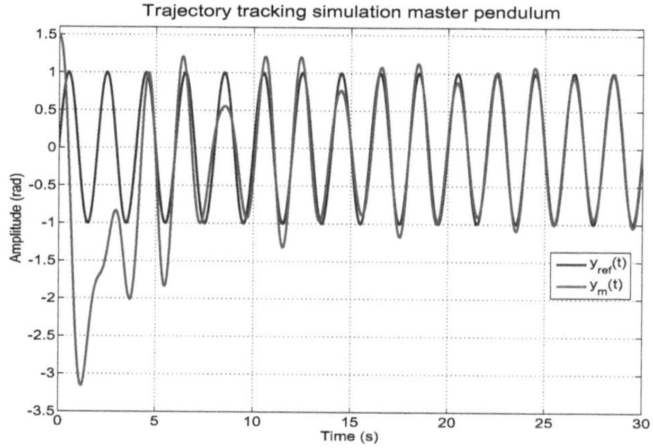

Fig. 3. Trajectory tracking for a sinusoidal function

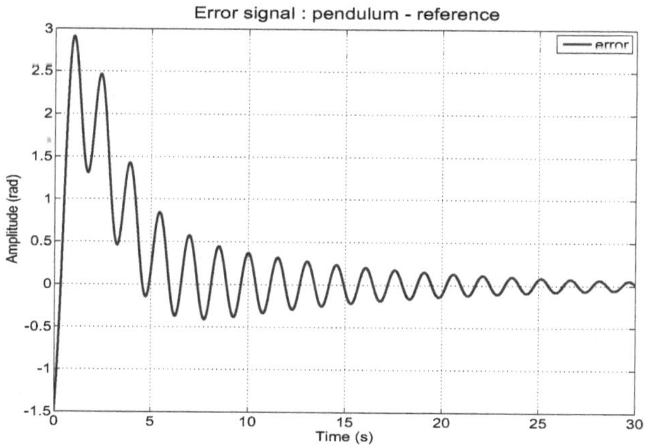

Fig. 4. Error signal for the master pendulum

Fig. 5. Synchronization scheme

$$e(t) = y_2(t) - y_1(t) \tag{23}$$

where $y_2(t)$ is the output of the slave, and $y_1(t)$ is the master output and acts as the reference trajectory. Again, we make $\gamma(t) = e(t)$ in (12); it yields

$$e^{(3)}(t) + a_s\ddot{e}(t-\tau) + b_s\dot{e}(t-\tau) + c_s e(t-\tau) = 0 \tag{24}$$

where $\{a_s, b_s, c_s\} \neq \{a_m, b_m, c_m\}$. Applying equation (23) in equation (24):

$$\begin{array}{r} y_2^{(3)}(t) - y_1^{(3)}(t) + a_s\big(\ddot{y}_2(t-\tau) - \ddot{y}_1(t-\tau)\big) + \\ +b_s\big(\dot{y}_2(t-\tau) - \dot{y}_1(t-\tau)\big) + c_s\big(y_2(t-\tau) - y_1(t-\tau)\big) = 0. \end{array} \tag{25}$$

and define

$$\begin{aligned} \omega_s = {} & a_s\big(\ddot{y}_2(t-\tau) - \ddot{y}_1(t-\tau)\big) + \\ & b_s\big(\dot{y}_2(t-\tau) - \dot{y}_1(t-\tau)\big) + c_s\big(y_2(t-\tau) - y_1(t-\tau)\big) \end{aligned} \tag{26}$$

Applying again the conditions given in [5], we obtain a set of coefficients for the zero solution of (25) to be asymptotically stable. In this case, we need the value of the signal $y_1^3(t)$, but we have a delay in the communication channel, so it is not possible to get it directly from the master. Instead, we can reconstruct this term from the equations (20) and (21) which yields

$$y_2^{(3)}(t) - \omega_m + \omega_s = 0 \tag{27}$$

ω_m depends on the term $y_{ref}^{(3)}(t)$. This is not a drawback since the reference is known for $-\infty < t < \infty$, so the third derivative of the reference signal is easily obtained at any time, if the delay is constant. We only need to change the phase to add the delay. Then, solving (27) for v_s we obtain as in (22)

$$v_s = \frac{\omega_m - \omega_s - \Psi(q,\dot{q})\big(R(q_1,q_2)\big) - \phi(q,\dot{q})}{\Psi(q,\dot{q})J\big(q_2(t)\big) + \xi(q,\dot{q})} \tag{28}$$

This causal control input for the slave pendulum solves the synchronization problem. The synchronization obtained is an asymptotic synchronization, as defined in (16).

In the case where $\{a_s, b_s, c_s\} = \{a_m, b_m, c_m\}$, the equation (27) will be

$$\begin{aligned} y_2^{(3)}(t) = {} & a\big(\ddot{y}_M(t-\tau) - \ddot{y}_M(t-\tau)\big) + b\big(\dot{y}_M(t-\tau) - \dot{y}_M(t-\tau)\big) \\ & +c\big(y_M(t-\tau) - y_M(t-\tau)\big) - a y_{ref}^{(3)}(t) + a\big(\ddot{y}_S(t-\tau) - \ddot{y}_{ref}(t-\tau)\big) \\ & +b\big(\dot{y}_S(t-\tau) - \dot{y}_{ref}(t-\tau)\big) + c\big(y_S(t-\tau) - y_{ref}(t-\tau)\big), \end{aligned} \tag{29}$$

and reduces to

$$\begin{aligned} y_2^{(3)}(t) = {} & a\big(\ddot{y}_S(t-\tau) - \ddot{y}_{ref}(t-\tau)\big) + b\big(\dot{y}_S(t-\tau) - \dot{y}_{ref}(t-\tau)\big) \\ & +c\big(y_S(t-\tau) - y_{ref}(t-\tau)\big) - a y_{ref}^{(3)}(t). \end{aligned} \tag{30}$$

This causes that the dependency of the master in the slave disappears, and the synchronization will be made with $y_{ref}(t)$ instead of $y_1(t)$. To avoid this singularity, differents coefficients in the master and slave have to be chosen.

4 Experiments

Now, the results obtained from the experimentation of the methodology developped above, are given using two distants mechanical pendulums. The Internet network was used to transfer data between both pendulums. The delay in Internet is not constant but it is considered to be bounded. A buffer was used and in practice the delay reduces to a constant delay equal to its upper bound! The results for the synchronization and the phase error are given in Figures 8 and 9.

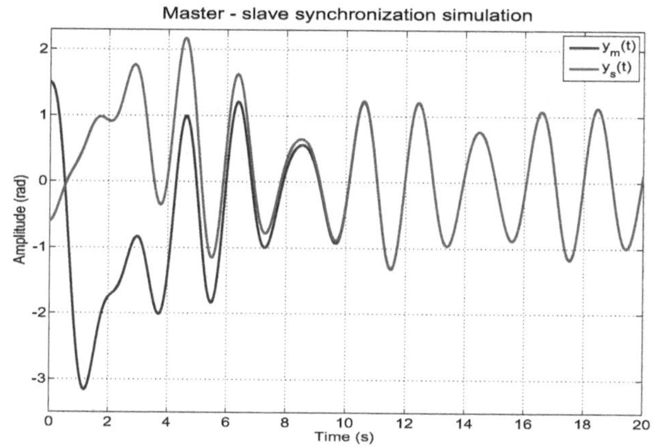

Fig. 6. Simulation of synchronization

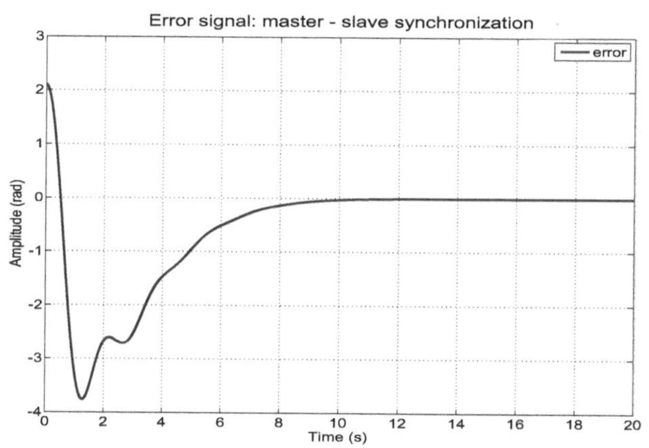

Fig. 7. Error signal for the slave pendulum

5 Conclusions

In this work a scheme for the synchronization of two mechanical systems with a constant transmission delay is presented. The principal characteristics of this result, is that it does not require any state prediction. The effects of the delay are not considered as disturbances and the control law which is obtained is causal. This scheme is based on the inclusion of terms with delays in the error equations used for the synthesis of the control law. When we have time delays in the communication path of the system we can maintain the stability of the synchronization law. These control schemes guarantee the stability of the synchronized system for almost any (constant) delay in the communication channel, the effect of time delay is observed in a large stabilization time, since the effect of delay is considered as a scale in time. This causes that the transitory response time will increase. It is worth to emphasize that the proposed method allows to prove stability of the synchronized systems for an arbitrary

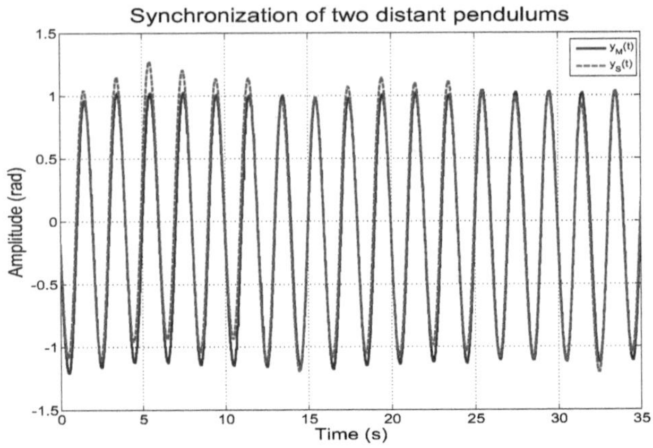

Fig. 8. Synchronization of two mecanical pendulums.

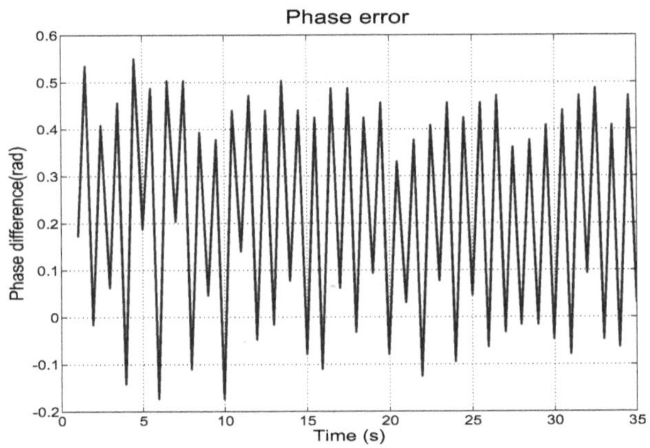

Fig. 9. Phase error.

constant communication delay and, in some sense, the stability is uniform with respect to the communication delay.

References

1. Alvarez-Gallegos, J., Rodriguez, D., de, C., Spong, M.W.: A stable control scheme for teleoperators with time delay. International Journal of Robotics and Automation 12(3), 73–79 (1997)
2. Anderson, R.J., Spong, M.W.: Bilateral control of teleoperators with time delay. IEEE Transactions on Automatic Control 34(5), 494–501 (1989)

3. Arcara, P., Melchiorri, C.: Control schemes for teleoperation with time delay: A comparative study. Robotics and Autonomous Systems 38(1), 49–64 (2002)
4. Blekhman, I.I., Fradkov, A.L., Tomchina, O.P., Bogdanov, D.E.: Self - synchronization and controlled synchronization: general definition and example design. Mathematics and Computers in Simulation 58, 367–384 (2002)
5. Cahlon, B., Schmidt, D.: Stability criteria for certain third-order delay differential equations. Journal of Computational and Applied Mathematics 188, 319–335 (2006)
6. Dugard, L., Verriest, E.I.: Stability and Control of Time-Delay Systems. Springer, Berlin (1998)
7. Gorecki, H., Fuksa, S., Grabowski, P., Korytowski, A.: Analysis and Synthesis of Time Delay Systems. Wiley, New York (1989)
8. Grizzle, J.W., Moog, C., Chevallereau, C.: Nonlinear Control of Mechanical Systems with an Unactuated Cyclic Variable. IEEE Transactions on Automatic Control 50(5), 559–576 (2005)
9. Gu, K., Kharitonov, V.L., Chen, J.: Stability of Time-Delay Systems, 1st edn. Birkhüser, Boston (2003)
10. Leung, G.M., Francis, B.A., Apkarian, J.: Bilateral controller for teleoperators with time delay via mu-synthesis. IEEE Transactions on Robotics and Automation 11, 105–116 (1995)
11. Moog, C., Castro-Linares, R., Velasco-Villa, M., Márquez-Martínez, L.A.: Disturbance decoupling for time-delay nonlinear systems. IEEE Trans. Automat. Control 45(2), 305–309 (2000)
12. Niculescu, S.I., Gu, K.: Advances in Time-Delay Systems. LNCSE, p. 38. Springer, Heidelberg (2004)
13. Niemeyer, G., Slotine, J.J.: Telemanipulation with time delays. International Journal of Robotics Research 23(9), 873–890 (2004)
14. Nise, N.S.: Control Systems Engineering, 3rd edn. Wiley, New York (2000)
15. Spong, M.W.: Partial Feedback Linearization of Underactuated Mechanical Systems. In: IROS, Munich, Germany, pp. 314–321 (1994)
16. Zhu, W.H., Salcudean, S.E.: Stability guaranteed teleoperation: An adaptive motion/force control approach. IEEE Transactions on Automatic Control 45(11), 1951–1969 (2000)
17. `http://www.ecpsystems.com/`

Index

Lecture Notes in Control and Information Sciences

Edited by M. Thoma, F. Allgöwer, M. Morari

Further volumes of this series can be found on our homepage:
springer.com

Vol. 388: Loiseau, J.J.; Michiels, W.; Niculescu, S-I.; Sipahi, R. (Eds.):
Topics in Time Delay Systems
418 p. [978-3-642-02896-0]

Vol. 387: Xia, Y.; Fu, M.; Shi, P.:
Analysis and Synthesis of Dynamical Systems with Time-Delays
283 p. 2009 [978-3-642-02695-9]

Vol. 386: Huang, D.; Nguang, S.K.:
Robust Control for Uncertain Networked Control Systems with Random Delays
159 p. 2009 [978-1-84882-677-9]

Vol. 385: Jungers, R.:
The Joint Spectral Radius
144 p. 2009 [978-3-540-95979-3]

Vol. 384: Magni, L.; Raimondo, D.M.; Allgöwer, F. (Eds.):
Nonlinear Model Predictive Control
572 p. 2009 [978-3-642-01093-4]

Vol. 383: Sobhani-Tehrani E.; Khorasani K.;
Fault Diagnosis of Nonlinear Systems Using a Hybrid Approach
360 p. 2009 [978-0-387-92906-4]

Vol. 382: Bartoszewicz A.; Nowacka-Leverton A.;
Time-Varying Sliding Modes for Second and Third Order Systems
192 p. 2009 [978-3-540-92216-2]

Vol. 381: Hirsch M.J.; Commander C.W.; Pardalos P.M.; Murphey R. (Eds.)
Optimization and Cooperative Control Strategies: Proceedings of the 8th International Conference on Cooperative Control and Optimization
459 p. 2009 [978-3-540-88062-2]

Vol. 380: Basin M.
New Trends in Optimal Filtering and Control for Polynomial and Time-Delay Systems
206 p. 2008 [978-3-540-70802-5]

Vol. 379: Mellodge P.; Kachroo P.;
Model Abstraction in Dynamical Systems: Application to Mobile Robot Control
116 p. 2008 [978-3-540-70792-9]

Vol. 378: Femat R.; Solis-Perales G.;
Robust Synchronization of Chaotic Systems Via Feedback
199 p. 2008 [978-3-540-69306-2]

Vol. 377: Patan K.
Artificial Neural Networks for the Modelling and Fault Diagnosis of Technical Processes
206 p. 2008 [978-3-540-79871-2]

Vol. 376: Hasegawa Y.
Approximate and Noisy Realization of Discrete-Time Dynamical Systems
245 p. 2008 [978-3-540-79433-2]

Vol. 375: Bartolini G.; Fridman L.; Pisano A.; Usai E. (Eds.)
Modern Sliding Mode Control Theory
465 p. 2008 [978-3-540-79015-0]

Vol. 374: Huang B.; Kadali R.
Dynamic Modeling, Predictive Control and Performance Monitoring
240 p. 2008 [978-1-84800-232-6]

Vol. 373: Wang Q.-G.; Ye Z.; Cai W.-J.; Hang C.-C.
PID Control for Multivariable Processes
264 p. 2008 [978-3-540-78481-4]

Vol. 372: Zhou J.; Wen C.
Adaptive Backstepping Control of Uncertain Systems
241 p. 2008 [978-3-540-77806-6]

Vol. 371: Blondel V.D.; Boyd S.P.; Kimura H. (Eds.)
Recent Advances in Learning and Control
279 p. 2008 [978-1-84800-154-1]

Vol. 370: Lee S.; Suh I.H.; Kim M.S. (Eds.)
Recent Progress in Robotics: Viable Robotic Service to Human
410 p. 2008 [978-3-540-76728-2]

Vol. 369: Hirsch M.J.; Pardalos P.M.; Murphey R.; Grundel D.
Advances in Cooperative Control and Optimization
423 p. 2007 [978-3-540-74354-5]

Vol. 368: Chee F.; Fernando T.
Closed-Loop Control of Blood Glucose
157 p. 2007 [978-3-540-74030-8]

Vol. 367: Turner M.C.; Bates D.G. (Eds.)
Mathematical Methods for Robust and Nonlinear Control
444 p. 2007 [978-1-84800-024-7]

Vol. 366: Bullo F.; Fujimoto K. (Eds.)
Lagrangian and Hamiltonian Methods for Nonlinear Control 2006
398 p. 2007 [978-3-540-73889-3]

Vol. 365: Bates D.; Hagström M. (Eds.)
Nonlinear Analysis and Synthesis Techniques for Aircraft Control
360 p. 2007 [978-3-540-73718-6]

Vol. 364: Chiuso A.; Ferrante A.; Pinzoni S. (Eds.)
Modeling, Estimation and Control
356 p. 2007 [978-3-540-73569-4]

Vol. 363: Besançon G. (Ed.)
Nonlinear Observers and Applications
224 p. 2007 [978-3-540-73502-1]

Vol. 362: Tarn T.-J.; Chen S.-B.; Zhou C. (Eds.)
Robotic Welding, Intelligence and Automation
562 p. 2007 [978-3-540-73373-7]

Vol. 361: Méndez-Acosta H.O.; Femat R.; González-Álvarez V. (Eds.):
Selected Topics in Dynamics and Control of Chemical and Biological Processes
320 p. 2007 [978-3-540-73187-0]

Vol. 360: Kozlowski K. (Ed.)
Robot Motion and Control 2007
452 p. 2007 [978-1-84628-973-6]

Vol. 359: Christophersen F.J.
Optimal Control of Constrained Piecewise Affine Systems
190 p. 2007 [978-3-540-72700-2]

Vol. 358: Findeisen R.; Allgöwer F.; Biegler L.T. (Eds.): Assessment and Future Directions of Nonlinear Model Predictive Control
642 p. 2007 [978-3-540-72698-2]

Vol. 357: Queinnec I.; Tarbouriech S.; Garcia G.; Niculescu S.-I. (Eds.):
Biology and Control Theory: Current Challenges
589 p. 2007 [978-3-540-71987-8]

Vol. 356: Karatkevich A.:
Dynamic Analysis of Petri Net-Based Discrete Systems
166 p. 2007 [978-3-540-71464-4]

Vol. 355: Zhang H.; Xie L.:
Control and Estimation of Systems with Input/Output Delays
213 p. 2007 [978-3-540-71118-6]

Vol. 354: Witczak M.:
Modelling and Estimation Strategies for Fault Diagnosis of Non-Linear Systems
215 p. 2007 [978-3-540-71114-8]

Vol. 353: Bonivento C.; Isidori A.; Marconi L.; Rossi C. (Eds.)
Advances in Control Theory and Applications
305 p. 2007 [978-3-540-70700-4]

Vol. 352: Chiasson, J.; Loiseau, J.J. (Eds.)
Applications of Time Delay Systems
358 p. 2007 [978-3-540-49555-0]

Vol. 351: Lin, C.; Wang, Q.-G.; Lee, T.H., He, Y.
LMI Approach to Analysis and Control of Takagi-Sugeno Fuzzy Systems with Time Delay
204 p. 2007 [978-3-540-49552-9]

Vol. 350: Bandyopadhyay, B.; Manjunath, T.C.; Umapathy, M.
Modeling, Control and Implementation of Smart Structures 250 p. 2007 [978-3-540-48393-9]

Vol. 349: Rogers, E.T.A.; Galkowski, K.; Owens, D.H.
Control Systems Theory and Applications for Linear Repetitive Processes
482 p. 2007 [978-3-540-42663-9]

Vol. 347: Assawinchaichote, W.; Nguang, K.S.; Shi P.
Fuzzy Control and Filter Design for Uncertain Fuzzy Systems
188 p. 2006 [978-3-540-37011-6]

Vol. 346: Tarbouriech, S.; Garcia, G.; Glattfelder, A.H. (Eds.)
Advanced Strategies in Control Systems with Input and Output Constraints
480 p. 2006 [978-3-540-37009-3]

Vol. 345: Huang, D.-S.; Li, K.; Irwin, G.W. (Eds.)
Intelligent Computing in Signal Processing and Pattern Recognition
1179 p. 2006 [978-3-540-37257-8]

Vol. 344: Huang, D.-S.; Li, K.; Irwin, G.W. (Eds.)
Intelligent Control and Automation
1121 p. 2006 [978-3-540-37255-4]

Vol. 341: Commault, C.; Marchand, N. (Eds.)
Positive Systems
448 p. 2006 [978-3-540-34771-2]

Vol. 340: Diehl, M.; Mombaur, K. (Eds.)
Fast Motions in Biomechanics and Robotics
500 p. 2006 [978-3-540-36118-3]

Vol. 339: Alamir, M.
Stabilization of Nonlinear Systems Using Receding-horizon Control Schemes
325 p. 2006 [978-1-84628-470-0]

Vol. 338: Tokarzewski, J.
Finite Zeros in Discrete Time Control Systems
325 p. 2006 [978-3-540-33464-4]

Vol. 337: Blom, H.; Lygeros, J. (Eds.)
Stochastic Hybrid Systems
395 p. 2006 [978-3-540-33466-8]

Kolibri 360 GmbH
Ferdinand-Jühlke-Straße 7
99095 Erfurt, Deutschland
produktsicherheit@kolibri360.de